Parascript

Smithsonian Series in Comparative Evolutionary Biology
V. A. Funk and Peter F. Cannell, Series Editors

The intent of this series is to publish innovative studies in the field of comparative evolutionary biology, especially by authors willing to introduce new ideas or to challenge or expand views now accepted. Within this context, and with some preference toward the organismic level, a diversity of viewpoints is sought.

Parascript

Parasites and the Language of Evolution

Daniel R. Brooks
and
Deborah A. McLennan

SMITHSONIAN INSTITUTION PRESS
Washington and London

Editor and typesetter: Peter Strupp/Princeton Editorial Associates
Production editor: Rosemary Sheffield
Designer: Janice Wheeler

Library of Congress Cataloging-in-Publication Data
Brooks, D. R. (Daniel R.), 1951–
Parascript : parasites and the language of evolution / Daniel R. Brooks, Deborah A. McLennan.
p. cm.—(Smithsonian series in comparative evolutionary biology)
Includes bibliographical references and index.
ISBN 1-56098-215-2 (cloth) ISBN 1-56098-285-3 (pbk.)
1. Parasitology. 2. Parasites—Life cycles. 3. Host-parasite relationships. 4. Parasites—Evolution. I. McLennan, Deborah A. II. Title. III. Series.
QL757.B785 1993
574.5′249—dc20 92.20822
CIP

British Library Cataloguing-in-Publication Data is available.

Manufactured in the United States of America
99 98 97 96 95 94 93 5 4 3 2 1

∞ The paper used in this publication meets the minimum requirements of the American National Standard for Permanence of Paper for Printed Library Materials Z39.48-1984.

CONTENTS

PREFACE

> There was no doubt about it
> It was the myth of fingerprints
> I've seen them all and man they're all the same
> —Paul Simon

We hope to accomplish several things in this book. First, we want to show what can be accomplished by a concerted effort to delineate the phylogenetic relationships among the members of large groups of species. Second, we want to use the phylogenetic data base to evaluate some of the "myths, metaphors, and misconceptions" about parasites and their evolution. And finally, we hope to demonstrate that parasite-host systems are excellent models for evolutionary studies and for use as biodiversity probes.

There will be two recurring themes in this book. The first is that most stories about parasite evolution focus on the host rather than on the parasite: what does the parasite do to the host, and what does the host do in response; how does the host limit, mold, and shape the evolution of the parasite? That this is currently the primary perspective about parasite evolutionary biology can be seen by reading Rennie (1992). We believe this viewpoint has retarded the development of understanding about parasite evolutionary biology and has led nonparasitologists to pursue research stemming from improper assumptions about parasites. We will focus on the parasite rather than the host throughout this book.

The second recurring theme is that every story about parasite evolution can be investigated fruitfully (if not fully) by using the rigor of modern phylogenetic and historical ecological analysis. Therefore, the empirical underpinnings of the evolutionary investigations presented in this book will flow from phylogenetic systematic studies.

It is our perspective that all phylogenetic studies currently published, including our own, are preliminary estimates, and that books such as this one serve to set the stage for the first concerted efforts in comparative biology and macroevolution, something that will become a mature part of evolutionary biology in the twenty-first century. The generalizations highlighted herein may not and need not be fundamental truths about some aspects of parasite evolution. Their primary purpose is to serve as working hypotheses functioning as springboards for future research. Just as with historical ecological studies (Brooks and McLennan 1991), our vested interest in studying the comparative evolutionary biology of parasites lies in an approach to discovering rather than in the defense of a priori hypotheses. Consequently, we will not be using all parasitic organisms in formulating a macroevolutionary perspective on parasite evolution, because there is a paucity of phylogenetic hypotheses available. This does not mean that we are disregarding the considerable history of thought about the evolution of parasites that predates the advent of phylogenetic systematics. On the contrary, we have tried to use the generalizations that have emerged from those previous studies as a set of hypotheses to be tested using the new methods and studies presented herein. In this way, we hope to demonstrate some continuity with traditional views and approaches.

We would like to express special thanks for insights, advice, and friendly discussions about the development of comparative evolutionary studies in parasitology to Susan Bandoni, Walter Boeger, Bill Font, Scott Gardner, David Glen, Eric Hoberg, Greg Klassen, Delane Kritsky, Cheryl Macdonald, Janice Moore, Richard O'Grady, Tom Platt, and Marilyn Scott. Many of these people also generously gave us access to unpublished material. Brent Mishler and Richard Norris provided invaluable insights for the discussion of parasite adaptive radiations.

Our thanks to Mary H. Pritchard (Harold W. Manter Laboratory, University of Nebraska State Museum, Lincoln, Nebraska, USA) for access to rare literature during the background research for this book, and to the myriad of parasitology curators throughout the world who have kindly made material available to DRB for study during the past twenty years. Some of this book was written at the Cedar Point Biological Station, University of Nebraska, during July 1991, and we thank John Janovy, Tony Joern, John Lynch, Brent Nickol, and the graduate students at CPBS during that time for their interest and generosity in sharing ideas, and for their forbearance during that period of gestation.

We thank Sherwin Desser, chairman, and the faculty, staff, and students of the Department of Zoology, University of Toronto, and Kathy Coates, Chris Darling, Mark Engstrom, Bob Murphy, and Rick Winterbottom of the Royal Ontario Museum for providing the superb scholarly and collegial atmosphere in which this book was conceived and written.

We dedicate this book to Harold W. Manter, who kept the light burning during the eclipse of parasite comparative evolutionary biology, and to Gerald D. Schmidt, whose light was lost to an untimely death that was a severe blow to parasitologists throughout the world.

1
INTRODUCTION

> The principles that govern the structure, life cycles, habitats and activities of free-living and parasitic animals are really the same.
> —Hegner et al., 1938

> Parasitologists have accumulated, during the past decades, an amount of information that is truly formidable. This information is not only valuable for the parasitologist, but is also a potential gold-mine for the evolutionist and general biologist. Yet, much of this information is hidden away in a widely scattered and highly technical literature.
> —Mayr 1957

Parasites are an enigma. To some people they are an unpleasant but unavoidable fact of life. To others they are, like Victorian ankles, an embarrassing topic to be avoided in polite conversation. But to a small subset of human beings parasites are glorious creatures, no less a part of Darwin's tangled bank than organisms more acceptable to anthropocentric idealization. Whatever the perspective, our fascination with these organisms is an old one (for interesting discussions about the history of parasitology, see Harant 1955 and Foster 1965).

The roots of modern parasitology lie deep within a nurturing infusion of ecology, developmental biology, and systematics. Given this legacy, parasitology should have been a leader in the recent renaissance of comparative biology (Brooks and McLennan 1991; Harvey and Pagel 1991 and references therein). Unfortunately, this has not been the case because, since 1940, evolutionary biology has passed by most of parasitology. For example, only one text on parasite evolutionary biology has appeared in the past generation (Price 1980) and current general evolution texts either pay only fleeting attention to parasitism (Dodson and Dodson 1985; Futuyma 1986) or do not mention it at all (Avers 1989). Because of this separation, many of our current attitudes about parasite evolution stem from concepts that were developed in the early part of the twentieth century, often from a non-Darwinian perspective. These archaic views have contributed to a perception among nonparasitologists that parasites are somehow aberrant, that the patterns and processes of parasite evolution are not a reflection of evolution in general. Consequently, parasites are not deemed adequate model systems for studies in evolutionary biology, and the gulf between parasitologists and evolutionary biologists widens.

One thing that history teaches us is that the disintegration of communication between individuals or groups is always detrimental in the long run. We believe that the corollary is also true, that increased communication between individuals or groups is always mutually beneficial at some level. Idealistic visions of reciprocal altruism aside, it is in parasitologists' own best interest to forge connections with other disciplines because these connections are conduits to new ideas.

DEFINITIONS OF PARASITISM

"Parasite," in its original sense, is wonderfully explicit. A *parasitos* (*para:* beside; *sitos:* grain or food) was a person who received free meals from a rich patron, in return for amusing, impudent, and flattering conversation; in other words, a sycophant (Harant 1955). Unfortunately, when biologists coopted the term, they had a much more difficult time finding a concise definition. Virtually every text about parasites begins with an apology for this problem followed, paradoxically, by the author's own special description of the phenomenon.

Definitions of parasitism have traditionally focused upon some ecological aspect of the parasite-host interaction. In general, researchers have stressed that parasites are distinguished from free-living organisms by foraging mode or by habitat preference, either singly or in combination. The first type of ecological character to be examined in any detail was foraging. According to this perspective, parasites were "all those creatures which find their nourishment and habitat on other living organisms" (Leuckart 1879 in Dogiel 1962: 2) "without destroying it [the host] as predators do their prey" (Caullery 1922 in Dogiel 1962: 3). Parasites, then, differed from their free-living brethren "only in their mode of feeding" (Leuckart 1879 in Dogiel 1962: 2). In modern terminology, feeding mode has been transformed into metabolic dependency on the host (Cheng 1986). Although intuitively pleasing, this definition does not provide an unambiguous way to distinguish parasites from nonparasites because it artificially groups species on the basis of a convergently evolved character, the type of feeding mode. The group "parasite" must therefore include certain organisms that most biologists would not consider to be parasites in the traditional sense, e.g., vampire bats, some mosquitoes, and all herbivores, while excluding organisms that we would consider to be parasites, e.g., nematodes of the family Trichostrongylidae, which feed on intestinal bacteria and protozoans, not host tissue.

Several authors, disillusioned with foraging but determined to retain an ecological flavor in the discipline, proposed that parasites could be distinguished by the unique habitat in which they live. They were interested in the observation that, for many parasites, the "environment was formed by another living animal" (Filipchenko 1937 in Dogiel 1962: 5) to which the parasite transferred "the burden of regulating its relationship with the external environment" (Moshkovski 1946 in Dogiel 1962: 5). The same problem inherent in the foraging definition is found here, only in this case habitat preference is the convergent trait. Membership in this group is equally problematical. Should it include bacteria like *Escherichia coli* or parasites for whom free-living stages form a major part of the life cycle? Do we include parasitic plants and, if so, insects that spend their entire lives in association with plants?

Attempts at clarification continued with the observation that some parasites, in particular the better-known medical and veterinary scourges, produced a harmful effect on their hosts. Given this emphasis, parasitism could be defined as the form of symbiosis in which one species lived "at the expense of the other" in the association (Chandler and Read 1961; Dogiel 1962; Henry 1966). However, as Noble and Noble (1961) pointed out, living "at the expense of the other" is pretty difficult to pin down, or even to establish for most species that we consider parasites (see also Whitfield 1979; Anderson and May 1982). For example, the amoeba *Entamoeba invadens* is pathogenic in snakes and is innocuous in tortoises (Dogiel 1962). The tapeworm *Schistocephalus solidus* adversely affects the health and behavior of its piscine intermediate hosts, while living in apparent harmony with its avian definitive hosts. Price (1980) argued that this definition forced us to consider up to 40% of all species on this planet to be parasites. He proposed that parasite-host relationships were more like herbivore-crop than predator-prey interactions. Recognizing that this step forward did not clear away all the ambiguity because the definition was still in essence ecologically based, he resorted to using the definition of parasitism listed in Webster's Third New International Dictionary ("an organism living in or on another living organism, obtaining from it part or all of its organic nutriment, commonly exhibiting some degree of adaptive structural modification, and causing some degree of real damage to its host"). In a final move of apparent desperation, parasitism was recently (Brusca and Brusca 1990) defined as "a particular lifestyle suited to a particular environment, requiring certain adaptations and endowing certain advantages."

In response to years of continued and frustrating failures in the search for characteristics that would unambiguously define parasitism, Cameron (1956),

Noble and Noble (1961, 1976), and Schmidt and Roberts (1985) concluded that there is no distinct ecology, function, evolution, or physiology that distinguishes all parasites from all nonparasites. This conclusion was based upon an effect: failure to find "the characteristics." Evolutionary theory provides us with the cause for that failure. In essence, there is no such thing as an unambiguous definition of parasitism, because only common ancestry is unambiguous in biology, and parasites do not represent a monophyletic group. What brought us to this impasse? One answer to this question may be that our recognition of "parasites" as a distinct group stemmed from our discovery that "parasites" caused diseases. Because the association between parasite and disease is so firmly entrenched in our history and our minds, we have accorded parasitic organisms the same taxonomic status as natural groups like mammals and birds. But, because parasites are not a natural group, our search for a defining characteristic for the "group" has produced only logically inconsistent results. If we define parasites by their foraging mode, and from this establish a separate discipline of parasitology, then we should also consider establishing the parallel disciplines of herbivology, predatology, and saprophology. Clearly, biologists have not separated themselves along these lines, nor are they likely to. A stronger case can be made for habitat preference. After all, we recognize "fish" as something different because of the environment through which they move. Like parasites, fish are not a monophyletic group, yet no one would suggest that ichthyology be either disbanded or revamped to include all the higher tetrapods. Nevertheless, the practice of defining parasites by the host environment, despite its illustrious history (see next section), is fraught with difficulties, as discussed in the preceding paragraphs.

Ichthyology and parasitology exist as separate disciplines because that is the way biology has traditionally been organized. One of the ramifications of organizing a discipline around a paraphyletic or polyphyletic group is that all useful definitions of that group (e.g., "parasite" or "fish") will be somewhat cryptic. This is not cause for despair. Virtually all of biology is filled with ambiguous working definitions because they serve to justify and direct productive research programs and to identify membership in research groups (Hull 1988). The objective reality of any definition is less important than the quantity and quality of research spurred by its formulation. Consequently, although we do not think there is any possibility that we will discover the one characteristic that delineates parasitism (and hence, by extension, parasitology), we do believe that numerous lines of productive research have been developed from attempts to do that very thing. This is as true in parasitology as it has been in ichthyology, herpetology, invertebrate zoology, microbiology, botany, and protozoology, and in areas such as studies of speciation, where we still argue about the definition of species. Working definitions of parasitism stress the adaptive radiation of parasites and their obvious evolutionary success, the ways in which parasitism and particular parasite-host systems originated, or the antagonistic aspects of parasite-host relationships, usually in those species that are regularly associated with disease syndromes in humans or in species important to human life.

Parasitology is strongly influenced by tradition so we have continued the custom of beginning a book about parasites with a definition of parasitism. We conclude, only somewhat tongue in cheek, that the only unambiguous definition is that parasites are those organisms studied by people who call themselves parasitologists (see also Schmidt and Roberts 1985).

A BRIEF HISTORY OF PARASITE EVOLUTIONARY BIOLOGY

> In much of classical parasitology, the evolutionary emphasis is on using taxonomy to deduce host phylogeny.
>
> —Anderson and May 1982

When Anderson and May wrote the above sentence, they were voicing the commonly held belief that parasitology has rather shallow conceptual roots in evolutionary biology. We believe that this is an overly pessimistic view of an old and rich tradition in parasite evolution. So why is it the most widespread view? In order to answer this, we must trace the history of the discipline, drawing on a recent historical analysis by Klassen (1992) beginning, as all evolutionary biologists do, with Darwin. On November 7, 1844, Darwin wrote to Henry Denny:

> Dear Sir,
>
> I am much obliged for your note & have been greatly interested by the facts you mention of the identical parasites on the same species of birds at immensely remote stations. . . . Are you aware whether the same parasites are found on any of our *land* birds in this country & in N. America. Some of the birds of Europe & N. America appear certainly identical; many form very closely related species or as some would think races: What an *interesting* investigation would be the comparison of the parasites of the closely allied & representative birds of the two countries.
>
> Should you chance to know anything of the parasites of *land*-birds of North America, perhaps, sometime you kindly would take the trouble to send me a line, as I am deeply interested in everything connected with geographical distribution, & the differences between species & varieties. . . .

Although Darwin had an early interest in the evolutionary significance of parasite-host relationships, there is little indication that he expressed any additional opinion about parasites, other than calling them disgusting (Cameron 1964). We must therefore look elsewhere for our roots.

The key element in Darwin's letter was his interest in the connection between geographic isolation and the production of new species, a position that he later abandoned in favor of sympatric speciation mediated by natural selection (Bowler 1989). Perhaps his interest in parasites disappeared at the same time. In any event, this separation of geographic isolation and natural selection set the stage for a scathing attack mounted by Wagner (1868), a champion of speciation by geographic isolation. Ironically, it is in this anti-Darwinian context that we find the origins of parasite evolutionary biology.

The First Generation

One of the biologists influenced by Wagner was von Ihering (1891, 1902), the first researcher to use parasite and host distribution patterns as evidence for continental drift. Von Ihering expressed strong views about two important biological issues of his day: (1) the relative importance of geographic isolation versus natural selection in speciation and (2) the degree of permanence of continental and oceanic shelves. As an adherent of Wagner's views, he felt that geographic isolation rather than natural selection accounted for species formation and the disjunct distribution of related species. In 1891, he wrote

> I am as much convinced of the erroneousness of this doctrine [fixity of continents] . . . as I am that the ideas of Darwin and Wallace on "natural selection" as the cause of the origin of species will have but a historical interest in the coming centuries.

Although not a parasitologist, von Ihering (1902) incorporated information from two aspects of parasite-host relationships into his research program. First, he felt that the relationships of parasitic helminths to their hosts would allow researchers to estimate the age of individual species and genera. This led von Ihering to grapple with the significance of host specificity. In his mind, hosts were either *autochthonous* (endemic), inhabited by highly specific parasites indicating coevolutionary relationships, or *allochthonous* (widespread), inhabited by parasites also found in other hosts in other regions, and indicating secondary dispersal. From his perspective, only autochthonous hosts and their host-specific parasites could provide unambiguous evidence for lineage age. Second, von Ihering believed that the close relationship between parasites and their hosts would lead to a parallel series of speciation events in both lineages if the hosts, and hence their parasites, were geographically isolated. He argued that the existence of this parallel speciation could only be explained by geographic isolation since it is difficult to envision a selective pressure that could be operating equally on such distantly related and ecologically distinct species. Overall then, von Ihering felt that the evidence for the importance of geographical isolation in species formation provided by correlated parasite-host phylogenies, combined with the information about lineage age potentially available within the parasite-host association, formed a powerful biogeographic tool. His insights laid the foundation for parasite evolutionary biology; however, the new research program lacked a critical component—a method for generating parasite and host phylogenies.

Although von Ihering was the first biologist to examine parasite-host data within an evolutionary framework, a number of his contemporaries were developing similar ideas independently. Vernon Kellogg (1896a,b, 1913a,b, 1914; Kellogg and Kuwana 1902), a student of avian Mallophaga and a collaborator with the ichthyologist David Starr Jordan (e.g., Jordan and Kellogg 1900, 1908; Jordan et al. 1909), tried to meld geographic isolation and natural selection into a more pluralistic view of parasite-host evolution. He recognized three categories of parasite-host relationships, based on geographic distribution and host specificity. The first class consisted of closely related highly host specific parasites occurring allopatrically on or in closely related hosts. Unlike many researchers of his time, Kellogg (1913a) attributed such correlated parasite-host phylogenies to genealogy, writing

> I do believe that it is a commonness of the genealogy rather than a commonness of adaptation that is the chief explanation of this restriction of certain parasite groups to certain host groups.

The second class was comprised of parasites occurring on or in two or more closely related hosts inhabiting different geographic regions. Kellogg believed that these relationships reflected instances in which both the hosts and the parasites had been geographically isolated but only the hosts had speciated. He hypothesized that if the environment was different enough in the geographically isolated regions to establish selection pressures leading to differentiation of the hosts, while the parasites' environment, consisting of parts of the host, did not change to the same extent, then the parasites would not speciate because they were not subjected to the same magnitude of selection (Kellogg 1896a).

Kellogg's third class consisted of parasites occurring on or in two or more distantly related host species, generally within the same geographic region. His observations about this class, the "stragglers," presaged an important analytical problem for students of geographic isolation, the potentially confounding influence of host specificity. Kellogg (1913a) suggested that three major factors affected patterns of host specificity: (1) physical contact between hosts (of the same or different, closely related or distantly related, species), (2) differential speciation between hosts and parasites, and (3) some inherent tendency for parasite species not to "straggle" (host switch) even when given the chance. He wrote:

> It is well to keep in mind, in treating this rather abundant parasitization, that the feeding habits of the birds [vultures] give some opportunity for the straggling of parasites from other bird or mammal kinds . . . [and] . . . it is interesting to note that no mammal-infesting Mallophaga have been taken from any vulture, despite the excellent chances for such straggling.

Having established these classes, Kellogg found himself in the same quandary as von Ihering. It was imperative for him to distinguish between the two classes in which the same species of parasite inhabited two or more host species, since one type (2 above) would indicate phylogenetic relationships among hosts whereas the other (3 above) would not. Imperative or not, such differentiation was difficult in the absence of a method for determining host and parasite phylogenies.

While von Ihering and Kellogg were pursuing their investigations in Brazil and California, respectively, two Australian parasitologists, S. J. Johnston (1912, 1913, 1914a,b, 1916), a helminthologist, and Lionel Harrison (1914, 1915a,b, 1916, 1922, 1924, 1926, 1928a,b, 1929), an expert with avian Mallophaga, were converging on similar ideas. Like von Ihering and Kellogg, Johnston and Harrison considered geographic isolation to be the primary causal agent for speciation:

> We must look for the nearest relatives of each of these worms [in a particular region], not in host-animals of other classes living in the same region, but in hosts of the same class living in different regions. [Johnston, 1913]

Harrison, in particular, was an early advocate of Wegener's (1912) theory of continental drift as the geological mechanism underlying geographical isolation. Like Kellogg, Johnston and Harrison believed that natural selection played a role in producing parasite-host relationships characterized by departures from a strictly one host : one parasite host specificity pattern:

> The members of the former group represent very old parasites of their hosts . . . [that] have become altered in their morphological structure much less than their hosts have. [Johnston 1914a]

And finally, like von Ihering and Kellogg, they did not have a method for reconstructing phylogenetic relationships beyond assuming that the phylogeny of host-specific parasite species must parallel the phylogeny of their hosts:

> The proposition I advance is:—That in the case of total obligate parasites, closely related parasites will be found to occur upon phyletically connected hosts, without regard to other ecological conditions . . . a study of such parasites may give valuable indications as to host phylogeny. [Harrison 1915b]

Although primarily interested in the effect of geographic factors on speciation, von Ihering and Kellogg wove the threads of host specificity in and out of their writing. Their interest, however, extended only to the recognition that the influences of such specificity might cloud the developing picture of geographic speciation.

Students of coevolution in Europe approached the problem of parasite-host evolution from the opposite direction: they concentrated primarily upon the host component of the system. The earliest of the European workers was Fahrenholz (1909, 1913), a German entomologist interested in lice. Fahrenholz (1909) asked (independently of any of the members of the "geographic" school) if the presence of the same or related parasite species on different host species would allow researchers to draw conclusions about the relationships among the hosts. Parasites, he argued, are as dependent on their habitats as free-living organisms, except that their habitat is a host. Because of this, one would expect members of the same host species to be inhabited by the same species of parasites, and members of different host species to be inhabited by parasites whose phylogenetic relationships would be as divergent from each other as the phylogenetic relationships of their hosts. In other words, if parasites, like any other organisms, speciate in response to their environments (something with which all Darwinians and many non-Darwinians of the day could agree), and if the host furnishes the environment, then it follows that parasites will speciate in response to differences among hosts.

From Fahrenholz's perspective parasites were highly specialized beings who were no longer masters of their own destiny. Their evolutionary fate was sealed; they were totally dependent upon and molded by their hosts (the environment). From this belief sprang the two related concepts that formed the cornerstones of his research program: (1) host specificity, because only the study of strongly host-specific parasites would reveal the causal mechanisms underlying parasite-host relationships, and (2) "true" or obligately permanent parasites, because only among the "true" parasites could one be certain that the environment to which the parasite responds evolutionarily is provided exclusively by the host. Ironically, the perspective engendered by this research program led in later years to the emergence of the view that parasite speciation may be mostly host-mediated sympatric speciation (e.g., Price 1980) at the same time that nonparasitologist Darwinians had mostly abandoned sympatric speciation in favor of modes requiring some degree of geographic isolation.

Equally noteworthy is the observation that Fahrenholz arrived at his conclusions about the environmentally determined fate of parasites via orthogenetic, rather than Darwinian, reasoning. The orthogenesis movement developed as a response to what many scientists saw as an overemphasis by Darwinians on the role of the environment in evolution. As an alternative to Darwinism, these researchers proposed that evolutionary change was internally, rather than externally or environmentally, driven. Among other things, this internal drive always led to progressive specialization, dependence on other species, a loss of evolutionary independence, and finally self-imposed extinction (for an extended discussion in a historical context, see Bowler 1983). No wonder Farenholz was intrigued with the theory; parasites—with their presumed evolutionarily degenerate nature, overspecialization, and dependence on their hosts—were key examples of the process.

So, by the early twentieth century at least five of the key players in the parasite evolution drama had made their entrances on the stage. Each of these players had independently realized the importance of the nonrandom distribution of parasite-host associations, in particular, the distribution of closely related species of parasites on closely related species of hosts. True to the history of most scientific issues, two separate schools of thought were developing in the search for a causal mechanism to explain that observation. In Australia and the New World, researchers focused their attention on the influence of geographic isolation on parasite speciation (the "geographic" school; see also Zschokke 1904, 1933; Ewing 1924, 1928; Kirby 1937; Hopkins 1942). Development of their ideas was hampered by the lack of a method for reconstructing parasite-host phylogenies that was independent of the evolutionary principle they were trying to document. This led to the tautological assertion that similar, host-specific parasites occurred in similar but geographically separate hosts because of parasite-host coevolution via geographical isolation, so one could assume that the parasites were related to each other phylogenetically in the same way that their hosts were related to each other. This, in turn, would produce congruent parasite-host phylogenies, which were used to support the existence of coevolution by geographic isolation. At the same time the geographic school was developing, researchers in Europe were turning their attention to the influence of the host on parasite evolution (the "host" school). This group, with its emphasis on the dependent, degenerative nature of parasites, turned its evolutionary focus away from the parasites themselves, toward characteristics of the host.

The Second Generation

The orthogenesis movement, flourishing in different forms in North America and Europe during the first third of the twentieth century, was to play a further role in parasite evolutionary biology through its influence on the second generation of parasite evolutionary biologists. This generation was led by three researchers: Maynard Metcalf, Lother Szidat, and Wolfdietrich Eichler. Metcalf (1920, 1922, 1923a,b, 1926, 1928a,b, 1929, 1934, 1940; see also Dunn 1925) championed the geographic school. Working simultaneously with a group of hosts (anurans) and a group of their parasites (opalinids), he dedicated his career to the study of

> relationships between groups of animals and plants, and their geographical distribution and migration routes, by means of comparison of their parasites and comparison of the distribution of the hosts with that of their parasites. [Metcalf 1920]

Metcalf consolidated the studies of von Ihering, Kellogg, Johnston, Harrison, and his own work into a research program unified by the "parasite-host" method (Metcalf 1923a,b; renamed the "von Ihering Method" by Metcalf 1929). Eager to provide a rigorous basis for this fledgling program, he adopted the Wegener hypothesis of continental drift to explain disjunct distributions of related species and invoked logical parsimony to argue that common distribution patterns of related hosts and related parasites were more likely the result of common ancestry than parallel evolution:

> It might be conceivable that the Australian and American "frogs" assigned to the Leptodactylidae may not be closely related and that their resemblances are due to convergent evolution, but it is not possible that both the Leptodactylidae and their Opalinid parasites have evolved in parallel or convergent lines on the American and Australian continents. Such a coincidence is altogether improbable. [Metcalf 1920]

and

> Parallel development, or convergent evolution of both hosts and their parasites is too large a dose for even the most credulous to accept. [Metcalf 1923a]

Invoking continental drift to explain parasite distributions provided an interesting causal mechanism, but it did not help Metcalf design a method for testing that mechanism. In order to do this, he needed a way to generate host and parasite phylogenies without relying upon assumptions about geographic isolation and coevolution. Parasite taxonomy had little to offer at that time because it was an authority-based system that relied solely upon the subjective judgements of powerful researchers.

During his search for a more objective approach, Metcalf was intrigued by the developmental patterns displayed by the protists he was studying. In 1926, he proposed that von Baerian recapitulation provided an independent source of phylogenetic information:

> Of course the recapitulative nature of these larval stages is of especial interest, helping us interpret the phyletic history of this the most archaic family of Ciliate Infusorians.

By 1928 Metcalf had encountered the protistologist H. Crawley, one of the members of the American orthogenesis movement (e.g., Crawley 1923). Following discussions with Crawley about "trends in evolution and such crucial evidence of their existence as is given by the distribution of certain characters along the stems and branches of some phylogenetic trees," Metcalf (1928b) wrote:

> If trends be real, they are highly significant in evolution. Internal factors then become relatively more important in evolution and environmental control is less dominant throughout the successive stages of progress.

Trends lay at the heart of the orthogenetic program. In America, neo-Lamarckism developed into orthogenesis, bringing with it the Lamarckian concept of recapitulation. In orthogenetic terms, evolution was internally driven along a regular, linear pathway that was mirrored (recapitulated) in the patterns of embryological growth (Bowler 1983). Here was the causal justification for Metcalf's use of an ontogenetic criterion to reconstruct phylogenies.

While Metcalf pursued his studies, Lother Szidat and Wolfdietrich Eichler were promoting Fahrenholz's views (the "host school"). Szidat (1940, 1956a–c, 1960), an Argentinian helminthologist working primarily with digenetic trematodes, was interested in using phylogenetic relationships of hosts and their parasites to test hypotheses about coevolution. Although a proponent of the concepts of Fahrenholz, he incorporated some ideas from the geographic school into his research. Like Metcalf, he believed that similarities in ontogenetic stages provided the crucial data for reconstructing phylogenies (Szidat 1940). Like Fahrenholz (1913), Kellogg (1913a), Odhner (1913), and Harrison (1914), he believed that parasites that evolved more slowly than their hosts must be strongly host specific. But Szidat (1940) carried this concept two steps further, arguing that host specificity was the inherent developmental end product (*Altersresistenz*) of an orthogenetic drive toward specialization that increased with both the age of the individual host and the age of the parasite-host association. Once again attention was deflected away from the parasite toward the host, placing the parasite in a subordinate position within its own evolutionary drama.

The most influential proponent of the "host" school was Wolfdietrich Eichler (1941a,b, 1942, 1948a,b, 1966, 1973, 1982; see also Baer 1948; Dougherty 1949; Stammer 1955, 1957), a German entomologist studying Mallophaga. Eichler (1941a) set the tone for his career with an early attack on the suggestion, made by Kéler (1938, 1939), that at least some Mallophaga exhibited reduced host specificity. He argued that explanations for the presence of the same parasite on different hosts fell into one of two categories: (1) the parasite had inhabited the common ancestor of the current host species (à la Fahrenholz) or, more likely, (2) there had been a mistake in species-level identification of the parasite. Claiming that many of the examples discussed by Kéler fell into the second category (and in so doing launching a secondary attack on Kellogg and his co-workers), Eichler concluded that one must base taxonomy on an understanding of the parasite-host relationship rather than on morphological or other characteristics of the parasites themselves. This belief entrenched the perspective that hosts were the primary determinants of parasite evolution (see also Osche 1958, 1960, 1963). From this theory were born the three parasitological rules that Eichler (1942) believed could be used to explain any set of observations about parasite-host associations:

1. **Fahrenholz's Rule:** *The historical development (anagenesis) and splitting (cladogenesis) of hosts are paralleled by the development and splitting of their parasites. The phylogenetic relationships of the parasites can therefore be used to draw conclusions about the (often obscured) phylogenetic relationships of their hosts.* As discussed previously, this was more a statement of belief than a "rule." Eichler (1941a) was adamant in his attribution of this principle to Fahrenholz. He was equally adamant in his claim that Kellogg had stolen the idea from Farenholz, even though Kellogg had published the idea in 1896 and Fahrenholz (1913) himself stated that he had not thought of it until 1907.
2. **Szidat's Rule:** *The orthogenetic trend toward higher development in hosts drives their (mainly permanent) parasites in the same direction. Because of this, host taxa with a relatively low level of organization (primitive hosts) harbor parasites with relatively low levels of organization (primitive parasites).* This rule, based upon the beliefs of the European orthogenesis movement, was proposed to provide a causal mechanism for Fahrenholz's

rule. In 1948, Eichler expanded upon the orthogenetic concept of linear trends encapsulated in Szidat's rule. He argued that nonspecific parasites had main (*Hauptwirt*) and secondary (*Nebenwirte*) hosts, and that such nonspecificity was indicative of younger parasites (i.e., younger parasites tend to have more secondary hosts) because, as Szidat's rule demonstrated, host specificity tends to increase with increasing age of the parasite-host association.

3. **Eichler's Rule:** *Among host groups of equal systematic rank, species-rich groups will have a more diverse (mainly permanent) parasite fauna than their species-poor relatives.* Although Eichler formulated this "rule" to explain the differential diversity that he observed among the mallophagan taxa, its implications go much deeper.

Eichler was unshakably convinced that the drive toward increasing host specificity was a universal trend among parasites. When faced with the suggestion that some parasites did in fact demonstrate widespread host preferences, he either dismissed the parasites as young and thus just starting out along the orthogenetic pathway to specialization, or challenged the basis for believing that the parasites were members of the same species. He based this challenge upon a reiteration of his previous attack on morphologically based taxonomy, only this time he supported his claim with information from population genetics:

> We really have here the opposite of geographic variation, where the . . . different parasites are not (clearly) distinguishable externally, but are not transferable. They are therefore undoubtedly genetically more different than would be acceptable for mere populations of the same form. [Eichler 1948b]

In other words, even though the widespread parasites looked the same, they were really distinct, host-specific species recognizable at the genetic, if not at the morphological, level.

By the late 1930s the two schools were well established in North America and Europe. The geographic school in America concerned itself with the search for a way to escape the tautology of phylogenetic reconstruction based upon assumptions of geographical isolation. Meanwhile, on the other side of the Atlantic, the host school was busy creating its own tautological albatross: parasites are highly host specific, so they coevolve, and because they coevolve, they become highly host specific. To this albatross was attached the specter of unfalsifiability: since host specificity was the *cause* of coevolution, rather than a function of the ecological and phylogenetic interaction between lineages, any conflicting or inconsistent observations were erroneous or irrelevant because they failed to conform to the doctrine of coevolution. Attempts to re-introduce Kellogg's notions about host-switching under the term "capture" (e.g., Chabaud 1965a) failed to improve the situation. And, although they differed in their fundamental beliefs about the factors which had influenced the evolution of parasite-host associations, both schools suffered from the same problem: they lacked a rigorous methodology for testing their beliefs.

The Third Generation

The demise of orthogenesis and continental drift, along with the untimely death of Maynard Metcalf in 1940, were severe blows to coevolutionary studies in North America. Fortunately, the efforts of Harold Manter (1940, 1954, 1955, 1963, 1966), a helminthologist at the University of Nebraska interested in the digenetic trematodes of marine fishes, kept the research alive over the next three decades. Manter was strongly influenced by Metcalf's (1923a) paper, which he called the "outstanding classical example" of the geographical school's approach to coevolution (Manter 1940). Following in Kellogg's footsteps, he incorporated many of David Starr Jordan's ideas about speciation into his theoretical framework. This approach, combined with his own observations on distribution patterns of hosts and their parasites, and the writings of von Ihering, Kellogg, and Metcalf, led to his (informal) formulation of the geographical school's three rules:

1. **Manter's First Rule:** *Parasites evolve more slowly than their hosts.*
2. **Manter's Second Rule:** *The longer the association with a host group, the more pronounced the specificity exhibited by the parasite group.* This principle represented a synthesis of Kellogg's views on the importance of host specificity and von Ihering's ideas that the taxonomic rank of the hosts and parasites in an association provided information about the age of that relationship. Since parasite speciation lagged behind host speciation (rule 1), Manter argued that specificity of a parasite taxon should be considered in light of the host at the next higher taxonomic level. This rule is, in essence, a restatement of the host school's assertion that parasites were driven, over time, to become more host specific (Szidat's rule).
3. **Manter's Third Rule:** *A host species harbors the largest number of parasite species in the area where it has resided longest, so if the same species, or two closely related species of hosts, exhibit*

> *a disjunct distribution and possess similar faunas, the areas in which the hosts occur must have been contiguous in the past.* This principle was the geographic school's answer to Eichler's rule. In this case the number of parasite species found on a given host species was linked to the host's relationship with geographical factors; in Eichler's case, the number of parasite species on a given host species was directly linked to some internal host mechanism that drove parasite evolution. In either case, the parasite's evolutionary fate was mediated through the host, not by characteristics specific to the parasites themselves.

It is remarkable that Manter managed to keep the flame of coevolutionary studies alive in North America at a time when the theoretical underpinnings of the research program, continental drift and orthogenesis, were being destroyed. Manter recognized the significance of geographical isolation in species formation, but did not outwardly pursue that kind of research because the scientific, especially the geological, community had judged the theory of continental drift and found it wanting (Leviton and Aldrich 1985). This judgment deprived parasite evolutionary biology of its strongest corollary evidence. Interestingly, Manter's biogeographic studies clearly indicate that he felt continental drift was a primary mechanism for isolating populations and producing geminate (twinned: Jordan and Kellogg 1908) species, although he never championed the idea in print until late in his career (Manter 1966).

Manter also recognized the importance of correlated parasite-host phylogenies, but did not pursue that line of research because there was no independent way to reconstruct phylogenies apart from Metcalf's orthogenetically based reliance on recapitulation. This reliance on orthogenetic processes to explain evolutionary patterns was being viewed with increasing disfavor, especially in North America (Mayr 1982). Neo-Darwinism emerged as the dominant evolutionary theory in the late 1930s to mid-1940s (Dobzhansky 1937; Mayr 1942; Simpson 1944), but parasitology did not keep pace by incorporating the new mechanisms into explanations for the coevolutionary patterns envisioned by von Ihering, Kellogg, Johnston, Harrison, and Metcalf. For example, Walton (1942) provided a taxonomic description of nematodes inhabiting giant tortoises on the Galapagos Islands without referring to the evolutionary significance of the hosts or the islands.

The decline of orthogenesis catapulted parasitologists back into the methodological quagmire that had existed before Metcalf appeared. No one could produce an objective, empirical method for reconstructing parasite and host phylogenies, and for comparing them with each other and with geographic distribution patterns. This meant that workers had to "fit" their host and parasite data to some set of preconceived theoretical guidelines about coevolution, making it impossible to test those theoretical guidelines.

And, if these theoretical disasters were not bad enough, students of parasite evolution had to deal with a more pragmatic problem. The fall of orthogenesis and the rise of neo-Darwinism coincided with the outbreak of World War II. As the war escalated, parasitologists either were encouraged to focus their attention on parasites of medical and veterinary significance or were displaced by the hostilities. This does not mean that parasitology had not had a significant medical and veterinary component prior to 1940. Indeed, the *Journal of Parasitology,* originally subtitled "A Quarterly Devoted to Medical Zoology," carried only two papers explicitly discussing phylogenetic relationships during the period 1914–1942 (Byrd and Denton 1938; Byrd and Reiber 1942). Prior to 1940, however, parasitologists were active in evolutionary biology, even if they were not publishing in parasitological journals, a theme common even today. Unfortunately, for the reasons outlined earlier, and perhaps more at which we can only guess, the study of parasite evolution became a low-priority item that did not recover its former professional vigor after the war ended. As a consequence, parasitologists became followers, rather than leaders, on the evolutionary playing field, relying more and more upon statements by nonparasitologists for their ideas about parasite evolution. The "schools" of parasite evolutionary biology quietly persisted in a holding pattern for more than thirty years.

Some parasite systematists of the day continued to discuss questions of correlated parasite and host phylogenies and the role of host specificity in parasite evolution, but their discussions were more often seen in special volumes (e.g., Baer 1957) than in mainstream journals. Between 1947 and 1959, the *Journal of Parasitology,* North America's leading parasitology journal, published no articles on adaptation, evolution, phylogeny, or speciation. During that same period, *Evolution* and *American Naturalist* each published only a single article on parasite evolution, neither of which was written by a parasitologist. *Systematic Zoology* published six articles written by parasitologists between 1951, when the journal began, and 1959; of those six, four were contained in a single issue of the journal devoted to intraspecific variation in parasite species.

The disjunction between evolutionary biology and parasitology was essentially complete by 1959, the centennial of the publication of *Origin of Species.* The years 1959–1961 produced a spate of books, collected works, journal issues devoted to the results of symposia, and individual contributions in many journals. This period heralded the triumph of the neo-Darwinian revolution, except in parasitology. Neither the *Journal of Parasitology* nor *Parasitology,*

Britain's principal parasitology journal, published a single article dealing with evolutionary concepts, nor did either offer any editorial comment about the centenary during that time.

Despite the explosion in theoretical and empirical studies of evolutionary biology stemming from the ascension of neo-Darwinism, only nine articles on adaptation, evolution, phylogeny, or speciation appeared in the *Journal of Parasitology* from 1960 to 1976. During the same period, two articles on parasite evolution appeared in *Evolution* (one of which was written by a parasitologist), five articles on parasite evolution appeared in *Systematic Zoology* (all written by parasitologists), and nine articles on parasite evolution appeared in *American Naturalist* (two of which were written by parasitologists). The crowning touch came when Price (1980), who was not a parasitologist, wrote the first book written exclusively about the evolutionary biology of parasites.

Despite the relative stasis of parasite evolutionary biology during the latter twenty-five years of his career, Manter (1966) never relinquished his belief that the study of parasite evolution had much to offer biology, if only a method for deciphering the wealth of data could be devised:

> Parasites . . . furnish information about present-day habits and ecology of their individual hosts. These same parasites also hold promise of telling us something about host and geographical connections of long ago. They are simultaneously the product of an immediate environment and of a long ancestry reflecting associations of millions of years. The messages they carry are thus always bilingual and usually garbled. As our knowledge grows, studies based on adequate collections, correctly classified and correlated with knowledge of the hosts and life cycles involved should lead to a deciphering of the message now so obscure. Eventually there may be enough pieces to form a meaningful language which could be called *parascript*—the language of parasites which tells of themselves and their hosts both of today and yesteryear.

In one of the strange coincidences usually sprinkled liberally throughout the history of any field, this optimistic paper was published at the same time that Willi Hennig, a German entomologist, was shaking the foundations of Eichler's "parasitological method." Hennig (1950, 1966) provided the first explicit method for reconstructing phylogenies that, because it was based upon analyzing observable attributes of the organisms, was independent of theories about their evolution. Thus, he was acutely conscious of the tautological assumptions that formed the basis of Eichler's research program. Unlike those of Manter, Hennig's (1966) conclusions about the utility of parasitological data in systematic analyses, and hence, by extension, in evolutionary studies, were not optimistic:

The theory of the parasitological method must be much better worked out than it is now . . . it is a question of the degree of phylogenetic relationships, of deciding whether the host species are more closely related to one another than to the species in which the parasites do not occur. Consequently, as with the morphological method, there must be criteria for determining when this assumption is justified . . . and when it is not. No such criteria are known.

Sixty-eight pages later, Hennig tempered his somewhat scathing view of the "present unsatisfactory state" of parasite evolutionary biology but did not offer much guidance for the future:

We cannot foresee the role that the parasitological method will play some day when relationships are more accurately known. . . . We can be sure that it will not be small.

Ironically, parasitologists were to use Hennig's own method beginning more than a decade later to launch a renaissance of parasite evolutionary biology and to provide the Rosetta Stone for Manter's Parascript.

2

PARASCRIPT

> Parasites . . . furnish information about present-day habits and ecology of their individual hosts. These same parasites also hold promise of telling us something about host and geographical connections of long ago. They are simultaneously the product of an immediate environment and of a long ancestry reflecting associations of millions of years. The messages they carry are thus always bilingual and usually garbled. Eventually there may be enough pieces to form a meaningful language which could be called *parascript*—the language of parasites which tells of themselves and their hosts both of today and yesteryear.
>
> —Manter 1966

Parascript studies are the modern descendants of the research program initiated by von Ihering, Kellogg, Harrison, Johnston, and Fahrenholz, synthesized by Metcalf, and sustained by Manter during the eclipse of parasite evolutionary biology. Students of parascript attempt to unravel the forces shaping parasite evolution by uncovering the patterns of *geographic* and *host* associations among parasite clades.

These associations may exhibit two types of patterns. In the first, a parasite species may occur in a particular geographical area because it evolved there (a phenomenon termed vicariance), or a parasite may inhabit a particular host because its ancestors, and the ancestors of its host, interacted with each other in the past. In each case, the species has inherited the association, and its phylogenetic history will be congruent with the geological history of the area or the phylogenetic history of its hosts. This type of interaction between parasites and geography or between parasites and their hosts is termed *association by descent* (Mitter and Brooks 1983; Brooks and Mitter 1984).

In the second pattern type, a parasite may be in a particular area or associated with a particular host because it evolved elsewhere and colonized the new area/host. This type of interaction between parasites and geography or parasites and their hosts is termed *association by colonization* (Mitter and Brooks 1983; Brooks and Mitter 1984). The dispersal or host switch may have occurred a long time ago or relatively recently. In either case, the history of the species will not be congruent with the history of the area into which it dispersed or with the history of its host. Of course, both types of patterns may be represented throughout the evolutionary history of interacting clades. For example, an association between a group of monogeneans and a group of minnows may have begun with a host switch, then persisted through a series of vicariance speciation episodes.

The theoretical underpinnings and goals of parascript studies are reasonably straightforward. What is required now is a rigorous method to test these ideas. Phylogenetic systematics, with its extensions into biogeography and coevolution, provides just such a method. In the remainder of this chapter we will discuss how the patterns of association by descent and the patterns of association by colonization can be uncovered and differentiated from one another. By the time we finish this chapter, we hope to have deciphered enough of the language of parasites to begin formulating some generalities about the influences of geography and hosts on parasite speciation and diversification.

THE ROOTS OF PARASITE SYSTEMATICS

Systematics is an old pursuit, arising out of our seemingly inherent need to organize the world into familiar units. To this end, systematists have traditionally wanted their classifications to be made up of "natural groups" delineated by unique or "diagnostic" features. In a perfectly ordered world each species would have at least one unique feature that indicated its existence as a distinct biological entity; groups of species would share unique features that indicated the existence of a natural group called a genus; groups of genera would be comprised of species sharing yet other unique features that indicated the existence of a natural group called a family; and so on. In pre-Darwinian times, these characters were thought of as guideposts to the natural order of life on this planet, a natural order that was the product of a logical and methodical Deity with an inordinate fondness for hierarchical arrangements. Darwin added an evolutionary twist to this scenario when he wrote that all true (=natural) classifications were genealogical. Systematic biologists attempted to incorporate this fundamental insight into their conceptual framework by asserting that (1) characters shared by members of natural groups were homologous characters, so natural groups were genealogical groups, and (2) if natural groups were indicated by uniformity (similarity) of (diagnostic) characters, and natural groups were genealogical groups, then homology equaled similarity (e.g., Cort 1917 and Stunkard 1940, among early parasitologists). Based on this perspective, it was asserted that pre-Darwinian classifications were in fact genealogical, even though the systematists producing them generally were not evolutionary biologists.

Such assertions worked when only one person studied a particular group. However, once several people began investigating the same organisms, differences of opinion rapidly emerged about the composition of the "true" natural groups and the identities of their diagnostic characters. This led to many arguments about "real" and "spurious" similarity, culminating in appeals to authority and claims that one's opponents could not discriminate homology from nonhomology, and often could not even prepare a specimen well enough to see the relevant characters. These exchanges actually stimulated many advances in the collection, fixation, and preparation of parasite specimens. They also highlighted the absence of any rigorous scientific method for determining genealogical relationships and dealt a serious blow to the belief that the pre-Darwinian systematists had been unconsciously documenting genealogy all along.

Parasitologists, like nonparasitologists of the day, found themselves in the midst of the Darwinian revolution with no way to evaluate evolutionary or coevolutionary hypotheses based on observable traits of their study organisms. Realizing that this was a serious problem, researchers like Kellogg and Metcalf sought alternative ways to reconstruct parasite phylogenies. Four major sources of data were investigated in this search: (1) geographic distribution, (2) host relationships, (3) morphological (originally orthogenetic) trends, and (4) ontogeny (von Baerian recapitulation). The use of geographic distribution and/or host data to infer parasites' identities and phylogenetic relationships was abandoned by many authors because of concerns about circular reasoning (see Chapter 1). This line of reasoning reached new heights in circularity when some researchers reconstructed host relationships based on parasite data, then classified new parasites according to those hosts. Circularity notwithstanding, the entire process rested on systematists' ability to determine host relationships, an ability that was seriously hampered by the lack of a rigorous method. One had to have Eichler's strength of conviction that parasite-host coevolution was overwhelmingly true to continue with this line of research. Most parasite systematists did not have such faith, although there were (e.g., Sandground 1926: "The specificity for their hosts is regarded as the best means of determining the specific standing of the parasite"; see also Stunkard 1940), and continue to be (e.g., Burt and Jarecka 1982; Humphery-Smith 1989) exceptions.

Metcalf championed the use of trends in adult parasite morphology and von Baerian recapitulation in parasite ontogenies as alternate sources of data for phylogenetic reconstruction. Parasitologists were apparently skeptical of the first approach. Only a few evolutionary trees based on morphological trends appeared in the literature (e.g., Metcalf 1923a; Byrd and Denton 1938; Byrd 1939; Dubois 1945; Byrd and Reiber 1942; Pritchard 1966; Gibson and Bray 1979). Many of those trees depicted whole groups of species in linear positions relative to other groups of species, rather than in the nested hierarchical arrangement that is characteristic of genealogical relationships. Moreover, most of the trees were based on a single character or character complex, provoking the criticism that someone choosing another character complex might see a different set of morphological trends. When multiple characters were used, the belief that species and groups of species could be arrayed from primitive to advanced in a linear sequence led to ambiguity in the placement of many "intermediate" taxa. With the demise of orthogenesis, parasitologists' skepticism was apparently justified; there was no longer any theoretical basis for believing that trees based on morphological trends actually represented phylogeny. Nevertheless, such diagrams represented the authors' bold attempts to delineate explicitly their evolutionary hypotheses so that further study and discussion could focus on the strength of the data, rather than on the authority of the writer.

Discouraged with the theoretical impasse, most parasitologists turned their attention to the production of more, and more detailed, descriptions of new species. A few, however, turned toward Metcalf's (1926) suggestion that the recapitulative nature of larval stages might be useful in uncovering phylogenetic relationships. This idea attracted Metcalf because it meshed with his orthogenetic belief that evolution was internally driven along a regular, linear pathway. It attracted other parasitologists because of a traditional fascination with life cycles in the discipline. At first parasitologists were unwilling to use life cycle data in phylogenetic reconstructions because they simply did not know enough about the biology of entire groups. But they were discouraged with hosts, geography, and morphological trends, and this appeared to be the only area remaining in parasitology that would not lead its adherents into the usual evolutionary dead end. The call went out for more descriptions of life cycles. Unfortunately, as the studies accumulated it became obvious that ontogenetic data did not delineate natural groups any more clearly than adult morphological data. So, although parasitologists now knew more about parasite life cycles, they were no closer to understanding parasite evolution.

This failure delivered a devastating blow to a discipline already reeling from a series of conceptual disappointments. Ironically, what had begun as a search for the fundamental source of phylogenetic data became the rationale for avoiding speculations about phylogeny in parasitological publications. Unlike its predecessors, this strategy appears to have been successful; during the period from 1925 to 1977 very few treelike diagrams based upon any type of data appeared in parasitological journals. Stunkard (1940) articulated the general state of parasite systematic biology when he wrote

> If taxonomy is to have meaning, it must express genetic and phylogenetic relationships. In the establishment of such a system, the principle of homogeneity of groups should receive first consideration. The well-known tendency for animals of diverse ancestry to assume a superficial resemblance and converge toward a common morphological type, after living for a long period in the same environment, is especially prevalent among parasitic forms. Thus, among species which have been greatly modified by parasitic adaptation, the morphological criteria applied to free-living animals lose much of their significance.

This statement reflected the duality of parasitologists' evolutionary beliefs. On the one hand, they realized that phylogenetic relationships were an important component of the evolutionary picture. On the other hand, they believed that the (putatively) extreme adaptive plasticity exhibited by parasites would obscure those relationships. The quest to overcome this problem heralded the beginning of a long-standing argument about potential sources of "evolutionarily conservative" characters. Some researchers advocated the use of larval data, arguing that adult stages were adaptively plastic, and thus subject to too much convergent evolution to be phylogenetically informative. Other researchers advocated the use of adult data, arguing that larval stages were adaptively plastic, and thus subject to too much convergent evolution to be phylogenetically informative (e.g., discussion in Cable 1974). In the end, the search for the magic characters that would indicate true phylogenetic relationships had returned parasitology to the unsatisfactory concept that the "uniformity of groups" was a valid criterion for recognizing natural groups.

PHYLOGENETIC SYSTEMATICS: THE ROSETTA STONE FOR PARASCRIPT

The state of parasite systematics did not change for almost forty years after Stunkard's statement. When that change finally came, it was not initiated by parasitologists or by the discovery of more informative data. It was initiated by the development of a more robust method for analyzing existing data bases developed by Willi Hennig (1950, 1966). Hennig suggested that systematists should focus their attention on a particular type of similarity, rather than on a particular type of data, in order to formulate a general reference system for comparative biology. He reasoned that this system should be based on reconstructing phylogenetic relationships from shared homologous traits because all homologies covary with each other and with phylogeny. Similarity owing to convergent or parallel evolution (homoplasy) is not phylogenetically informative because homoplasies need not covary with anything. Although this insight was fundamentally important to the study of evolutionary biology, it was not immediately accepted because in traditional approaches to systematics homology was at once defined by, and required to reconstruct, phylogeny. Hence, a systematist, in true Orwellian fashion, needed to *know the phylogeny* in order to *determine the homologies to build the phylogeny.* Knowledge of the phylogeny, in turn, tended to be determined by the experience and intuition of authority figures in systematics, rendering systematic biology less than scientifically rigorous.

One aspect of Hennig's perspective that is critical for parasite systematics is his assertion that the evolutionary process ensured that all species, including the ancestral species that gave rise to groups we now recognize as higher taxa, are mosaics of traits reflecting varying degrees of ancestry. That is, *not all traits change at the same time evolutionarily.* Consequently, we should not believe that there are any such things as groups defined absolutely by uniformity of characteristics. Hennig believed that genealogical influences in evolution are so pronounced that phylogenetic relationships could be reconstructed using empirical observations and formal rules for decision making rather than appeals to authority and intuition. He suggested that researchers begin with the initial assumption that all characters that conform to nonphylogenetic criteria for homology are homologous (e.g., Remane 1956; Wiley 1981). In some cases this will lead to the incorrect identification of homoplasious traits as homologues. However, when a phylogeny is reconstructed by grouping taxa according to their shared putative homologies, these misidentifications will be revealed because the homoplasious characters will not covary with the majority of the other characters. *This freed systematics from the specter of circularity because in the Hennigian system homologies, which indicate phylogenetic relationships, are hypothesized without a priori reference to a phylogeny, whereas homoplasies, which are inconsistent with phylogeny, are determined as such by reference to the phylogeny.* In other words, it is impossible to determine whether a character is homologous or homoplasious until it has been tested against a phylogenetic hypothesis based on a multitude of characters. So, years after the pioneering work had been done, a method was finally available that allowed researchers to arrive at a phylogenetic hypothesis based on the overall weight of evidence rather than on the favorite characters of authority figures or on theories about the effects of biogeography, hosts, or morphological and developmental trends (see Brooks and McLennan 1991 and Wiley et al.

1991 for recent detailed discussions of phylogenetic systematics). The first study using phylogenetic systematic methods for a group of parasites appeared in 1977 (Brooks 1977), and the number of such studies has been growing since that time. The Appendix provides the phylogenetic data base from which the studies in this book were drawn.

Every parasite-host system has an evolutionary story to tell (Table 2.1). Some systems, however, are more amenable to parascript studies than others. These systems can be recognized by (1) *host groups* that are ancient in origin, are species rich and widespread geographically, have been analyzed phylogenetically, and support a diverse parasite fauna, and (2) *parasite groups* that are ancient in origin, are species rich and widespread geographically, have been analyzed phylogenetically, and for which a wealth of life cycle and other ecological information is available. In the following examples, we hope to convey an initial sense of the breadth of systems that can be encompassed by parascript studies, and to stimulate interest in the multitude of systems that have not yet been scrutinized. None of the stories has been fully deciphered yet, but all of them contain interesting evolutionary information.

Parascript studies come in a variety of forms. In some cases, phylogenetic information about the parasites is relatively incomplete, but can still indicate interesting avenues for future research. In other (rare) cases, information about the phylogenies of both the parasites and their hosts, and information about the geological history of the areas in which the parasites reside, is available. Researchers working on these groups chose wisely. At the moment we usually have detailed phylogenetic information about the parasites but not about their geographic distributions or hosts. There are three ways around this problem. The first involves convincing someone, with offers of fame and funding for "cutting edge" research, to perform a phylogenetic systematic analysis for the host group of interest. Laughter on the part of the presumptive collaborator leads us to the second alternative: infer the history of the association by examining the parasites' phylogenetic tree side by side with the most current host classification scheme. This has always been, at best, a weak option. The third option was provided by Brooks (1981c, 1985, 1988, 1990b; Brooks and McLennan 1991), who developed a more robust method based upon reconstructing the host's phylogenetic relationships with *more than one* of its parasite groups. If we have phylogenies for five parasite groups inhabiting, say, gasterosteid fishes, then we have five independent estimates of gasterosteid relationships. We can compare these estimates, and search for points of congruence, indicating a common history of association between members of the host group and members of the parasite groups, and points of incongruence, highlighting possible cases of host switching. If there is a robust phylogeny for the

Table 2.1
Studies using phylogenetic systematics in studies of the evolution of parasite-host systems, listed by parasite group with host groups and references following

Protists

Coccidians in cricetid rodents (Reduker et al. 1987).

Sporozoeans in a variety of hosts (Barta 1989).

Diplomonads in a variety of hosts (Siddall et al. 1992).

Helminths

Platyhelminths

Digeneans in vertebrates (Brooks 1979a; Brooks and Macdonald 1986); tetrapods (Brooks and Overstreet 1978); anurans (Brooks 1977); crocodilians (Brooks 1981a,b); North American freshwater turtles (Platt 1988, 1992; Macdonald and Brooks 1989b).

Aspidobothreans in vertebrates (Brooks et al. 1989a).

Monogenea on elasmobranchs (Boeger and Kritsky 1989); North American ictalurid fishes (catfish) (Klassen and Beverly-Burton 1987); North American centrarchid fishes (basses and sunfish) (Klassen and Beverly-Burton 1988b); African cyprinid fishes (genus *Labeo*); South American characid fishes (piranhas) (Van Every and Kritsky 1992).

Gyrocotylidea in chimaeroid fishes (Bandoni and Brooks 1987a).

Amphilinidea in vertebrates (Bandoni and Brooks 1987b).

Eucestoda in tetrapods (Brooks 1978a,b); neotropical catfish (Brooks and Rasmussen 1984); carnivore mammals (Moore and Brooks 1987); alcid birds (Hoberg 1986); pinnipeds (Hoberg 1992); neotropical mammals (Gardner and Campbell 1992).

Nematoda

Oxyuroids in Old World primates (Brooks and Glen 1982).

Ascaridoids in neotropical rodents (Gardner 1991); North American amphibians (Richardson 1988).

Strongyloids in Old World primates (Glen and Brooks 1985).

Trichostrongyloids in North American ruminants (Lichtenfels and Pilitt 1983).

Metastrongyloids in North American cervids (Platt 1984).

Digeneans + nematodes in crocodilians (Brooks and O'Grady 1989).

Digeneans + eucestodes + nematodes in hominoid primates (the great apes) (Glen and Brooks 1986).

Digeneans + eucestodes + monogeneans + nematodes in neotropical freshwater stingrays (Brooks et al. 1981b; Brooks and McLennan 1991; Brooks 1992a).

host group, parasite phylogenetic relationships can be optimized onto the host tree to provide a detailed accounting of the historical context of each parasite-host association. We will discuss this approach, dubbed Brooks parsimony analysis by Wiley (1988), and studies using it, later in this chapter.

Brooks's method requires that we have, minimally, a phylogenetic tree and comprehensive host or area distribution patterns for the parasite groups of interest. What happens if we cannot even meet that minimal requirement? The long-term answer, of course, is "keep collecting more data." As we will demonstrate in the following three studies, however, even preliminary analyses can uncover interesting and significant information, as well as helping to focus the data collecting process.

BLOOD FLUKES AND TURTLES: A STORY OF CONTINENTAL DRIFT

Many people have heard of schistosomes, the blood flukes causing schistosomiasis or bilharziasis in humans throughout the tropics. Fewer people have heard that these digenean platyhelminths apparently originated with avian hosts and switched at least twice into mammals (Carmichael 1984). Even fewer people realize that spirorchids, the sister group of the schistosomatids, inhabit the circulatory system of freshwater and marine turtles worldwide (Platt 1992). Although they were overlooked by most, questions about the systematics of this small group provoked fifty years of debate among the few trematode taxonomists who found them interesting (e.g., MacCallum 1921, 1926; Stunkard 1921, 1923; Ward 1921; Price 1934; Byrd 1939; and Yamaguti 1971). Perhaps this debate was sparked by the spirorchids' phylogenetic proximity to the medically important schistosomes. Or perhaps it was a measure of each researcher's fascination with the organisms. Whatever the reason, there was no way to resolve the problem objectively until the advent of a data-based method for reconstructing phylogenetic relationships.

Using phylogenetic systematic methodology, Platt (1988, 1992) has begun to provide a robust phylogenetic framework for the spirorchids (Fig. 2.1). Blood flukes have been reported from turtle species representing approximately 25% (21/80) of the recognized turtle genera. As with so many other parasite studies, it is likely that the "absence" of spirorchids in some turtle species reflects the absence of parasitologists interested in, or capable of, collecting blood flukes, rather than the actual absence of the flukes themselves. The framework, therefore, is far from complete (Platt 1992). In addition, although turtles have been subjected to phylogenetic systematic scrutiny (see Gaffney and Meylan 1988

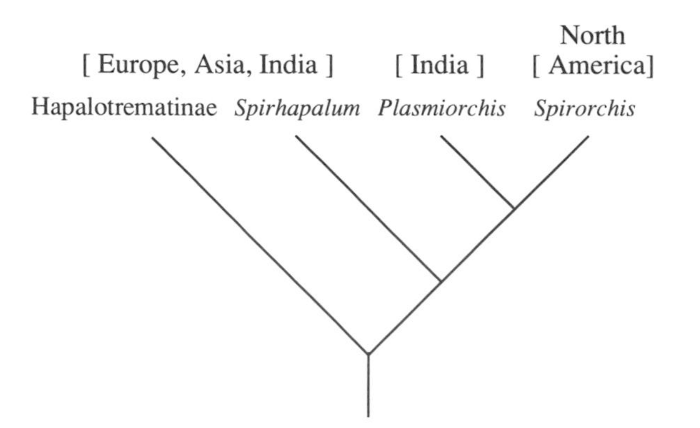

Figure 2.1. Phylogenetic tree for the genera of freshwater turtle blood flukes (family Spirorchidae), showing the genera of the Spirorchinae and the possibly paraphyletic Hapalotrematinae.

and references therein), they have not yet been examined within a historical biogeographic context. What little information is available, however, raises enough intriguing possibilities to warrant the attention of inquiring parasitologists.

Turtles are an ancient group with roots extending to at least the Permian (Colbert and Morales 1991). They are split into two major groups, the Pleurodira, or side-necked turtles, and the more derived, "modern" Cryptodira, or hidden-necked turtles (Fig. 2.2). Fossil evidence from early Jurassic deposits (Gaffney et al. 1987) indicates that the Cryptodira existed before the breakup of Pangaea. As that supercontinent began to drift apart, the hidden-necked turtles experienced a burst of speciation events leading to the appearance of all modern families during the Cretaceous and early Tertiary (Gaffney 1975; Bickham 1981).

Species of the Hapalotrematinae, which probably contains the basal spirorchids (and which is possibly paraphyletic; see Platt 1992 and Appendix), inhabit a wide range of turtle hosts throughout North America, Africa, India, Asia, and the former USSR (Fig. 2.2). *Hapalorhynchus* species are most often reported from relatively basal groups, including African pleurodires (Platt 1991) and from relatively basal cryptodires, chelydrids (snapping turtles), trionychids (soft-shelled turtles), and kinosternids (mud turtles and stinkpots). *Hapalotrema,* the other member of the parasite subfamily, inhabits members of the marine family Cheloniidae, a relatively basal cryptodire group. These observations suggest a long period of association between spirorchids and turtles.

The freshwater species of the Spirorchinae exhibit a narrower host range than the Hapalotrematinae, but are no less interesting (Fig. 2.2). They associate

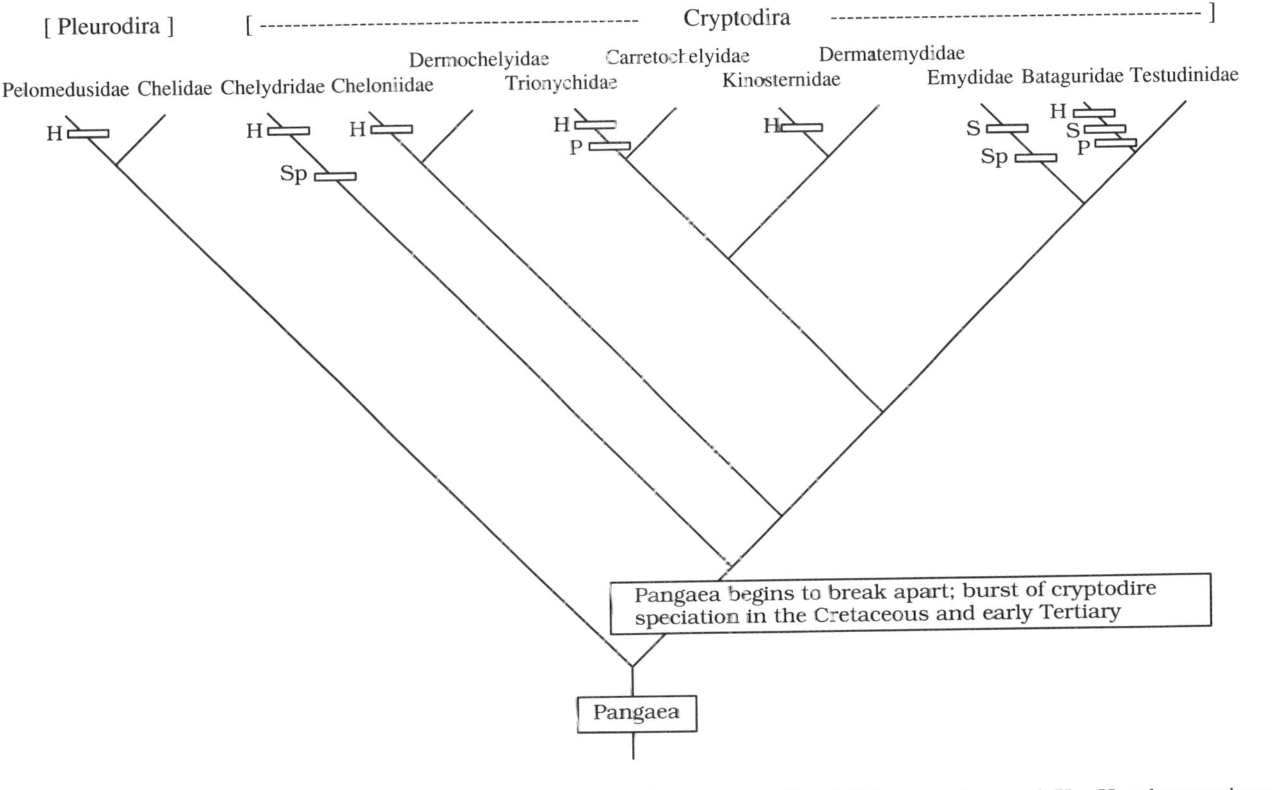

Figure 2.2. Phylogenetic tree for extant turtle families with distributions for some blood flukes superimposed. H = Hapalotrematinae; Sp = *Spirorchis* and close relatives (*Henotosoma, Haematotrema,* and *Aphanospirorchis*); S = *Spirhapalum;* P = *Plasmiorchis* (and *Hemiorchis*). Redrawn and modified from Platt (1992).

primarily with turtles in the families Emydidae (including sliders, cooters, and painted, chicken, spotted, and wood turtles) and Bataguridae. The emydids comprise 34 species occurring in Europe, Northern Africa, the former USSR, North and Central America, and the Caribbean. The possibly paraphyletic batagurids (Gaffney and Meylan 1988) comprise 54 species distributed in Central America, northern South America, the Mediterranean portions of Europe and Africa, and Asia. *Spirhapalum* species have been reported from both emydid and batagurid hosts: *S. polesianum* from *Emys orbicularis* in Poland and the far eastern portion of the former USSR (Ejsmont 1927; Scharpilo 1960) and *S. elongatum* from *Cyclemys amboinensis* in Malaysia and Borneo (Rohde et al. 1968; Brooks and Palmieri 1979). *Plasmiorchis* species inhabit batagurid and trionychid hosts in India, whereas *Spirorchis* species inhabit predominantly emydid hosts in North America, the exceptions being the nominal species *S. haematobium, S. magnitestis,* and *S. minutum,* which inhabit the snapping turtle, *Chelydra serpentina* (Chelydridae). Examination of Figure 2.2 shows that the Spirochinae are noticeably absent in the Testudinidae. It seems likely that tortoises have secondarily lost the ancestral association with spirorchids because their highly terrestrial habits would not bring them into contact with cercariae on a regular basis, if at all. (However, *Terrapene ornata* and *T. carolina,* two species of North American box turtles, are known occasionally to host *Telorchis robustus,* a digenean species that uses anuran tapdoles as second intermediate hosts, indicating that these terrapenes have some contact with aquatic habitats.)

We can begin to detect two coevolutionary patterns from this preliminary data base. The first pattern is concerned with the evolutionary diversification of the Hapalotrematinae. These flukes appear to have enjoyed a long relationship with their hosts extending into the early Mesozoic prior to the breakup of Pangaea. They continued to diversify while their cryptodire hosts were speciating during the Cretaceous following the breakup of the continents. Since they were well established throughout Asia, Eastern Europe, North and Central America, Africa, and India by the time the ancestors of the Emydidae and Bataguridae appeared in the early Tertiary, their association with Indian batagurids appears to represent a secondary invasion of those hosts (Fig. 2.3). If this scenario is accurate, then the spirorchid fauna of pleurodires, comprising the South American and African pelomedusids and the South American and Australian chelids, should be dominated by the Hapalotrematines and/or a yet to be discovered sister group.

The second pattern focuses upon the diversification of the freshwater Spirorchinae (Fig. 2.3). This younger fluke group is hypothesized to have originated in Laurasia with the ancestors of the emydids plus batagurids and testudinids.

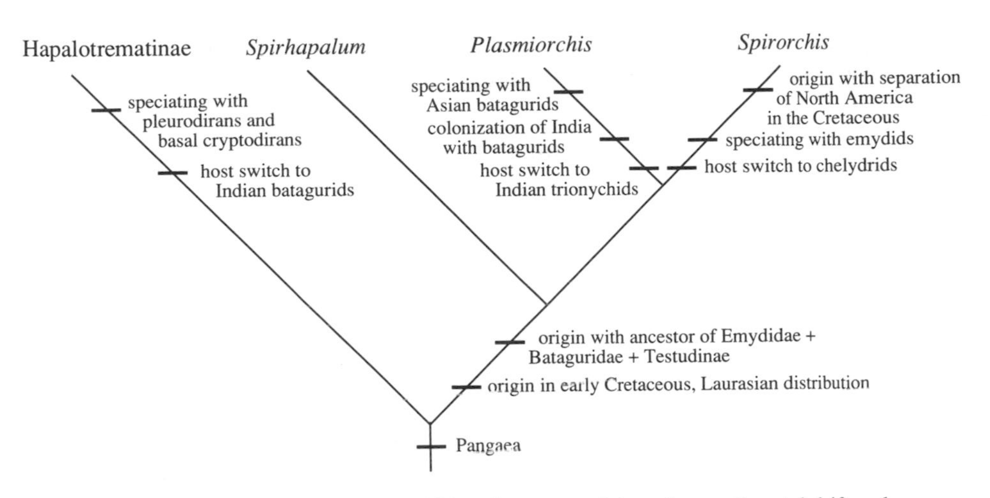

Figure 2.3. Evolution of the Spirorchidae. Summary of the roles continental drift and host switching have played in the diversification of these blood flukes.

Spirhapalum maintains a primitive host distribution pattern, occurring in batagurids and emydids throughout Europe and Asia. The origin and diversification of *Spirorchis* and the North American emydid fauna occurred following the separation of North America from Europe in the Cretaceous, suggesting that the *Spirorchis* species found in the snapping turtle represent episodes of host switching. Given this hypothesis, the only *Spirorchis* species we would expect to find in neotropical turtles are those derived from emydids that colonized South America relatively recently from North America. Finally, the evolution of *Plasmiorchis* appears to be closely tied to the evolution of batagurids, whose diversity is concentrated in Asia. *Plasmiorchis* and the Indian batagurines would therefore be descended from Asian species that colonized India during the early Tertiary (ca. 60 mya) when India came into contact with Asia (Owen 1983; Patterson and Owen 1991). If this scenario is accurate, we would expect to find *Plasmiorchis*-like species in Asian batagurids, but not in turtles endemic to South America, Africa, or Australia.

At the moment, there are not enough data to perform a more thorough investigation of these flukes and their turtles. This study, however, clearly demonstrates that a phylogenetic analysis of even a poorly known group can produce some tantalizing hypotheses and predictions. These, in turn, help a researcher to determine which pieces of information, among an often overwhelming absence of data, must be collected in order to corroborate or modify that evolutionary explanation.

ALCID BIRDS, PINNIPED MAMMALS, AND TAPEWORMS: DEATH OR DIVERSITY IN THE ARCTIC?

Ask most biologists about the tropical regions of the world and they will respond with enthusiastic descriptions of almost overwhelming species richness. Those same biologists, when questioned about high-latitude (Arctic and Antarctic) biotas, will paint a bleak picture of limited diversity and of simple trophic interactions. Since there is substantial fossil evidence that the high-latitude regions of the world once supported a richer biodiversity than exists there today, the current paucity of species is thought to reflect widespread extinctions in those areas. The story is straightforward: the advent of global cooling fragmented populations of ancestral species both in the tropics and at high latitudes. In the tropics, this isolation resulted in numerous episodes of peripheral isolates allopatric speciation. Upon the return to balmier and wetter conditions, the fledgling species dispersed and intermixed with each other, producing the complex interactions that characterize high-diversity food webs. In the polar areas, on the other hand, the effects of cooling were more severe. Rather than speciating, most isolated populations failed to adapt and became extinct. What we see today, then, are the remnants of a preexisting flora and fauna that managed to survive the inclement weather.

This view of the Arctic as a place where clades go to die is being challenged, in part, by the intriguing results of Hoberg's (1986, 1987, 1992; Hoberg and Adams 1992) work with the tapeworms of alcids (puffins, murres, auks, and their relatives) and pinnipeds (seals, sea lions, walruses, and their relatives). Hoberg recognized that the parasite-host assemblages in Arctic habitats exhibited some important features. First of all, the cestode fauna in alcids and pinnipeds is depauperate, existing within relatively confined geographical boundaries; that is, the associations are restricted to the Arctic, even though members of the host group and the closest relatives of the parasite groups live outside the Arctic (Fig. 2.4b). Second, comparison between parasite and host phylogenies demonstrates substantial incongruence. For example, mapping hosts above the phylogenetic tree for the cestode genus *Anophryocephalus* indicates that the phocids and the otariids are sister groups (Fig. 2.4a). When the relationships among the pinnipeds based upon characters of the hosts themselves are examined (Fig. 2.4b) we discover that, contrary to what the parasites tell us, otariids are the oldest, and phocids among the youngest, members of the clade. These two hypotheses are obviously substantially incongruent.

The picture is similar for alcids, in which parasite relationships derived from the cestode genus *Alcataenia* (Fig. 2.5a) place *Uria* and *Cepphus* as sister groups, *Aethia* as the sister group to *Uria + Cepphus,* and so on, in marked

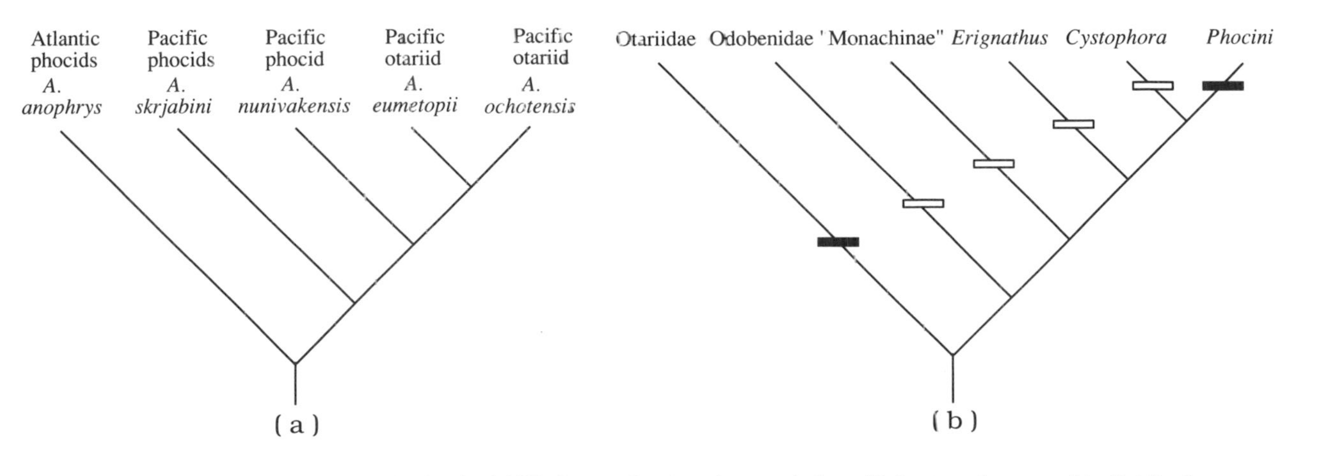

Figure 2.4. Speciation of tapeworms in pinnipeds. (a) Phylogeny for *Anophryocephalus* with hosts and geographic distributions superimposed above each species. (b) Phylogeny for the pinnipeds. Open bars indicate no association with *Anophryocephalus* ; black bars indicate an association with *Anophryocephalus*. Examination of the two phylogenetic trees reveals marked incongruences. For example, the two youngest species of tapeworm (*A. ochotensis* and *A. eumetopii*) are associated with the basal member of the pinnipeds (Otariidae). Redrawn and modified from Hoberg (1992).

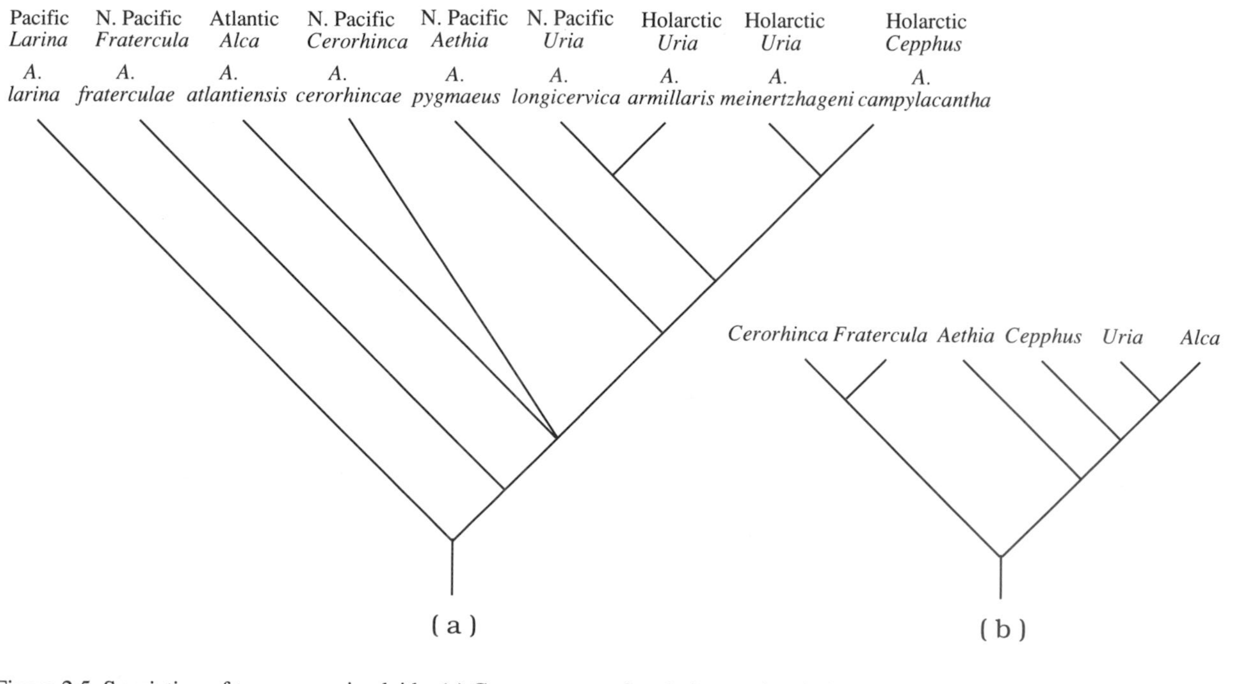

Figure 2.5. Speciation of tapeworms in alcids. (a) Consensus tree for phylogenetic relationships within *Alcataenia* with hosts and geographic distributions superimposed above each species. (b) Phylogeny for the Alcidae, simplified to highlight points of incongruence between the host phylogeny and the sequence of host appearance on the parasite's tree. Redrawn and modified from Hoberg (1992).

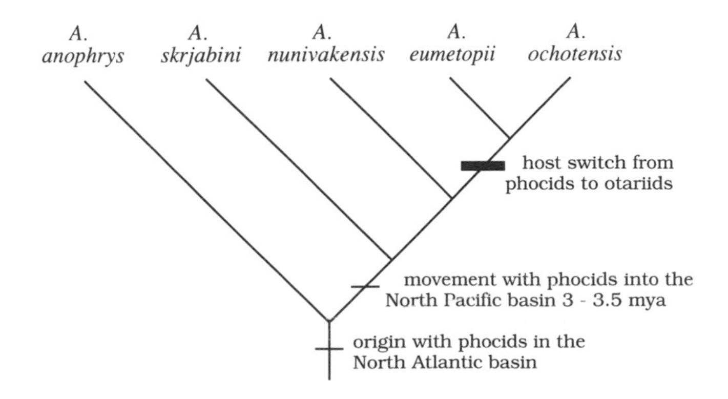

Figure 2.6. Phylogeny for *Anophryocephalus* highlighting important events in its speciation. Redrawn and modified from Hoberg (1992).

contrast to the relationships proposed for the birds (Fig. 2.5b). Interestingly, speciation by host switching in both parasite groups appears to have occurred among hosts of similar trophic ecological requirements, suggesting that ecological determinism could have also played a role in the diversification of these tapeworms.

These observations led Hoberg to conclude that the parasite-host assemblages in the Arctic were not remnants of a previous biota, but were structured by sequential host switching and geographical colonization and extreme climatic variation during the Pleistocene. As the glaciers advanced, sea levels dropped, trapping populations of seabirds, sea mammals, and their parasites in refugia scattered around the Arctic, Subarctic, and Boreal zones. This increased the potential for parasite speciation via both vicariance and host switching. Periodic glacial retreats and rises in ocean levels allowed the confined species to expand their ranges, increasing opportunities for more host switching. Patterns of speciation within *Anophryocephalus* and *Alcataenia* are compatible with this theory of refugial expansion and contraction. For example, the basal member of *Anophryocephalus, A. anophrys,* is hypothesized to have originated in the North Atlantic basin with a phocid host (Fig. 2.6). Following the movement of phocids through the Bering Strait into the North Pacific Basin (3.5–3 mya), the tapeworms experienced both cospeciation with their seal hosts and speciation via host switching within phocids and to stellar sea lions (for details see Hoberg and Adams 1992). So both the seals and their tapeworms diversified during this harsh period in the earth's history.

At approximately the same time that seals were moving through the Bering Strait, an *Alcataenia* species was switching from larids to horned puffins

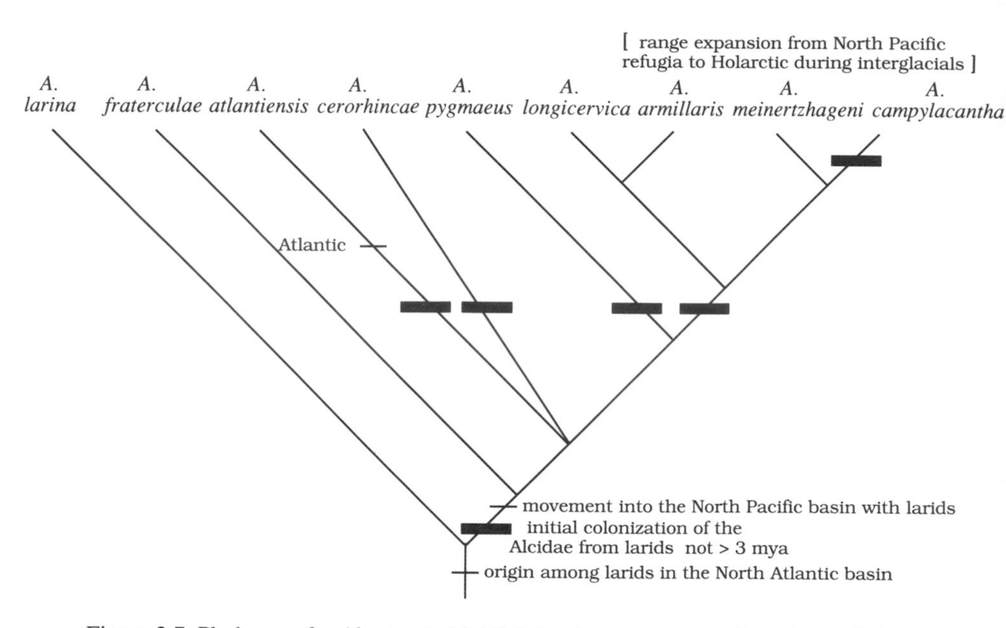

Figure 2.7. Phylogeny for *Alcataenia* highlighting important events in their speciation. Black bars indicate instances of host switching. Redrawn and modified from Hoberg (1992).

(Fig. 2.7: producing *A. fraterculae*). This first switch, also in the North Pacific basin, presaged these cestodes' traveling tendencies, and further speciation occurred following jumps to puffins (*A. cerorhinca*), razorbills (*A. atlantiensis*), auklets (*A. pygmaeus*), and guillemots (*A. campylacantha*). The three tapeworm species in murres are more enigmatic. That association began with a host switch from auklets, and may have involved either further switching within or cospeciation with the murres. In order to differentiate between these two possibilities, we need a phylogeny for *Uria*.

Hoberg's (1992) *boreal refugium hypothesis* suggests that the dramatic environment fluctuations characterizing the Pliocene-Pleistocene glaciation promoted the diversification, not the extinction, of some arctic and boreal organisms (see also Siegel-Causey's [1991] discussion of the biogeography of shags in the western North Pacific Ocean). Recent phylogenetically based studies of the effects of Pliocene-Pleistocene refugia in Amazonia and the eastern highlands of North America (see Brooks and McLennan 1991 and references therein) demonstrated that much of the current diversity predates the origin of those refugia. Paradoxically, then, it appears that while species in the tropics were experiencing extinction and stasis, the flora and fauna in the Arctic were busily becoming extinct or speciating. So, rather than being simply the place where clades went to die, the Arctic may also have been one of the few places where evolutionary diversification took place during the last glacial period.

TREMATODES AND FROGS: THE IMPORTANCE OF LIFE CYCLES

Frogs are an old, widespread, and diverse group hosting a bewildering variety of parasites. The relationships between frogs and their parasites have provided support for ideas about continental drift and evolution (Johnston and Metcalf: see Chapter 1) and have been used as ecological indicators of host habitat preference (Brandt 1936; Prokopic and Krivanec 1975; Brooks 1976). This system would appear, at first glance, to provide an excellent base for parascript studies; however, only a few of the many parasite groups inhabiting frogs have been analyzed phylogenetically. One of those groups is the digenetic trematode genus *Glypthelmins* (Appendix; Fig. 2.8).

Phylogenetic analysis combined with information about geographic distribution, hosts, and life cycles highlights three major evolutionary components in this group. First, superimposing the geographic distribution of the genus on a map of Pangaea (Fig. 2.9) uncovers two collections of neotropical species. One clade, occurring in Gondwana (South America) is the sister group of the rest of the species in the genus, all of which occur in Laurasia. The second group is divided into a clade of neotropical species, a clade comprising three species occurring in Eurasia, and a clade restricted to North and Central America. These observations suggest that *Glypthelmins* is a relatively ancient lineage that was widespread across South America into Laurasia before Pangaea drifted apart. This breakup isolated an assemblage of neotropical species from the Laurasian species, then the further splitting of Laurasia isolated distinct Eurasian and North American groups.

Next we concentrate less on the abiotic and more on the biotic influences on these digeneans' evolution. If the diversification of *Glypthelmins* was associated with the diversification of their anuran hosts, we should find evidence of Pangaean origins for the frogs. This is, in fact, exactly what we find. Fossil evidence indicates that anurans were alive and well in the Triassic, and that extant anuran families existed during the turbulent beginnings of Pangaea's demise in the Jurassic (Duellman and Trueb 1986). That same evidence, however, also indicates that anurans were widespread and diversified throughout Gondwana. Since *Glypthelmins* is not currently distributed throughout the

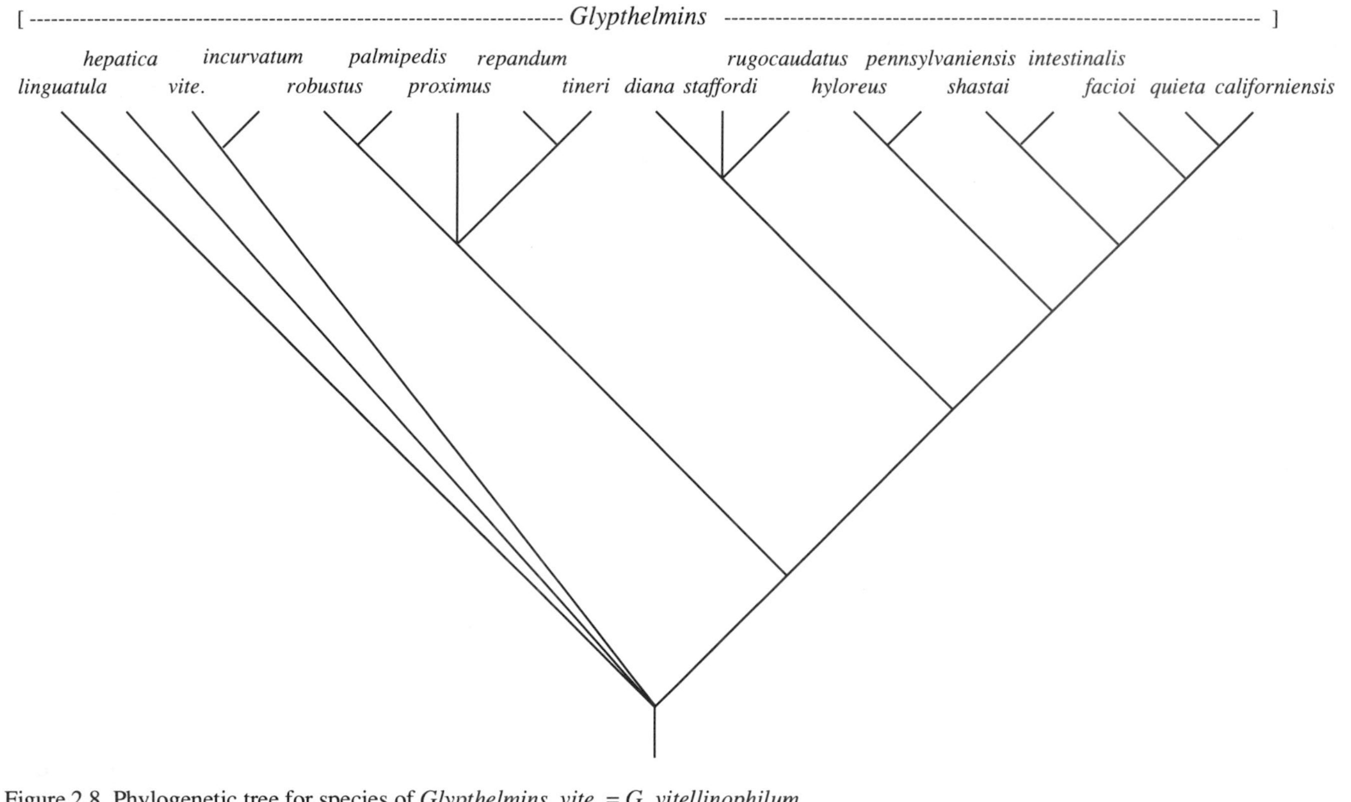

Figure 2.8. Phylogenetic tree for species of *Glypthelmins*. *vite*. = *G. vitellinophilum*.

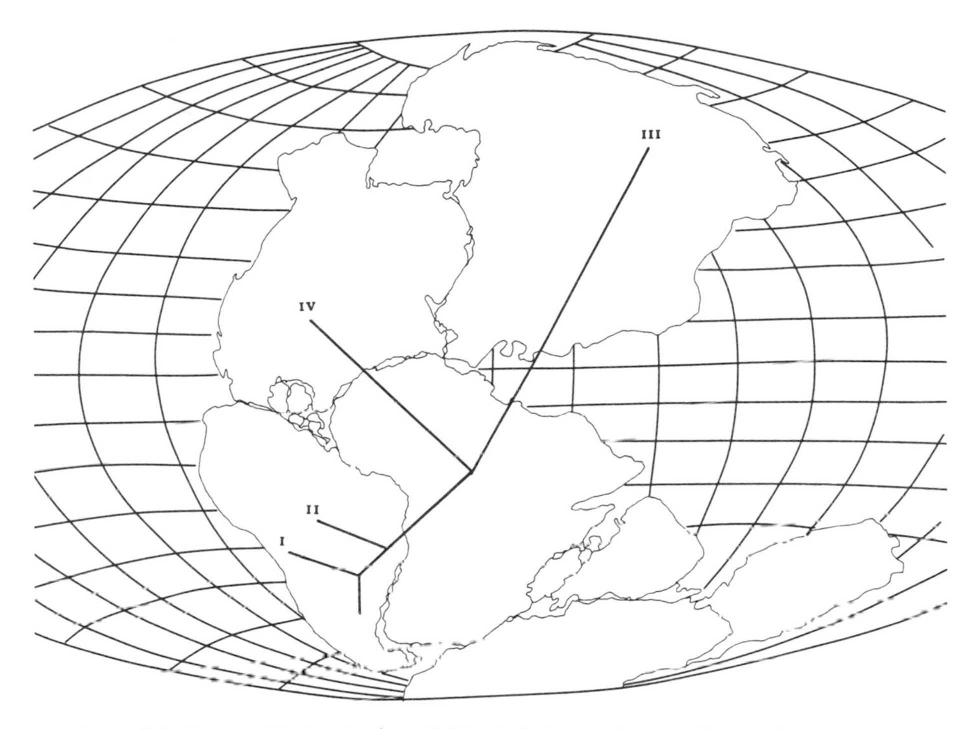

Figure 2.9. Geographic distribution of *Glypthelmins* species superimposed on a map of Pangaea. I = *G. linguatula, G. hepatica, G. vitellinophilum,* and *G. incurvatum;* II = *G. robustus, G. palmipedis, G. proximus, G. repandum,* and *G. tineri;* III = *G. diana, G. staffordi,* and *G. rugocaudutus;* IV = *G. hyloreus, G. pennsylvaniensis, G. shastai, G. intestinalis, G. facioi, G. quieta,* and *G. californiensis.*

fragments of that supercontinent, we need to explain why the pattern of parasite-host diversification in Gondwana differs from the pattern in Pangaea. There are a number of potential reasons for this. For example, *Glypthelmins* may have originally been more widely distributed throughout Gondwana than its current restriction to South America indicates. Or perhaps we have discovered only a small fraction of the Gondwanian *Glypthelmins,* a finding that further research will clarify. Finally, it is possible that "*Glypthelmins,*" which lacks a synapomorphy (Appendix), is paraphyletic (i.e., does not include all of the descendants of a common ancestor). If some of the many generalized plagiorchiform digeneans inhabiting the intestinal tracts of anurans throughout Africa, India, and Eurasia (e.g., *Astiotrema, Haplometroides, Laiogonimus, Metaplagiorchis, Opisthioglyphe, Ostioloides,* some members of *Plagiorchis, Tremiorchis,* and *Xenopodistomum*) are eventually found to be members of *Glypthelmins,* then the distribution pattern of the genus throughout Gondwana

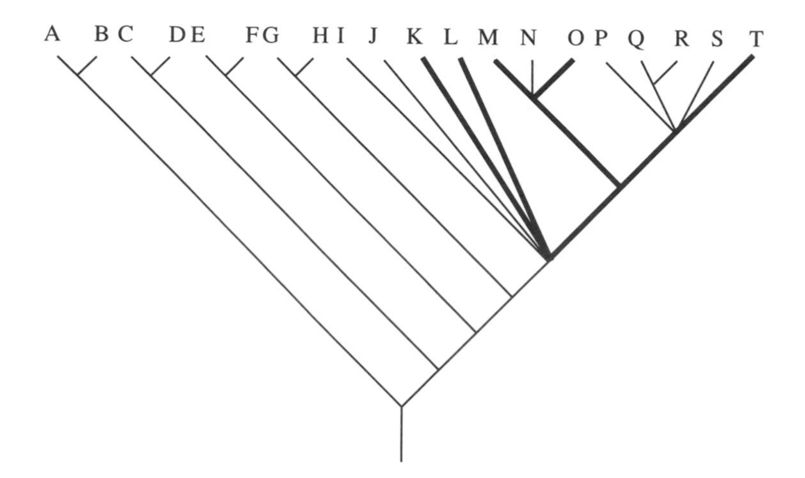

Figure 2.10. Phylogenetic tree for families of extant anurans. A = Leiopelmatidae; B = Discoglossidae; C = Rhinophrynidae; D = Pipidae; E = Pelobatidae; F = Pelodytidae; G = Sooglossidae; H = Myobatrachidae; I = Brachycephalidae; J = Rhinodermatidae; K = Leptodactylidae; L = Bufonidae; M = Pseudidae; N = Centrolenidae; O = Hylidae; P = Dendrobatidae; Q = Hyperoliidae; R = Rhacophoridae; S = Microhylidae; T = Ranidae. Redrawn and modified from Duellman and Trueb (1986). Bold lines denote families known to host species of *Glypthelmins*.

would increase dramatically. In order to clarify this problem we need to collect more data from Gondwanian areas, then establish the monophyly of the group based upon extensive taxonomic revisions and rigorous phylogenetic analysis.

One thing we can do at this stage is compare the phylogenetic relationships of the known and included members of *Glypthelmins* with those of their hosts. Figure 2.10 is a phylogenetic tree for families of extant anurans, taken from Duellman and Trueb (1986). Figure 2.11 is the phylogenetic tree for *Glypthelmins* with the host families mapped onto the branches representing each species. Although there is evidence of much host switching, there is a fair amount of apparent concordance between the host and parasite phylogenies (Fig. 2.12). This supports the idea that *Glypthelmins* is a rather ancient group.

Let us turn our attention away from the "big picture" to the North American *Glypthelmins* clade. These worms are the most diversified group outside of South America. This diversification is correlated with geographic changes, host switches and host diversification, and changes in life cycle patterns (O'Grady 1987). Optimization of hosts onto the phylogenetic tree for the Laurasian members of *Glypthelmins* indicates that ranids are the plesiomorphic host type for the clade (Fig. 2.13). Two sister groups of *Glypthelmins* occur in North America. One of these comprises two species (*G. pennsylvaniensis* and *G.*

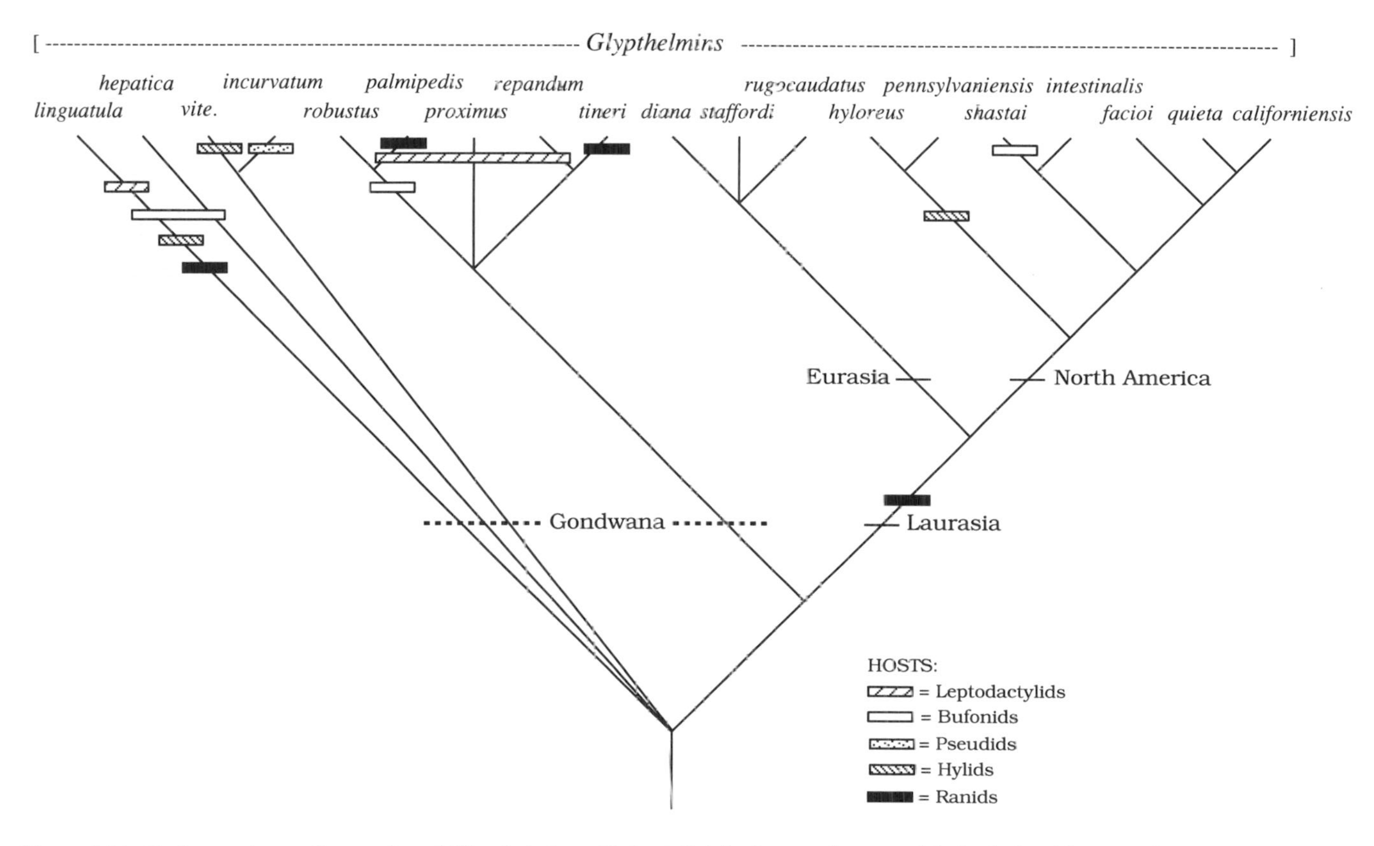

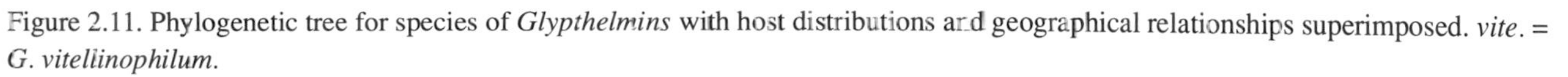

Figure 2.11. Phylogenetic tree for species of *Glypthelmins* with host distributions and geographical relationships superimposed. *vite.* = *G. vitellinophilum.*

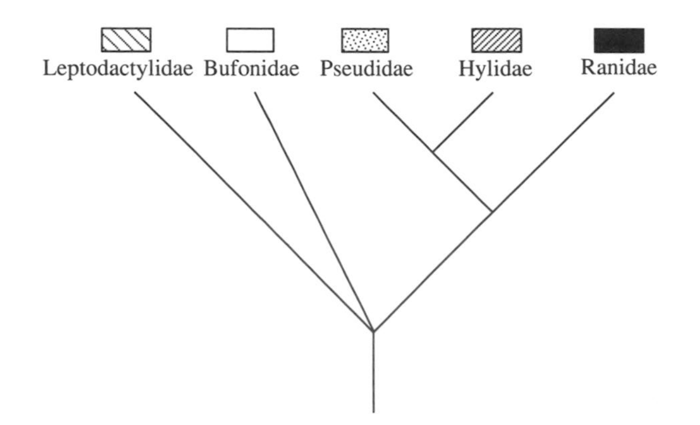

Figure 2.12. Relationships of some anuran families based on phylogenetic relationships of *Glypthelmins* species. Note congruence with boldfaced portion of Figure 2.10.

hyloreus) inhabiting chorus frogs, members of the hylid genus *Pseudacris*. This appears to be the result of a host switch from ranids to hylids by the common ancestor of *G. pennsylvaniensis* and *G. hyloreus*. Like the older members of the Laurasian group, the second North American clade (*G. intestinalis* + *G. shastai* + *G. facioi* + *G. quieta* + *G. californiensis*) is predominantly restricted to ranids. Within this group, *G. shastai*, known only from *Bufo boreas* in northern California, USA, and Alberta, Canada, represents another apparent host switch.

Perhaps details of the life cycles of these worms might provide some clues to the mystery of how such host switches occurred. Life cycles are known for *G. hyloreus*, *G. pennsylvaniensis*, *G. intestinalis*, *G. quieta*, and *G. californiensis*. All five species display two characters that are plesiomorphic for plagiorchiform digeneans: the miracidium (1) hatches from the egg when ingested by a snail host and (2) undergoes a developmental sequence from a mother sporocyst to rediae lacking pharynges and guts (traditionally termed "daughter sporocysts") to cercariae. In this group, evolution appears to have focused on the cercariae. The plesiomorphic behavior of plagiorchiform cercariae is "emerge from the snail host, penetrate, assisted by a sharp stylet (hence the name 'xiphidiocercaria'), and encyst in an arthropod host, and transform into juveniles called metacercariae." The cercariae of *Glypthelmins* species encyst in anurans rather than arthropods, an apomorphic behavior for the plagiorchiforms, and a plesiomorphic behavior for the North American *Glypthelmins* clade. Cercariae of *G. intestinalis*, *G. quieta*, and *G. californiensis* have penetration stylets, and encyst in the epidermis of the same frogs in which they become adult worms. The frogs become infected when they ingest shed epidermis containing encysted infective metacercariae. The life cycle pattern exhib-

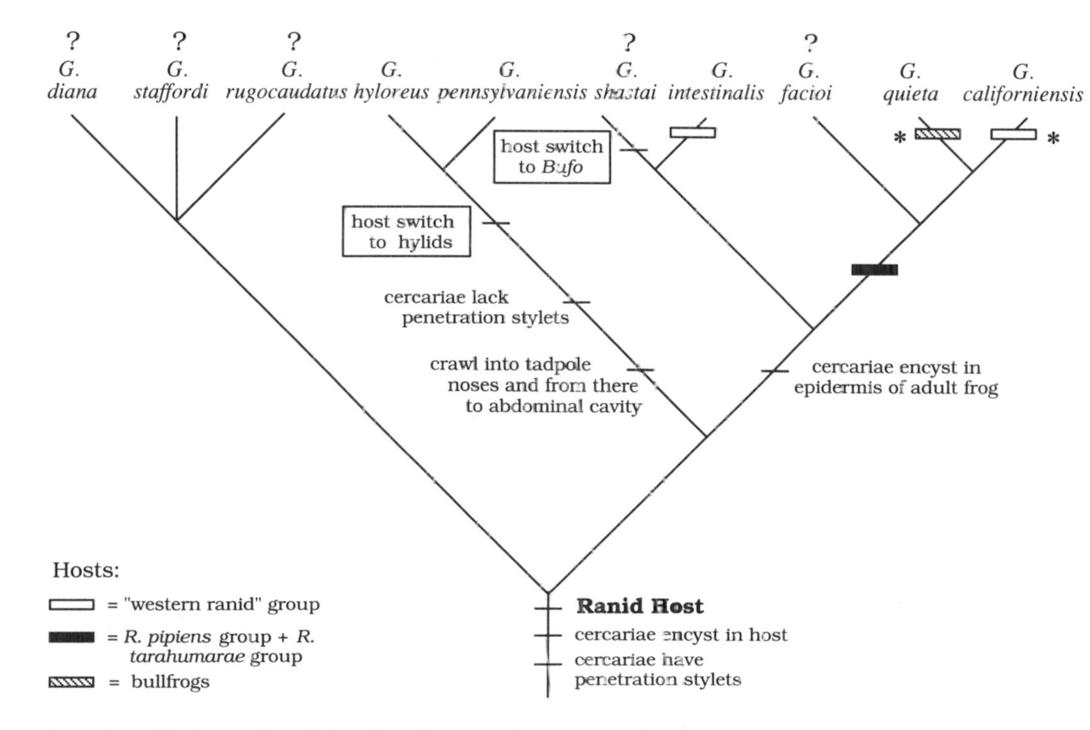

Figure 2.13. Phylogenetic relationships for Laurasian species of *Glypthelmins,* with host superimposed. * = Host switches from a member of the *Rana pipiens* group. ? = Life cycle information not known.

ited by the two species inhabiting hylids appears to be derived relative to the pattern exhibited by the preceding three species. Cercariae of *G. pennsylvaniensis* and *G. hyloreus* lack penetration stylets, and crawl into the nares of their anuran hosts when they are still tadpoles. They migrate into the abdominal cavity where they become unencysted metacercariae, then migrate into the intestine of their hosts and mature when the tadpoles metamorphose into frogs.

These derived characters within the North American clade represent a potentially important correlation between the evolution of a novel life cycle pattern and the colonization of a new anuran host group. The hylids that host *G. pennsylvaniensis* and *G. hyloreus* have palustrine habits, the adults coming to water only to breed. Once they metamorphose from tapdoles into young adults, members of *Pseudacris* and their relatives (e.g., *Hyla regilla* and *H. crucifer*) leave their natal ponds and do not return until the following spring, when they breed for a very short time. In addition, chorus frogs do not live very long when compared with the ranid hosts for the other members of *Glypthelmins* (hylids are essentially annual species). Consequently, there is a much narrower window in time for the hylids to become infected with *Glypthelmins* than for the ranids, many of whom spend a great deal of time in and around water, and may live for several years. That window of time could be extended by the infection of tadpoles rather than adults, and this would suggest that the infection of tadpoles is a derived trait. Although it sounds intuitively plausible, until we have life cycle information for the more basal members of *Glypthelmins,* we cannot determine whether this scenario is supported by data or by imagination.

The members of the ranid-inhabiting clade of North American *Glypthelmins* exhibit an interesting pattern of association with respect to the evolution of some of their anuran hosts and the geological evolution of western North America and Central America. The group including *G. intestinalis* and *G. shastai* exhibits a montane distribution in the central Rocky Mountains and coastal mountain ranges. *Glypthelmins shastai* inhabits *Bufo boreas* and is known only from two localities (northern California and southern Alberta). As we suggested previously, this distribution appears to reflect a host switch from ranids to a bufonid. *Glypthelmins intestinalis* inhabits *Rana pretiosa,* the Spotted frog, throughout the frog's range. *Rana pretiosa* is a member of a group of North American frogs called the "western ranids" by anuran systematists (Fig. 2.14; see Case 1978). This group includes *Rana aurora,* the Red-legged frog; *R. cascadae,* the Cascades frog; *R. boylii,* the Foothills Yellow-legged frog; and *R. muscosa,* the Mountain Yellow-legged frog—species that are rapidly approaching extinction owing to habitat destruction and the introduction of predatory fish to their remaining habitats in the California mountains.

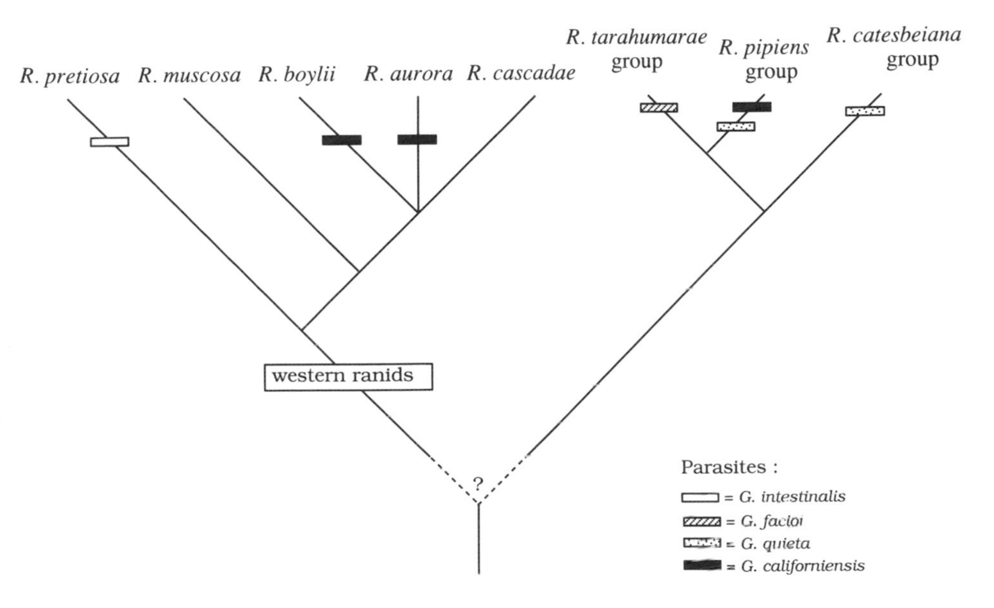

Figure 2.14. Distribution of *Glypthelmins* species superimposed on a phylogenetic tree depicting relative phylogenetic relationships of ranid groups occurring in North and Central America.

Glypthelmins facioi, G. quieta, and *G. californiensis* inhabit a variety of ranid hosts. The type host for *G. facioi* is listed as *Rana pipiens*. When *G. facioi* was described, *Rana pipiens* was the name given to frogs called leopard frogs that ranged from northeastern Canada to Panama. This is now recognized as a group of species, including the "*Rana pipiens* group," that may number close to 20 distinct species (Fig. 2.15; Hillis et al. 1983; Hillis and Davis 1986). The host for *G. facioi* is most likely *R. palmipes,* which also hosts *G. palmipedis* in northern South America. *Glypthelmins quieta* has been reported from numerous members of the *R. pipiens* group throughout North America east of the Rocky Mountains, including *R. pipiens,* the Northern Leopard frog; *R. blairi,* the Western Leopard frog; *R. sphenocephala,* the Southern Leopard frog; and *R. palustris,* the Pickerel frog. It also occurs commonly in *R. catesbeiana,* the Bullfrog, and *R. clamitans,* the Green or Bronze frog, two members of a clade of ranid frogs that is more closely related to the *R. pipiens* group than either is to the western ranids. Although *G. californiensis* was described originally from a member of the western ranids, *R. aurora,* in northern California, it is also known from a member of the *R. pipiens* group, *R. montezumae,* in Mexico. From this pattern it appears that this group of three *Glypthelmins* species arose in association with the common ancestor of the *R. pipiens* group, and

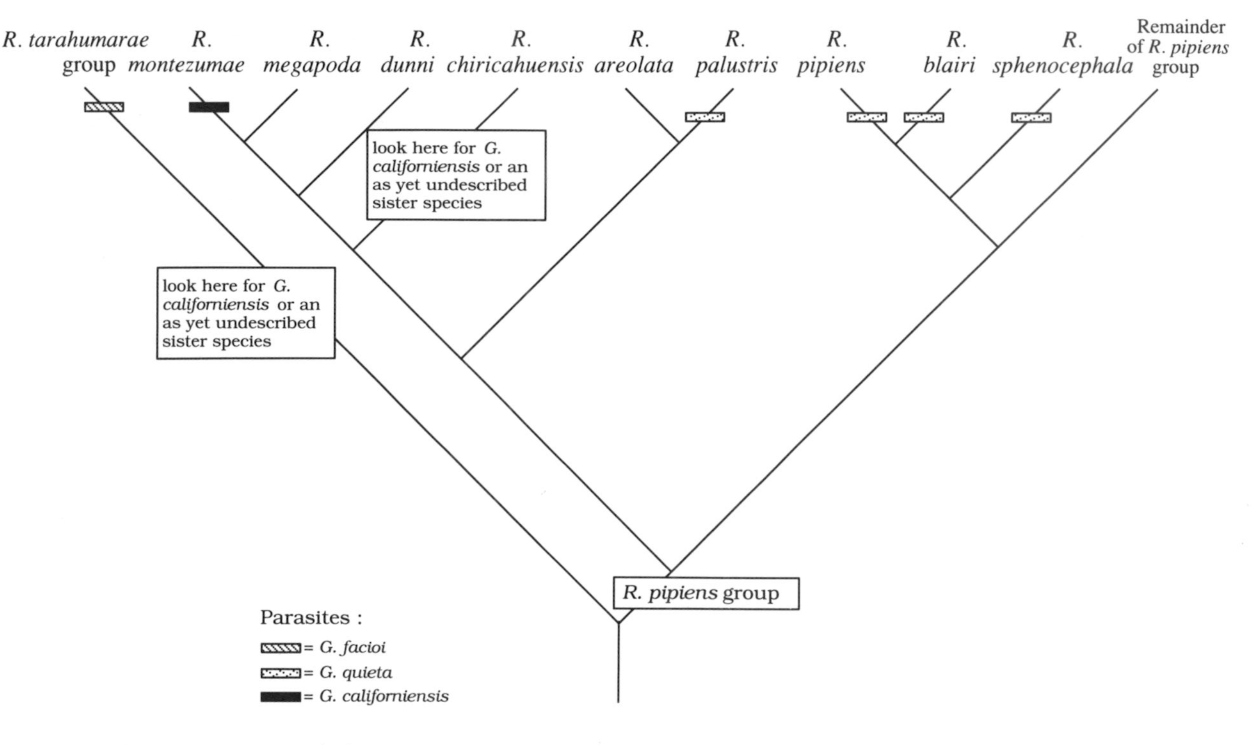

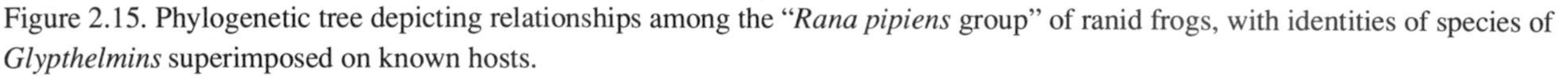

Figure 2.15. Phylogenetic tree depicting relationships among the "*Rana pipiens* group" of ranid frogs, with identities of species of *Glypthelmins* superimposed on known hosts.

that the occurrence of *G. californiensis* in *R. aurora* is the result of a host switch.

Putting the preceding bits of information together reveals that the two subgroups of *Glypthelmins* inhabiting ranids and bufonids in North America each exhibit a different geographic distribution pattern, one northern and one southern. Case (1978) suggested that (1) the western ranids are most closely related to ranids found in Eurasia, and (2) the *Rana pipiens* group, its sister group the *Rana tarahumarae* group, and the *Rana catesbeiana* group are most closely related to species of ranids occurring in Laurasian areas to the southeast of present-day Central and North America. The *R. pipiens* group appears to have originated in the southern part of North America/Central America and moved northward, contributing to a common vicariant pattern exhibited by certain plants, fishes, salamanders, snakes, birds, and mammals, and dating from mid-Cenozoic orogenic and climatic changes (see Rosen 1978). It would also seem that members of the *R. pipiens* group must have come into contact with members of the western ranids, whereupon *G. californiensis* continued moving north by hopping from *R. pipiens* group hosts to *R. aurora* and *R. boylii,* two species of western ranids. This would explain the occurrence of two different, nonsister, species of *Glypthelmins* (*G. californiensis* and *G. intestinalis*) in western ranids.

Is there any evidence to support the hypothesis that the relevant ranid hosts were sympatric at some point in the past? *Rana montezumae* is not sympatric with either *R. aurora* or *R. boylii.* On the other hand, *R. tarahumarae,* the sister species of the *R. pipiens* group, and *R. chiricahuensis,* a member of the *R. pipiens* group, both occur in areas lying between the present-day distribution of *R. montezumae* and *R. aurora* and *R. boylii.* At the moment neither species has been examined for parasites. If the preceding scenario is correct, we would expect to find either *G. californiensis* or an as yet undescribed sister species of *G. californiensis* inhabiting those frogs.

Gylpthelmins represents a potential gold mine of information about the interplay of coevolution, host switching, geographical isolation, climatic change, host biology, and parasite life cycle diversification during the course of evolution. It also represents an ideal system for collaborative efforts among systematists and ecologists interested in both the hosts and the parasites.

PARASCRIPT AND BROOKS PARSIMONY ANALYSIS: THE PARASITES, THE HOSTS, THE METHOD

Studies that make use of Brooks parsimony analysis (BPA: Wiley 1988) are based upon the belief that parasites can provide valuable information about the evolution of close ecological associations. This perspective is not new. Recall

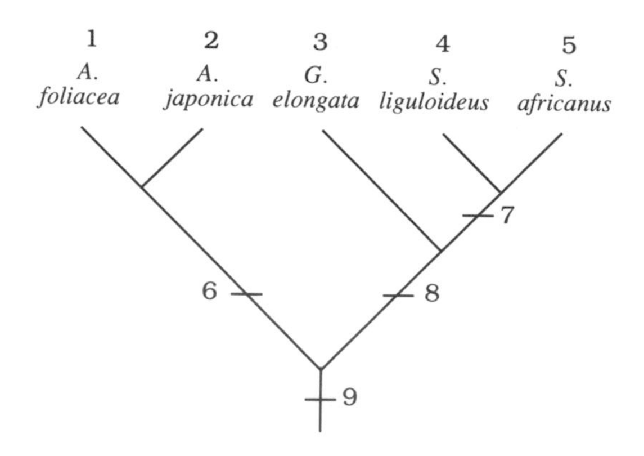

Figure 2.16. Phylogenetic relationships for five species of amphilinids based upon morphological characters. Redrawn and modified from Brooks and McLennan (1991). *A.* = *Amphilina; G.* = *Gigantolina; S.* = *Schizochoerus.*

that the geographic school was based upon the assumption that, since most parasites were so highly host specific, their phylogenetic relationships would parallel the phylogeny of their hosts. What is new in this approach is the recognition that the phylogenetic relationships of parasites can be reconstructed based upon characters of the parasites themselves, thus freeing evolutionary biologists from the specter of circular reasoning in assessing the relationships among hosts and geography during the course of parasite evolution.

We will begin this next level of investigation by demonstrating BPA using a group of rare flatworms, the Amphilinidea, as an example (for a more detailed discussion of the method, see Brooks and McLennan 1991 and Wiley et al. 1991). Amphilinids, the sister group of the species-rich true tapeworms, are a small (eight known species) but widespread group of parasites that live in the body cavities of freshwater and estuarine ray-finned fishes and in one species of freshwater turtle. The first step is to reconstruct the phylogenetic relationships for the parasite group based upon characters of the parasites (Fig. 2.16). To simplify the demonstration, we will begin with only five of the eight species (we will include the other three later). Second, reconstruct the historical connections among the study areas (an area cladogram) using geological evidence (e.g., Dietz and Holden 1966; Owen 1983). Next, prepare a list placing the species of amphilinids with the areas in which they occur. Now, treat the phylogenetic relationships of the five amphilinid species as if they were a completely polarized multistate transformation series, in which each taxon and each internal branch of the tree is numbered. Each species of amphilinidean now has a code that indicates both its identity and its common ancestry.

Table 2.2
Data matrix listing binary codes indicating phylogenetic relationships among five species of amphilinid flatworms

Taxon	Binary code
Amphilina foliacea	100001001
A. japonica	010001001
Gigantolina elongata	001000011
Schizochoerus liguloideus	000100111
S. africanus	000010111

Construct a data matrix for these codes by using the convention that the presence of a number in the species code is listed as 1 and the absence of a number in the species code is listed as 0. The phylogenetic relationships of the study group are now encased within the binary codes. You can check this by performing a phylogenetic systematic analysis for species 1–5 using the binary codes from Table 2.2. If nothing is amiss, the analysis will reproduce the original phylogenetic tree for the group.

Next, replace the species names in Table 2.2 with their geographic distributions (Table 2.3). Finally, construct a new area cladogram based, this time, on the phylogenetic relationships of the species. This produces a picture of the areas' historical involvement in the parasites' evolution. In this example the area cladogram based upon geological evidence (Fig. 2.17) and the area cladogram reconstructed from the phylogenetic relationships of the taxa occurring in

Table 2.3
Data matrix listing binary codes indicating phylogenetic relationships among five species of amphilinid flatworms and the geographic areas in which they occur

Area	Binary code
Eurasia	100001001
North America	010001001
Australia	001000011
South America	000100111
Africa	000010111

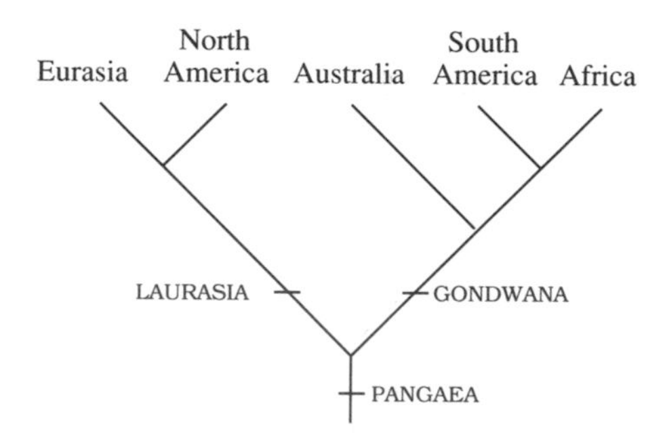

Figure 2.17. Area cladogram based upon geological evidence for five geographical areas inhabited by amphilinideans. Redrawn and modified from Brooks and McLennan (1991).

each region (Fig. 2.18) are identical. In addition, the consistency index for the area cladogram is 100%, indicating that all the speciation events postulated by the phylogenetic tree are congruent with the area cladogram. We can hypothesize, therefore, that the occurrence of the study species in the study areas is a result of a long history of association between amphilinids and the areas in which they now occur.

Dispersal of organisms is a common phenomenon in nature, so real data sets will generally show less than the 100% congruence depicted in the preceding example. The complete phylogenetic tree for the Amphilinidea, including *Gigantolina magna* (taxon 6), *Schizochoerus paragonopora* (taxon 7), and *Schizochoerus janickii* (taxon 8), is shown in Figure 2.19. The three additional species of amphilinids inhabit South America (*S. janickii*) and Indo-Malaysia

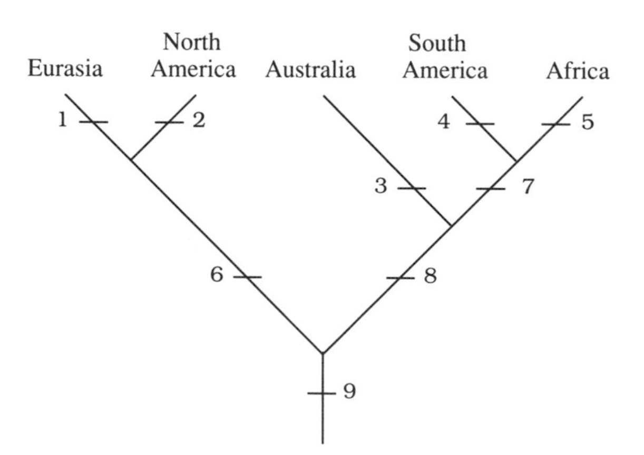

Figure 2.18. Area cladogram reconstructed from the phylogenetic relationships of the amphilinidean taxa occurring in each region. Compare this with the area cladogram depicted in Fig. 2.17. Redrawn and modified from Brooks and McLennan (1991).

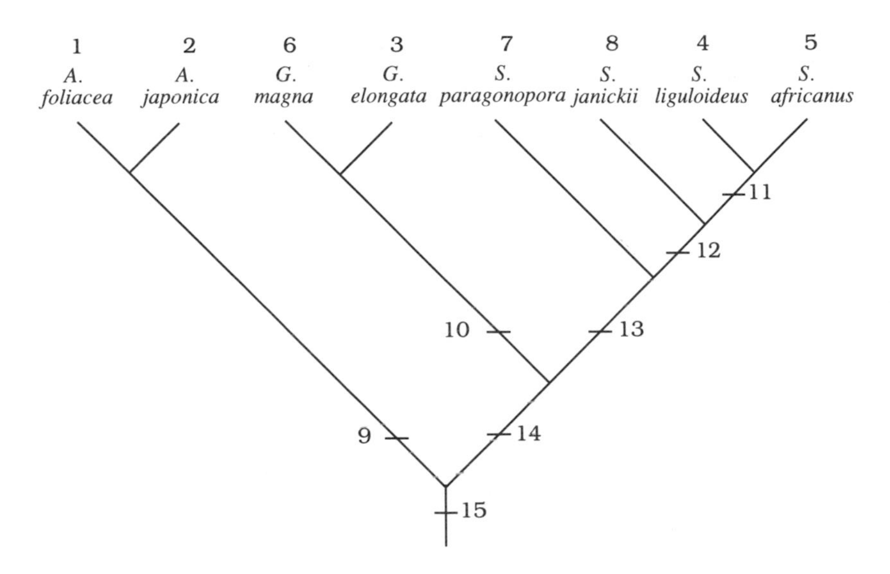

Figure 2.19. Phylogenetic relationships for all eight species of amphilinids numbered for BPA. *A.* = *Amphilina*; *G.* = *Gigantolina*; *S.* = *Schizochoerus*. Redrawn and modified from Brooks and McLennan (1991).

(*G. magna* and *S. paragonopora*). Species codes are converted to binary codes and listed for each area in Table 2.4.

The area cladogram reconstructed from the new data matrix (Fig. 2.20) has a consistency index of 93.8%. Note that 10 appears twice on the tree. This indicates that the common ancestor (species 10) of *Gigantolina elongata* (species 3) and *G. magna* (species 6) occurred in both Australia and Indo-Malaysia.

Table 2.4
Data matrix listing binary codes for species of amphilinid flatworms inhabiting six geographic areas

Area	Binary code
Eurasia	100000001000001
North America	010000001000001
Australia	001000000100011
South America	000100010011111
Africa	000010000011111
Indo-Malaysia	000001100100111

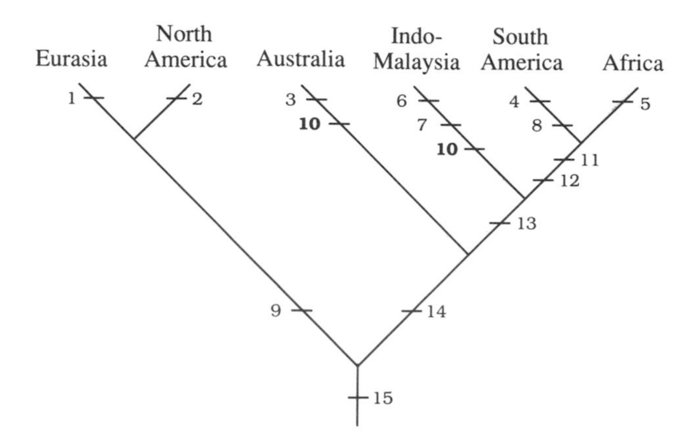

Figure 2.20. Area cladogram reconstructed from the phylogenetic relationships of all eight species of amphilinids. Note the dual placement of ancestral taxon 10. Redrawn and modified from Brooks and McLennan (1991).

Its occurrence in Australia coincides with the geological history of the areas, so we explain this by saying that *G. elongata* evolved in the same place as its ancestor. On the other hand, the occurrence of 10 in Indo-Malaysia does not coincide with the geological history of the areas. We explain this by hypothesizing that at least some members of ancestor 10 dispersed to Indo-Malaysia, where the population evolved into *G. magna*. Hence, the occurrence of *Schizochoerus paragonopora* (species 7) in Indo-Malaysia is due to common history,

Table 2.5
Data matrix listing binary codes for species of amphilinid flatworms inhabiting six geographic areas

Area	Binary code
Eurasia	100000001000001
North America	010000001000001
Australia	001000000100011
South America	000100010011111
Africa	000010000011111
IM_1	000001000100011
IM_2	000000100000111

Note: Indo-Malaysia is listed as two different areas, one for species 6 (IM_1) and the other for species 7 (IM_2).

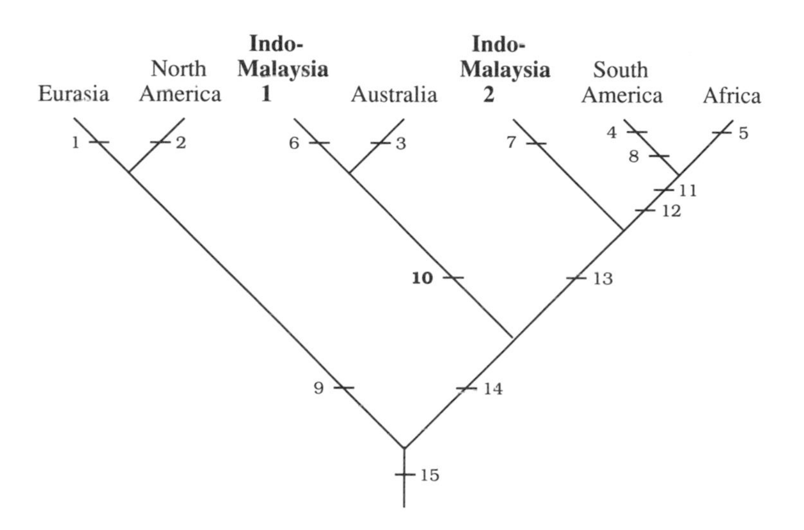

Figure 2.21. New area cladogram obtained when Indo-Malaysia is listed as two areas. Redrawn and modified from Brooks and McLennan (1991).

whereas the occurrence of *G. magna* in Indo-Malaysia is due to the dispersal of its ancestor into that area. If this is true, then what we have called "Indo-Malaysia" is, from historical a perspective, two different areas for those species.

We can test this possibility and further examine the question of ancestor 10's dispersal by recoding the data matrix in Table 2.4, listing *Gigantolina magna* and *Schizochoerus paragonopora* in different subsections of Indo-Malaysia (Table 2.5). When we perform a phylogenetic analysis using this new matrix, we obtain the area cladogram depicted in Figure 2.21.

We now find the two Indo-Malaysian areas in different parts of the geographic cladogram, with IM_1 connected to Australia and IM_2 associated with South America and Africa. The placement of IM_2 is in accordance with the patterns of continental drift, but the placement of IM_1 is not. This "misplacement" strengthens our hypothesis that ancestor 10 did disperse (from Australia into Indo-Malaysia). Interestingly, this dispersion involved a movement into a different habitat than that occupied by *Schizochoerus paragonopora;* IM_1 encompasses estuarine Indo-Malaysian habitats, whereas IM_2 represents freshwater, nuclear Indian subcontinent habitat.

As noted earlier, the species of Amphilinidea live in the body cavities of freshwater and estuarine ray-finned fishes and in one species of freshwater turtle. By replacing geographical areas with hosts, we can use BPA to obtain a picture of the history of involvement of various types of hosts in the evolution-

Table 2.6
Data matrix of hosts for amphilinid flatworms and the binary codes for the parasites and their phylogenetic relationships

Host	Parasite no.	Binary code
Acipenseriformes	1	100000001000001
Acipenseriformes	2	010000001000001
Perciformes	3	001000000100011
Chelonia	4	000100000100011
Siluriformes	5	000010000000111
Osteoglossiformes	6	000001000001111
Osteoglossiformes	7	000000100011111
Osteoglossiformes	8	000000010011111

ary diversification of amphilinideans. Table 2.6 is the data matrix produced when the binary codes for the parasite species are listed with their associated hosts. Phylogenetic analysis of this data matrix produces one host cladogram with a consistency index of 100% (Fig. 2.22). The species of parasites inhabiting acipenseriforms are a monophyletic group, as are those inhabiting osteoglossiforms. However, unless there is something amiss with current theories of evolution, the turtle *Chelodina longicollis* is not the sister group of perciform teleostean fishes, so we must interpret the presence of *Gigantolina magna* in

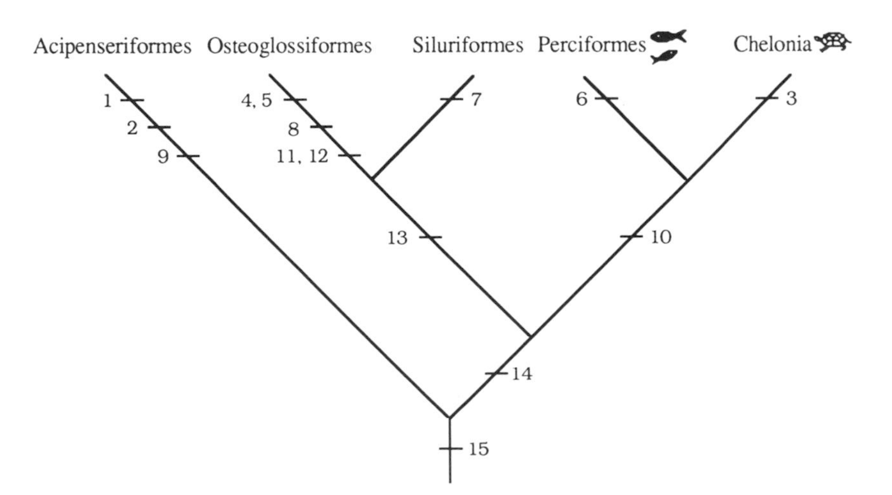

Figure 2.22. Host cladogram based on phylogenetic relationships of amphilinideans. Redrawn and modified from Brooks and McLennan (1991).

this turtle as being the result of a host switch. In addition, contrary to the current phylogenetic analysis of the actinopterygians, this cladogram places siluriform fishes with the osteoglossiforms rather than with the perciforms; therefore, the presence of *Schizochoerus paragonopora* in a siluriform host must be the result of a host switch as well. Mapping the amphilinidean phylogeny onto the relative phylogenetic relationships for the fish and turtle hosts produces a fit between the parasite and host phylogenies of 75% (estimated by the consistency index: see Chapter 3 for a definition). This reemphasizes the *interaction between cospeciation and host switching* (two events) in the evolution of the amphilinideans.

NEOTROPICAL FRESHWATER STINGRAYS AND THEIR PARASITES: A TALE OF TWO OCEANS

Earlier in this chapter, we suggested some attributes of host groups that would make them excellent systems for parascript studies. Does this mean that parasitologists must wait until the evolutionary history of a particular host group is established before studying a given group of parasites? By no means. The publication of phylogenetic systematic studies for different parasite groups, based strictly on characteristics of the parasites themselves, opens the door for parasitologists to lead the way in investigating the history of host groups for which minimal evolutionary information is known.

Elasmobranchs are an old and diverse group of fishes inhabiting areas ranging from marine to freshwater. Marine elasmobranchs retain urea and other organic substances in their blood and tissue fluids, creating an internal osmotic environment that is similar to the surrounding seawater. The rectal gland supplements kidney function, secreting salt in a fluid that has twice the osmotic concentration of body fluids. Euryhaline elasmobranchs function like marine species under conditions of high salinity; however, in less saline waters their urea concentration drops by 50–80%, and rectal gland function is either reduced or stopped (see references in Thorson et al. 1983). Some of these species, such as the bull sharks and sawfish of Lake Nicaragua, may even spend extended periods of time in fully freshwater habitats.

Members of the stingray family Potamotrygonidae, the only elasmobranchs permanently adapted to freshwater habitats, occur throughout the major river systems of eastern South America. They cannot concentrate urea, although they produce some of the necessary enzymes, and their rectal glands are small and apparently nonfunctional. Not even physiological conditioning can induce these rays to concentrate urea or excrete salt (see references in review by

Thorson et al. 1983). The highly modified nature of stingrays relative to sharks and skates, and the absence of totally freshwater species in any other elasmobranch group, led biologists to assume that potamotrygonids are derived from marine ancestors. The presence of a rectal gland and of some of the enzymes for concentrating urea have been accepted as evidence that potamotrygonids are not a particularly old group. Given this conclusion, and the observation that potamotrygonids are restricted to rivers that empty into the Atlantic Ocean, ichthyologists have tended to assume that the ancestor of the potamotrygonids was an Atlantic marine or euryhaline stingray that dispersed into freshwater.

Brooks et al. (1981b) proposed an alternative perspective on the origins of the potamotrygonids, based on a series of studies using the helminth parasites that inhabit them. They began by establishing a sequence of questions to be answered:

1. Are potamotrygonids inhabited by helminth parasites that are related to helminths inhabiting other elasmobranchs? If not, then parasite evidence would shed no light on the evolutionary origins of potamotrygonids. If the helminths inhabiting potamotrygonids were related to those found in other elasmobranchs, however, they, their phylogenetic relationships, and their host and geographic affinities might provide evidence about potamotrygonid origins.
2. Do the helminths inhabiting potamotrygonids form monophyletic groups? If so, then the most parsimonious hypothesis is that the potamotrygonids themselves are monophyletic, and that their common ancestor brought the ancestors of the various helminth groups into freshwater with it. If not, then the most parsimonious explanation is that potamotrygonids are derived from more than one marine ancestor.
3. What is (are) the source(s) of the helminths? The geographic distribution of their marine sister groups would provide an estimate of the general geographic area from which the ancestral potamotrygonid and its helminth fauna came.
4. How long ago did potamotrygonids arrive in freshwater habitats? Correlated biogeographic and phylogenetic patterns can often provide a robust estimate of age of groups, even in the absence of a fossil record; in fact, as we will show later, this study provides just such an example. In addition, the degree of complexity of a geographic distribution pattern for a group of species is often correlated with the length of time the group has been in an area.

5. What has been the pattern of diversification for the helminth fauna in potamotrygonids? For example, if they arrived relatively recently in freshwater and speciated as a result of independent dispersal, potamotrygonids and their parasites should not show correlated patterns of speciation with organisms that evolved in the freshwater habitats.

Question 1: Are potamotrygonids inhabited by helminth parasites that are related to helminths inhabiting other elasmobranchs? Twenty-four species of parasitic worms have been found inhabiting potamotrygonids thus far. Four of the 24 parasite species listed in Table 2.7 inhabit either teleosts (*Paravitellotrema overstreeti, Terranova edcaballeroi*) or crocodilians (*Leiperia gracile, Brevimulticaecum* sp.) and local potamotrygonids appear to have picked them up, at least on occasion (they are not commonly encountered in potamotrygonids). The remaining 19 species of parasites are restricted to stingray hosts. The closest relatives of these species inhabit marine stingrays (with the exception of *Megapriapus ungriai,* the only acanthocephalan known to inhabit elasmobranchs of any kind, and whose relationships to other acanthocephalans is uncertain). Hence, it seems likely that most of the parasite groups inhabiting potamotrygonids were brought into neotropical freshwater habitats with the ancestor of the stingrays themselves.

Question 2: Do the helminths inhabiting potamotrygonids form monophyletic groups? Fourteen of these species are members of three clades (Fig. 2.23), while the remaining ten species each represent a different clade. This supports the hypothesis that the potamotrygonids are monophyletic, and that their common ancestor brought the ancestors of the various helminth groups into freshwater with it.

Question 3: What is (are) the source(s) of the helminths? Phylogenetic hypotheses postulating the sister groups for some, but not all, of the helminths inhabiting potamotrygonids indicate two things: (a) the helminth fauna of potamotrygonids arose in the Pacific Ocean, and (b) many of the closest relatives of the helminth fauna of potamotrygonids inhabit urolophid stingrays. This suggests that the ancestor of the potamotrygonids was a marine urolophid living in the Pacific Ocean. This observation, by itself, is troublesome, because no potamotrygonids are known from river systems that empty into the Pacific Ocean, and because no urolophid species has ever been reported from freshwater habitats (all reports of marine stingrays in freshwater involve members of the [possibly paraphyletic] Dasyatidae). Consideration of the pattern and de-

Table 2.7
Geographic distribution of 24 species of parasitic worms inhabiting South American freshwater stingrays

Parasite species	Locality					
	1	2	3	4	5	6
1. *Acanthobothrium quinonesi*	0	0	0	0	+	+
2. *A. regoi*	0	0	0	+	0	0
3. *A. amazonensis*	0	0	+	0	0	0
4. *A. terezae*	+	0	0	0	0	0
8. *Potamotrygonocestus magdalenensis*	0	0	0	0	0	+
9. *P. orinocoensis*	0	0	0	+	0	0
10. *P. amazonensis*	0	0	+	+	+	0
13. *Rhinebothroides moralarai*	0	0	0	0	0	+
14. *R. venezuelensis*	0	0	0	+	+	0
15. *R. circularisi*	0	0	+	0	0	0
16. *R. scorzai*	+	0	0	+	0	0
17. *R. freitasi*	0	+	0	0	0	0
18. *R. glandularis*	0	0	0	+	0	0
19. *R. mclennanae*	+	0	0	0	0	0
26. *Eutetrarhynchus araya*	+	+	0	+	0	0
27. *Rhinebothrium paratrygoni*	+	+	0	+	0	0
28. *Paraheteronchocotyle tsalickisi*	0	0	+	0	0	0
29. *Potamotrygonocotyle amazonensis*	0	0	+	0	0	0
30. *Echinocephalus daileyi*	0	0	+	+	0	0
31. *Paravitellotrema overstreeti*	0	0	0	0	0	+
32. *Terranova edcaballeroi*	0	0	0	+	0	0
33. *Megapriapus ungriai*	0	0	0	+	0	0
34. *Leiperia gracile*	+	0	0	0	0	0
35. *Brevimulticaecum* sp.	+	0	0	0	0	0

Note: Locality 1 = upper Parana River, including the lower Mato Grosso; 2 = mid-Amazon, near Manaus; 3 = upper Amazon, near Leticia; 4 = delta of the Orinoco; 5 = Lake Maracaibo tributaries; 6 = mid- to lower Magdalena River. Species are numbered for phylogenetic and biogeographic analysis.

gree of complexity of the distribution of helminth species inhabiting potamotrygonids provided additional information that clarified this evidence.

The geographic distribution patterns for the parasites of potamotrygonids are complex: some of the parasite species appear to be restricted to single river systems, while others are more widespread (Table 2.7).

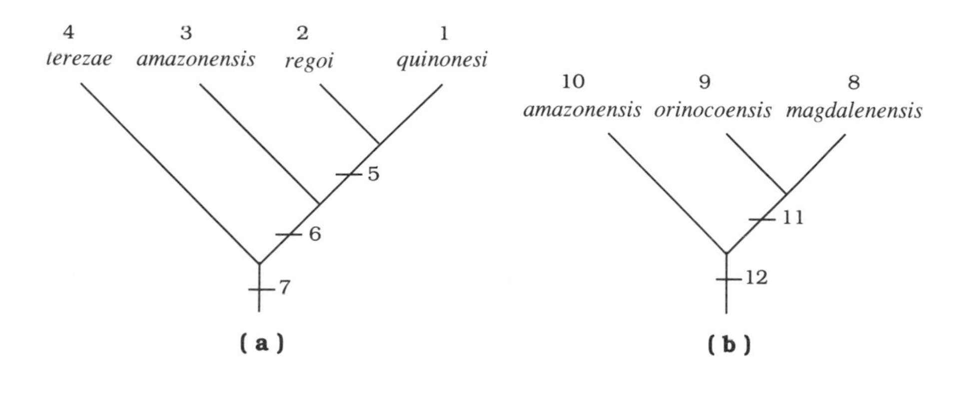

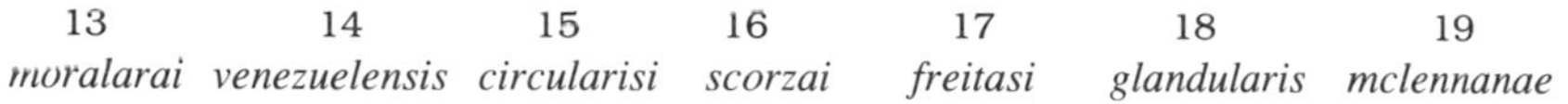

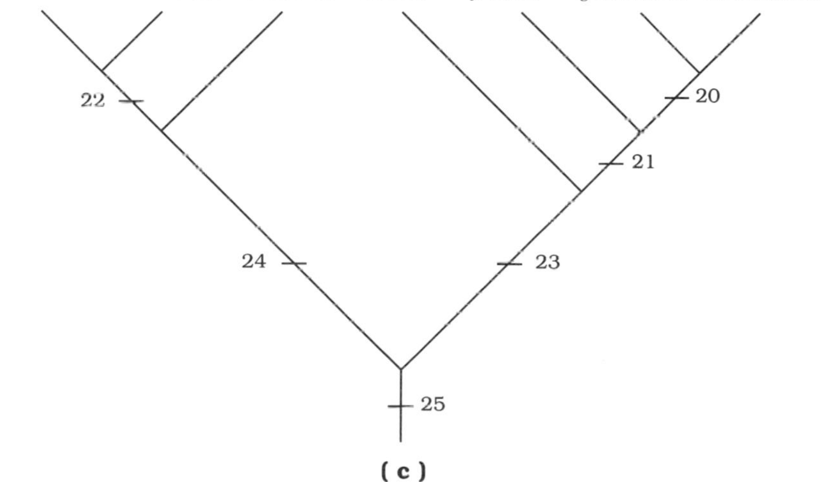

Figure 2.23. Phylogenetic relationships of three parasite groups inhabiting neotropical freshwater stingrays. (a) *Acanthobothrium*, (b) *Potamotrygonocestus*, (c) *Rhinebothroides*. From Brooks (1992a).

Table 2.8 is a data matrix for six localities based on the phylogenetic relationships and distributions of the stingray helminths. Phylogenetic analysis of this data matrix produces an area cladogram with a consistency index of 85% (Fig. 2.24). This finding supports a qualitative assessment that approximately 85% of the species occur in these areas as a result of common phylogenetic history. This observation is more consistent with a history of vicariant allopatric speciation than with a history of speciation via dispersal. Brooks et al. (1981b) pointed out that the areas of endemism for the helminths inhabiting

Table 2.8
Data matrix listing six river systems in eastern South America and the binary codes for members of the helminth parasite groups inhabiting freshwater stingrays residing in those areas

Area	Binary code
Leticia, upper Amazon	00100110010100100000000110011100000
Manaus, middle Amazon	00000000000000001000101011100000000
Upper Parana River	01011110101101010011111111101000111
Orinoco Delta	01001110111101010100111111100101100
Lake Maracaibo	10001110010101000000010110000000000
Magdalena River	10001111001110000000010110000010000

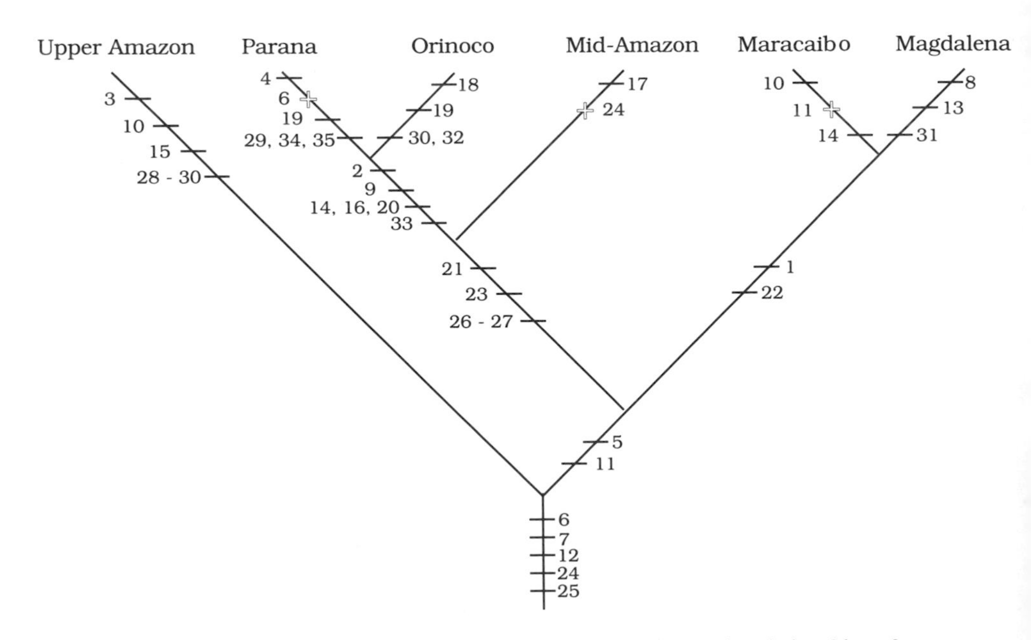

Figure 2.24. Area cladogram reconstructed from the phylogenetic relationships of three parasite clades inhabiting neotropical freshwater stingrays. Crosses indicate secondary absence. From Brooks (1992a).

potamotrygonids are also areas of endemism for species of fish groups that appear, on the basis of phylogenetic, biogeographic, and paleontological evidence, to have originated in neotropical freshwater systems.

The area cladogram is also congruent with the hypothesized geological history of the region, which links the origins of the major South American river systems with the uplifting of the Andes beginning early in the Cretaceous (see Brooks et al. 1981b and Windley 1986 for references to the geological evidence supporting this interpretation). It appears, then, that potamotrygonids are not relatively recent invaders of the neotropics.

A large portion of this phylogenetic component in parasite community composition may be obscured when attention is focused on individual communities. For example, the Parana, mid-Amazon, upper Amazon, Orinoco, and Magdalena systems all contain species whose phylogenetic relationships correspond to the geological history of the areas in which they occur. In addition to species endemic to the area, the Orinoco community contains species that have colonized from three other systems: the upper Amazon (species 10 and 28), the Parana (species 16, 24, and 25), and the mid-Amazon (species 19, the ancestor of species 18). The Orinoco community thus has the highest diversity, although it is not the oldest. The Maracaibo community also has representatives from three different source areas: the Magdalena (species 1), the Orinoco (species 14), and the upper Amazon (species 10). Finally, the Parana community contains species that have colonized from the Orinoco via two different routes (O_2: species 19, 26, 27; O_1: species 2, 9, 14, 29, 33, 34, 35). Clearly, these communities have been assembled in different ways historically, and in ways that could not be deduced from observations of contemporaneous species distributions alone. Such complex distribution patterns also support the conclusion that potamotrygonids have been in the neotropics for a long time.

On the basis of this initial analysis, the Orinoco, Maracaibo, and Parana localities appear to be composite areas. So, we recode the data matrix, separating the Maracaibo and Parana each into three areas and the Orinoco into four (Table 2.9). Phylogenetic analysis of this matrix produces a new area cladogram (Fig. 2.25) with a consistency index of 97%. This new area cladogram highlights five evolutionary components that have contributed to the evolution of this helminth fauna. The first component is the historical geological, or vicariant, backbone linking the upper Amazon, Parana, Orinoco, and Magdalena areas. These areas all contain species whose phylogenetic relationships correspond to the geological history of the regions. The remaining components of the parasite distribution patterns involve three sequences of dispersal from these areas along the following routes: (1) from the Parana to the mid-Amazon to the Orinoco; (2) from the Orinoco to the Upper Parana; (3) from the upper

Table 2.9
Recoded version of data matrix in Table 2.8

Area	Binary code
Leticia, upper Amazon (MA)	00100110010100100000000110011100000
Upper Parana River (P_1)	01001110101101000000000110001000111
Upper Parana River (P_2)	0001001?????00010000000101000000000
Upper Parana River (P_3)	????????????00000011101011100000000
Manaus, mid-Amazon (UA)	????????????00001000101011100000000
Orinoco Delta (O_1)	010011101011?????????????0000001100
Orinoco Delta (O_2)	???????00101?????????????1100000000
Orinoco Delta (O_3)	????????????00000101101010000100000
Orinoco Delta (O_4)	????????????01000000001011000000000
Lake Maracaibo (Mar_1)	1000111??????????????????0000000000
Lake Maracaibo (Mar_2)	???????00101?????????????0000000000
Lake Maracaibo (Mar_3)	????????????01000000001011000000000
Magdalena River (Mag)	10001111001110000000010110000010000

Note: Matrix is based on duplicating the Upper Parana River (P) and Lake Maracaibo (Mar) areas twice and the Orinoco Delta (O) three times, following the method described by Brooks and McLennan (1991).

Amazon into the Orinoco; and (4) from the upper Amazon, the Orinoco, and the Magdalena, forming the Maracaibo fauna (Fig. 2.26).

Questions 4 and 5: The questions of how long ago the ancestor of the potamotrygonids arrived in neotropical freshwater habitats and what the patterns of diversification have been since that invasion can be approached from several perspectives. Assumptions about the evolutionary sequence of osmoregulatory modifications and observations of current geographic distribution led many biologists to think that potamotrygonids were the recently derived descendants of marine stingrays. The most common evolutionary scenario postulated that the potamotrygonid ancestor moved from the Atlantic Ocean into the Amazon Basin during the Pliocene marine ingression. Subsequent to this invasion, a population was isolated from the ancestor, progressively adapting to freshwater and spreading throughout South America by stream capture during the past 3–5 million years.

The congruence among the area cladogram based on the phylogenetic relationships of the parasitic worms inhabiting potamotrygonids, areas of endemism for ostariophysan fishes inhabiting the neotropics, and the geological

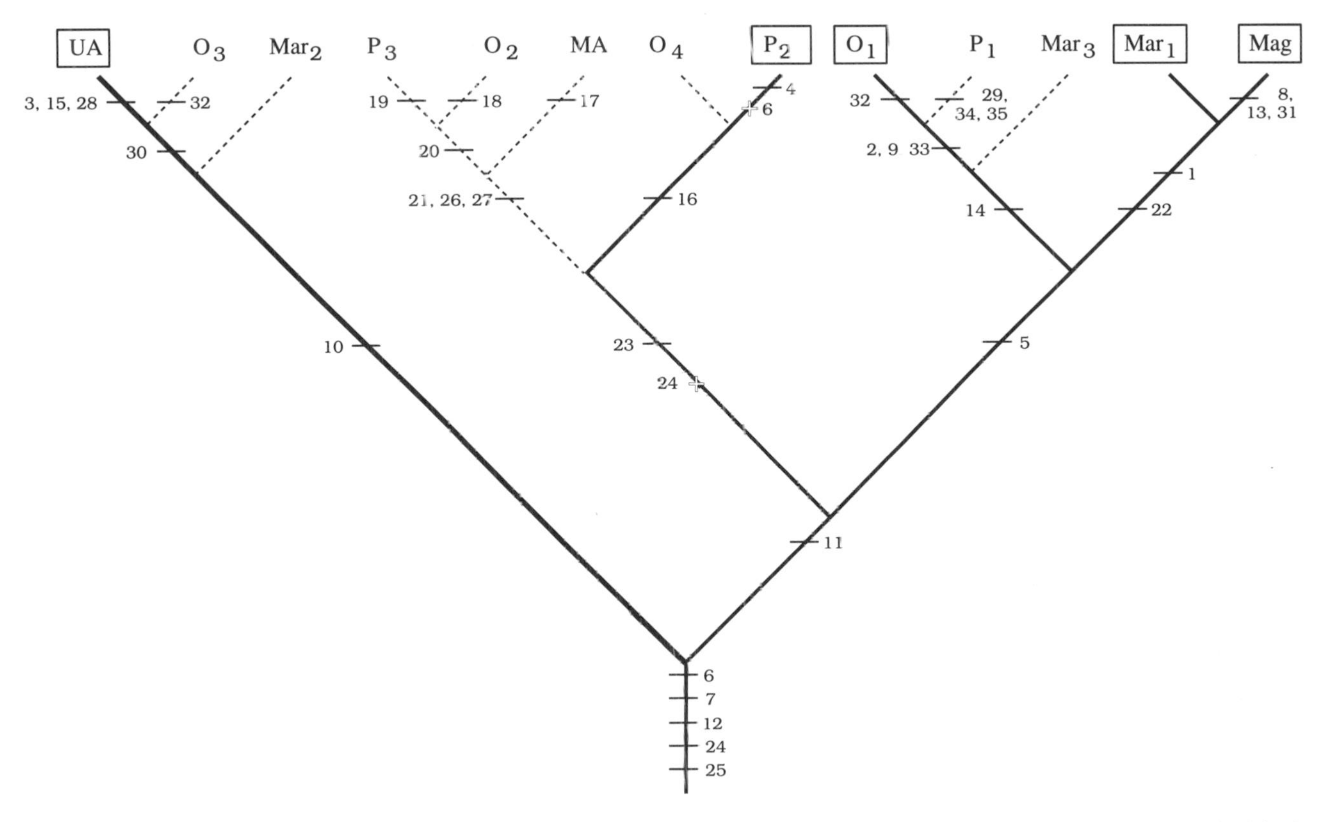

Figure 2.25. New area cladogram reconstructed from the phylogenetic relationships of three parasite clades inhabiting neotropical freshwater stingrays based upon replicating areas having more than one historical influence. From Brooks (1992).

Figure 2.26. Patterns of geographical distribution for related species of helminths inhabiting neotropical freshwater stingrays superimposed on a map of South America. From Brooks (1992a).

history of the river systems in which they occur pushes the origin of the parasite communities back to at least the mid-Miocene. Given that the parasite fauna is an old one, there are two alternative explanations for the origins of the parasite/stingray associations. First, the parasitic fauna that now inhabits the potamotrygonids occurred in South America prior to the appearance of these stingrays. In this case, the parasites or their closest relatives should be found in freshwater organisms, such as ostariophysan fishes, that were living in South America during the mid-Miocene or earlier. This explanation supports the proposal that the potamotrygonids are recently derived. Alternately, the freshwater stingrays may be older than biologists once believed, and their parasites reflect that ancient ancestry. If this is the case, the parasites inhabiting potamotrygonids, or their closest relatives, should inhabit marine stingrays whose geographic distribution is consistent with a hypothesis that the group originated as a result of marine invasion of South America no later than the mid-Miocene.

The geography of South America prior to the mid-Miocene differed in three significant ways from what we see today: (1) Africa and South America were not completely separated (i.e., there was no Atlantic Ocean at the mouth of the Amazon), (2) the Andes began sweeping upwards from the south beginning in the early Cretaceous and moving northward, and (3) the Amazon River flowed into the Pacific Ocean until no later than the mid-Miocene, when it was blocked by Andean orogeny, becoming an inland sea and eventually opening to the Atlantic Ocean. Thus, if potamotrygonids are a relatively old component of neotropical freshwater diversity east of the Andes, they must have come from the Pacific Ocean, which is today west of the Andes.

Enlarging the spatial scale of this study to include the geographic distribution of the marine relatives of the parasites inhabiting potamotrygonids provides additional support for the hypothesis that these stingrays and their parasites originated from marine ancestors that were isolated in South America from the Pacific Ocean by the Andean orogeny. The closest relatives of the parasites inhabiting potamotrygonids occur in Pacific marine stingrays and those members of two groups for which phylogenetic hypotheses exist appear to exhibit exhibit circum-Pacific rather than trans-Pacific distribution patterns (Fig. 2.27). A similar Pacific origin has been suggested for Amazonian freshwater anchovies (Nelson 1984) and possibly for neotropical freshwater needlefish (Collette 1982). In addition, each of the parasite species inhabiting potamotrygonids requires a mollusk or arthropod intermediate host, so it seems likely that mollusk and arthropod species derived from marine ancestors also moved into neotropical freshwater habitats along with the ancestor of the potamotrygonids. As a consequence, we now recognize the possibility that a

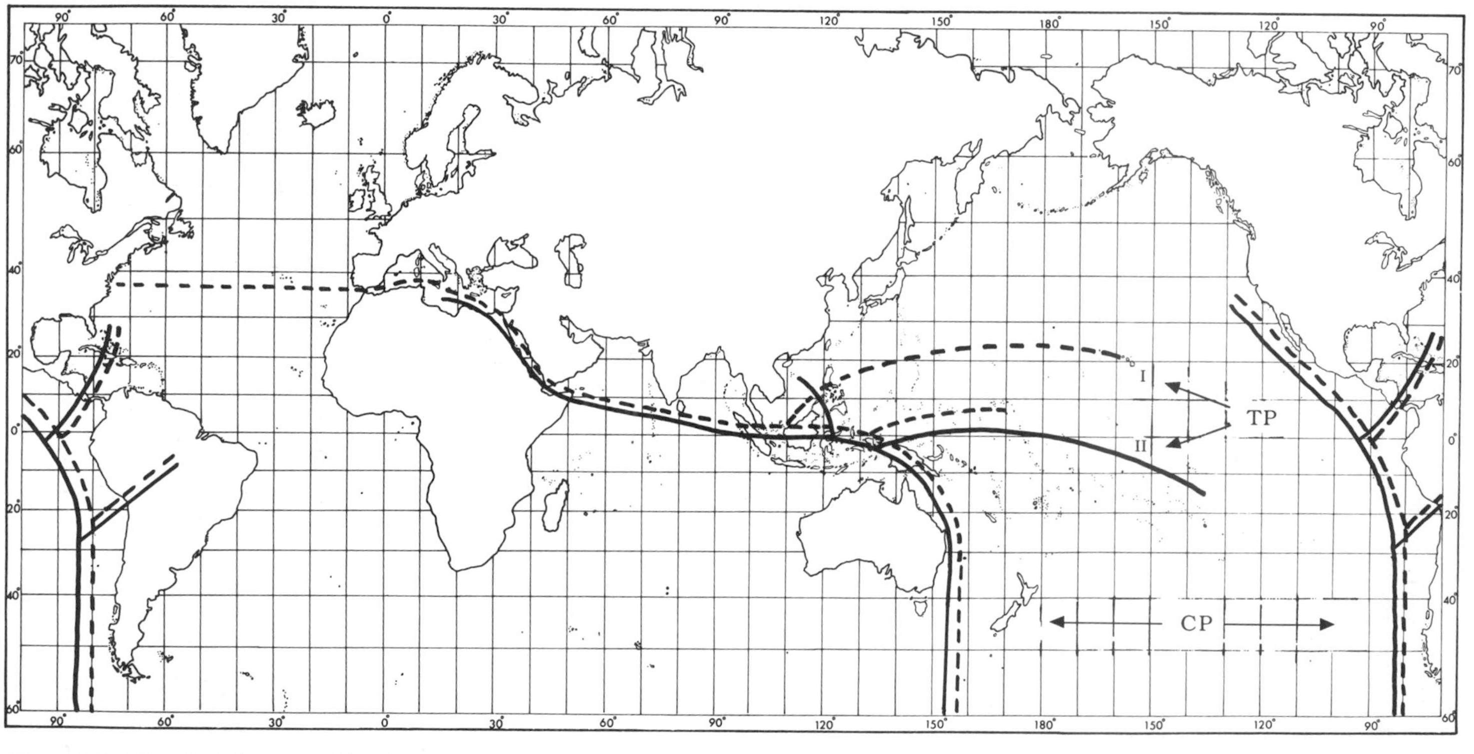

Figure 2.27. Historical biogeographic relationships of helminth parasites inhabiting neotropical freshwater stingrays and their closest relatives, based on phylogenetic trees for species groups in the tapeworm genera *Acanthobothrium, Eutetrarhynchus,* and *Rhinebothrium* (dashed lines) and in the roundworm genus *Echinocephalus* (solid lines). Note both circum-Pacific and trans-Pacific distribution patterns, with species most closely related to those occurring in freshwater stingrays being part of the circum-Pacific pattern. From Brooks and McLennan (1991). Copyright © 1991 by The University of Chicago.

sizable component of current neotropical freshwater diversity might be derived from Pacific marine ancestors.

We suggested earlier that correlations of geographic distribution patterns and phylogenetic relationships of helminths could provide a robust estimate of the age of origin for a group, even in the absence of a fossil record. A decade after the first proposal of Pacific origins of the potamotrygonids, potamotrygonid fossils have been collected from mid-Miocene freshwater deposits in the Amazonian portions of Peru and Colombia (Frailey 1986; J. Lundberg, personal communication). This particular study, therefore, represents a graphic empirical example of the power of combining parasite phylogeny and historical biogeography to answer questions about evolutionary origins in the manner first envisioned by von Ihering, Kellogg, Harrison, and Johnston, amplified by Metcalf, and kept alive by Manter until the advent of modern phylogenetic systematics.

CROCODILIANS AND HELMINTHS: EVOLUTIONARY PATTERNS IN A RELICTUAL HOST GROUP

Living crocodilians inhabit a variety of estuarine to freshwater habitats throughout the tropical and subtropical regions of the world. They prey on a wide variety of vertebrates, mostly fish, and some invertebrates. Today, crocodilians are known chiefly for two reasons. They are a symbolic group for discussions about conserving biodiversity because of their relictual status (and their value in the exotic leather trade) and, strangely enough, they are a symbolic group for theoretical discussions in systematic and evolutionary biology. Crocodilians play an important role in systematic and evolutionary biology because of their phylogenetic relationships with birds and dinosaurs. In the past, the sister group relationship between crocodilians and birds was not reflected in vertebrate classifications. Birds were given their own group, while crocodilians were placed in the Reptilia, a paraphyletic assemblage. This made it difficult to understand why crocodilians hosted so many parasites that appeared to be closely related to parasites in birds (Brooks 1979b; Brooks and O'Grady 1989).

From the fossil record, it is clear that crocodilians have seen better days, having once been more species rich and comprised larger representatives, including some monsters marauding throughout the Cretaceous seas. In addition, the earliest known crocodilian fossils suggest a bipedal stance and a terrestrial origin for the group. Today, crocodilians are reduced to about 22 living species, only one of which, *Crocodylus porosus,* even approaches 10 meters in total length. Hence, the current diversity of crocodilians represents only a fraction of the species number and ecological diversity once encompassed by the group (for reviews and references see Densmore and Owen 1989; Frey et al. 1989; Taplin and Grigg 1989; Tarsitano et al. 1989).

Because crocodilians are themselves evolutionary relicts (we will discuss the question of relicts again and in more detail in chapter 3) studying their helminth parasite fauna presents some interesting challenges. Stunkard (1970) suggested that relictual host groups should host parasite faunas made of relictual parasite groups. It is also intuitively reasonable to think that relictual host groups should be inhabited by fewer groups of parasites than their more species-rich sister groups. Although intuition and science often find themselves at cross purposes, in this case both ways of looking at the world appear to agree. Crocodilians are notable because they lack tapeworms and acanthocephalans and host almost no members of many nematode groups, such as filarids and spirurids, parasite groups that are commonly found in avian hosts. They do, however, host a wide variety of digeneans, the phylogenetic breadth of which (Echinostomiformes, Strigeiformes, Opisthorchiformes, and Plagiorchiformes) is consistent with the long history of the crocodilians themselves. In accordance with the revised status of the now defunct "reptiles," the digeneans inhabiting crocodilians include taxa whose sister groups inhabit birds, such as the echinostomes, schistosomes, proterodiplostomes, liolopids, and clinostomes. Crocodilians also host a number of parasite groups, most notably the acanthostomes, which appear to have been acquired originally by host switching from species that used piscivorous fish as definitive hosts.

The challenge of working with the crocodilian parasite fauna is to determine whether there is enough left in this ancient diversity to give us a picture of its evolutionary origins, in terms of both historical biogeographical relationships and host phylogenetic relationships. We can use BPA to examine these two components, concentrating on the well-represented and (fortunately) well-studied digenean platyhelminths and ascaridoid nematodes (Figs. 2.28 and 2.29).

The Historical Biogeographical Picture

If the current parasite diversity represents fragments of a once more diverse, widespread, and coevolved fauna, we should find that a BPA matrix for geographical areas will contain many "missing data" entries. In addition, if the current diversity represents an evolutionarily coherent fragment of that once more diverse, widespread, and coevolved fauna, BPA should provide strong evidence of ancient geographic configurations. Tables 2.10 and 2.11 are the BPA matrices for the parasite fauna; as anticipated, both contain many

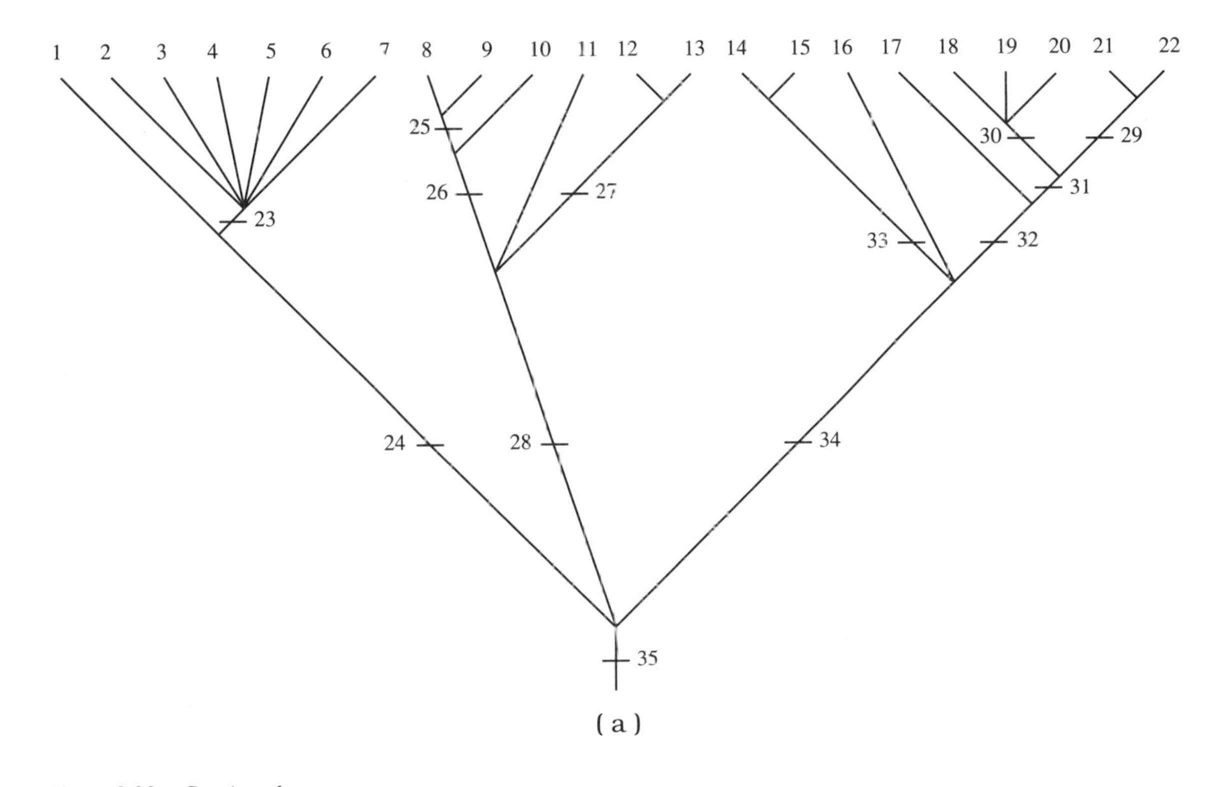

Figure 2.28—*Continued on next page.*

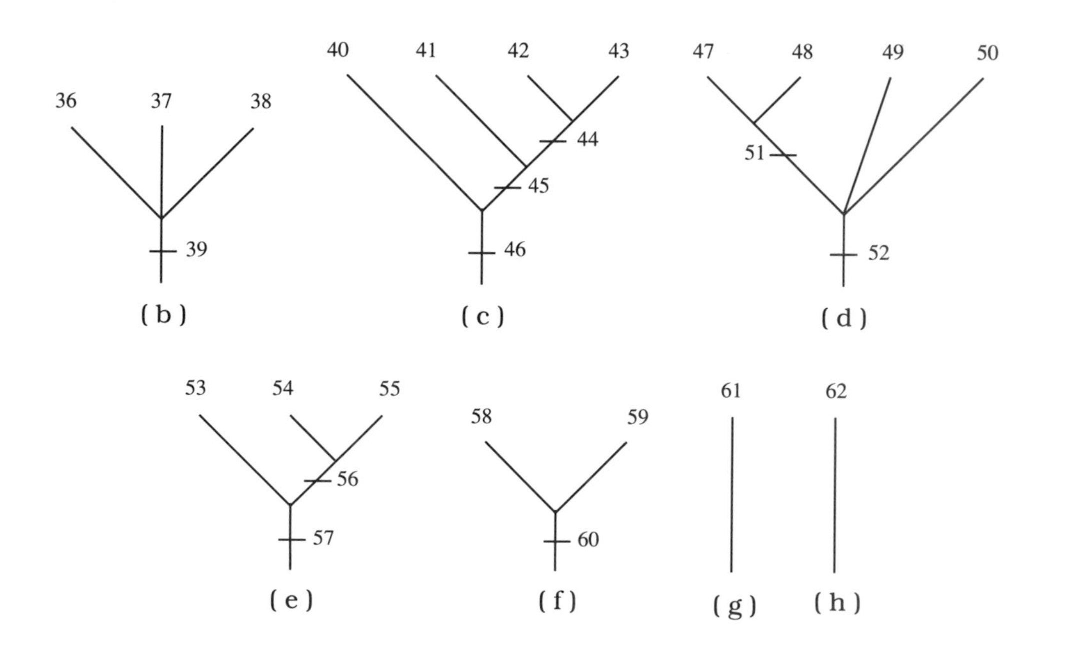
36
37
38
39
(b)
40
41
42
43
44
45
46
(c)
47
48
49
50
51
52
(d)
53
54
55
56
57
(e)
58
59
60
(f)
61
(g)
62
(h)

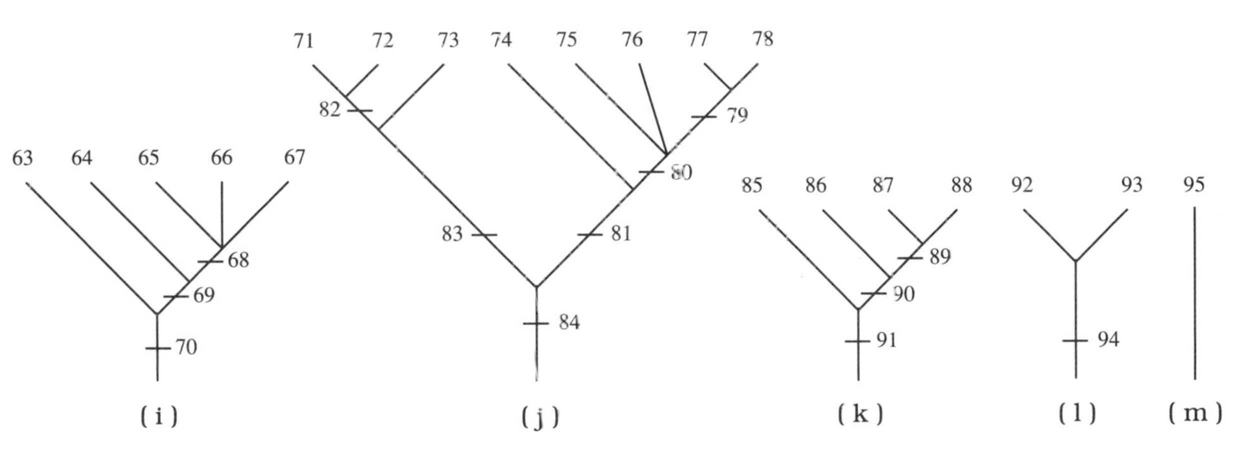

Figure 2.28. Phylogenetic trees for various digenean groups inhabiting crocodilians, numbered for BPA. (a) Proterodiplostomidae (1 = *Neelydiplostomum;* 2–7 = *Pseudoneodiplostomum;* 8–9 = *Crocodilicola;* 10 = *Archaeodiplostomum;* 11 = *Polycotyle;* 12–13 = *Pseudocrocodilicola;* 14 = *Cystodiplostomum;* 15 = *Prolecithodiplostomum;* 16 = *Mesodiplostomum;* 17 = *Massoprostatum;* 18–20 = *Proterodiplostomum;* 21 = *Herpetodiplostomum;* 22 = *Paradiplostomum*). (b) *Allechinostomum* (including part of *Stephanoprora*) (36 = *jacaretinga;* 37 = *ornata;* 38 = *crocodili*). (c) *Dracovermis* (40 = *occidentalis;* 41 = *nicolli;* 42 = *brayi;* 43 = *rudolphii*). (d) *Odhneriotrema* (47 = *incommodum;* 48 = *microcephala*), *Tremapoleipsis* (49) and *Nephrocephalus* (50). (e) *Cyathocotyle* (53 = *C. brasiliensis;* 54 = *C. fraterna;* 55 = *C. crocodili*). (f) *Exotidendrium* (58) and *Renivermis* (59). (g) *Pachypsolus sclerops* (61). (h) *Gryphobilharzia amoenae* (62). (i) *Timoniella* (63 = *incognita;* 64 = *ostrowskiae;* 65 = *unami;* 66 = *loossi;* 67 = *absita*). (j) *Proctocaecum* (71 = *coronarium;* 72 = *vicinum;* 73 = *gonotyl;* 74 = *productum;* 75 = *elongatum;* 76 = *crocodili;* 77 = *atae;* 78 = *nicolli*). (k) *Caimanicola* (85 = *pavida;* 86 = *caballeroi;* 87 = *marajoara;* 88 = *brauni* [inhabits turtles]). (l) *Acanthostomum* (92 = *scyphocephalum;* 93 = *americanum*). (m) *Acanthostomum slusarskii* (95).

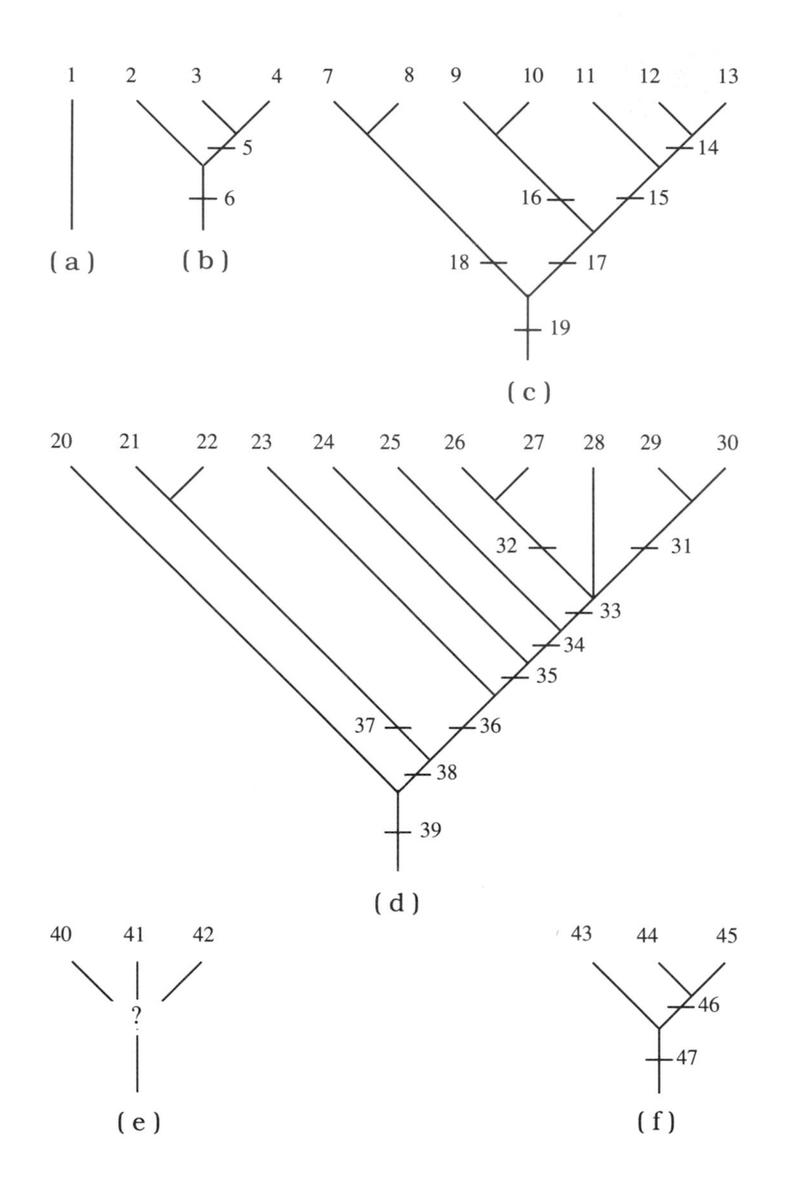

Figure 2.29. Phylogenetic trees for various ascaridoid nematode groups inhabiting crocodilians, numbered for BPA. (a) *Multicaecum* (*agile* [1]). (b) *Ortleppascaris* (*antipini* [2]; *nigra* [3]; *alata* [4]). (c) *Gedoelstascaris* + *Brevimulticaecum* (*vanderbrandeni* [7]; *australiensis* [8]; *gibsoni* [9]; *stekhoveni* [10]; *tenuicolle* [11]; *pintoi* [12]; *baylisi* [13]). (d) *Dujardinascaris* (*chabaudi* [20]; *gedoelsti* [21]; *puylaerti* [22]; *woodlandi* [23]; *longispicula* [24]; *taylorae* [25]; *dujardini* [26]; *waltoni* [27]; *madagascariensis* [28]; *helicini* [29]; *mawsonae* [30]). (e) *Goezia* (*lacerticola* [40]; *gavialidis* [41]; *holmesi* [42]). (f) *Terranova* (*caballeroi* [43]; *crocodili* [44]; *lanceolata* [45]).

Table 2.10
BPA matrix for digeneans inhabiting crocodilians and major geographic areas where they occur

Area	Binary code
India$_1$	10000000000000000000000100000000001 ???? ??????? ?????? ????? ??? ? ? ???????? ?????????????? ??????? ?? 1
Africa$_1$	01100000000000000000001100000000001 0101 ??????? 011011 01011 ??? ? ? ???????? ?????????????? ??????? ?? ?
Australo-Oceania$_1$	00011110000000000000001100000000001 0011 ??????? ?????? 00111 ??? ? 1 ???????? ?????????????? ??????? ?? ?
South America	00000000000001111111110000001111111 1001 ??????? 010011 10001 ??? 1 ? 11000011 ?????????????? 0011111 10 ?
North America	00000001111110000000000011110000001 ???? 1000001 100011 ????? ??? ? ? 00110111 10000000000111 1100011 01 ?
Africa$_2$	??????????????????????????????????? ???? ??????? ?????? ????? ??? ? ? ???????? 01100000000111 ??????? ?? ?
India$_2$	??????????????????????????????????? ???? 0100011 ?????? ????? ??? ? ? ???????? ?????????????? ??????? ?? ?
Africa$_3$	??????????????????????????????????? ???? 0010111 ?????? ????? ??? ? ? ???????? 00010000001001 ??????? ?? ?
Australo-Oceania$_2$	??????????????????????????????????? ???? 0001111 ?????? ????? ??? ? ? 00001111 00001111111001 ??????? ?? ?

Note: See Figure 2.28 for identities of taxa indicated by codes. There is a break between the codes for each taxon.

Table 2.11
BPA matrix for ascaridoid nematodes inhabiting crocodilians and major geographic areas where they occur

Area	Binary code
India	1 ????? ????????????? 0001000000000001011 010 ?????
Africa$_1$	1 ????? 10?????????11 ??????????????????? 000 ?????
Australo-Oceania$_1$	1 ????? 01?????????11 ??????????????????? 001 ?????
South America	? 10001 ??110111111?1 1000100000000011011 000 01111
North America	? 00111 ??010111111?1 0000001010111111011 100 10001
Africa$_2$	? 01111 ????????????? 0110001010001111111 000 ?????
Australo-Oceania$_2$	? ????? ????????????? 0000100001101110111 000 ?????

Note: See Figure 2.29 for identities of taxa indicated by codes. There is a break between the codes for each taxon.

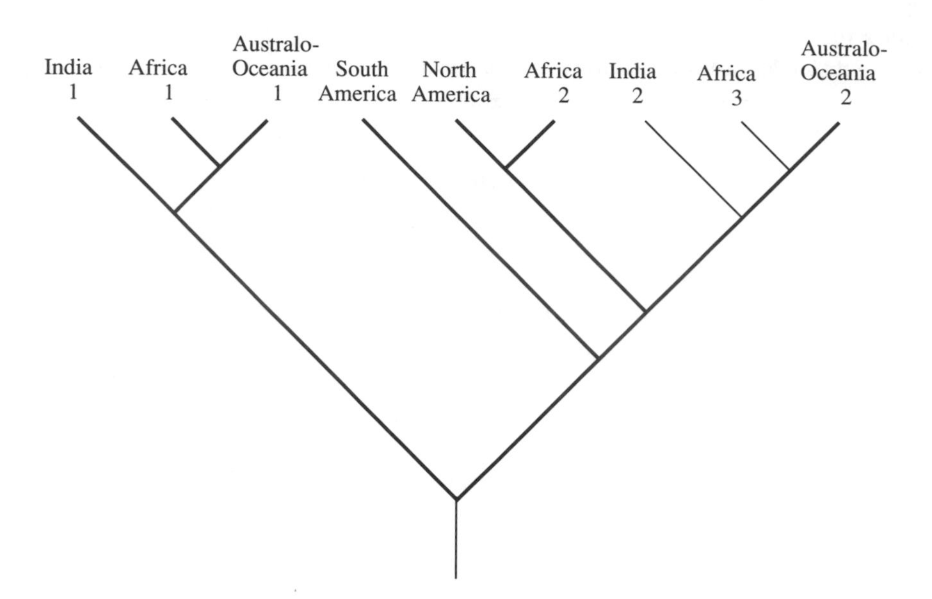

Figure 2.30. Area cladogram based on phylogenetic relationships of digeneans inhabiting crocodilians.

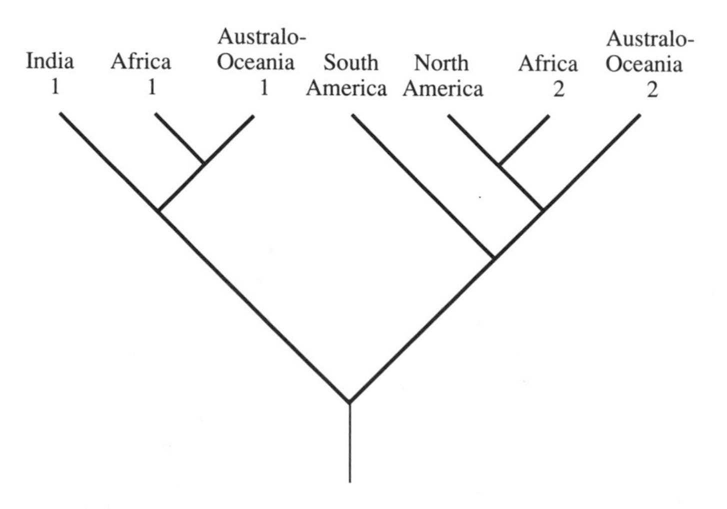

Figure 2.31. Area cladogram based on phylogenetic relationships of ascaridoid nematodes inhabiting crocodilians.

"missing data" entries, presumably indicating extinctions in the parasite fauna concomitant with extinctions in the host fauna.

Figures 2.30 and 2.31 are area cladograms based on the digenean and nematode faunas, respectively. Despite the fact that each is reconstructed from a fragment of a fragment of a past parasite fauna, the two cladograms show remarkable concordance with each other. In fact, the only difference between the two cladograms is the presence of I-2 and Af-3 for the digeneans (or their absence for the nematodes). If we superimpose Figure 2.30 on a map of Pangaean continental configurations, we find apparent concordance with ancient geographic configurations coincident with the known age of crocodilians derived from fossil evidence (Fig. 2.32). Even though they originated long ago, and a good portion of them have probably become extinct, the helminth faunas of crocodilians inhabiting India, Africa, and the Australo-Oceania region appear to be evolutionary composites, because those areas appear twice on the area cladograms, whereas the

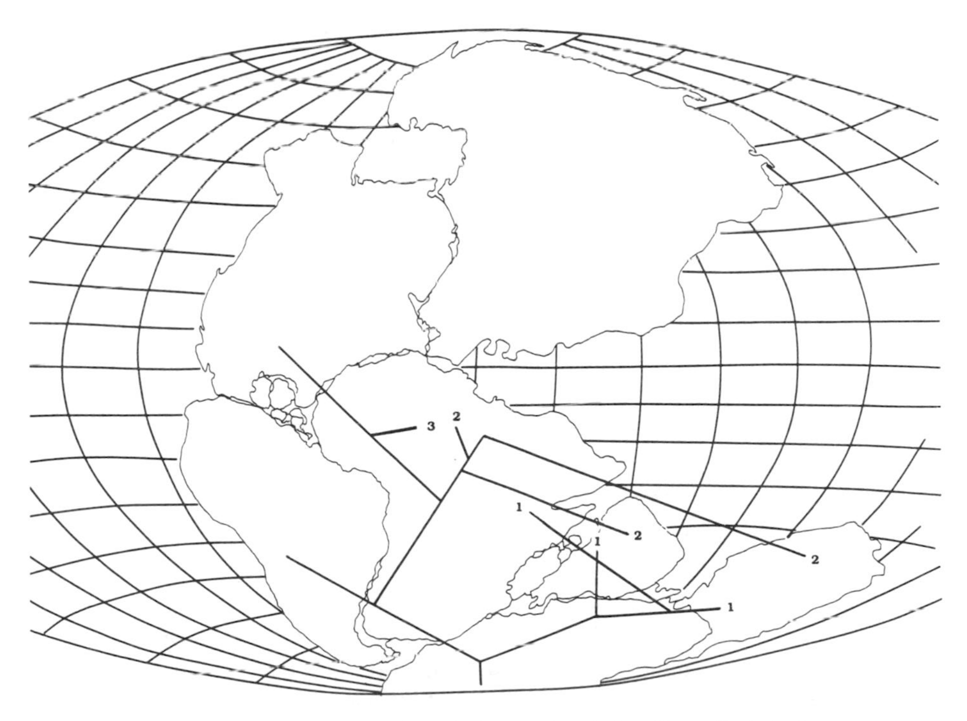

Figure 2.32. Area cladogram based on phylogenetic relationships of helminth parasites inhabiting crocodilians superimposed on a map of Pangaea.

helminth faunas of South America and North America appear to have single origins.

The relationships among India, Africa, and Australo-Oceania on the area cladograms are intriguing, because we do not understand them completely. In each area cladogram, these three areas appear twice, and in each case India appears as the "sister area" of Africa plus Australo-Oceania. The three-area configuration is an interesting pattern in itself, partly because India is now, and has apparently always been, situated *between* Africa and Australo-Oceania, and partly because the appearance of the same pattern repeated in the area cladograms suggests two possible interpretations. In one interpretation, the repeating pattern occurs because there was substantial early evolutionary divergence among crocodilians, allowing time for two lineages to evolve and become distributed throughout eastern Gondwana prior to the breakup of Pangaea. In the other interpretation, the sister area relationships with North America may indicate a later set of vicariance events that happened to be congruent with the area relationships resulting from an earlier set of vicariance events. In both interpretations, the sister area relationships for each of the repeating three-area components appear to be different, one having clear affinities with North America (reminiscent of the distribution of the turtle blood fluke group Spirorchinae) and the other appearing to have more affinities with South America.

The configuration is also interesting because the geological history of India plays a key role in the debate that has become known as the "expanding earth" hypothesis (Owen 1983; Patterson and Owen 1991). This hypothesis stems from the observations that (1) bottom sediments in the Pacific Ocean, Atlantic Ocean, and Indian Ocean all are of similar ages, even though the Pacific Ocean is considerably older than the other two, and (2) the continental margins fit together better if one assumes an ancient earth with a diameter approximately 25% smaller than it is today. Opponents of the hypothesis respond that sea floor spreading and subduction have obliterated the older Pacific sea floor and eroded the continental margins to such an extent that we have lost some of the record of geological history. In a manner reminiscent of the geologists' objections to continental drift in the first half of the twentieth century, opponents also decry the lack of an accepted mechanism by which the earth could have expanded by 25%. One of the differences between the "traditionalists" and the "expanders" concerns the relative amount of time during which India is thought to have been geologically isolated from Africa and Asia. The "expanders" believe that India was never out of contact with the other continents for very long, if at all, and that this situation explains the lack of a highly endemic and basal Indian flora and fauna. (The response to this is that India became submerged during its long drift from Gondwanian Antarctica to the underbelly of Asia, killing off its endemic terrestrial and freshwater flora and fauna.) If the congruence of the second set of India–Africa–Australo-Oceania patterns reflects evolutionary events following the breakup of Pangaea, they (along with the turtle blood flukes) might provide evidence that India was not isolated for a very long time. In the light of the biogeographic patterns exhibited by the parasites of crocodilians, we have to conclude that something interesting and nonrandom appears to be trying to explain itself to us.

In the late 1800s and early 1900s, parasitologists like von Ihering (1891, 1902), Metcalf (1920), and Harrison (1929) used parasite and host distribution patterns as evidence for the controversial theory of continental drift (Chapter 1). It will be ironic indeed if, over half a century later, their academic descendants once again lead the way in providing evidence for another controversial theory about earth history.

Parasite-Host Relationships

A wide variety of analyses, based on anatomical, physiological, ecological, and molecular data, have been used to tackle the question of crocodilian evolution. Nevertheless, their phylogenetic relationships remain controversial and this, of course, makes comparisons of host and parasite phylogenies a more dynamic exercise than most evolutionary biologists would like. In the most recent review of the problem, Densmore and White (1991) suggested four major areas of uncertainty: (1) Do all living crocodilians belong to the same inclusive monophyletic group? The key question is, where does *Gavialis* fit? Is it the sister group of crocodylids plus alligatorids, of the alligatorids, of the crocodylids, or is it a separate lineage not closely related to the other crocodilians at all? (2) To which crocodilians are false gavials (*Tomistoma*) most closely related—are they the sister group of *Gavialis* or are they crocodylids? (3) If living crocodilians represent an inclusive monophyletic group, what are the sister-group relationships among the gavialids, crocodylids, and alligatorids? (4) What are the relationships among the various species of the genus *Crocodylus*? Fortunately, there are some areas of agreement: (1) Alligatorids are monophyletic. (2) *Alligator* is the sister group of the caimans. (3) Within the caimans *Paleosuchus* is the sister group of *Melanosuchus* plus *Caiman*. (4) Crocodylids are a monophyletic group. (5) *Osteolaemus* (the dwarf crocodile) is the sister group of *Crocodylus*. Purely by chance, the crocodilians whose relationships are most in

Table 2.12
BPA matrix for digeneans inhabiting crocodilians and the hosts they inhabit

Host	Binary code
Gavialis	1000000000000000000000100000000001 ???? 0100011 ?????? ????? 101 ? ? ????? ?????????????? ??????? ??? ?
Alligator	0000000111111000000000001111000001 ???? 1000001 100011 ????? ??? ? ? 00010111 1000000000111 100001 ??? ?
Melanosuchus	0000000000000000100011000000001111011 ???? ??????? ?????? ????? ??? ? ? ???????? ?????????????? ?????? ??? ?
Paleosuchus	0000000000000000000010000000000111011 ???? ??????? ?????? ????? ??? ? ? ???????? ?????????????? ?????? ??? ?
Caiman crocodilus	00000000000001100110110000001111111 1001 ??????? 010011 10001 ??? 1 ? 10000001 ?????????????? 0010111 ??? ?
C. fuscus	00000000000000100110110000001111111 ???? ??????? 010011 ????? ??? ? ? 01000011 ?????????????? 0110111 101 ?
C. latirostris	00000000000001000000110000001111111 ???? ??????? ?????? ????? ??? ? ? ???????? ?????????????? ?????? ??? ?
Osteolaemus	0000001000000000000000001100000000001 ???? ??????? ?????? ????? ??? ? ? ???????? ?????????????? ?????? ??? ?
Crocodylus niloticus	0000011000000000000000001100000000001 0111 ??????? ?????? 01011 ??? ? ? ???????? 01110000001111 ??????? ? ? ? ???????? 01110000001111 ??????? ??? ?
C. cataphractus	?????????????????????????????????? ???? 0010111 ?????? 01011 ??? ? ? ???????? ?????????????? ??????? ??? ?
C. palustris	?????????????????????????????????? ???? ??????? ?????? ????? ??? ? ? ???????? ?????????????? ?????? ??? 1
C. siamense	0110000000000000000000001100000000001 ???? ??????? ?????? ????? ??? ? ? ???????? ?????????????? ?????? ??? ?
C. johnsoni	01000000000000000000000001100000000001 ???? ??????? ?????? ????? ??? ?0 1 ???????? 00000001111001 ?????? ??? ?
C. porosus	0001100000000000000000001100000000001 ???? 0001111 ?????? 00111 011 ? ? 00001111 00001110111001 ??????? ??? ?
C. rhombifer	?????????????????????????????????? ???? ??????? ?????? ????? ??? ? ? 00010111 ?????????????? 0110111 011 ?
C. acutus	?????????????????????????????????? ???? ??????? ?????? ????? ??? ? ? 00110111 ?????????????? 0110111 011 ?
C. intermedius	?????????????????????????????????? ???? ??????? ?????? ????? ??? ? ? 00010111 ?????????????? 0010111 ??? ?
C. moreletii	0000000010000000010000000110100001011 ???? ??????? ?????? ????? ??? ? ? ???????? ?????????????? ?????? ??? ?

Note: See Figure 2.28 for identities of taxa indicated by codes. There is a break between the codes for each taxon.

Table 2.13
BPA matrix for ascaridoid nematodes inhabiting crocodilians and the hosts they inhabit

Host	Binary code
Gavialis	1 ????? ????????????? 00010000000000001011 001 ?????
Alligator	? 00111 0000101110101 00000001000011111011 100 10001
Melanosuchus	? ????? 0011001111101 ???????????????????? 000 ?????
Caiman crocodilus	? 10001 0001011111101 10001000000000011011 000 00111
C. latirostris	? ????? 0000010110101 ???????????????????? 000 01011
Crocodylus niloticus	1 01011 1000000000011 01100010100011111111 000 ?????
C. cataphractus	1 01011 1000000000011 00000000100001111011 000 ?????
C. palustris	1 ????? ????????????? ???????????????????? 000 ?????
C. novaeguineae	? ????? ????????????? 00000000001101111011 000 ?????
C. johnsoni	1 ????? 0100000000011 00000100000000111011 000 ?????
C. porosus	? ????? 0100000000011 00000100001101111011 010 ?????
C. rhombifer	? 00111 ????????????? 00000000010101111011 000 ?????
C. acutus	? 00111 ????????????? 00000000010101111011 000 ?????

Note: See Figure 2.29 for identities of taxa identified by codes. There is a break between the codes for each taxon.

doubt are also the species for which the fewest parasites have been reported. There are no helminths known from *Tomistoma*, and relatively few from *Gavialis*. Efforts to apply data from parasite phylogenies to this group of hosts have proceeded at a slower pace than studies of the crocodilians themselves (Brooks 1979b; Brooks and O'Grady 1989) and have produced a mosaic of results. The following represents yet another update in this continuing saga.

Once again, despite numerous "missing data" entries (Tables 2.12 and 2.13), there is still marked concordance between the crocodilian relationships reconstructed from the phylogenetic relationships of their helminths (Figs. 2.33 and 2.34) and current estimates of crocodilian phylogeny. At the generic level, there is almost complete concordance with only two points of possible conflict: the placement of *Gavialis* as the sister group of the crocodylids and the placement of *Osteolaemus* as the sister group of *Crocodylus niloticus* + *C. cataphractus* rather than as the sister group of *Crocodylus* as a whole (Fig. 2.35). This "misplacement" indicates either that the

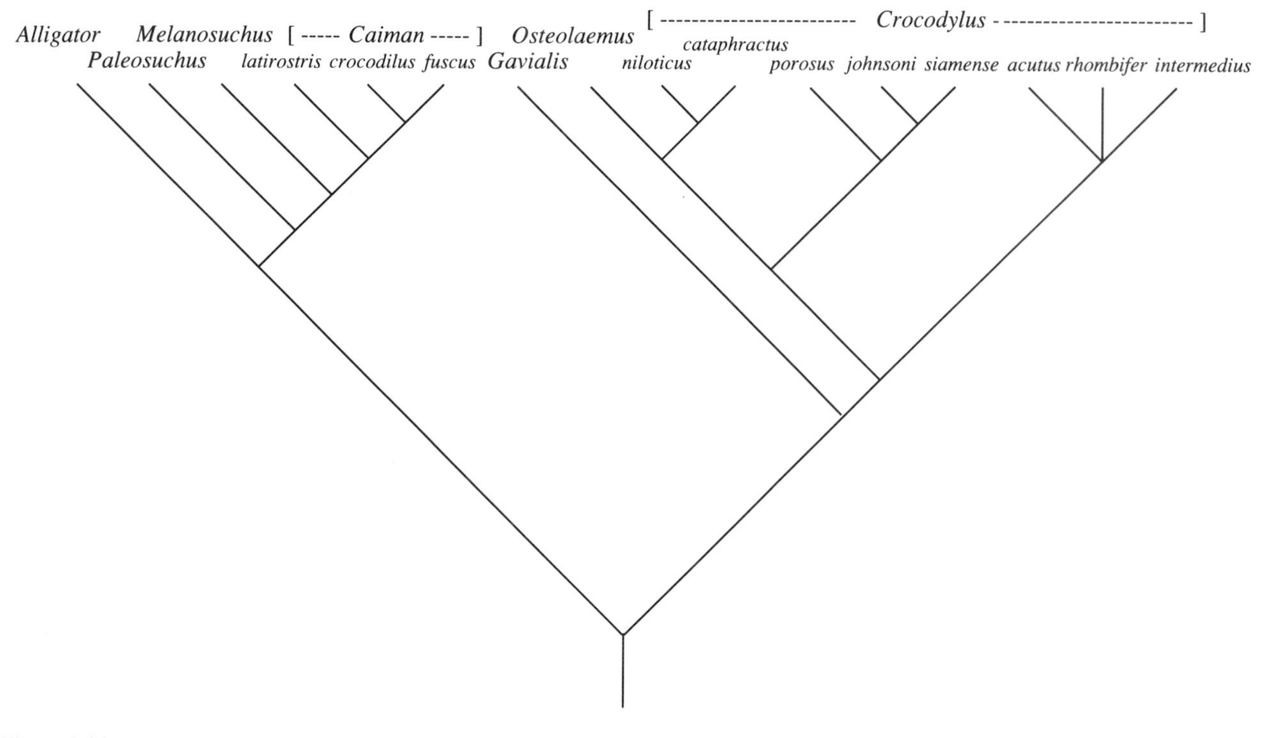

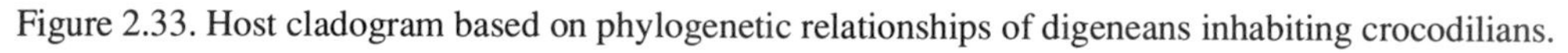
Figure 2.33. Host cladogram based on phylogenetic relationships of digeneans inhabiting crocodilians.

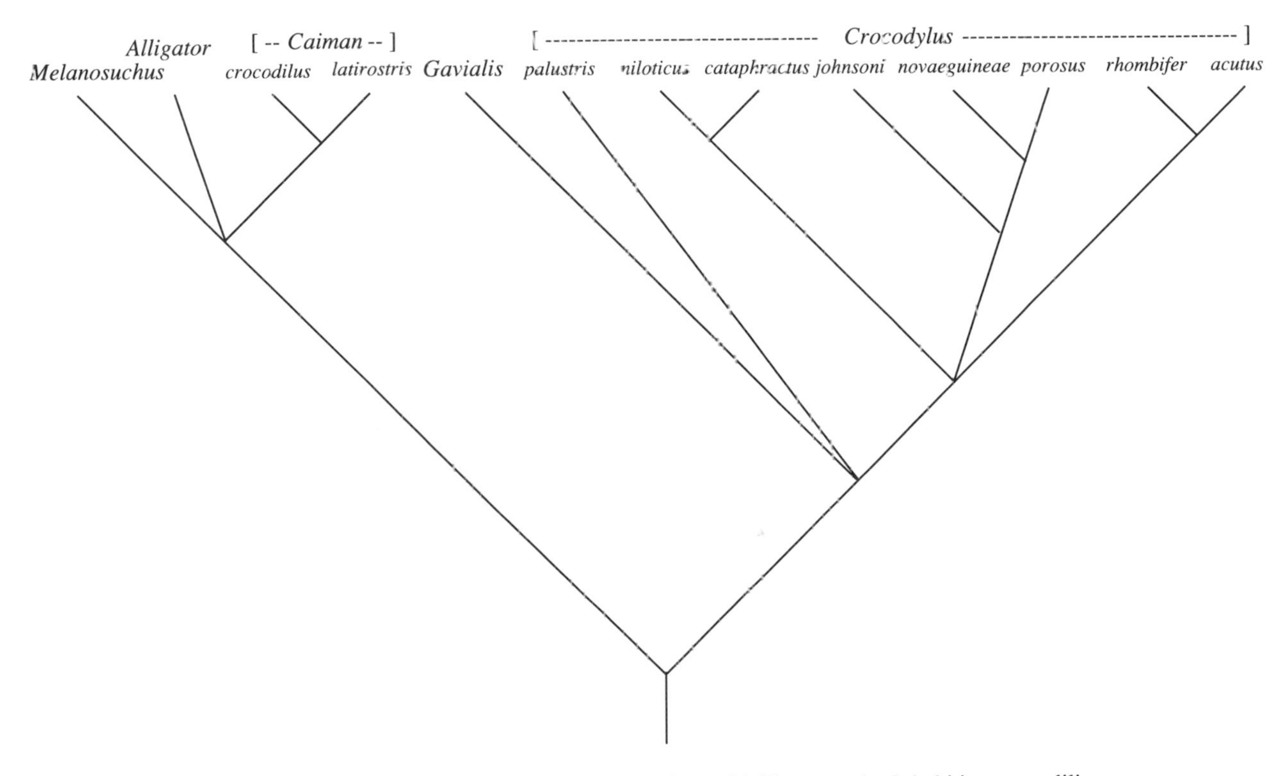

Figure 2.34. Host cladogram based on phylogenetic relationships of ascaridoid nematodes inhabiting crocodilians.

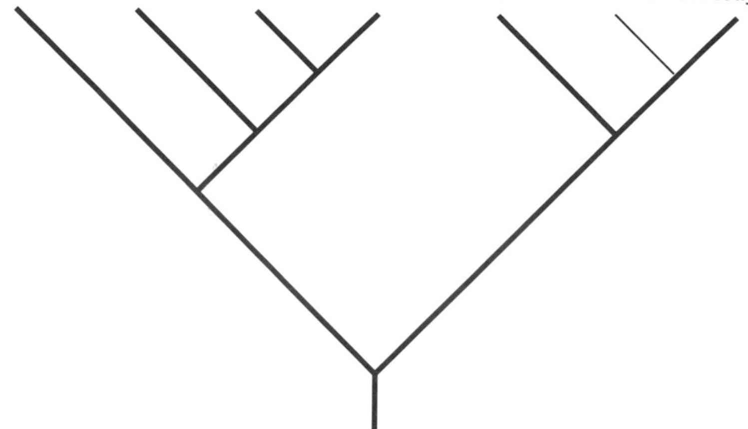

Figure 2.35. Current estimate of generic relationships among living crocodilians. Relationships corroborated by the phylogenetic relationships of helminth parasites inhabiting crocodilians are in boldface. Redrawn and modified from Densmore and White (1991).

parasites known from *Osteolaemus* have been acquired by host switching from African species of *Crocodylus,* or that *Osteolaemus* is not actually the sister group of the entire genus *Crocodylus.* Within the genus *Crocodylus,* ambiguity rules both the host and the parasite data, although even in this morass we can catch a glimpse of the "ghost of concordance future" (Fig. 2.36a,b). Densmore and White (1991) discussed the recurring hypothesis that the species of *Crocodylus* inhabiting each geographic region are not necessarily each other's closest relatives. This possibility is particularly intriguing if we refer back to the area cladograms, which suggest that the helminth fauna of crocodilians inhabiting India, Africa, and the Australo-Oceania region was assembled in at least two evolutionary waves.

This study is important from a parasitological perspective because it indicates that even a relictual parasite fauna in relictual hosts may preserve substantial indications of long (co)evolutionary history among parasite groups and between the parasites and their hosts. From a more general standpoint, these data indicate the way in which studies of host and parasite phylogenies— conducted independently but with the common goal of integrating all information into a robust picture of evolutionary history—can reinforce, inform, and stimulate each other.

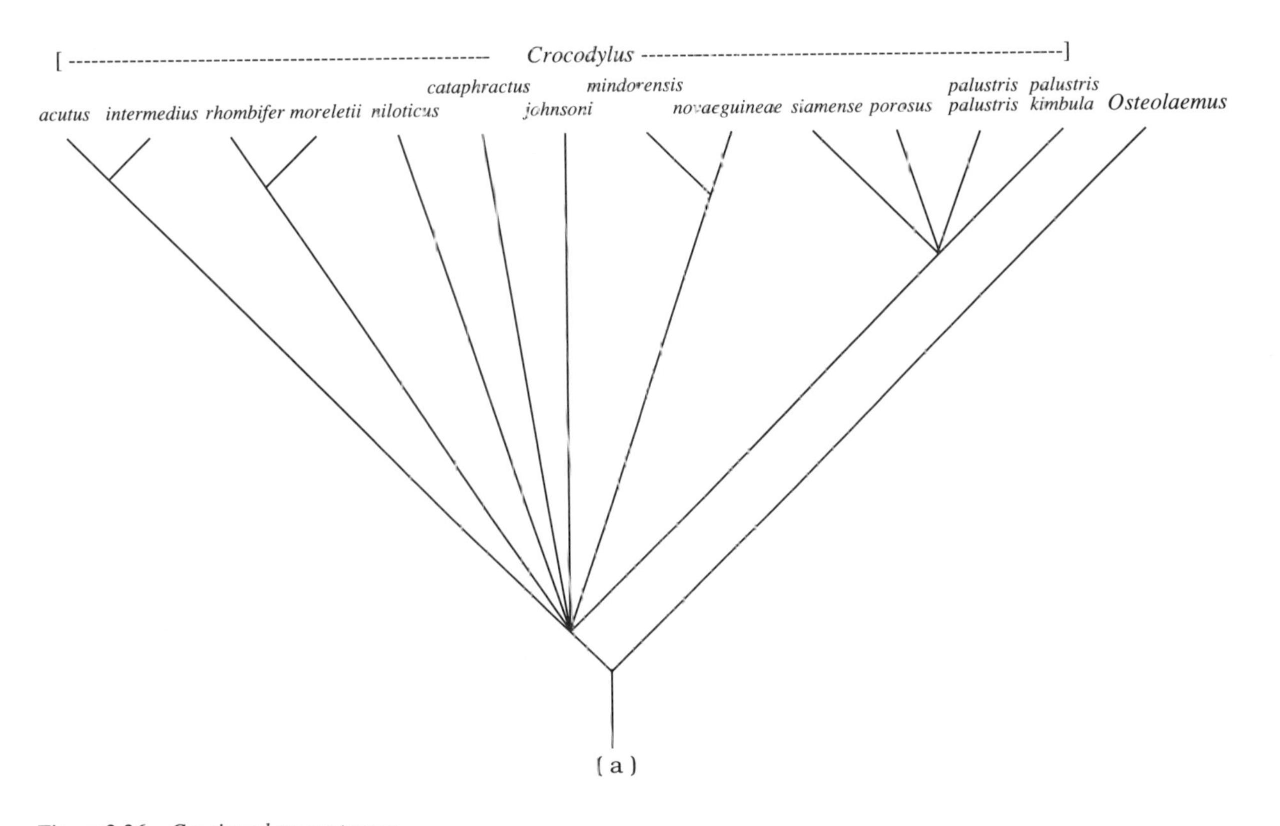

Figure 2.36—*Continued on next page.*

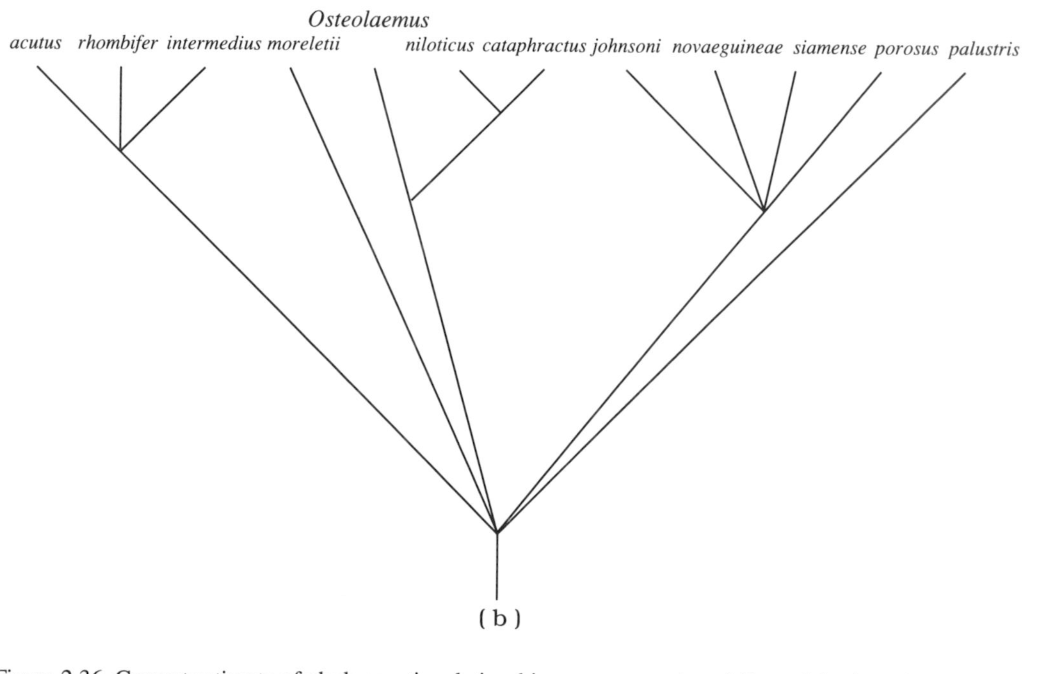

Figure 2.36. Current estimate of phylogenetic relationships among species of *Crocodylus* based on molecular analysis of the hosts (a) and from phylogenetic relationships of their parasites (b). Part (a) redrawn and modified from Densmore and White (1991).

NEMATODE PARASITES OF GREAT APES: PARASITOLOGICAL TRACES OF HUMAN ORIGINS AND HISTORY

Recognition of the profound similarities between humans and primates, especially the ones we have named the great apes, predates formal evolutionary theory. Given the progressive nature of parasite evolutionary biology in the late nineteenth and early twentieth centuries, it is not surprising that a number of parasitologists attempted to decipher what, if anything, parasites could tell them about human origins and genealogical relationships. For example, Fahrenholz (1913) offered evidence from biting lice to support the hypothesis that humans were more closely related to great apes than to any other primate group. Cameron (1929) was among the first helminthologists to propose that parasitological evidence indicated a close relationship between humans and the African great apes, gorillas (*Gorilla*) and chimpanzees (*Pan*).

As with all other areas of parasite evolutionary biology, these studies suffered from the lack of a method for generating explicit phylogenetic hypotheses. So, when Kuhn (1967) supported Fahrenholz, when Sandosham (1950) and Inglis (1961) provided corroborating evidence based upon pinworm data for Cameron's conclusions, or when Garnham (1973; see also Peters et al. 1976) suggested that human-infecting species of *Plasmodium* (malaria) were most closely related to *Plasmodium* species infecting gorillas and chimpanzees, their assessments were based more on intuition than on rigorous data analysis:

> The relationship of the malarial parasites of man and their African pongids are extremely close; the plasmodia of *Pongo* [orangutans] and the gibbons [hylobatids], however, are well defined separate species. [Dunn 1966]

Dunn (1966) made the first attempt to synthesize evidence from many groups of parasites. He collected information for representatives of 34 genera inhabiting gibbons, humans, orangutans, and chimpanzees, and performed a simple phenetic clustering analysis based upon shared parasites. From this, he concluded that humans were more closely related to (i.e., shared more parasites with) chimpanzees than to the Asian great apes. Glen and Brooks (1986) added more parasite taxa and information from cercopithecid monkeys (Old World monkeys) and gorillas to Dunn's data base. The results of the new phenetic analysis were startling: humans clustered with cercopithecids. Using cercopithecids as outgroups to determine whether the presence or absence of a parasite taxon in a great ape taxon was plesiomorphic or apomorphic, then analyzing those data phylogenetically, did not change the outcome (see Glen and Brooks 1986 for details). The conclusion: either parasites have nothing to tell us about human origins or we have not tapped all the information available in the parasite data.

The early 1980s was a time for reevaluating the robustness of many phylogenies. Despite Washburn's (1973) assertion that the close relationship between "man . . . [and] the African apes may now be considered a fact," the great apes did not escape this period unscathed. One of the pioneers of North American phylogenetic systematics, the noted herpetologist Arnold Kluge, examined the morphological and molecular data base for the great apes within a rigorous phylogenetic systematic context (Kluge 1983). He concluded that the molecular data were neither as robust nor as unambigious as previously thought, and that both data sets supported more than one phylogenetic tree. The consensus view of these alternate trees suggested that humans were the sister group of gorillas plus chimpanzees (the "troglodytian hypothesis"). Echoing Washburn, Pilbeam (1984) concluded that the debate between physical anthropologists and molecular biologists over the pattern and timing of hominoid evolution was basically settled. Kluge (1983), however, noted that phylogenetic systematic analysis of anatomical data actually favored the old (Schultz 1930; Keith 1931), generally discarded hypothesis that humans were the sister group of orangutans plus gorillas plus chimpanzees (the "hylobatian hypothesis"). To add fuel to the controversy, Schwartz (1984) proposed that humans were the sister group of orangutans (the "pongoian hypothesis").

The first attempt at a phylogenetic systematic analysis for a group of parasites inhabiting great apes used the pinworm genus *Enterobius,* which Cameron (1929), Sandosham (1950), and Inglis (1961) unanimously agreed showed support for the consensus troglodytian hypothesis. Phylogenetic systematic analysis (Brooks and Glen 1982), however, placed *Enterobius vermicularis,* a cosmopolitan parasite of humans that also inhabits gibbons, as the sister group of *E. buckleyi,* inhabiting orangutans, *E. lerouxi,* inhabiting gorillas, and *E. anthropopitheci,* inhabiting chimpanzees, supporting the hylobatian hypothesis. Subsequently, Glen and Brooks (1985) performed a phylogenetic systematic analysis of a group of hookworms, *Oesophagostomum* (*Conoweberia*), and discovered that a single species, *O. stephanostomum,* inhabits chimpanzees, gorillas, and human beings, while its sister species inhabit gibbons and orangutans, a finding that supports the consensus troglodytian hypothesis. When both groups were combined using BPA, the consensus troglodytian hypothesis was supported, but not unambiguously (Glen and Brooks 1986).

During this time, a French parasitologist specializing in the systematics of pinworms collected specimens from a human, assuming them to be *E. vermicularis.* To his surprise, the specimens represented a previously unrecognized and undescribed species of *Enterobius,* which was named *E. gregorii* after the parasitologist's son (Hugot 1983; Hugot and Tourte-Schaefer 1985). Using this new information, let us redo the BPA analysis for *Enterobius* and *Oesophagostomum (Conoweberia)* (Fig. 2.37) to see if any new light can be shed on the perplexing state of human origins.

BPA of the data presented in Table 2.14 results in two equally parsimonious host cladograms each having a consistency index of 96.7% and differing only in their placement of orangutans relative to hylobatids (Fig. 2.38). *Both of these trees are congruent with the troglodytian hypothesis.* Geographic distribution data for the various helminth species suggest complex, and as yet incompletely resolved, movements of primates between Africa and Asia, correlated with the numerous host switches depicted in Figure 2.38. While in Africa, humans were busy coevolving with (at least) one species of pinworm (*Enterobius gregorii*) and one species of hookworm (*Oesophagostomum stephanostomum*). They were also adding to their parasite repertoire, through host switches, the hookworms *O. brumpti* (from Old World monkeys) and *O. bifurcum* (may have been acquired originally by the ancestor of humans plus chimpanzees plus gorillas, or may have been acquired independently by humans and by chimpanzees). During the journey to Asia, they lost *E. gregorii, O. stephanostomum,* and *O. brumpti,* and acquired an Asian pinworm (*E. vermicularis*) that is phylogenetically associated with hylobatids. (This conclusion is, of course, somewhat tentative until we can disentangle cases in which the newly recognized *E. gregorii* has been confused with *E. vermicularis.*) Somehow *O. bifurcum* managed to survive the trip, to be introduced to the marvels of the Orient.

In addition to the evidence provided by phylogenetic comparisons of parasites and their hosts, we can search for an understanding of our more immediate roots in the data being uncovered by paleo- (e.g., Ferreira et al. 1988a) and archaeoparasitologists (e.g., Horne 1985; Reinhard 1990). For example, *E. vermicularis* eggs have been discovered in New World human coprolites dating from A.D. 375 to A.D. 1275 (Colorado Plateau: Samuels 1965; Fry and Hall 1969, 1975; Hall 1972; Fry 1977; Stiger 1977; Reinhard et al. 1987), from approximately 10,000 years ago (Utah: Fry and Moore 1969), from 1075 B.C. to A.D. 1140 (Utah: Fry and Hall 1969, 1975; Moore et al. 1969), and from 200 B.C. (Tennessee: Faulkner et al. 1989). Further south, *E. vermicularis* eggs have been found in mummies and coprolites dating from 2277 to 400 B.C. in Argentina (Zimmerman and Morilla 1983), Peru (Patrucco et al. 1983), and Chile (Araujo et al. 1985; Ferreira et al. 1989). These discoveries set at least a minimal time frame for the appearance of human travelers in the New World. By at least 8000 B.C. they had colonized middle North America, and, moving at

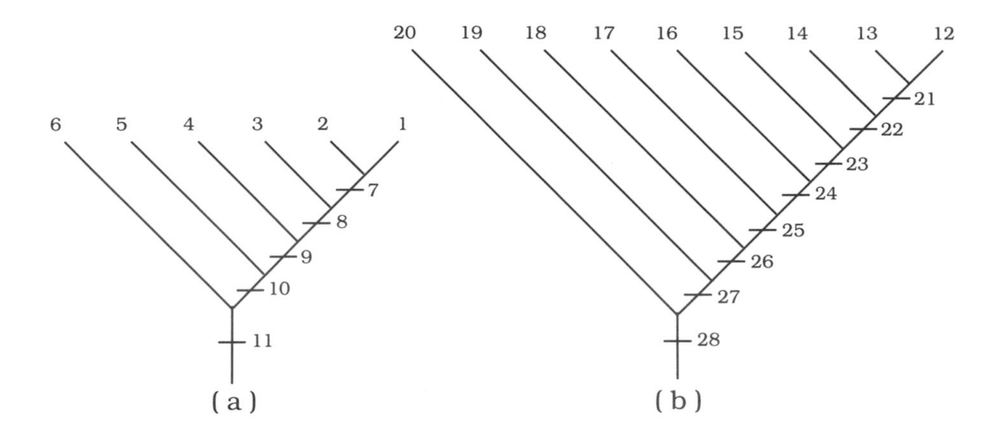

Figure 2.37. Partially modified phylogenetic trees for (a) *Enterobius* and (b) *Oesophagostomum* (*Conoweberia*), numbered for BPA.

Table 2.14
BPA matrix for pinworms (*Enterobius,* 1–11) and hookworms (*Oesophagostomum* [*Conoweberia*], 12–28) inhabiting Old World monkeys (cercopithecids) and great apes (gibbons [*Hylobates*], orangutans [*Pongo*], humans [*Homo*], chimpanzees [*Pan*], and gorillas [*Gorilla*])

Host	Binary code
Cercopithecid 1	0000010000100000100000001111
Cercopithecid 2	???????????00000010000000111
Cercopithecid 2	???????????00000001000000011
Cercopithecid 3	???????????00000000100000001
Cercopithecid 4	???????????01000000011111111
Hylobatid 1	0000100001100001000000011111
Hylobatid 2	???????????00110000001111111
Pongo 1	0001000011100001000000011111
Homo 1	0010000111110000000011111111
Homo 2	00001000011?????????????????
Homo 3	???????????00000010000000111
Homo 4	???????????00000001000000011
Pan 1	1000001111110000000011111111
Pan 2	???????????00000001000000011
Gorilla 1	0100001111110000000011111111

Note: See Figure 2.37 for identities of taxa indicated by codes.

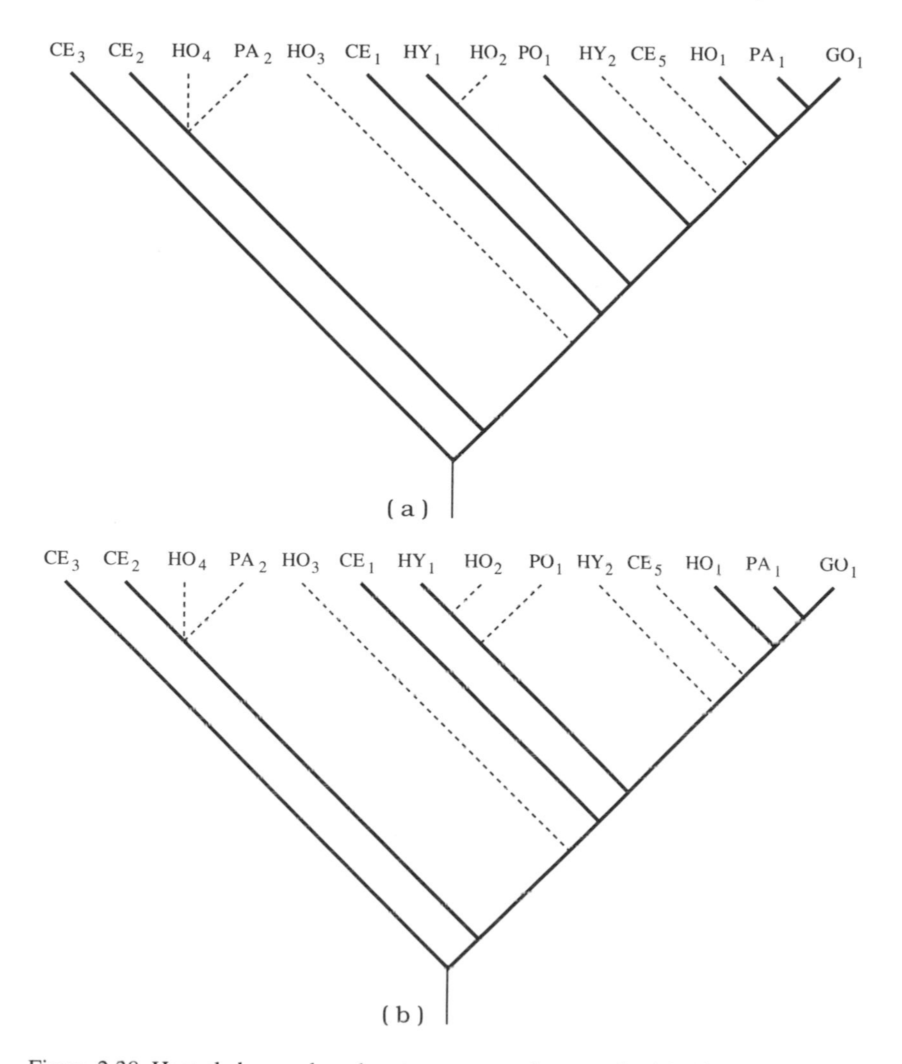

Figure 2.38. Host cladogram based on two groups of nematodes inhabiting Old World monkeys and great apes. CE = Cercopithecidae; HY = *Hylobates;* PO = *Pongo;* HO = *Homo;* GO = *Gorilla;* PA = *Pan.* Bold lines indicate phylogenetic congruence; dotted lines indicate host switching.

a seemingly leisurely pace, had established themselves throughout South America by 2277 B.C.

Where might these colonizers have come from? Darling (1920, 1921) believed that distribution patterns of two hookworm species, *Necator americanus* and *Ancylostoma duodenale,* could shed some light on the problem. *N. americanus* is native to Africa south of the Sahara and to southern Asia, while

A. duodenale is native to northern Africa, to Eurasia, and to Asia north of the Himalayas. Darling hypothesized that the presence of only *N. americanus* in South American Indians would indicate an African or southern Asian origin for the first peoples of South America, while the presence of only *A. duodenale* in the native inhabitants would indicate a northern Eurasian origin. Since hookworm eggs and larvae are not very cold resistent, if the first peoples of South America migrated to the New World via the Bering land bridge, it is unlikely that *Ancylostoma* and *Necator* would have survived the trip unless the migration was rather rapid (recall the evidence from *Enterobius vermicularis*). Therefore, Darling suggested, the occurrence of either *Ancylostoma* or *Necator* in South American Indians would indicate that the hookworms were introduced, after the initial peopling of the New World, from someplace where they both occurred, such as Japan and China. In recounting the history of these ideas, Darling spoke of the possibility that "storm-tossed fishermen" from eastern Asia might have introduced the parasites.

Soper (1927) found that *Ancylostoma duodenale* was 13 times more prevalent than *Necator americanus* in a group of South American Indians living in western Paraguay that apparently had a limited history of contact with Europeans or Africans. Assuming that such a discrepancy in prevalence patterns reflected different ages of association between the humans and the two hookworm species, Soper concluded that *Ancylostoma* infections predated the European invasion of South America, and that *Necator* was a more recent arrival, presumably from eastern Paraguay, where Europeans and Africans lived.

For nearly forty years these ideas remained interesting, but untestable, parasitological fancy, until Meggers and Evans (1966) discovered evidence that Japanese fishermen had reached and become established in Ecuador about 3000 B.C. Estrada and Meggers (1961) had previously documented the introduction of southeast Asian culture into Ecuador around 200 B.C. Manter (1967), always aware of the implications of new findings, suggested that the Japanese fishermen who arrived in 3000 B.C. were probably infected with *A. duodenale*, and that the later arrivals in 200 B.C. might have brought *Necator* with them, explaining the relatively higher prevalence of *Ancylostoma* among South American Indians. Thus, he suggested, both *Necator* and *Ancylostoma* might have arrived in South America after that continent's colonization by its first people, but before the arrival of Europeans and Africans. This possibility has gained credence recently; *Necator* has been reported from human coprolites in Minas Gerais, Brazil, in surroundings dated from 360 to 3610 years ago (Ferreira et al. 1988a), while *Ancylostoma* has been reported from human coprolites (and in some cases from mummified humans) in Lake Tiahuanaco, Peru dating from A.D. 900 (Allison et al. 1974), from Minas Gerais, Brazil, dating from 430 to 3490 years ago (Ferreira et al. 1988a), and from Piaui State, Brazil, dating from 7230 years ago (Ferreira et al. 1988b). It would appear that the next step in unraveling this part of the human parascript story requires the application of some of today's modern methods of molecular genetic analysis to assess the evolutionary status of different strains of *Ancylostoma duodenale* and *Necator americanus.*

Overall then, these findings are interesting for a variety of reasons. First of all, they support the consensus view of great ape phylogeny, providing an additional source of evidence in our quest to provide as detailed a reconstruction of human origins and relationships as possible. Second, they provide empirical rigor for conclusions that parasitologists had drawn using less formal means, attesting to the robustness of the coevolutionary picture sometimes afforded by parasites. Finally, the story is just beginning; there are still so many parasites and so little time. For example, what information can be gleaned from the five species of *Plasmodium,* at least an equal number of *Schistosoma,* and the numerous parasites acquired from animals that were deliberately (e.g., dogs, cats, sheep, cattle, camels, horses, pigs) or accidentally (e.g., many species of rats and mice) domesticated? The parascript interpretation of human history is already exciting. It can only become more so as we improve our ability to decipher those "always bilingual and usually garbled" messages.

SPECIATION: DO PARASITES DIFFER FROM FREE-LIVING ORGANISMS?

Mayr (1963) recognized three general classes of speciation: (1) *reductive speciation,* in which two existing species fuse to form a third; (2) *phyletic speciation,* in which a gradual progression of forms through a single lineage (anagenesis) are assigned species status at different points in time, and (3) *additive speciation,* which involves lineage splitting (cladogenesis) and reticulate evolution. The majority of examples of speciation represent cases of additive speciation. In addition, the concept of phyletic speciation has been criticized on empirical and theoretical grounds. On the empirical side, we recognize that, although the endpoints of any anagenetic continuum or lineage may be recognizably "different," separation of the intermediate forms into distinct species is an inherently arbitrary exercise (Hennig 1966; Wiley 1981). On the theoretical side, additive speciation appears to be closely analogous to reproduction in organisms, while anagenesis appears to be more analogous to ontogeny than to reproduction. Thus, because ontogeny does not produce new organisms, but only changes them, anagenesis refers to the evolution of single species and not to the production of new species.

There are three major additive speciation models (for detailed discussions see Wiley 1981; Wiley and Mayden 1985; Brooks and McLennan 1991; and references therein). *Vicariant speciation* (allopatric mode I) occurs when an ancestral species is geographically separated into two or more relatively large and isolated populations, with subsequent lineage divergence by the isolated descendant populations. The speciation rate will depend on the degree of variation in the ancestral species prior to isolation and the rate of origin of evolutionary novelties in the subdivided populations. The *peripheral isolates allopatric speciation* model (allopatric mode II) postulates that a new species arises from a small, isolated population usually, but not always, on the periphery of the larger central ancestral population. Gene flow between the peripheral and central populations contributes to species cohesion, because it is initially sufficient to keep novel traits from being fixed; however, it is not strong enough to prevent the establishment of novel phenotypes in the peripheral population. Once geographic separation is complete, gene flow from the central population is stopped. This could happen rapidly, as with founder effect phenomena (Carson 1975, 1982; Templeton 1980; Lande 1981; Carson and Templeton 1984; Goodnight 1987; Charlesworth and Rouhani 1988; Barton 1989) or it could be a relatively gradual process, such as a gradual environmental change in the peripheral area (Mayr 1954, 1963, 1982; Patton and Smith 1989). *Sympatric speciation* (Maynard Smith 1966; Dickinson and Antonovics 1973; Felsenstein 1981; Gittenberger 1988) occurs when one or more new species arise without geographical segregation of populations. Unlike the allopatric models, which postulate that gene flow between populations is initially severed by factors extrinsic to the biological system, sympatric speciation requires the involvement of biological processes intrinsic to the system, e.g., hybridization, the evolution of asexual or parthenogenetic populations, or a change in mate recognition systems. Additionally, differentiation must occur "within the dispersal area of the offspring of a single deme [the cruising range]" (Mayr 1963). Although this was the mode preferred by Darwin (1859) after he (perhaps ironically) abandoned earlier support for speciation by geographic isolation, support for sympatric speciation wavered when population geneticists demonstrated that the effects of gene flow among populations would tend to swamp out or homogenize any novel traits arising within a population. If gene flow were restricted or interrupted, as in allopatric or peripheral isolates speciation models, the novel trait would have a better chance of becoming fixed within a deme, and the whole process would operate much more smoothly. The work of the population geneticists, coupled with the earlier recognition that most "related" species (this usually meant members of the same genus) exhibited allopatric distributions (e.g., Mayr 1942; Wallace 1955), provided a strong foundation for the hypothesis that most speciation was allopatric. In recent years, however, there has been a revival of interest in sympatric speciation, as researchers have intensified investigations of phenotypic plasticity, disruptive selection, and chromosomal divergence (see discussion and references in West-Eberhard 1989; chapters in Otte and Endler 1989).

To date, very few studies have been published that examine speciation using phylogenetic evidence. Development of a more robust data base is vitally important to the future of speciation research because that is the only known way to assess the relative frequencies of different speciation modes based on evidence rather than on theory. Lynch (1989) produced the first such investigation using free-living organisms and with intriguing results. He examined species ranges for members of a number of clades for which phylogenetic trees and extensive distributional data were available, estimating ancestral ranges by the sum of all descendants' ranges. Based on an analysis of 66 documented cases of vertebrate speciation, he suggested that 71% of the speciation events were due to vicariant allopatric speciation, 15% of the cases resulted from peripheral isolates allopatric speciation, and 6% of the evolutionary divergence fulfilled the requirements of sympatric speciation. In the other 8% of the cases, Lynch discovered dichotomies buried deep within the phylogenetic trees that explained significant geographical overlap between more highly derived sister groups, but which could not be interpreted in terms of speciation modes. Lynch's study, combined with other examples drawn from a wide variety of vertebrate species (see examples and references in Brooks and McLennan 1991) supports the major contention of evolutionary theorists, such as Mayr (1963) and Futuyma and Mayer (1980), that sympatric speciation does not seem to occur very often among vertebrates.

In the past, researchers have assumed that parasites are so different from the majority of free-living organisms that it is difficult, if not impossible, to generalize about them from other studies. This assumption was reflected in a proposal by Price (1980) that the evolution of new parasite-host relationships was driven predominantly by sympatric speciation via colonization of new hosts, the hosts involved in the colonization events not necessarily being closely related to each other. This proposal has two parts: (1) unlike the patterns uncovered by Lynch for vertebrates, sympatric speciation has been as important as allopatric speciation within parasite groups and (2) sympatric speciation can be initiated by a parasite switching to a different species of host (ecological segregation).

Let us examine the case for sympatric speciation, before delving into the realm of host switching. Price believed that the small and relatively discrete nature of parasite demes facilitated sympatric speciation. The deme concept is an important component of speciation theory (Mayr 1963). In general, evolutionary biologists view demes as local inbreeding aggregates whose "stability" through time is due to inbreeding and whose "variation" is due to immigration from other demes. The relative magnitude of inbreeding versus immigration determines whether a species population is polymorphic (lots of immigration) or polytypic (lots of inbreeding). This characteristic, in turn, is thought to play a role in determining the probability of speciation because demes can serve as units that can be reinforced by inbreeding to such an extent that they produce incipient species. Local dispersion (including vagility and home range size), habitat loyalty and patchiness, mate recognition systems, and social structure all affect inbreeding in free-living organisms. *The more these factors tend to enhance inbreeding and inhibit immigration in individual demes, the more cohesive that deme will be, and the higher the likelihood that incipient speciation will occur.* When these factors do not inhibit immigration and/or enhance inbreeding, we expect to find a panmictic polymorphic species.

In parasitology, demes are defined as the groups of adult (sexually reproducing) parasites inhabiting an individual host organism. Although this definition portrays parasite demes as discrete entities at the population level, it creates the following paradox when extended to views of sympatric speciation: *parasite demes have a tendency to be ephemeral.* In the vast majority of cases, they must be reassembled each generation by the relatively random immigration of larval and/or juvenile stages, usually from a much larger gene pool (the parasite species population occurring in multiple hosts) than that represented by the members of the original deme. Parasite demes thus generally have no spatiotemporal continuity, and because they have no continuity, there is no opportunity for pre- or postmating isolating mechanisms to reinforce differentiation via inbreeding. Such mechanisms are an essential element of sympatric speciation. *So, rather than promoting that speciation mode, the structure of many parasite demes appears to increase the influence of immigration and decrease individual deme cohesion, ensuring that sympatric speciation will occur even less frequently in parasites than it does in free-living organisms.*

Does this mean that sympatric speciation is totally superfluous to discussions of parasite evolution? Fortunately, the answer to this question is "no" because factors do exist that enhance inbreeding in some species. For example, parasites that autoinfect their individual hosts and inhabit long-lived hosts can produce more than one generation on the same host organism. This reduces the ephemerality of the deme structure for those species of parasites, increasing the potential that differences appearing within a deme could be maintained by inbreeding. If sympatric speciation has played a role in parasite evolution, we might expect to find it concentrated within groups like the monogenean platyhelminths or oxyurid nematodes (pinworms) that show this type of life cycle. Many fish species play host to more than one species of congeneric monogeneans, making us suspicious that sympatric speciation might have occurred. Two recent phylogenetic studies (Van Every and Kritsky 1992, in a study of monogeneans inhabiting piranha hosts in the middle Amazon; Guegan and Agnese 1991, in a study of monogeneans inhabiting mud-sucking minnow hosts in Africa) suggest that there is an apparent phylogenetic "signal" in this noise of evolutionary divergence. In neither study was an inclusive monophyletic group of hosts analyzed, but the results are tantalizing for studies of speciation modes. In Chapter 3, we will return to monogeneans in a comparative study of diversification among the species-rich groups of parasitic platyhelminths.

The second part of Price's proposal is that speciation by host switching is a form of sympatric speciation, a common assumption (Diehl and Bush 1989; Grant and Grant 1989; Tauber and Tauber 1989). If we consider this a mode of sympatric speciation by ecological segregation, we create a twofold paradox. First, as we discussed previously, the opportunity for sympatric speciation is highest in parasites that autoinfect and have direct life cycles. This mode of life, however, actually *decreases the likelihood of host switching* because the parasites are relatively immobile. Increasing the number and the dispersal of larval stages facilitates host switching; however, this mode of life *decreases the likelihood of sympatric speciation* because it increases immigration and decreases the cohesion among individual demes. The second part of the paradox appears following the putative host switch. Once host switching has occurred, the probability that the new host (resource) will exert strong directional selection pressure on the colonizing population should be higher if the colonizers display pronounced host specificity. Such species are more tightly coupled to their resource bases and thus should be more sensitive evolutionarily to changes in that component of their environment than their generalist counterparts. The likelihood of a habitat change occurring in the first place, however, is decreased for species that only respond to a small number of cues. *Therefore, specialist species are least likely to colonize new hosts, but are the ones most likely to speciate as a result of any such switch, while generalist species are most likely to colonize new hosts, but are the least likely to speciate as a result of the interaction.*

Is there a way out of this problem that does not involve abandoning host switching as a potential influence in parasite speciation? We believe that there is, but that it requires us to reexamine our concept of host switching, this time from the parasite's perspective rather than the host's perspective. A new host may or may not represent a new kind of resource to a parasite, but it definitely represents a new geographical distribution. Thinking of hosts as geographical areas eliminates the host specificity paradox because, once a species has moved into a new area (a new host species), the potential for speciation is created by the effects of geographical isolation on gene flow, even if the host does not represent a new kind of environment provoking an adaptive response by the colonizing parasites (although such responses can accelerate the speciation process). *Speciation by host switching is thus an example of* peripheral isolates allopatric speciation *(peripatric speciation of Mayr 1942, 1954), not sympatric speciation.* In free-living organisms, the speed with which gene flow is finally severed between the colonizers and their ancestral population depends upon the magnitude of the geographical barrier and the dispersing capabilities of the organisms. This equation is repeated in parasites with one substitution: gene flow severance is dependent upon the dispersing capabilities of the parasites and upon the extent of sympatry between the old and new host species.

We believe that the confusion about modes of speciation in parasites was created in the first place because the speciation process was viewed from the host's perspective. It is time for a new view of parasite speciation and the origin of parasite-host associations. This view must be based on the notion that parasites have their own individual evolutionary tendencies and fates, and thus may exhibit geographic distribution patterns that can be determined, evaluated, and explained without reference to a priori theories about host evolution or the influence of hosts on parasite evolution. When we view speciation from the parasite's perspective, we discover a set of speciation modes analogous to those proposed for free-living organisms (Table 2.15). As we discussed in the preceding paragraph, *host switching involves the movement of a small subset of a species into a new "geographical area."* Once there, the colonizers will either simply add the new host to the species range of preferred habitats or speciate (via a peripheral isolates mode). The interactions depicted in Table 2.15 highlight two interesting, and unique, aspects of host switching: (1) the parasite is not speciating in association with its "normal" host and (2) speciation is initiated by an active movement of the parasite. New parasite-host associations can also arise without the movement of a parasite population onto a new host species. For example, an ancestral parasite species inhabiting an ancestral host population can be subdivided geographically with the host population, resulting in the evolution of new associations (vicariant allopatric cospeciation: Brooks and McLennan 1991). There are three possible outcomes of such separation: the parasite will speciate and the host will not, the host will speciate

Table 2.15
Comparison of host, geographic, and phylogenetic correlates for different speciation modes in parasites

Distribution of parasite sister species	What initiates speciation (what initiates the severance of gene flow)?	Host distribution
Allopatric	Geographical changes subdivide an ancestral species (passive movement of parasites)	Allopatric (the same host or sister species)
	Geographical isolation of a small subset of the host population (passive movement of parasites)	Allopatric (the same host or sister species)
	Host switching (active movement of parasites)	Sympatric (different host species)
	Behavioral or ecological segregation of hosts via sympatric speciation (passive movement of parasites)	Sympatric (sister species)
Sympatric	Interactions between two diverging populations (no movement of parasites)	Sympatric (same host species)

and the parasite will not, or both the parasite and the host will speciate. Speciation can also be initiated by character changes within one parasite deme that result in reproductive isolation of the new parasite phenotype from another sympatric parasite deme without involving any changes around or to the host; this is sympatric speciation from the parasite's perspective. This type of speciation process can be recognized by (1) the presence of sister species in the same host species that (2) differ in some apomorphic ecological or genetic characteristics that could, in themselves, produce independent species. Both allopatric cospeciation and sympatric speciation in parasites have one thing in common: the new parasite-host associations did not originate as a result of active colonization by the parasite because both the host and the parasite maintain the ancestral association throughout the speciation process.

Speciation does not just produce species, it produces *sister species*. Since this irreversible production of groups that are each other's closest relatives introduces a historical component into the process, speciation cannot be studied without first determining the sister group relationships within the system of interest. Assuming that two species are or are not "closely related," and basing hypotheses of the speciation model involved in their production on this assumption, will, in most cases, ultimately lead to confusing and contradictory results. Bearing this is mind, let us turn our attention to a study delineating the potential importance of the three speciation modes within a group of parasitic flatworms.

TREMATODES AND TURTLES (AGAIN): GEOGRAPHY, HOST SWITCHING, AND MORPHOLOGICAL INNOVATIONS IN SPECIATION

We have already examined one group of trematodes inhabiting freshwater turtles, the blood flukes of the family Spirorchidae. In this example, we will move out of the bloodstream and into the intestine, where we will encounter trematodes of the genus *Telorchis.* Macdonald and Brooks (1989b) examined geographical, host, and morphological correlates of the phylogenetic diversification of the 11 *Telorchis* species found in North America. Many species of these flukes cooccur geographically, very often in the same species of hosts, and often in the same host individual. In order to disentangle the relative roles of geography and the host in the evolution of these flukes, we must first determine whether any of the sympatric species are each other's sister species. Sympatric congeners that are not each other's closest relatives provide evidence of secondary contact subsequent to speciation, thus ruling out a role for sympatric or peripheral isolates speciation in the origin of those two species. We need two pieces of evidence to address this problem: a phylogenetic tree for *Telorchis* (Fig. 2.39) and information about species distributions.

There are four clades of North American telorchids: (1) *T. sirenis* and *T. stunkardi;* (2) *T. corti, T. dollfusi, T. auridistomi,* and *T. angustus;* (3) *T. chelopi* and *T. scabrae;* and (4) *T. robustus, T. singularis,* and *T. attenuatus.* The distributions of these species are depicted in Figures 2.40a–d. *Telorchis stunkardi* (clade 1: Fig. 2.40a) is a relatively widespread species that appears to occur sympatrically with its more resticted sister species, *T. sirenis.* The phylogenetic relationship and sympatric distribution provide evidence that either sympatric or peripheral isolates allopatric speciation may have played a role in the origin of this species pair. In order to examine this hypothesis further, we need detailed information about the biological characteristics of these flukes that may have contributed to the initiation of speciation. *Telorchis chelopi* and *T. scabrae* (clade 3: Fig. 2.40c) have been collected from only a single and two confirmed localities, respectively, but inhabit hosts having more widespread distributions and presumably are more widespread themselves. At the moment

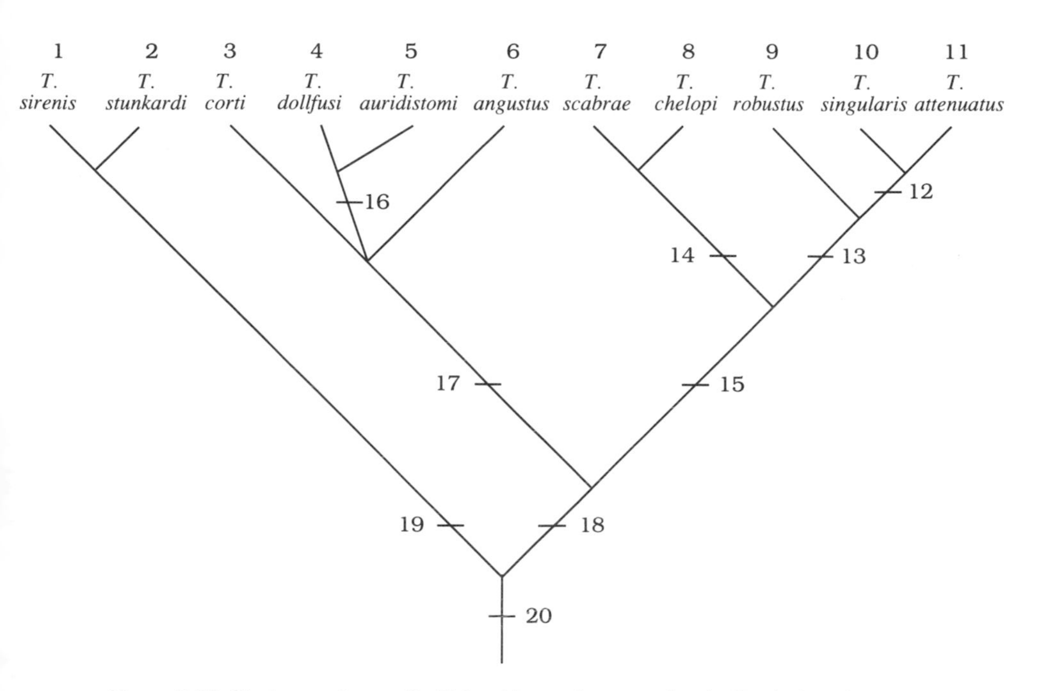

Figure 2.39. Phylogenetic tree for *Telorchis* species occurring in North America, numbered for BPA.

their distribution suggests vicariant speciation of a more widespread ancestor, followed by extinction of populations in the middle of that ancestor's range.

The fourth clade (clade 4: Fig. 2.40d) is composed of three species occurring primarily in a north-south range east of the Appalachian mountains (*T. robustus*), an east-west range in the southern United States (*T. singularis*), and an east-west range in the northern United States and southern Canada (*T. attenuatus*). Examination of the distribution patterns of these flukes indicates that the two putative speciation events within this clade have been vicariant, via the geographical division of (1) the ancestor denoted by 13 to produce *T. robustus* and ancestor 12 and (2) ancestor 12 to produce *T. singularis* and *T. attenuatus* (Fig. 2.39). Although suggestive of sympatric speciation, the observation that *T. attenuatus* overlaps the range of *T. robustus* does not support a role for that speciation mode because the two groups are not sister species.

The second clade presents us with a very complex picture. Evidently the ancestor of the clade, ancestor 17, was widespread throughout central North America (Fig. 2.40b). The trichotomy among *T. corti*, *T. angustus*, and ancestor 16 indicates that either sympatric or peripheral isolates allopatric speciation

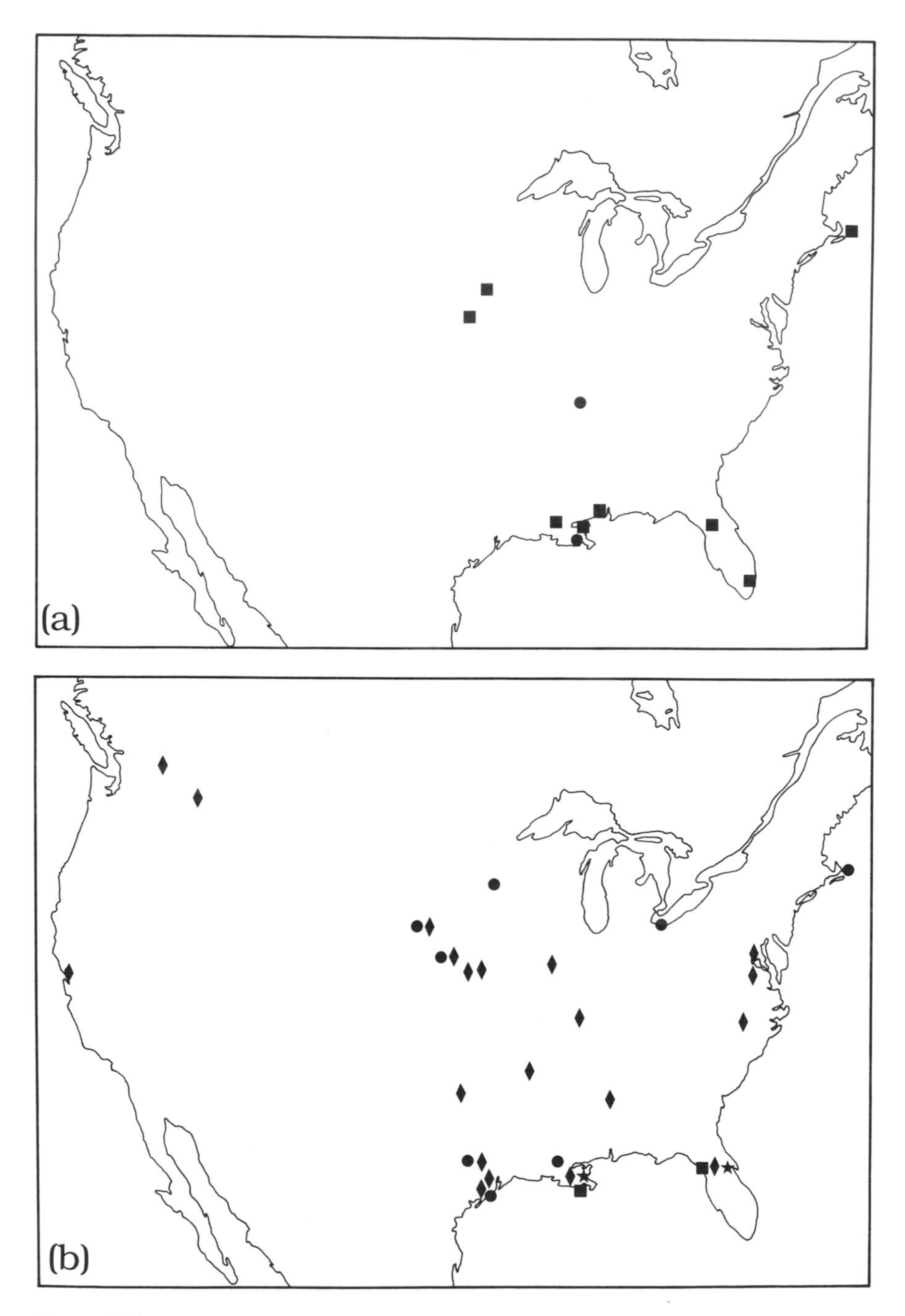

Figure 2.40. Reported geographic distribution records for North American *Telorchis* species. (a) ● = *Telorchis sirenis*, ■ = *T. stunkardi*. (b) ♦ = *Telorchis corti*, ★ = *T. dollfusi*, ■ = *T. auridistomi*, ● = *T. angustus*.

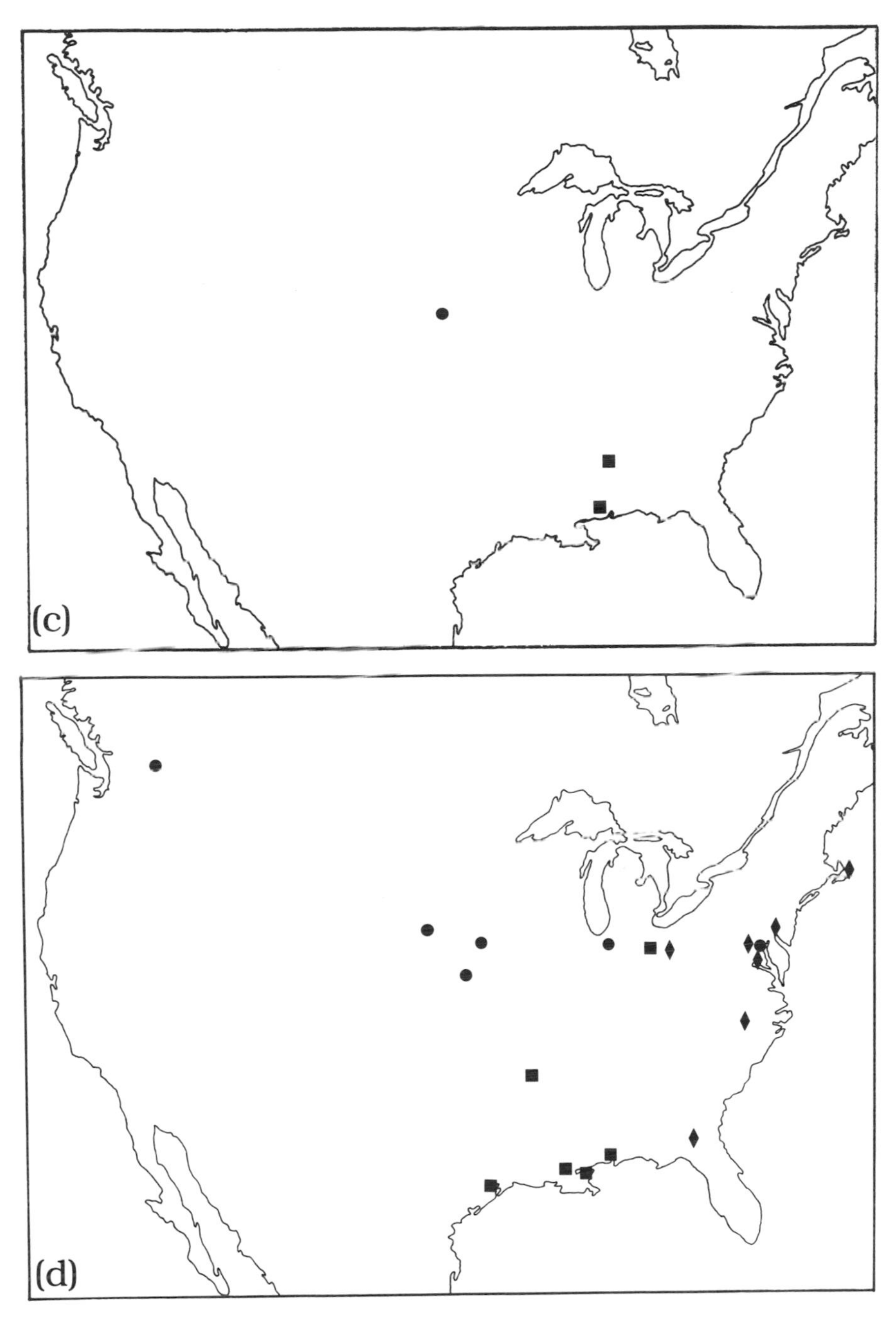

(c) ■ = *Telorchis scabrae*, ● = *T. chelopi*. (d) ♦ = *Telorchis robustus*, ■ = *T. singularis*, ● = *T. attenuatus*. Redrawn and modified from Macdonald and Brooks (1989b).

was important in the formation of these three species. The observation that both *T. angustus* and ancestor 16 are located almost entirely within the range of the more widespread *T. corti* supports a role for two episodes of either sympatric or peripheral isolates speciation. Interestingly, *T. corti* does not have an autapomorphy, suggesting that it may be the surviving ancestor 17. In the same vein, the overlapping ranges of the sister species *T. dollfusi* and *T. auridistomi* also provide evidence for the potential involvement of sympatric speciation in their origin. Once again, further evidence is required to corroborate the macroevolutionary suggestion that sympatric speciation has played a role in the diversification of these flukes. Specifically, we need to search for autapomorphies in two sister species that could have driven their evolutionary divergence.

This part of the analysis indicates that the North American telorchids have experienced three cases of vicariance speciation and four cases of peripheral isolates allopatric or sympatric speciation. Of course, the situation is more complex than this because we have not discussed the potential modes involved in the speciation of the ancestors of each of the four clades. In order to resolve this problem fully, we need more detailed distribution patterns, information about the geological events that could have initiated the vicariant speciation episodes, and information about the biological characters that could have initiated the sympatric speciation episodes.

Now let us turn our attention to the relationship between the flukes and their hosts. The relative phylogenetic relationships for the North American turtles inhabited by species of *Telorchis* are depicted in Figure 2.41. Figure 2.39 depicts the phylogenetic tree for North American *Telorchis* species with the taxa and their putative phylogenetic connections numbered for BPA. The two most basal North American telorchid species, *T. sirenis* and *T. stunkardi,* are not included in the analysis because they commonly inhabit salamanders, although *T. stunkardi* has been reported from turtles. All other members of the genus inhabit turtles or snakes. If analysis of the entire genus supports the placement of *T. sirenis* and *T. stunkardi* as a basal group, it is possible that their occurrence in salamanders indicates a long evolutionary history for the genus. If this is the case, then the association between *T. stunkardi* and the map turtle, *Graptemys pseudogeographica,* originated from a host switch.

Telorchis auridistomi and *T. dollfusi* are also not included in this analysis because they inhabit aquatic snakes. Their putative common ancestor was a sister species of *T. corti,* which has been reported from aquatic snakes on occasion. Given this, it seems likely that the speciation event producing that ancestor involved a host switch from turtles to snakes. *Telorchis dollfusi* inhabits *Regina grahami* and *Regina alleni* whereas *T. auridistomi* dines with *Faran-*

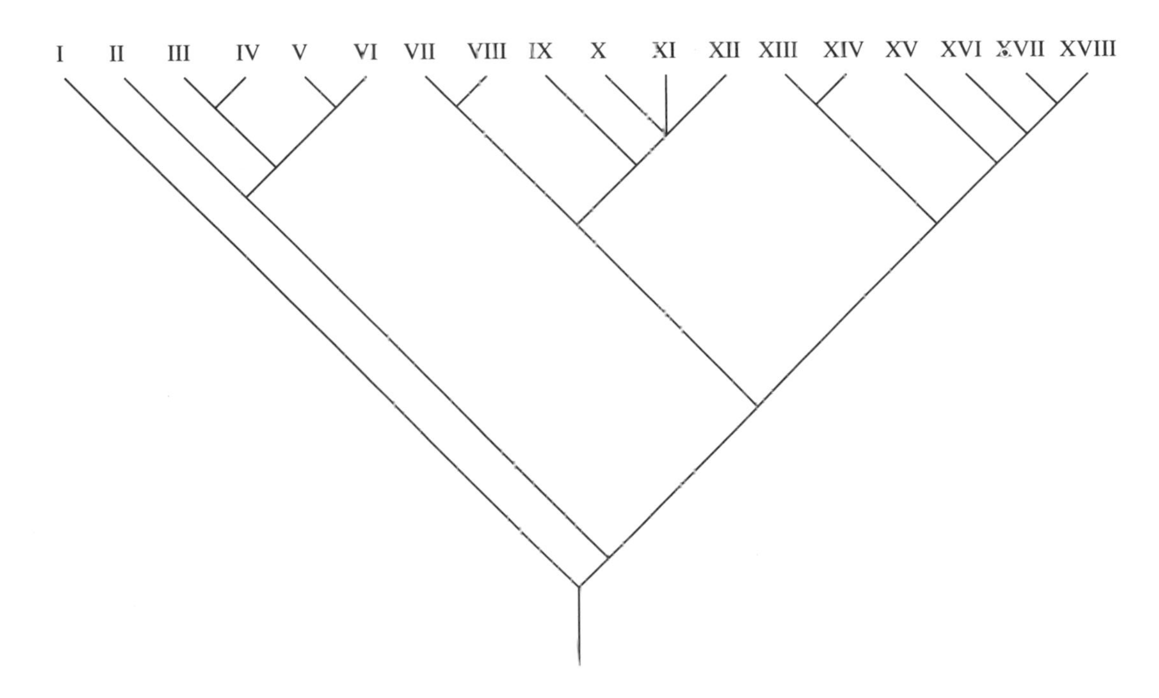

Figure 2.41. Phylogenetic relationships for North American turtles inhabited by *Telorchis* species. I = *Chelydra;* II = *Trionyx;* III = *Kinosternon flavescens;* IV = *K. subrubrum;* V = *Sternotherus odoratus;* VI = *S. carinatus;* VII = *Deirochelys;* VIII = *Emydoidea;* IX = *Terrapene;* X = *Clemmys guttatta;* XI = *C. marmorata;* XII = *C. insculpti;* XIII = *Graptemys pseudogeographica;* XIV = *G. geographica;* XV = *Chrysemys picta;* XVI = *C. concinna concinna;* XVII = *C. scripta scripta;* XVIII = *C. s. elegans.*

cia abacura. Since *Farancia* is not the sister species of *Regina,* further host transfers among aquatic snakes might have been involved in the divergence of *T. dollfusi* and *T. auridistomi* from their common ancestor.

Figure 2.42 depicts the results of the BPA. A historical backbone, highlighted in bold lines, indicates that *T. chelopi* (8), *T. singularis* (10), and *T. attenuatus* (11) have cospeciated with their turtle hosts. The placement of *T. corti* (3) is more problematical because this fluke inhabits 19 different host species, including members of 10 different chelonian genera. The most parsimonious representation based upon the data at hand suggests that *T. corti* cospeciated with *Chelydra.* From that host, the fluke colonized the ancestor of the remaining turtle genera (the sister group of *Chelydra*). This latter colonization is interesting because it implies that the turtles underwent extensive radiation past that point in time, while *T. corti* remained unchanged. Finally, *T. corti* was "lost" from *Chrysemys concinna concinna* and *C. scripta scripta*—although such "absences" are often due to poor collection records. The presence of *T. corti* in salamanders and snakes is best explained, according to this scenario, as the result of host switching from turtles. This, in combination with the preceding suggestion that *T. corti* is apparently buffered against host speciation, further highlights the propensity for that fluke to accumulate definitive hosts. The combination of colonizations and "losses" makes a total of eight departures from strict phylogenetic association between *T. corti* and its turtle hosts.

In contrast to *T. corti,* which demonstrated extensive host switching without speciation, three hypothesized cases of speciation via host switching are depicted in Figure 2.42. The first of these involves the association of *T. angustus* (6) with *Chrysemys picta* and *C. scripta elegans.* Since these two turtles are not each other's closest relatives, ancestor 17 must have colonized one of them and speciated, producing *T. angustus,* which switched to the other turtle without speciating again. It is impossible to tell from this analysis whether the speciation/host switching sequence might have been *C. picta* to *C. scripta elegans* or vice versa. It is, however, interesting to note that *C. scripta elegans* was already associated with *T. singularis* and *T. scabrae,* while *C. picta* was hosting either *T. attenuatus* or its ancestor (12) and possibly *T. scabrae* and *T. corti* at the time ancestor 17 was attempting a landing on a new host. In both cases, with the possible exception of ancestor 12, none of the resident parasite species speciated as a result of the appearance of the new telorchid. The second case involves *T. scabrae* (7) and its association with *C. concinna concinna, C. scripta scripta,* and *C. scripta elegans.* These three turtles are each other's closest relatives (a clade). So, in contrast to the preceding scenario, the origin of *T. scabrae* involves a switch by members of its ancestor (14) to the ancestor

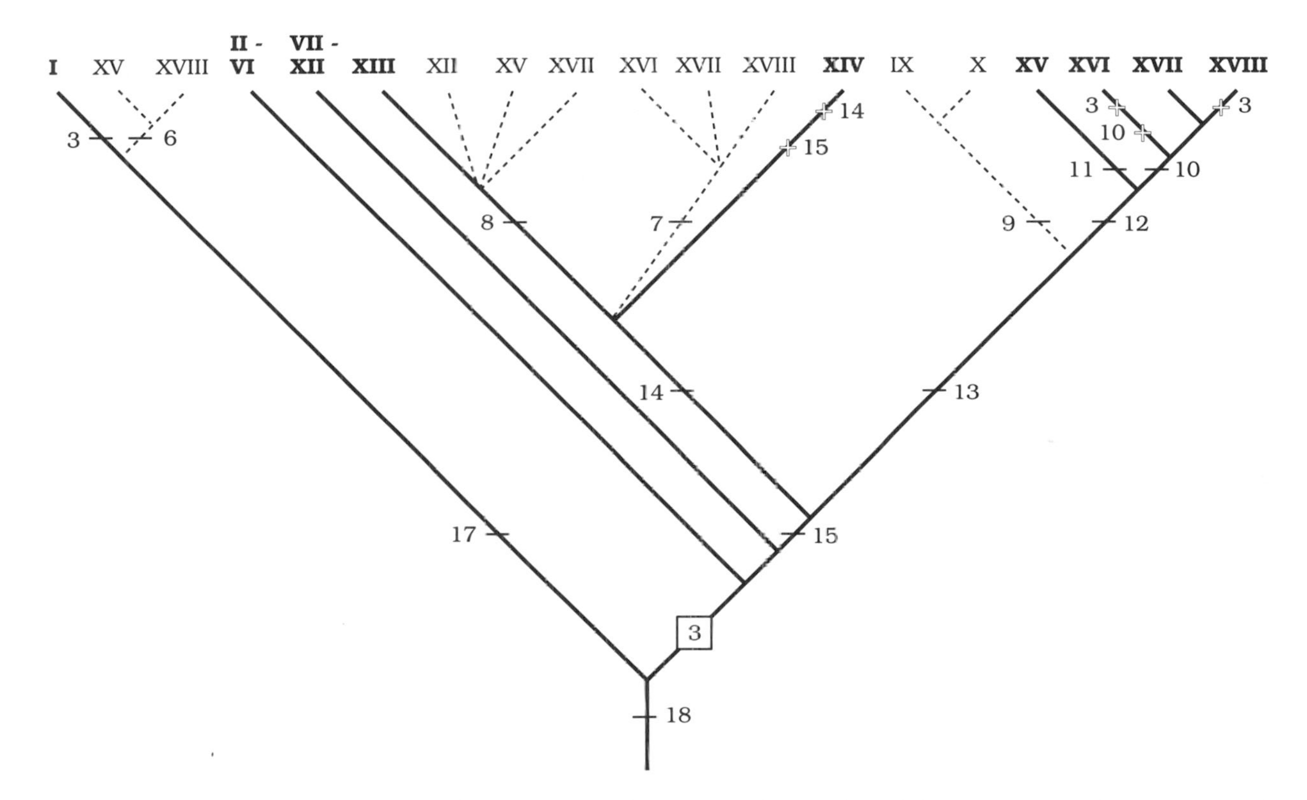

Figure 2.42. Host cladogram for a variety of North American freshwater turtles based on phylogenetic relationships of species of *Telorchis* inhabiting them. Bold lines indicate phylogenetic congruence; dotted lines indicate host switching. Arabic numbers denote BPA codes for *Telorchis;* crosses denote secondary absence of a parasite lineage or species. Roman numerals denote turtle taxa (see Fig. 2.41 for explanation).

of the turtle clade, with subsequent speciation in that turtle producing *T. scabrae*. Past that point, the turtles continued to speciate, while *T. scabrae* remained unchanged. Interestingly, this event occurred in the presence of another telorchid species (*T. singularis:* 10), raising the possibility that speciation, in one or both of these flukes, was either initiated or strengthened by an interaction between the two populations. Finally, we consider *T. robustus* (9) and its relationships with *Terrapene* and *Clemmys guttata*. Once again, these turtles are members of a clade, so it appears that ancestor 13 colonized the ancestor of the turtle clade, then speciated to produce *T. robustus*. Past that point, the fluke did not speciate even though the turtles continued to diverge. Unlike the situation for *T. scabrae,* though, the ancestor of the turtle clade did not already host a telorchid species, so ancestor 13 moved into untouched intestine.

Three cases of host switching without speciation are also highlighted by this analysis. We have already discussed the rambling proclivities of *T. corti* and the transfer of *T. angustus* from *Chrysemys picta* to *C. scripta elegans* or vice versa. *Telorchis chelopi* has also undergone some speciation-free host switching, originating with *Graptemys pseudogeographica,* then switching to *Clemmys insculpti, C. picta,* and *C. scripta scripta.*

Now let us combine the information from geographic and host distribution patterns with information from morphology to develop a more complete picture of speciation in the North American telorchids. The parasite-host relationships do not throw any light on the hypothesized influence of sympatric speciation in clade 1. We need more details about the salamander hosts, as well as an analysis expanded to include all the members of *Telorchis,* before we can pursue this question further.

In contrast, the parasite-host relationships do not support the two cases of sympatric speciation proposed for clade 2. *Telorchis angustus,* which resides completely within the range of one of its sister species, *T. corti,* originated following a host switch by its ancestor into one of two turtles. The origins of the sister species *T. corti* and *T. angustus* involve two separate influences on their ancestor; some populations of ancestor 17 cospeciated with the turtle host to produce *T. corti,* while other populations switched to a different host, then speciated. Since the populations producing *T. corti* and its sister species *T. angustus* seem not to have interacted evolutionarily, we can rule out sympatric speciation. Instead, this is an example of host switching leading to peripheral isolates allopatric speciation. Now, recall that both of the turtles already hosted at least one species of telorchid prior to the arrival of ancestor 17. It is possible, therefore, that an interaction between the residents and the colonizer reinforced the fledging speciation process initiated by the reduction in gene flow owing to geographic isolation. If this occurred we would expect to find an autapomorphy for the colonizing species that could have been involved in driving it along the pathway to "specieshood." That is exactly what we find. Hermaphroditic digeneans mate by aligning their genital pores and exchanging sperm. The plesiomorphic condition of the genital pore for species of *Telorchis* is medial or slightly sinistral and immediately preacetabular. Members of *T. angustus,* however, have genital pores that are dorsolateral to the left cecum. This change in location makes it difficult, if not mechanically impossible, to align genital pores with other species of *Telorchis*. *Telorchis angustus* is therefore characterized by an autapomorphic trait that could act as a premating isolating mechanism relative to other members of the same genus occurring in the new host. This characteristic, combined with the geographic distribution and host switch, provides extremely strong evidence for the influence of peripheral isolates speciation in the origin of *T. angustus.*

The second presumptive case of sympatric speciation in this clade involves the sister species *T. auridistomi* and *T. dollfusi*. Information from parasite-host relationships indicates that these species originated as the result of a host switch from turtles to semiaquatic or aquatic snakes that prey on tadpoles. The distribution of the two flukes among their hosts, however, does not support the sympatric speciation scenario because, although the ranges of the two species overlap geographically, they are not located in the same host (they are "allohospitalic"). Once again, the parasite-host relationships, and the limited distribution of the two sister species, provide evidence for the influence of either random or sequential peripheral isolates speciation (see Brooks and McLennan 1991 for a discussion). Moving to the third clade, geographical distributions indicated that *T. scabrae* and *T. chelopi* were formed by some form of vicariance speciation. The analysis of parasite-host relationships confirmed the abiotic environments story, opting for the peripheral isolates form of vicariance (microvicariance of Lynch 1989) for *T. scabrae*. Members of ancestor 14 that did make the switch to the new host encountered another telorchid, *T. singularis,* in that host's intestine. At this point, a sequence of events occurred paralleling the events leading to the origin of *T. angustus*. *Telorchis scabrae* have genital pores that are ventral to the left cecum, another configuration that makes it difficult, if not mechanically impossible, to align genital pores with other species of *Telorchis*. This change may have hastened the reproductive isolation of *T. scabrae* that had begun with the renegade population's shift to a new turtle host. Evidence of a morphological change correlated with a resticted geographic distribution and host switch provides an extremely strong argument for the influence of peripheral isolates speciation on the origin of *T. scabrae*.

Telorchis chelopi, on the other hand, appears to have cospeciated with its turtle host. Combining information from the two sister species provides us with the following picture: ancestor 14 was originally more widespread than the current distributions of its descendants would imply. When this ancestor, together with the ancestor of *Graptemys pseudogeographica* and *G. geographica,* was divided, both subsequently speciated. During this process members of ancestor 14 were busy either jumping to yet another turtle ancestor, and from there producing *T. scabrae,* or going extinct in *G. geographica.* This state of affairs, of course, raises the possibility that the gap in the distribution of these two species is due to that extinction. How can this scenario explain the occurrence of both species in the same host, *C. scripta scripta* (although they have not been reported from the same host populations yet)? It would appear, from Figure 2.42, that this host sharing has arisen from the propensity of these two flukes to remain with the same hosts through a series of speciation events, or to switch hosts, without themselves speciating. Specifically, *C. scripta scripta* "inherited" the association with *T. scabrae* and gained *T. chelopi* through a host switch. Neither fluke speciated as a result of the interaction with *C. scripta scripta,* so, even if sympatric populations are discovered, we do not have evidence of sympatric speciation. If sympatric populations are found, we do, however, have an interesting model system for testing MacArthur's dictum that closely related species should demonstrate high levels of competition with subsequent resource partitioning.

Finally, we come to the fourth clade of flukes. BPA supports the proposition, based upon geographic distributions, that the sister species *T. singularis* and *T. attenuatus* were formed by vicariant speciation. The origins of *T. robustus* are more problematical. Geographical data indicate a case of vicariance, whereas host data indicate a case of host switching. *Telorchis robustus,* however, is too widespread to support a hypothesis of peripheral isolates vicariant speciation. Is there a way out of this conundrum? Examination of Figure 2.42 shows that *T. robustus* is one of the flukes that has not speciated even though its original host has undergone numerous speciation events. Thus it is possible that the current distribution of *T. robustus* reflects the increasing diversity of its host clade. We will therefore tentatively assign this host switch to the peripheral isolates category.

SUMMARY

It is evident from the studies presented in this chapter that the evolution of parasite systems involves a complex and intriguing variety of factors. Of these, perhaps the most surprising is that we can make limited generalizations about the role of hosts in parasite speciation. This is surprising because one of the few things the traditional "geographic isolation school" and "host school" of parasite speciation had in common was a belief that the degree of host specificity was an unfailing indicator of evolutionary association (cospeciation) between host and parasite group. That is, *the more pronounced the host specificity, the greater the degree of phylogenetic congruence between host and parasite phylogenies.* This dictum was based on the assumption that parasites had no evolutionary abilities of their own; hence, they were at the mercy of their host environments. The more specific the environment, the more direct its effect on the evolution of the parasites, and the more evolutionarily dependent upon that environment would be the parasite group. Phylogenetic analysis of biogeographic and host relationships gives us reason to question that simple syllogism because most cases in which the host appears to have played a causal role in parasite speciation are examples of peripheral isolates allopatric speciation via host switching. *This is the mode of speciation that produces incongruence between host and parasite phylogenies.*

For example, parasite groups that are capable of producing a viable deme from a single colonization event would more likely speciate as a result of colonization than those requiring a larger founding population. Because of this likelihood, we would expect these organisms to form clades comprised of many host-specific species rather than clades comprised of a few generalists. Monogeneans are excellent examples of both these points. Most of them exhibit direct development and have generation times much shorter than those of their single, vertebrate hosts; hence, it is possible for a single monogenean to establish a viable deme, and produce colonizing offspring, while residing on one host. *As a result, the monogeneans as a group exhibit many highly host-specific species, most of which have evolved as a result of host switching, not as a result of cospeciation.* Likewise, each of the ten species comprising the Gyrocotylidea resides in a single species of chimaeroid fish, but 50% of the observed parasite-host associations in that system are due to host switching (Bandoni and Brooks 1987b; Brooks and McLennan 1991).

The converse of the traditional assumption about host specificity and congruence between host and parasite phylogenies rests on equally shaky ground, and yet is an equally widely held belief. Aho (1990) recently studied the parasites of a number of salamanders in the southeastern United States. Having found that none of the parasite species was restricted to a single host species, he concluded that phylogenetic association with hosts had played no role in the evolution of the parasite fauna (similar sentiments were expressed by Holmes 1990; Kennedy 1990; Pence 1990). *A lack of pronounced host specificity, however, does not rule out the possibility that any given parasite has evolved with one of the hosts it inhabits and has colonized the others without speciating*. Even generalists must evolve in association with hosts. Thus, an assertion that there is no congruence between a parasite phylogeny and the phylogeny of one of its hosts requires phylogenetic analysis and comparison, not just an enumeration of hosts.

Our observations suggest that host specificity is not generally a direct causal component of parasite speciation. This does not rule out a role for specificity in the adaptive response of parasite species to their host environments, according to the evolutionary options available to, and abilities of, the parasites themselves. Brooks and McLennan (1991) considered adaptation to be an evolutionary phenomenon that is historically correlated with, and complementary to, speciation, but is not necessarily causally linked with the production of new species. This would appear to be true for both parasites and free-living species, and we will investigate the adaptive significance and consequences of host specificity in the next chapter.

Host specificity aside, what have we discovered about parasite speciation? Our first discovery is that parasites are amenable to phylogenetic analyses of speciation. As for free-living organisms, we can produce a set of phylogenetic patterns that are associated with each type of speciation mode. So points of congruence between host and parasite phylogenies indicate common evolutionary histories (host speciation in the absence of parasite speciation, cospeciation of hosts and parasites, parasite speciation in the absence of host speciation, including sympatric speciation of parasites), whereas points of incongruence between host and parasite phylogenies indicate parasite speciation by host switching. Our second discovery is that, according to the current data base, host and parasite phylogenies are congruent approximately half the time. When combined with information about geological history and the distribution of parasitic sister species, this fact indicates that most parasite speciation is allopatric, either vicariant or peripheral isolates. More importantly, it suggests that each parasite-host system has followed its own unique speciation pathway, in terms of both the sequence and relative importance of different speciation modes.

Although allopatric modes reign supreme in a general overview, sympatric speciation may have played an important role in the evolution of some parasite groups, in particular within the Monogenea. As we discussed previously, this expectation is based upon the biology and population structure of monogeneans. However, the result may also reflect the paucity of species-level phylogenetic analyses of parasites, their hosts, and their geographic distributions. For example, we think it likely that sympatric speciation has played a significant role in the evolution of species of blood flukes in the genus *Spirorchis* inhabiting North American freshwater turtles. Many of these species occur in the same host species, and each seems to have occupy a different site within the host. No species-level phylogeny for the group is yet available, although phylogenetic analysis of the family Spirorchidae has begun (Platt 1988, 1991, 1992), and we can anticipate a most interesting outcome in the near future. Likewise, Schad (1963) reported that the members of a species flock of nematodes in the genus *Tachygonetria* inhabited different, albeit overlapping, portions of the intestine of a European turtle. As with the species of *Spirorchis*, there is no phylogenetic tree for these nematode species.

Overall, in terms of the importance of allopatric versus nonallopatric modes, the results from parasite systems do not depart notably from findings based on studies of free-living species (Lynch 1989; Brooks and McLennan 1991). There is, however, one difference between the two: parasites appear to demonstrate higher levels of peripheral isolates speciation than free-living groups (at least, groups of vertebrates). This is an important finding for students of speciation theory. There are potential difficulties in distinguishing between a vicariant division of an ancestral population into two roughly equal-sized sister species and a peripheral isolates event followed by widespread dispersal of the descendant. Parasite systems provide us with a way around this problem because episodes of peripheral isolates speciation will be identified by a phylogenetic systematic analysis (incongruent portions of the parasite-host phylogenies) *independently of any assumptions about dispersal following the host switching event.* So not only can parasite speciation be studied within the wider framework of abiotic and biotic influences reserved, until recently, for free-living species, but those studies can also contribute useful model systems and valuable information toward our understanding of speciation and evolution.

3

ADAPTATION AND ADAPTIVE RADIATION OF PARASITES

> . . . unpleasant occupant in its strange dwelling-place . . .
>
> —Lockwood 1872

> It is to the credit of the Vertebrata that they have almost completely avoided the dirty corners of parasitism and degeneracy.
>
> —Morely Davies 1920

> Parasites as a whole are worthy examples of the inexorable march of evolution into blind alleys.
>
> —Noble and Noble 1976

> Visually stimulating organisms, the large, the colorful, the active, the aggressive, command our attention, while the secretive and insidious remain largely ignored.
>
> —Price 1980

> The development of the parasitic habit almost always involves degenerative evolution.
>
> —Dodson and Dodson 1985

Clearly, parasites have never enjoyed a positive reputation among biologists. This reputation, conceived in a pre-Darwinian framework and nurtured on orthogenesis, has created an abundance of "myths, metaphors, and misconceptions" about parasite biology that persist even in this age of evolutionary synthesis and unification. In this chapter we will remove the veils from these myths and, we hope, discover that empirical data are no less amazing, and infinitely more useful, than fantasy. In so doing we hope to reinforce Price's (1980) belief that parasite-host systems are excellent models for studies of adaptive evolutionary processes.

We will address four issues concerning patterns of adaptive evolution in parasite systems, concentrating primarily on protist and helminth groups:

(1) analysis of host specificity, with particular emphasis on the reality of evolutionary trends in host specificity; (2) morphological adaptations, specifically the degree of homoplasy and the degree of secondary character loss; (3) the evolution of life cycles; and (4) the adaptive radiation of parasites (see Kim 1985 for similar discussions based on parasitic arthropods and their mammalian hosts).

MYTH I: HOST SPECIFICITY IS A UNIQUE AND IMPORTANT FEATURE OF PARASITE EVOLUTION

> High specificity indicates a long history of parasitism. . . . In ethological and ecological specificity the range of hosts is wide and this is said to indicate a more recent association between the parasite and the host.
>
> —Rogers 1962

> The environment must also have the properties necessary for the organism to survive. This is the basis of host specificity which, in its essence, is no different from that underlying the survival and ecology of free-living terrestrial, freshwater, and marine organisms.
>
> —Cameron 1964

This chapter begins where the last chapter ended, with a discussion about host specificity and its role in parasite evolution. As can be seen from these two quotations, researchers hold widely disparate views about the importance of evolutionary specialization and host specificity in parasite evolution. And, like many scientific debates, these disparate views arise because the concepts, as currently stated, lead the adherents into a logical paradox. Host specificity, the hypothesis goes, plays an important role in parasite diversification because the extent to which a parasite can be expected to colonize new hosts is dependent upon its degree of specialization on the original host resource, and upon the evolutionary diversification of the hosts (see O'Grady in Noble et al. 1989 for an excellent review of host specificity in a phylogenetic context; see Futuyma and Moreno 1988 for a review of theories about the evolution of specialization, including host specificity). This hypothesis is at once theoretically attractive and perplexingly paradoxical. The attraction lies in its invocation of adaptive processes to drive speciation. The paradox is twofold. First, once colonization of a new type of resource (host) within the ancestral species range has occurred, the probability that the new resource will exert strong directional selection pressure on the colonizing population should be higher for species displaying pronounced host specificity. These species are more tightly coupled to their resource bases and thus should be more sensitive evolutionarily to changes in that component of their environment than their generalist counterparts. However, the likelihood of a host change occurring in the first place is decreased for species that only respond to a small number of cues. Therefore, the species least likely to colonize new hosts (specialists) are the ones most likely to speciate as a result of any such switch, whereas the species most likely to colonize new hosts (generalists) are the least likely to speciate as a result of the interaction. Recognizing this paradox was one of the factors that led us to the conclusion in Chapter 2 that host specificity was not directly involved in parasite speciation.

The paradox is created, in part, by the belief that parasites are resource utilization specialists (Price 1980). Although this is perhaps the most popular generalization about parasites, authors do not agree on the evolutionary origin(s) of those specialized capabilities. Cameron (1956, 1964) proposed that becoming a specialist was a necessary precursor to becoming a parasite, indicating that parasites are derived from specialist free-living ancestors. Read (1972), by contrast, asserted that the free-living ancestors of parasitic groups were probably generalists, and that the emergence of specialist tendencies was an adaptive response to being parasitic. Rogers (1962) presented a compromise between the previous two views by suggesting that "Many of these features were present in the free-living ancestors of parasites; their increased development is an adaptation which allows parasites to exploit a new environment effectively." This placed the origin of resource specialization in the free-living ancestors, while allowing for the enhancement of those traits within the parasite groups.

Evolutionary problems aside, just how do we recognize a resource utilization specialist? Host specificity may be obligate or facultative, and may stem from ecological, behavioral, biochemical, or geographical phenomena affecting either the hosts or the parasites. A given species of parasite may be quite capable of inhabiting a wide range of hosts in the laboratory, but will occur in only a small number of them in nature. Alternatively, there are a large number of parasite species that seem to be obligately restricted to a single host species, although the majority of parasites have some range of host tolerance. There are also parasites that are specialists during each developmental stage and yet inhabit a broad range of hosts during the course of a life cycle. For example, consider the larch. Now consider any digenean in the genus *Haematoloechus*. During the course of its life cycle, the worm will reside in a freshwater snail, in the musculature of a dragonfly, and in the lungs of a frog. At each stage, the resource utilization appears to be quite specialized, and yet the organism has the genetic capability over its lifetime to utilize molluscs, arthropods, and vertebrates as habitat patches. In terms of mechanisms, increasing host specificity might reflect an internal property of the parasite (Smiley 1978). If, however, the association is maintained by the host group's possession of a particular resource that is necessary for the survival of the parasite, then the specificity may reflect the resource's distribution among sympatric or parapatric host species and the opportunistic behavior of the parasite.

The second component of the specificity paradox arises from the belief that host specificity "causes" speciation. It was this belief that allowed the "host school" of Eichler and his followers (see Chapter 1) to reconstruct parasite phylogenies using only host relationships. Today, it manifests itself in the assertion that the degree of host specificity and the relative phylogenetic relationships of the hosts indicate something about parasite phylogeny (e.g., Burt and Jarecka 1982; Humphery-Smith 1989; Holmes 1990). Given this belief, we need to ask whether "phylogenetic specificity" (Humphery-Smith 1989) is a "cause" or an "effect" of parasite diversification. The fact that we did not need to make any assumptions about host specificity in Chapter 2 in order to determine patterns of parasite speciation made us suspicious that host specificity may be a component of parasite anagenesis, but may not be causally related to parasite speciation. That is, host specificity may determine the survival of parasite species in the event of some host extinctions, but does not necessarily affect the probability of parasite speciation. If host specificity is decoupled from speciation, then we would not expect to find any particular general pattern of specificity emerging over the evolutionary diversification of parasite groups. Theoretically, there are three possibilities, not one. First, host specificity can increase during phylogenesis as historical effects progressively constrain host preferences (Fig. 3.1a). This is consistent with hypotheses of progressive specialization in coevolving lineages (Smiley 1978) and is the pattern expected if host specificity and cospeciation are causally linked. Second, host specificity can decrease during phylogenesis (Fig. 3.1b). This is consistent with hypotheses about the evolution of extreme opportunists, in

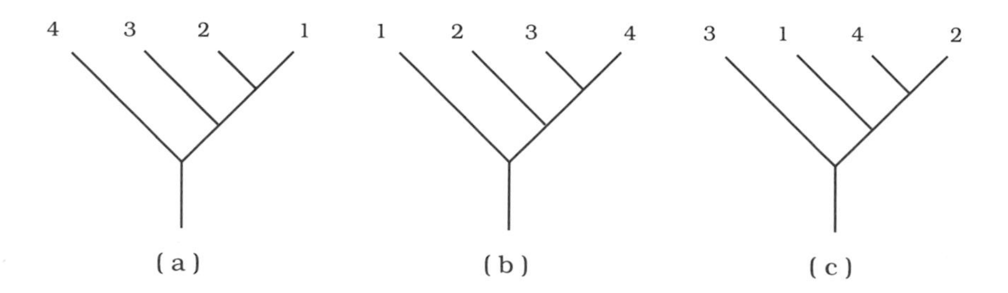

Figure 3.1. Phylogenetic patterns of host specificity. From Brooks and McLennan (1991).

Table 3.1
Comparison between the number of reported definitive host species and the number of published reports of collections of each species of acanthostome digeneans

Name of parasite species	No. of reported hosts	No. of reports of parasite
Timoniella imbutiforme	3	5
T. praeterita	3	4
T. ostrowskiae	3	4
T. incognita	2	1
T. loossi	4	4
T. unami	1	1
T. absita	1	1
Gymnatrema gymnarchi	1	2
Proctocaecum coronarium	1	4
P. productum	1	4
P. vicinum	1	3
P. gonotyl	1	1
P. atae	1	1
P. crocodili	1	1
P. elongatum	1	1
P. nicolli	1	2
Caimanicola marajoara	5	6
C. brauni	2	1
C. caballeroi	4	3
C. pavida	1	1
Acanthostomum scyphocephalum	2	1
A. spiniceps	4	5
A. absconditum	2	5
A. knobus	1	1
A. niloticum	1	1
A. minimum	1	1
A. astorquii	1	1
A. gnerii	4	4
A. megacetabulum	2	2
A. americanum	2	3
A. burminis	2	6
A. indicum	1	1
A. slusarskii	1	1
A. simhai	1	2
A. pakistanensis	1	1
A. asymmetricum	1	1
A. proctophorum	1	1
A. lobacetabulare	1	1
A. cerberi	1	1
A. marinum	1	1

Note: Nineteen of the 25 species for which only a single host species is known are themselves known only from a single collection.

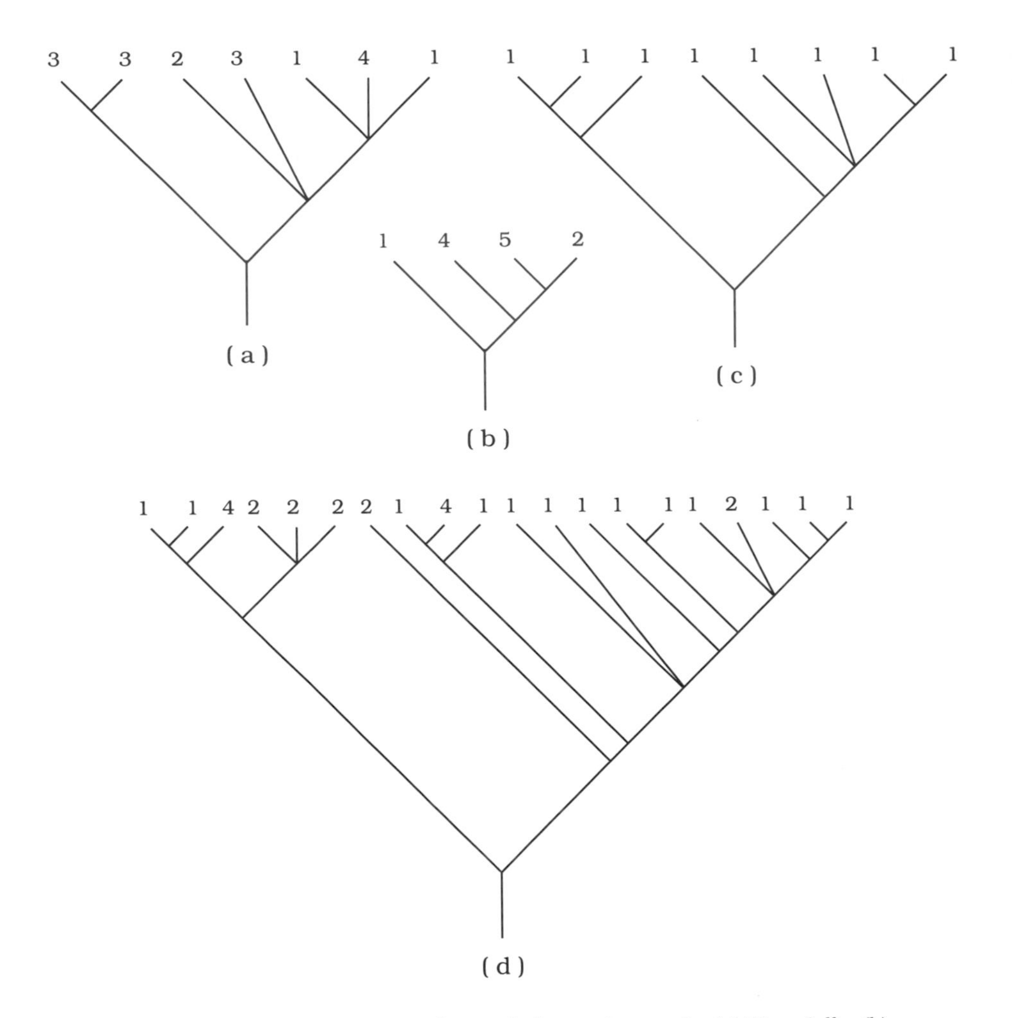

Figure 3.2. Host specificity mapped onto phylogenetic trees for (a) *Timoniella;* (b) *Caimanicola;* (c) *Proctocaecum;* and (d) *Acanthostomum.*

which traits that facilitate opportunistic behavior are optimized during evolution. Finally, there may be no macroevolutionary regularities in patterns of host specificity (Fig. 3.1c).

One caveat is required here. We cannot begin to search for the processes underlying patterns of host specificity until we have examined more species within a phylogenetic framework. And we cannot begin to reconstruct that framework until we have reliable estimates of host preferences under natural conditions. Most relatively species-rich clades contain several species that have been reported only once, others that have been reported only rarely, and a few that have been reported commonly. The majority of parasite species are originally described from a single host species (and from a single geographic location, for that matter), making assessments of pronounced host specificity ambiguous at best. Overall, a survey of host and parasite records will generally uncover a strong correlation between the number of reported hosts and the amount of attention accorded a particular parasite species or group (Table 3.1). Nevertheless, according to the current data base, the predominant macroevolutionary trend in changes in host specificity within a clade corresponds to the pattern shown in Figure 3.1c: some members of any given parasite clade are highly host specific, others are not, and the distribution of these two within the clade presents no apparent regularities (Fig. 3.2).

MYTH II: PARASITES ARE SIMPLE AND DEGENERATE COMPARED WITH FREE-LIVING ORGANISMS

> The practice of terming some evolutionary changes "progressive" and others "regressive," especially prevalent in parasitology, is another example of the intrusion of anthropocentrism into evolutionary studies. . . . The loss or reduction of physical attributes is hardly restricted to those organisms we call parasites. One need only consider dermal bones in tetrapods, limbs in snakes, wings in kiwis, pelvic appendages in whales, digits in horses—and the body hair, olfactory lobes, the coccyx, and the appendix in humans.
>
> —O'Grady in Noble et al. 1989

The perception that parasites show extreme simplification is generally based upon a comparison between a given parasite and its (usually) vertebrate host. We agree that parasitic protists and helminths do not demonstrate the degree of complexity present in vertebrates; however, we do not believe that this difference is sufficient evidence to postulate that parasites have experienced an evolutionary drive toward simplification. If it were, then we would have to postulate that among free-living species most invertebrates and all protistans have experienced the same drive. This is clearly a ludicrous proposal, but it serves to illustrate an important point: in order to study the directions evolution has taken it is necessary to compare sister groups. So the only way to pin down whether parasites have become secondarily simplified is to compare the amount and type of character change within a parasite group with the amount and type of character change within its free-living sister group. Once sister groups have been identified, we need to investigate both character losses *and* character innovations. Such identification has two components. The first is the recognition that loss of a structure might be correlated with the evolution of compensating mechanisms (e.g., microvilli on the body surface of neodermatans allowed the gut to be lost in cestodarians: Rogers 1962; Brooks et al. 1985a). The second is the realization that our initial identification of a "character" might be biased by our own perceptions of what is, and what is not, important, and by our ability to detect characters. For example, Rogers (1962) stated that "the loss of sense organs is a common feature of parasitism." When that statement was investigated using sophisticated electron microscopic techniques, Rohde (1989) discovered eight types of sense receptors in a single species of endoparasitic flatworm! One is left with the speculation that there actually has been a net gain of sense organs in the evolution of these parasites.

At the moment, the data base available for the free-living sister groups of the parasites discussed in this book is not developed enough to allow meaningful comparisons. In addition, no researcher has undertaken an analysis of any free-living group in order to determine what constitutes "unusual" secondary simplification, beyond noting that some groups appear to exhibit more secondary character loss than others (e.g., osmeroid fishes: Begle 1991). Thus, we have no baseline for comparison for the parasite information. Nevertheless, we believe that a preliminary investigation will provide us with some interesting patterns and highlight the necessity for nonparasitologists to gather a data base comparable to that available for parasites.

Focusing our attention on the parasitic flatworms, we began our investigation with two questions: (1) How much character loss has occurred within the Cercomeria? and (2) Are particular types of characters more susceptible to being lost? In order to examine the second question, we divided characters into four categories: male reproductive, female reproductive, adult nonreproductive, and larval. The results of our investigation are summarized in Table 3.2. We did not include separate columns for the Aspidobothrea, Gyrocotylidea, or Amphilinidea because so few characters have been described for these taxa that analysis of each group is uninformative. We did, however, include the data from those groups in the "totals" column.

The first result that leaps to the eye is that only 10.6% of the 1882 character changes displayed by the parasitic flatworms involve secondary loss, or, to put it another way, *the overwhelming majority (nearly 90%) of evolution in these organisms has involved character innovation rather than character loss.* The interpretation of this result, as mentioned previously, must be taken with a grain of salt because we do not have any information about rates of loss in free-living organisms. Nevertheless, we believe that this result sheds some serious doubt on the proposition that parasites are "extremely" simplified and degenerate. If this were true, we would have expected to see an unambiguously high rate of character loss, coupled with the observation that such losses obscured our

Table 3.2
Comparison of character loss for four different character categories among major groups of parasitic flatworms

	Amount of character loss within each clade				
Character category	Cercomeria	Digenea	Monogenea	Eucestoda	Total[a]
Male	4/20 (20%)	31/130 (23%)	27/188 (14%)	5/74 (5%)	67/427 (16%)
Female	3/26 (11%)	6/230 (3%)	14/98 (14%)	2/109 (2%)	25/493 (5%)
Adult nonreproductive	4/56 (7%)	31/222 (14%)	22/263 (8%)	6/128 (5%)	66/691 (9%)
Larval	8/55 (14%)	20/157 (13%)	10/38 (26%)	7/19 (37%)	45/271 (17%)
Total	19/157 (12%)	88/739 (12%)	73/587 (12%)	20/330 (6%)	200/1882 (10.6%)

Note: Numerator denotes number of losses; denominator denotes number of apomorphies in that category.
[a] Information from the Aspidobothrea, Gyrocotylidea, and Amphilinidea included in totals.

ability to reconstruct phylogenetic relationships. Clearly, given the results presented in Chapter 2 and the Appendix, this is not the case.

Interestingly, the total amount of character loss is not distributed among the three major cercomerian groups in proportion to the amount of character change in each group ($\chi^2 = 8.95$; $p < 0.025$; Table 3.3). In essence, character loss in the Digenea and Monogenea occurs in proportion to change, whereas character loss is lower than expected within the Eucestoda. This is a striking result because, outside of parasitology at least, tapeworms are usually cited as the best example of parasite degeneracy. This analysis suggests the exact opposite: tapeworms have a reduced rate of character loss compared to digeneans and monogeneans. How can we reconcile this low level of character loss with the perception that tapeworms are simplified? At the moment the data base suggests that eucestodes have not undergone as much character diversification as their relatives. Thus they look simplified relative to digeneans and monogeneans because they have been evolutionarily conservative.

Although the general trend within the Cercomeria appears to be one of increasing complexity, there are occasional lapses into extreme simplification. We interpret this observation to mean that it is not necessary to be highly simplified to be a parasite, but, if you are highly simplified, being a parasite might increase your chances of survival. This represents a reversal from the

Table 3.3
Summary of results of χ^2 analysis for character loss

Comparison of all character categories within clades (losses)	χ^2 value and level of significance	Comparisons of each character category among clades (losses)	χ^2 value and level of significance
Digenea	33.07 $p < 0.001$	Male reproductive	9.22 $p < 0.01$
Monogenea	10.31 $p < 0.025$	Female reproductive	21.83 $p < 0.001$
Eucestoda	35.28 $p < 0.001$	Adult nonreproductive	7.97 $p < 0.025$
Total	8.95 $p < 0.025$	Larval	8.16 $p < 0.025$

Note: Results are presented for character loss among all character categories within major groups of parasitic flatworms (columns 1 and 2) and among all major groups of parasitic flatworms within each character category (columns 3 and 4).

traditional orthogenetic view that there is an inherent drive toward degeneracy within parasites (a negative perspective) to a more positive view that being "parasitic" is one option that allows the survival of species exhibiting evolutionary simplification.

Can we make any generalizations about the types of characters that are lost in these organisms? In order to answer this question we investigated the proportion of character types lost versus the proportion of change for all characters within groups (columns) and for individual character categories among groups (rows). The results of χ^2 analyses for those data are given in Table 3.3. Since this is a preliminary analysis, we will content ourselves with the observation that the three major groups appear to have followed different evolutionary "character loss" pathways. In the Digenea fewer female characters and more male characters have been lost than would be expected by the total amount of male and female characters in that group, and more male and more nonreproductive characters have been lost in proportion to their distribution across groups. In the Monogenea fewer nonreproductive and more larval characters have been lost than would be expected, and the loss of female characters is high relative to that in the other groups. Finally, in the Eucestoda fewer female and more larval characters have been lost than would be expected, and the loss of male and nonreproductive characters is low, while the loss of larval characters is high, compared to the other groups. These descriptions can be summarized as follows: the monogeans show no strong overall trends, the digeneans have a strong propensity to lose male characters, and the eucestodes have a tendency to lose larval characters and to preserve adult characters. These complex patterns of character loss demonstrate that no group shows an overall trend toward secondary simplification. In addition, as discussed previously, the group that shows the strongest resistance to evolutionary loss is the Eucestoda. We will return to this point, as well as some speculation about what all this might mean, in our discussion about adaptive radiations.

So far we have discovered that character loss has not been rampant within the Cercomeria and that the particular types of characters that are lost varies among the major cercomerian groups. We can take this investigation one step further by turning our attention to the amount of homoplasious loss within each group and character category. In essence, we are attempting to uncover whether particular characters have a propensity to be lost over and over again. The results of this analysis are shown in Table 3.4. Overall, there is a trend toward decreasing homoplasious loss from the digeneans to the eucestodes, which reemphasizes the conservative nature of tapeworms. Among character categories, the amount of homoplasious loss appears to hover around 50%, but the distribution of that homoplasy differs among clades. In digeneans, the highest proportion of homoplasious loss is found in male reproductive and adult nonreproductive characters, and the lowest proportion is found in female reproduc-

Table 3.4
Summary of homoplasious character loss in four character categories for the major groups of parasitic flatworms

Character category	Cercomeria	Digenea	Monogenea	Eucestoda	Total[a]
Male	0/4	22/31 (71%)	11/27 (40%)	3/4 (75%)	37/70 (53%)
Female	1/3	2/6 (33%)	8/11 (73%)	0/2	10/20 (50%)
Adult nonreproductive	0/4	20/31 (65%)	10/20 (50%)	0/6	30/61 (49%)
Larval	2/7 (28%)	9/20 (45%)	4/10 (40%)	4/7 (57%)	19/44 (43%)
Total	3/18 (17%)	53/88 (60%)	33/68 (48%)	7/19 (37%)	97/200 (49%)

Note: Numerator denotes number of homoplasious losses; denominator denotes total number of apomorphies in that category.

[a] Information from the Aspidobothrea, Gyrocotylidea, and Amphilinidea included in totals.

tive characters. Of the male characters, the genital sac and male intromittent organs are lost most often. On the nonreproductive side, the homoplasy is divided between the body surface (the acetabulum has been lost at least eight times and body spines have been lost at least five times) and the digestive system (the pharynx has been lost at least four times, the esophagus at least four times, and one of the two ceca at least four times). Notice that the distribution of homoplasy parallels the distribution of overall losses depicted in Table 3.2. The digenean pattern is reversed in monogeneans, which show the highest proportion of homoplasious loss in female characters and the lowest proportion in male and larval traits. Within the female characters, the vagina and egg filaments appear to have been lost most often. The moderate proportion of nonreproductive homoplasy centers around loss of hooks and hooklets. Eucestodes show very few homoplasious losses. Of these, the loss of an axoneme in the sperm's tail (male) and the loss of cilia on hexacanths and tails on plerocercoids (larval) occur more than once. Interestingly, the types of characters that show homoplasious losses within each of these groups differ among the groups.

We can summarize this discussion with three general statements: (1) these parasites do not show high levels of character loss; (2) the types of characters that are lost show different distribution patterns among the three major cercomerian clades; and (3) these patterns are partially the result of differences in homoplasious loss among groups. The discussion of homoplasious loss leads us into our third myth.

MYTH III: PARASITES EXHIBIT HIGH LEVELS OF ADAPTIVE PLASTICITY IN MORPHOLOGY

The belief that parasites are secondarily simplified and degenerate is problematical because parasites appear to have enjoyed a long history of speciation on this planet. By their persistence and numbers, they are an evolutionary success story, a point emphasized by Price (1980). Abandoning the myth of simplicity leads us to another commonly held view about parasite evolution: parasite success is due, in part, to their ability to adapt rapidly to an ever-changing set of challenges posed by the environment. If parasites are as adaptively plastic as our intuition tells us, then we should have extraordinary trouble reconstructing phylogenetic relationships using the characters of the parasites because we would expect the amount of homoplasy to obscure, and possibly to swamp, the genealogical information available in homologous characters. Once again, in order to investigate this problem we need information on two parameters: the amount of homoplasy within free-living groups in general, and that within the free-living sister groups of the parasitic clades in particular. As we mentioned earlier, we do not have the necessary sister group data. We do, however, have evidence of rampant homoplasy among a variety of groups of free-living organisms, including angiosperms (Crane 1985; Bremer 1987; Freire 1987), insects (Throckmorton 1965; Saether 1977), vertebrates (Hecht and Edwards 1976; Butler 1982; Begle 1991), opisthobranch molluscs (Gosliner and Ghiselin 1984), fungi (Høiland 1987; Crisci et al. 1988), amphipods (Myers 1988), prokaryotes (Bremer and Bremer 1989), and nemerteans (Sundberg 1989). Thus, although we cannot make the precise comparisons we would like, we can assess whether parasites (in this case the parasitic platyhelminths) fall into the "high-homoplasy" group of taxa.

We have used two measures for estimating the amount of homoplasious evolution in our data base, and for evaluating its significance. The first of these is a convenient measure called the consistency index (CI: see Wiley et al. 1991 for a discussion). This measure is a ratio of the number of character transitions needed to account for the different character states in a data set divided by the total number of character changes on the final phylogenetic tree. Each episode of homoplasious evolution adds a character change that was not recognized in the original matrix. That is, the CI is an indicator of the proportion of mistaken hypotheses of homology made by the researcher prior to performing a phylogenetic analysis of an entire data set. There are problems with using the CI as a very strong indicator. For example, the inclusion of autapomorphies or symplesiomorphies in the calculation of the CI inflates the value of the measure based on characters that do not in fact affect the robustness of the phylogenetic hypothesis under consideration; hence, it has been suggested that the CI should be calculated using only putative synapomorphies. In addition, the CI is not strictly comparable across data sets, since it is dependent on the number of characters and the number of taxa. This leads us to perhaps the greatest problem with the measure: we do not have any good indication of its statistical properties, so we do not have any means of comparing our results with any form of expected results.

In the past few years, a number of systematic theorists have begun investigating the problem of providing a statistically robust means of assessing the results of phylogenetic analyses (e.g., Archie 1989; Farris 1989; Sanderson and Donoghue 1989; Meier et al. 1991). These studies have produced some interesting, and in some cases initially counterintuitive, findings, although there is still much to do. For example, it appears that the minimum significant value for a CI actually drops as one adds taxa and characters to a study; for example, a study using 50 characters for 20 taxa and reporting a CI of 65% may actually be more robust than one using 10 characters for 7 taxa and reporting a CI of 80%. This happens because there are often apomorphic character changes occurring once within a given taxon that also occur once in another taxon. If one were to expand the scope of a study to include both those taxa, the estimate of homoplasy would increase (and the CI would drop) even if the hypothesized phylogenetic relationships of the (now) subgroups did not change. Or, to use current jargon, we would say that a "global" phylogenetic analysis had discovered homoplasy that the two "local" analyses had failed to recognize. Of course, in some cases such global homoplasy could affect the hypotheses of relationships, so recognizing "global homoplasy" may play an important role in determining robust character polarities during outgroup comparisons at the inception of a phylogenetic analysis (e.g., Maddison et al. 1984; Wiley et al. 1991).

Among the recent proposals for additional measures of homoplasy is the homoplasy slope ratio (HSR), a measure that is independent of number of characters or number of taxa and that compares the CI of a real data set, taking the number of characters and taxa used, with the results of a randomly generated data set of the same number of characters for the same number of taxa (see Meier et al. 1991 for details of the calculations). An HSR of 0.35 means that the observed homoplasy is 35% of the homoplasy that would be expected for a randomly generated data set of the same number of characters for the same number of taxa. This measure behaves in a manner similar to the homoplasy excess ratio proposed by Archie (1989) but is easier for a working systematist to implement, since it does not require one to perform a series of randomizations of the data set that he or she has so diligently tried to nonrandomize.

In addition to producing a standard of comparison based on simulations of randomly generated data sets, Meier et al. (1991) performed a regression analysis based on a set of published phylogenetic studies, so that one could begin to assess the expected significant HSR values for real data sets. They also noted that the HSR does not take into account the internal structure of an individual data set, which may have a pronounced influence on the robustness of a phylogenetic tree, but this is a problem for all such measures at the moment (for example, the HSR may be overly sensitive to multistate characters: A. Kluge, personal communication). Thus, despite the progress systematic theorists have made in assessing the statistical robustness of phylogenetic trees, the best way to evaluate a tree is to accumulate more characters and perform new phylogenetic analyses based on all those characters.

We have calculated both the CI and the HSR for the major groups of cercomerians, based on both local and global homoplasy searches (Table 3.5). The only group for which the HSR value falls above the regression line for HSR values generated from real data sets is the Aspidobothrea, a group that has

Table 3.5
Comparison and summary of character evolution within the parasitic flatworms

Groups	No. of TS (local)	No. of characters	CI	Local HSR	No. of TS (global)	No. of characters	CI	Global HSR
Cercomeria (higher groups)	157	154	98.1%	0.014	157	154	98.1%	0.014
Cercomeridea	1725	1466	84.9%		1725	1305	75.6%	
Trematoda	749	681	90.9%		749	537	71.8%	
Aspidobothrea	10	10	100%	0	10	8	80%	0.359
Digenea	739	671	90.8%	0.002	739	529	71.7%	0.005
Cercomeromorphae	976	785	80.4%		976	768	78.7%	
Monogenea	587	449	76.5%	0.008	587	442	75.3%	0.010
Cestodaria	389	336	86.3%		389	324	83.5%	
Gyrocotylidea	19	15	78.9%	0.184	19	15	78.9%	0.184
Cestoidea	370	321	86.7%		370	309	83.7%	
Amphilinidea	40	37	92.5%	0.066	40	37	92.5%	0.066
Eucestoda	330	284	86.0%	0.004	330	272	82.7%	0.004
Total	1882	1620	86.1%	0.001	1882	1459	77.5%	0.002

Note: Columns 2–5 list estimates based on local considerations; columns 6–9 list estimates based on global considerations. TS = character transformations on trees; CI = consistency index (number of characters/number of TS); HSR = homoplasy slope ratio.

a high CI value but a very small number of characters. All other values fall well below the regression line, suggesting that we are dealing with a highly nonrandom data base; in fact, the lowest HSR value reported for a real data set by Meier et al. was 0.03. The value for the total data base, using global homoplasy considerations, is 0.002. This study also provides us with an opportunity to highlight one of the positive attributes of the HSR. In their paper, Meier et al. reported the HSR for an early study of the relationships among the major groups of cercomerians by Brooks et al. (1985a). That study was based on 39 characters, two of which were homoplasious, giving a CI of 95%. The current data base for this set of taxa is 154 characters, three of which are homoplasious, giving a CI of 98%. Intuitively, we would think that adding 115 characters, only one of which is homoplasious, would increase the robustness of our phylogenetic hypothesis by more than 3%. The HSR value for the original study was 0.04, whereas the current HSR value for the expanded data base is 0.014, or 65% better than the previous estimate. This accords more with our intuitions about the increase in robustness of the phylogenetic hypothesis resulting from a substantial increase in characters that corroborate the original tree.

The results of our investigation of homoplasy levels in the cercomerians support the theoretical predictions: homoplasy on a global level is higher than that seen on a more local scale of investigation (Table 3.5). It is interesting, however, that most of the increase in homoplasy that resulted from the global search occurs within the digeneans, which experienced a dramatic drop in the overall consistency index from 90.8% to 71.7%. The consistency indices of the monogeneans, gyrocotylideans, amphilinideans, and eucestodes did not change appreciably when character evolution was examined among higher taxonomic levels. Despite the drop in CI resulting from the increased homoplasy found in global considerations, the consistency indices for the parasitic platyhelminth groups remain high. In addition, the HSR value for the digeneans indicates that the data are not substantially less robust when considered globally. Most of the digenean homoplasy appears to be spread among groups, whereas with the other cercomerian taxa most of the homoplasy occurs within groups (see Sanderson and Donoghue 1989 for a discussion of patterns of distribution of homoplasy among taxa on phylogenetic trees). This is an example of global homoplasy not having much of an influence on the robustness of local phylogenetic trees (see also Wilkinson 1991). It would appear then that "adaptive plasticity" in morphological evolution has not been unusually high in these groups of parasites. Furthermore, we believe the same can be said for the other groups of parasites reported in the Appendix, although the data bases for those taxa are not as extensive as for the platyhelminths.

The myth of rampant homoplasy, like the myth of rampant character loss, has fallen, but in its wake a number of interesting problems have appeared. For example, let us investigate the distribution of homoplasious characters among our four character classes within and among the major groups of cercomerians (Table 3.6). The total amount of character homoplasy is not distributed among the three major cercomerian groups in proportion to the amount of character change in each group ($\chi^2 = 11.30$; $p < 0.005$; Table 3.7). The picture is similar to that depicted for evolutionary character losses: homoplasy within the monogeneans occurs in proportion to overall character change, whereas homoplasy is slightly higher than expected in the digeneans and much lower than expected within the eucestodes. Once again, this finding supports the hypothesis that eucestodes are the most evolutionarily conservative group within the Cercomeria.

The results of χ^2 tests examining the distribution of homoplasious characters among different categories of characters are presented in Table 3.7. In general, the trends shown for homoplasy parallel those uncovered for losses, with some notable exceptions. Homoplasy in the digeneans occurs less often than ex-

Table 3.6
Comparison of character homoplasy for four different character categories among major groups of parasitic flatworms

Character category	Distribution of homoplasious characters				
	Cercomeria	Digenea	Monogenea	Eucestoda	Total[a]
Male	0/20	39/130 (30%)	45/188 (24%)	21/74 (28%)	106/427 (25%)
Female	1/26 (4%)	81/230 (35%)	27/98 (27%)	16/109 (15%)	129/493 (26%)
Adult nonreproductive	0/56	76/222 (34%)	50/263 (19%)	13/128 (10%)	141/691 (20%)
Larval	2/55 (4%)	14/157 (9%)	16/38 (42%)	8/19 (42%)	40/271 (15%)
Total	3/157 (2%)	210/739 (28%)	138/587 (24%)	58/330 (17%)	416/1882 (22.1%)

Note: Numerator denotes number of homoplasious changes; denominator denotes total number of apomorphies in that category.
[a]Information from the Aspidobothrea, Gyrocotylidea, and Amphilinidea included in totals.

Table 3.7
Summary of results of χ^2 analysis for character homoplasy

Comparisons of all character categories within clades (homoplasy)	χ^2 value and level of significance	Comparisons of each character category among clades (homoplasy)	χ^2 value and level of significance
Digenea	29.37 $p < 0.001$	Male reproductive	1.14 (NS)
Monogenea	8.40 $p < 0.05$	Female reproductive	11.11 $p < 0.005$
Eucestoda	15.65 $p < 0.005$	Adult nonreproductive	23.65 $p < 0.001$
Total	11.30 $p < 0.005$	Larval	26.06 $p < 0.001$

Note: Presented are both results among all character categories within major groups of parasitic flatworms (columns 1–2) and results among all major groups of parasitic flatworms within each character category (columns 3–4).

pected in larval characters within and among groups, and more often than expected in nonreproductive characters within and among groups. Flukes do not show an unusual amount of homoplasy in male characters. They do show a high level of male character loss (see earlier), most of which was homoplasious loss. Monogeneans show unusually high levels of homoplasy in larval characters both within and across groups. In conjunction with the observed trend toward increased loss, it appears that larval characters are extremely flexible within the Monogenea. Finally, the saga of tapeworm conservatism continues. Within the Eucestoda, there are fewer homoplasious male and nonreproductive, and more homoplasious larval, characters than would be expected. When eucestodes are compared with other cercomerian groups, levels of homoplasious larval characters are higher and levels of homoplasious female and nonreproductive characters are lower than expected.

We can combine this information with the data on character losses to begin developing a picture of morphological character evolution within the three major groups of parasitic platyhelminths. In general, the "digenean profile" is one of constrained larval characters (average levels of loss and lower than expected homoplasy), constrained female characters (average homoplasy balanced by very low levels of loss), flexible male characters (above average loss, average homoplasy), and very flexible nonreproductive characters (higher than expected levels of both loss and homoplasy). The "monogenean profile" presents constrained nonreproductive characters (lower than expected levels of loss and average homoplasy), unremarkable male characters (average amount of loss and homoplasy), flexible female characters (higher than expected levels of loss and average homoplasy) and very flexible larval characters (higher than expected levels of loss and high homoplasy). Finally, the "eucestode profile" shows tightly constrained female and nonreproductive characters (lower than expected levels of loss and homoplasy), constrained male characters (lower than expected levels of loss, lower than expected homoplasy), and very flexible larval characters (higher than expected levels of loss and homoplasy).

We do not have an extensive set of explanations for all the patterns that we observe, because the theories of parasite morphological evolution established prior to this book were based on assumptions about homoplasious and reductive evolution that are not true. The discovery that levels of character loss and homoplasy are not unusually high within the parasitic flatworms is exciting on its own, because it causes us to rethink our theories about parasite evolution, and because it provides us with more evidence that the evolution of parasites does not differ from that of free-living species. This suggests that when we rethink our theories about parasite evolution we should pay attention to current theories in general evolutionary biology. For example, it would be interesting to investigate questions such as the types of characters that show the highest level of homoplasy and the degree to which the origins of any of those homoplasious characters might be correlated with changes in the environment—a contemporary approach to modern studies of adaptation (Brooks and McLennan 1991; Harvey and Pagel 1991 and references therein). At the moment we have evidence for the existence of homoplasy, but we have no evidence about which of those characters might serve an adaptive function. In the future, we should also delve deeper into the organism, searching for the mechanisms that actually cause homoplasious evolution to occur, as opposed to mechanisms that preserve homoplasious characters once they appear. At the moment, we know very little about such mechanisms, and most of what we do know concerns the influence that alterations in ontogeny may have on character expression.

Heterochrony as an Explanation for Some Cases of Homoplasy

The past decade has witnessed a resurgence of interest in a collection of developmental phenomena termed heterochrony (Gould 1977; Alberch et al. 1979; Bonner 1982; Fink 1982; Raff and Raff 1987; McKinney 1988). Biologists studying these phenomena are engaged in a search for changes in the timing of development that can be used to explain changes in morphology—the classical "how" question of evolution (e.g., Emerson 1986 [frogs]; Alberch 1980 [amphibians in general]; Griffith 1991 [lizards]; Winterbottom 1990; Strauss 1990 [teleostean fishes]; McNamara 1986; Mooi 1987 [echinoids]; Shea 1983; Gould 1977 [primates]; Ambros and Horvitz 1984 [nematodes]; O'Grady 1985 [digeneans]; Guerrant 1982 [plants]). Such changes in morphology may also be accompanied by changes in the life history and ecology of an organism (McNamara 1982; McKinney 1986). For example, Alberch and Alberch (1981) proposed that the salamander *Bolitoglossa occidentalis* was able to assume an arboreal existence because it matured at a smaller size (earlier age) than its close relatives, which are terrestrial. So it appears that alterations in development can initiate a series of cascading effects that manifest themselves on all levels from morphology, through ecology, to behavior.

Gould (1977) attempted to quantify and describe the developmental mechanisms underlying morphological change with a "clock" model that reduced all changes to either an acceleration or a retardation of developmental rate. Recognizing that heterochrony was more complex than Gould's description suggested, Alberch et al. (1979) developed a model based upon the idea that the development of any part of an organism can be represented by a positive trajectory with a starting point, a (growth starts); an endpoint, b (growth stops); and a rate of growth, K, for changes in shape, g, or size, S. Development could thus be tracked graphically by plotting a and b on the x axis (time), and g on the y axis (increasing complexity), and by varying the onset of growth, the rate of growth, and the ending of growth (see Funk and Brooks 1990). Based upon these manipulations, Alberch et al. (1979) recognized two general outcomes of heterochrony arising from changes in the rate and the timing of development: paedomorphosis and peramorphosis.

Paedomorphosis results in the production of adult descendant morphologies that are less complex than those of their immediate ancestor. This does not necessarily imply that the descendant will have a morphology comparable to that of a juvenile or even a larval ancestor; it may resemble a less developed ancestral adult (McKinney 1988). Paedomorphosis can be accomplished in three ways: growth onset can be delayed (postdisplacement), growth can be terminated earlier (progenesis), and development can proceed at a slower rate (neoteny). Peramorphosis, on the other hand, results in the production of adult descendant morphologies that are more complex than those of their immediate ancestor. Since this process produces a morphological trait in an organism that passes beyond the condition found in its ancestor, it will result in a recapitulation of the ancestral ontogeny during development. Not unexpectedly, there are three ways in which peramorphosis can happen: growth onset can begin earlier (predisplacement), growth can continue for a longer period of time (hypermorphosis), and development can proceed at a faster rate (acceleration: see McNamara 1986 for formal definitions).

The model of Alberch et al. has been modified slightly over the years. McKinney (1988) argued that shape and size were not independent variables. Rather, shape was a subset or by-product of size. Although this hypothesis may oversimplify our descriptions of a particular structure's development, it does allow us to reach a first approximation based upon two easily quantified parameters, size and time. The other modification arose from the realization that ontogeny is a widespread phenomenon, encompassing all the changes an organism undergoes from conception to death. In other words, the effects of heterochrony should be recognizable on more levels than just changes in adult morphology. McNamara (1986) recognized that this global perspective was of particular importance to parasitologists, who studied organisms that frequently passed through a sequence of discrete and identifiable developmental stages. For example, since all digeneans have an ontogenetic sequence that includes an egg, a miracidium, a sporocyst, a cercaria, and an adult, these stages are equivalent, or, more to the point, homologous. Any modification of this sequence, whether by the addition or the deletion of stages, may therefore be interpreted as an outcome of heterochrony (but see Alberch 1985). Based upon this line of reasoning, McNamara expanded the concept of paedomorphosis to include a descendant passing through fewer ontogenetic stages than the ancestor (e.g., the deletion of the metacercaria), and the concept of peramorphosis to include a descendant passing through more ontogenetic stages than the ancestor (e.g., the addition of a redial stage).

As the theory developed, researchers were faced with two problems. The first arose from the practical aspects of identifying and measuring changes in development. From this were generated purely technical questions like "how can you recognize environmentally induced heterochrony?" and more metaphysical questions like "how do you measure time on a biologically relevant scale?" and "when does the growth of a structure begin, and when does it end?" (Alberch and Alberch 1981; Reiss 1989). These questions can only be solved through collaboration with developmental biologists.

The second problem was created by the lack of a way to identify the ancestral ontogeny, since that ancestor was presumably no longer available for direct observation. This methodological stumbling block was removed by Fink (1982), who proposed that the ancestral ontogeny could be reconstructed in the same manner that the plesiomorphic condition for any character was delineated: by outgroup comparison within a phylogenetic context. This proposal requires two pieces of information: (1) a well-supported phylogeny for the study group and (2) details about the ontogenetic patterns within the study group and the outgroups. As with all studies cast within a phylogenetic framework, our confidence in the identification of heterochronic phenomena is only as strong as our confidence in the framework (e.g., Winterbottom 1990).

Heterochronic processes might be an important explanation for some of the homoplasy we see throughout the parasite phylogenies. For example, let us consider some characters of larval digeneans. The plesiomorphic digenean ontogeny includes three larval stages: the miracidium, sporocyst, and cercaria (Fig. 3.3). Early in the phylogeny of digeneans a fourth stage, the redia, was intercalated between the sporocyst and the cercaria. Not all descendants of the ancestral digenean in which the redia arose, however, are characterized by the presence of a redial stage. Many of them exhibit a stage termed a daughter sporocyst that occurs, like the redia, between the sporocyst (now called a mother sporocyst) and the cercarial stages. The presence of a daughter sporocyst coincides with the absence of the redia (Pearson 1972; Brooks et al. 1985b) and appears convergently among digeneans (Brooks et al. 1985b, 1989a; Appendix).

Rediae and daughter sporocysts are both derived from mother sporocyst germinal tissue; both have a birth pore and both give rise to cercariae. The difference between a redia and a daughter sporocyst is morphological: a redia has a pharynx and gut while a daughter sporocyst does not. This morphological

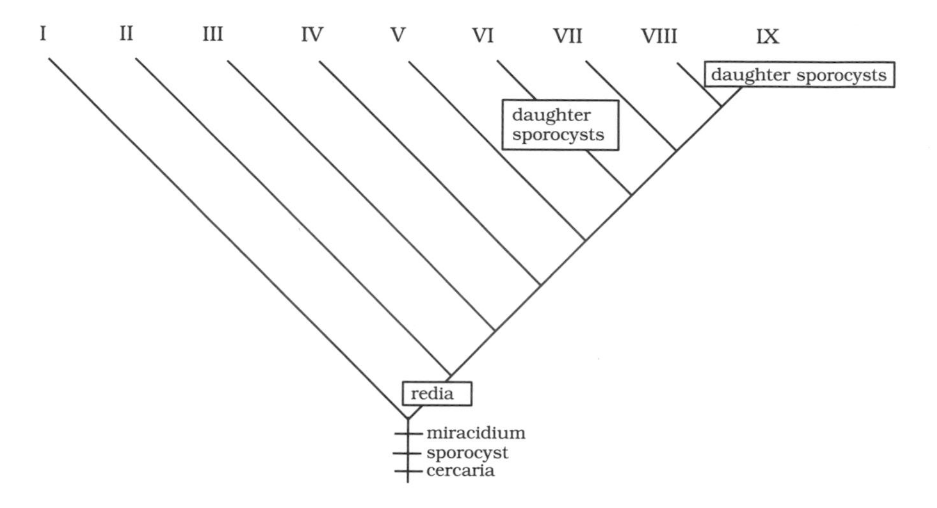

Figure 3.3. Life cycle stages mapped onto a simplified tree for the digeneans. I = Heronimiformes; II = Paramphistomiformes; III = Echinostomiformes; IV = Haploporiformes; V = Hemiuriformes; VI = Strigeiformes; VII = Opisthorchiformes; VIII = Lepocreadiiformes; IX = Plagiorchiformes.

difference, coupled with their mutually exclusive presence in ontogenies and the fact that both give rise to cercariae that generally possess pharynges and guts, indicates the presence of heterochronic phenomena leading to the later than expected expression of the pharynx and gut in ontogeny. In the absence of a phylogeny and experimental studies of developmental rates, we cannot tell which of the three categories of paedomorphosis were involved (or even that the same process was involved in each case), but we can narrow the field down to two. Either the daughter sporocyst is a paedomorphic redia, in which case the redia must have evolved before the daughter sporocyst, or a redia is a peramorphic daughter sporocyst, in which case the daughter sporocyst must have arisen first. The patterns uncovered by phylogenetic analysis support the first hypothesis: rediae arose before daughter sporocysts, so daughter sporocysts appear to be paedomorphic rediae (see also Pearson 1972; Brooks et al. 1985b, 1989; Gibson 1987).

The study of evolutionary effects caused by heterochrony is in its infancy, in parasitology in particular and in evolutionary biology in general. Parasitologists, and evolutionary biologists addressing questions of phenotypic plasticity in parasitic organisms, have both generally assumed that parasites will exhibit rampant homoplasy. We have demonstrated that parasitic platyhelminths, at least, do not exhibit high levels of homoplasy. In addition, we have raised the possibility that at least some of the homoplasy that does occur is caused by changes in the timing of events during parasite ontogeny.

MYTH IV: LARVAL AND ADULT FEATURES REPRESENT INDEPENDENT ADAPTIVE RESPONSES TO DIFFERENT ENVIRONMENTS AND INDICATE DIFFERENT PHYLOGENETIC RELATIONSHIPS

The evolutionary relationship between larval forms and their adult counterparts has provided the basis for plausible adaptive scenarios that, as they unfold, lead us to our fourth myth. Each scenario begins with the assumption that any stage added to the ontogeny of a parasite species is an adaptive response to some problem set by the environment, to which the parasite must respond or face extinction. So far, so good; the fitness benefits of enhanced transmission potential are straightforward. Following the origin of a particular stage, any evolutionary changes in its characteristics reflect independent adaptive responses to the environment in which the larval form finds itself. That "environment" may be dramatically different from the "environment" of its parents. For example, the miracidium of many species of digeneans escapes from the egg and swims about searching for a molluscan host (environment number 1); the mother sporocyst and rediae develop within that mollusc's tissues (environment number 2); the cercariae leave the mollusc and swim about in water looking for a second intermediate host (environment number 1); the cercariae encyst and develop into metacercariae in a second intermediate host, which may be an invertebrate or a vertebrate (environment number 3); the second intermediate host is ingested by the final definitive host, normally a vertebrate, where the adults mature and reproduce (environment number 4). In this case, a particular individual is exposed to at least four different environments during its lifetime. Since adaptation involves a direct interaction between an organism and the environment, this individual is also potentially exposed to four different sets of selection pressures. It is not, therefore, unreasonable to assume that each stage will exhibit adaptive evolutionary responses to each type of selection pressure. At this point, however, the scenario takes a wrong turn. An emphasis on adaptation, in the absence of phylogenetic information, allowed researchers to assert that similarities in the characteristics of different ontogenetic stages of different digenean species tend to reflect convergent adaptive responses rather than phylogenetic relationships. This wrong turn leads us away from a strong hypothesis into the realm of mythology. In this case the myth states that *systematic analysis of each developmental stage for a group of species will produce a different set of relationships.*

By now, the reader will not be surprised if we suggest that this view about the evolution of ontogenies is an old one. In this case, it grew out of research by the German Darwinian, Ernst Haeckel. Haeckel (1866) maintained that ontogenies evolved by *palingenetic* or by *cenogenetic* changes. Palingenetic changes, also known as recapitulation, result when phylogenetic changes are added to the ontogeny of a descendent species. These changes could come in the form of terminal additions (*Haeckelian recapitulation:* Fink 1982), or terminal substitutions to ancestral ontogenies (*von Baerian recapitulation:* Fink 1982). Cenogenetic changes, on the other hand, result from nonterminal changes in ontogenetic programs. Such changes were thought to disrupt the parallel between evolutionary history and individual developmental sequences. Russell (1916), among others, considered cenogenetic changes to be the result of adaptive responses to particular natural selective pressures, so the historical precedent for the scenarios is old.

Parasitologists were drawn to Haeckel and Russell's ideas because of the strong adaptationist component of discussions about the function and evolution of life cycles. For example, Chitwood and Chitwood (1974) considered that certain larval features of parasitic nematodes were cenogenetic changes associated with selection pressures specific to the larval stages, and thus not indicative of phylogenetic relationships. Goodchild (1943) presented similar arguments when discussing cercarial features of some gorgoderid digeneans. Cable (1974) proposed that the larval stages of digeneans (especially the cercarial stage) were recapitulations of ancestral adult free-living stages and that the adult stages were cenogenetic adaptations to inhabiting different groups of vertebrates; therefore, he argued, phylogenetic relationships could be correctly inferred from larval morphology but not from adult morphology. Gibson (1981), in direct contrast to Cable but using exactly the same theoretical argument, proposed that the larval stages of digeneans were cenogenetic adaptations, so phylogenetic relationships could be correctly inferred from adult morphology but not from larval morphology.

We believe that discussions about the evolution of life cycles became sidetracked because authors were so preoccupied with phenotypic variation within an individual. This preoccupation eclipsed the fact that the same genotype was responsible for the production of all those phenotypic manifestations. Since genealogical information ultimately resides in the genotype, there is no a priori reason to assume that any one stage is a better reflection of genealogical relationships than another. Incorporating genealogy into our evolutionary perspective moves us toward the view that the morphological diversification of different ontogenetic stages is an example of the "constraints" view of evolution. That is, the ways in which, and the extent to which, particular ontogenetic stages adapt to selective pressures are constrained by the fact that the entire ontogenetic program is a coherent and integrated sequence whose parts are correlated primarily with phylogenetic history and secondarily with selective pressures peculiar to portions of the ontogeny (see also Brooks and Wiley 1988; Brooks et al. 1989a; Brooks and McLennan 1990; Maurer and Brooks 1991). It does not appear, therefore, that parasite life cycles can be cobbled together as an assortment of independent adaptive responses, giving us little or no hint of phylogeny. This conclusion does not negate the importance of adaptation at every step of the process. It just means that the origin and elaboration of such adaptations must be examined within a phylogenetic framework.

Modern phylogenetic systematic methods allow us to assess the myth that juvenile and adult characters do not provide congruent reconstructions of phylogenetic relationships because the analysis is not based on any assumptions about recapitulation (or its lack). In order to do this, we need both a group for which nonterminal changes in ontogeny have arisen during the course of evolution and an assessment of the degree of congruence between phylogenetic trees derived from larval and from adult characters. As the reader might expect from the opposing views of Cable (1974) and Gibson (1981) mentioned previously, we have exactly this type of example in the digeneans.

Brooks et al. (1985a) discovered that the phylogenetic relationships postulated by the larval and by the adult characters *are completely congruent with each other.* Forty percent of the nodes on the tree were supported by both adult and larval characters. Interestingly, of the remaining nodes, there appeared to be a tendency for adult characters to resolve relationships among families within orders and for larval characters to be buried more deeply in the phylogeny, resolving relationships among orders. Overall, the 90 larval characters in their analysis clarified 74% of the phylogenetic tree and produced no conflicting groupings, whereas the 113 adult characters clarified 76% of the tree with no conflicts. In addition to showing that larval and adult features indicate the same phylogenetic relationships, this finding suggests that information from both forms of data is necessary to build robust phylogenies. Furthermore, both O'Grady (1987), in a study of one clade within the macroderoidid genus *Glypthelmins,* and Caira (1989), in a study of some papillose allocreadiids, found that data from larval and adult morphology produced congruent phylogenies at the species level of investigation. So evidence from both a family and a species-level analysis supports the proposition that there is no a priori reason to prefer one form of data over another. Not only does this evidence reinforce current views of the evolution of developmental programs, it is also consistent with the results of similar studies using free-living groups (see examples in Brooks and Wiley 1988).

Now that we have data about the morphological characters of larval and adult stages, let us put those data into a larger context to study the evolution of life cycle patterns.

MYTH V: PARASITES EXHIBIT ADAPTIVELY PLASTIC LIFE CYCLE PATTERNS

There are two fundamental components to parasite life cycles. The first is based on aspects of the association between a parasite and its host/resource. It is ecological in focus and raises questions about the species of host, the number of hosts, and the part of the host in or on which the parasite resides. Ecological studies have traditionally concentrated on life cycle patterns as a source of information about the possibility of host switching and convergent evolution, as well as a source of evidence about the evolution of adaptive stages that increase the likelihood of infection. The second component is based on aspects of the parasite's biology. It is developmental in focus and raises questions about the origin and number of developmental stages and in some cases the degree of asexual amplification of those stages. Like ecology, parasite ontogenies have been studied as a source of evidence about the evolution of adaptive stages that increase the likelihood of infection.

The existence of these two components provides us with an opportunity to examine the difference between traits that are historically correlated and those that are causally linked in evolution. There are strong ecological and developmental correlates of each parasite species's life cycle, which might seem to indicate a tight evolutionary relationship between the two components. Is this in fact the case? Until the advent of phylogenetic analysis, it was not possible to tease apart the relative patterns of diversification in each component in a rigorous manner. We now, however, have at our fingertips a method that will allow us to disentangle these components and study their relationship, if any, throughout the evolutionary history of various parasite groups.

Parasitologists have traditionally concentrated on three primary ecological features when discussing the evolution of life cycle patterns: (1) the sequence and number of hosts (e.g., Dougherty 1951; Price 1980; Sprent 1982; Holmes 1983), (2) the type of host in which each stage develops (e.g., Chabaud 1954, 1955, 1965b, 1982; Anderson 1957, 1958, 1982, 1984; Price 1980; Holmes 1983; Durette-Desset 1985), and (3) the site within the host in which each stage develops (e.g., Williams 1960, 1966, 1968; Schad 1963; Williams et al. 1970; Crompton 1973; McVicar 1979; Rohde 1979, 1981, 1984; Price 1980; Holmes 1983; Rohde and Hobbs 1986). There are, of course, other factors involved in the evolution of life cycle patterns. For example, we know very little about the evolutionary context of the parasite's behavior when it moves from one host to another or the effect of the parasite on the behavior of its hosts (e.g., Holmes and Bethel 1972; Moore 1984). Nevertheless, researchers have accumulated enough information about a variety of ecological and developmental characters for us to begin asking questions about the evolution of life cycles.

Digeneans exhibit an astounding diversity of life cycles, marked by complex ontogenetic and ecological attributes. Consequently, they should serve as excellent model systems for the study of life cycle evolution. Brooks et al. (1985a) investigated five classes of life cycle characters: (1) developmental changes in invasive larvae that increased the numbers of such colonizing stages; changes in preferences for (2) the first intermediate host, (3) the second intermediate host, and (4) the final host; and (5) changes in the mode of infection of the second intermediate host (a reflection of juvenile colonization ability). The outcome of this investigation is depicted in Figure 3.4. Derived changes in life cycle characters for the digeneans are interspersed throughout the more basal (suprafamilial) branches of the phylogenetic tree, not concen-

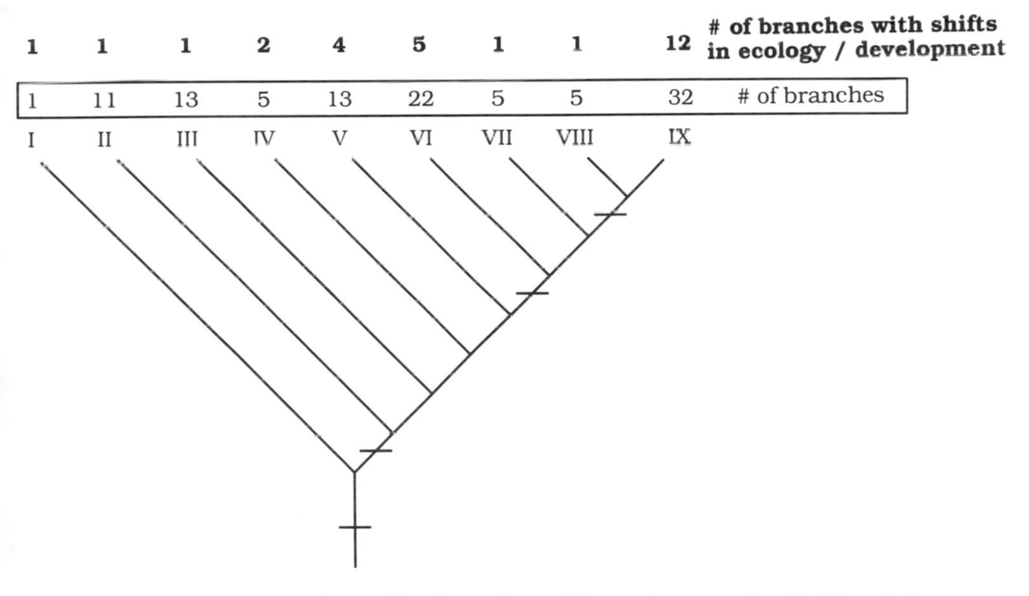

Figure 3.4. Phylogenetic tree for nine orders of digenetic trematodes. I = Heronimiformes (1 family); II = Paramphistomiformes (6 families); III = Echinostomiformes (7 families); IV = Haploporiformes (3 families); V = Hemiuriformes (7 families); VI = Strigeiformes (12 families); VII = Opisthorchiformes (3 families); VIII = Lepocreadiiformes (3 families); IX = Plagiorchiformes (23 families). *Numbers* above each order indicate the number of evolutionary changes in five classes of life cycle traits, discussed in the text, and the total number of terminal (family) and nonterminal branches within each order. *Slash marks* on nonterminal branches of the phylogenetic tree indicate additional points of diversification in life cycle traits for this group. Of the 107 branches on the phylogenetic tree, 32 (30%) are associated with some form of diversification in life cycle patterns. From Brooks and McLennan (1991).

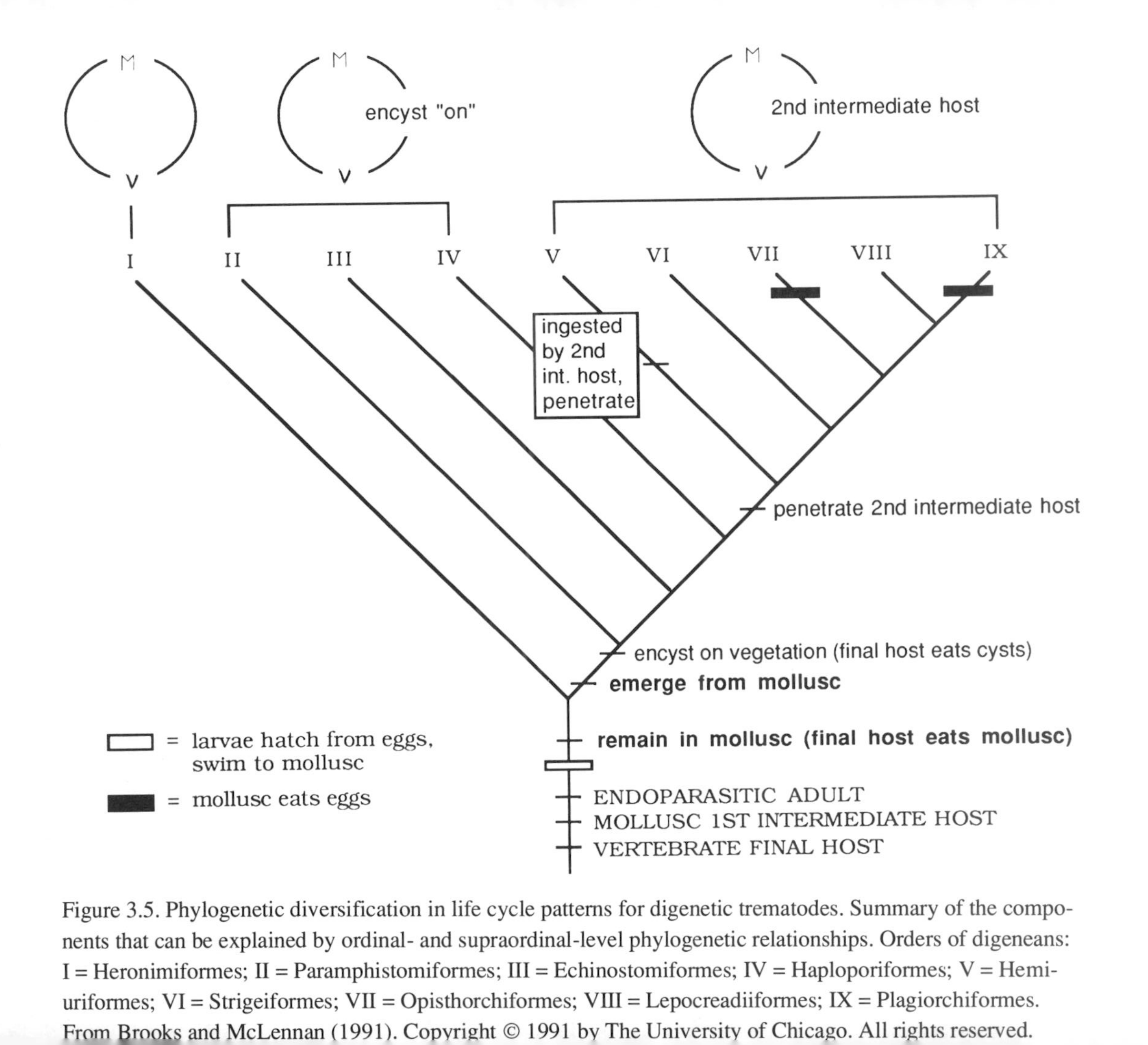

Figure 3.5. Phylogenetic diversification in life cycle patterns for digenetic trematodes. Summary of the components that can be explained by ordinal- and supraordinal-level phylogenetic relationships. Orders of digeneans: I = Heronimiformes; II = Paramphistomiformes; III = Echinostomiformes; IV = Haploporiformes; V = Hemiuriformes; VI = Strigeiformes; VII = Opisthorchiformes; VIII = Lepocreadiiformes; IX = Plagiorchiformes. From Brooks and McLennan (1991).

trated within family groups. Since only a small proportion (about 30%) of the branchings on the phylogenetic tree are correlated with any diversification in the life cycle characters, the modification of life cycle patterns is more conservative than phylogenetic diversification at the family level.

Furthermore, changes to life cycle characters do not "come and go" all over the tree as would be expected if they were adaptively plastic. Their patterns of diversification are instead closely correlated with the hypothesized phylogenetic relationships of their bearers. In the language of phylogenetic systematics, we would say that these characters display little homoplasy throughout their evolutionary history. Because of this, there is a historically coherent sequence of life cycle elaboration at this level of phylogenetic analysis that explains much of the diversity of digenean life cycle patterns (Fig. 3.5).

Interestingly, most of the diversification appears to have been initiated by evolutionary changes in the cercarial stage. The cercaria is a juvenile stage that develops in the molluscan intermediate host and becomes infective to the second intermediate host (or sometimes to the final, definitive, host). When cercariae began emerging from their molluscan hosts and encysting on vegetation and animal exteriors (most often on the exoskeletons of aquatic arthropods), the range of potential vertebrate hosts was enlarged greatly. No longer would trematodes be restricted to molluscivores. When the cercariae began penetrating particular intermediate hosts and encysting within them, a high degree of specificity in second intermediate host type emerged, possibly enhancing adaptive modes of speciation. The evolution of cercarial emergence and the evolution of cercarial encystment and penetration therefore had significant adaptive consequences. The general evolutionary trend in the case of the digeneans appears to have been from relatively simple to relatively complex life cycles.

Departures from phylogenetic congruence are due primarily to the reappearance of plesiomorphic life cycle patterns in relatively derived groups, generally involving the secondary loss of a host (Fig. 3.6, Table 3.8). A digenean three-host life cycle may secondarily become a two-host life cycle by dropping any one of the three hosts. There are no reports of the mollusc first intermediate host being dropped from the life cycle (although it reportedly can be replaced by a polychaete for at least one species of fish blood flukes). The second intermediate host may be lost when the cercariae remain in the molluscan first intermediate host (e.g., members of the Microphallidae and Dicrocoeliidae). By far the most common mechanism for truncating a life cycle is deletion of the definitive vertebrate host. This occurs because individuals become reproductively mature while they are still in the intermediate host (in a very rare case, *Microphallus abortivus* passes through its entire ontogeny in the molluscan first intermediate host: Deblock 1974). Note that cases of truncated life cycles are concentrated in the Plagiorchiformes and the

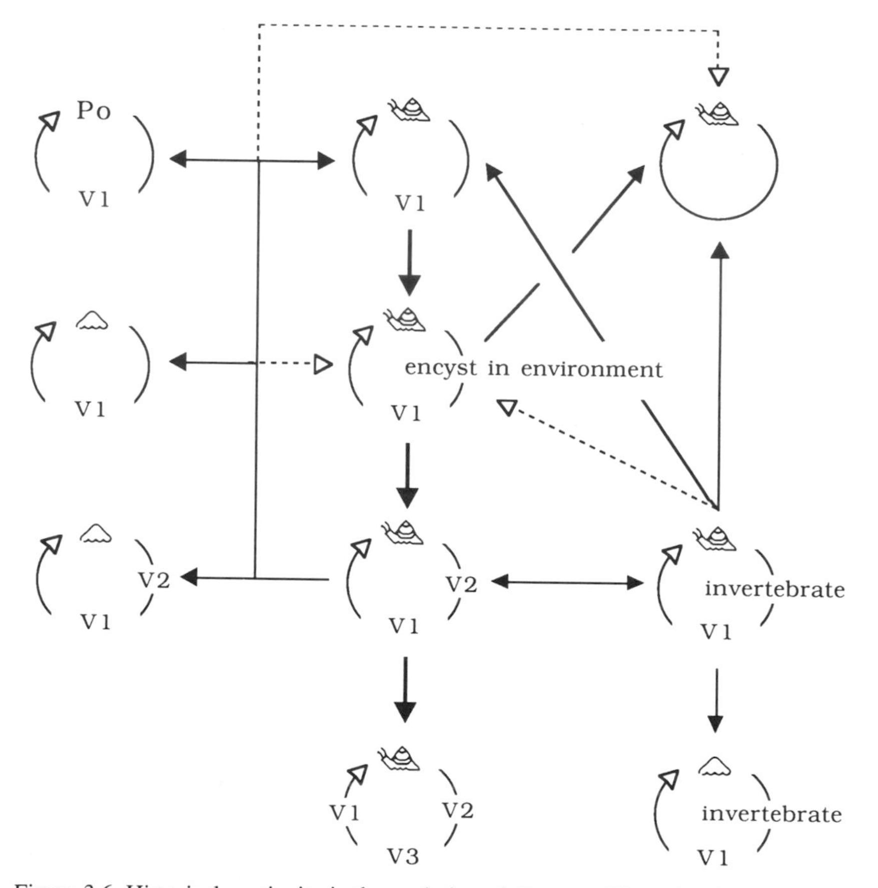

Figure 3.6. Historical continuity in the evolution of digenean life cycles. Snails and pelecypods are indicated with tiny figures. Po = polychaete, V = vertebrate host. Broken lines refer to possible transitions that have not yet been documented. Redrawn and modified from Brooks et al. (1985b).

Strigeiformes, the groups in which the paedomorphic change in rediae (to "daughter sporocysts") took place (Fig. 3.7).

Explanations for departures from phylogenetic congruence, like the reversals to a primitive life cycle pattern discussed previously, usually invoke either heterochronic development or ecological adaptation (e.g., McIntosh 1935; McMullen 1938; Buttner 1951; Stunkard 1959; Grabda-Kazubska 1976; Shoop 1988; Brooks et al. 1989a). Heterochronic processes need not affect the ecological components of a digenean life cycle. Theoretically, however, there are a number possible ecological outcomes to heterochronic changes in terminal (sexual maturity) and nonterminal development. The interactions between development and ecology listed in Table 3.8 highlight an interesting problem.

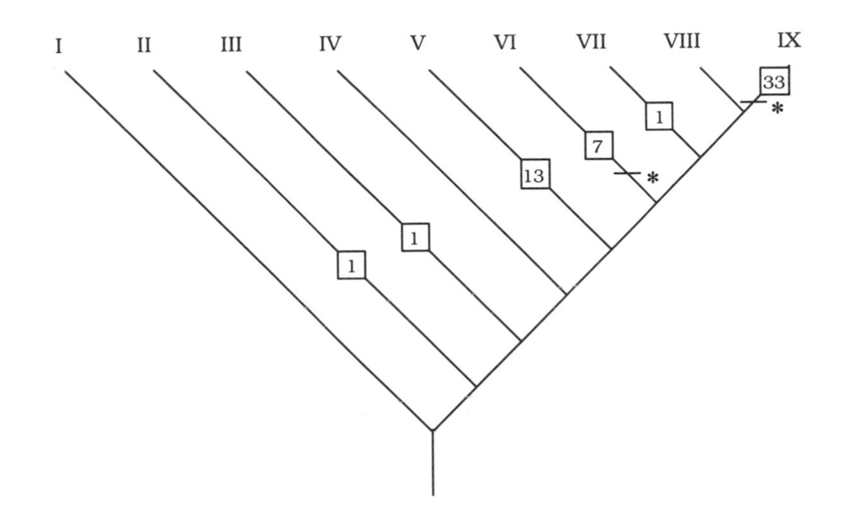

Figure 3.7. Instances of truncated life cycles mapped onto a simplified phylogenetic tree for the digeneans. Orders of digeneans: I = Heronimiformes; II = Paramphistomiformes; III = Echinostomiformes; IV = Haploporiformes; V = Hemiuriformes; VI = Strigeiformes; VII = Opisthorchiformes; VIII = Lepocreadiiformes; IX = Plagiorchiformes. * = Orders in which daughter sporocysts have evolved.

Table 3.8
Hypothetical effects that changes in two components of development can have on the ecological component of parasite life cycles

Heterochronic process	Nonterminal ontogeny	Terminal ontogeny
Acceleration[a]	Drop an intermediate host	Drop a terminal host
Pre-displacement	Add an intermediate host	Add a terminal host
Hypermorphosis	Add an intermediate host	Add a terminal host
Neoteny[a]	Add an intermediate host	Add a terminal host
Postdisplacement	Drop an intermediate host	Drop a terminal host
Progenesis	Drop an intermediate host	Drop a terminal host

[a]Processes whose ecological outcomes are opposite to the developmental outcome.

Two peramorphic changes in development result in peramorphic changes in ecology, whereas two paedomorphic changes in development result in paedomorphic changes in ecology. This predictability, however, vanishes when we consider the ecological outcomes of changes in developmental rate. In these cases, peramorphic development results in paedomorphic ecology, and paedomorphic development results in peramorphic ecology. Because of this, *it is impossible to determine which of the six developmental processes has been involved in the evolution of a particular life cycle pattern by examining the changes in host utilization alone, even if that examination occurs within a phylogenetic framework.* In order to understand the evolution of life cycle diversity data about parasite development are also required. Or, to return to a by now familiar theme, *parasite life cycles are a property of the parasite, not of the host.*

At the moment, no one has investigated changes in rates and timing of development within a phylogenetic context. So we cannot yet combine the theoretical data from Table 3.8 with the life cycle patterns depicted in Figure 3.6. Nor can we determine the relative influences that each of these processes has had in the evolution of life cycle patterns. What we can say is that, based upon the data collected to date, change in the development of both terminal and nonterminal ontogenetic stages has been the driving force in the evolution of digenean life cycle diversity. Although heterochrony in ontogeny need not have an ecological outcome, change in the number of hosts in a life cycle requires ontogenetic heterochrony. This suggests the possibility of an interaction between ontogenetic and ecological phenomena in the evolution of life cycle diversity. Parasite groups, like the digeneans, are thus excellent model systems for investigating hypotheses about the relationship among ontogeny, sexual maturation, and ecology in evolution.

For example, consider the nine species forming the macroderoidid digenean genus *Alloglossidium.* These digeneans are unusual because they display a range of definitive host types from catfish through crustaceans to leeches. Font (1980) attempted to explain that range from a developmental perspective. He proposed that (1) the species inhabiting catfish exhibited the primitive life cycle pattern (gastropod first intermediate host, arthropod second intermediate host, vertebrate definitive host) for the group, (2) the vertebrate definitive host was lost through precocial maturation of the metacercarial stage in what was primitively the second intermediate host, and (3) the developmental change had arisen independently in the "crustacean" and "leech" lineage. Riggs and Ulmer (1983a,b) took an ecological stand. They proposed that (1) the two-host life cycle is primitive and (2) some ancestral flukes in leeches switched to catfish, remaining and flourishing in the new host because of similarities between leech intestines and catfish intestines.

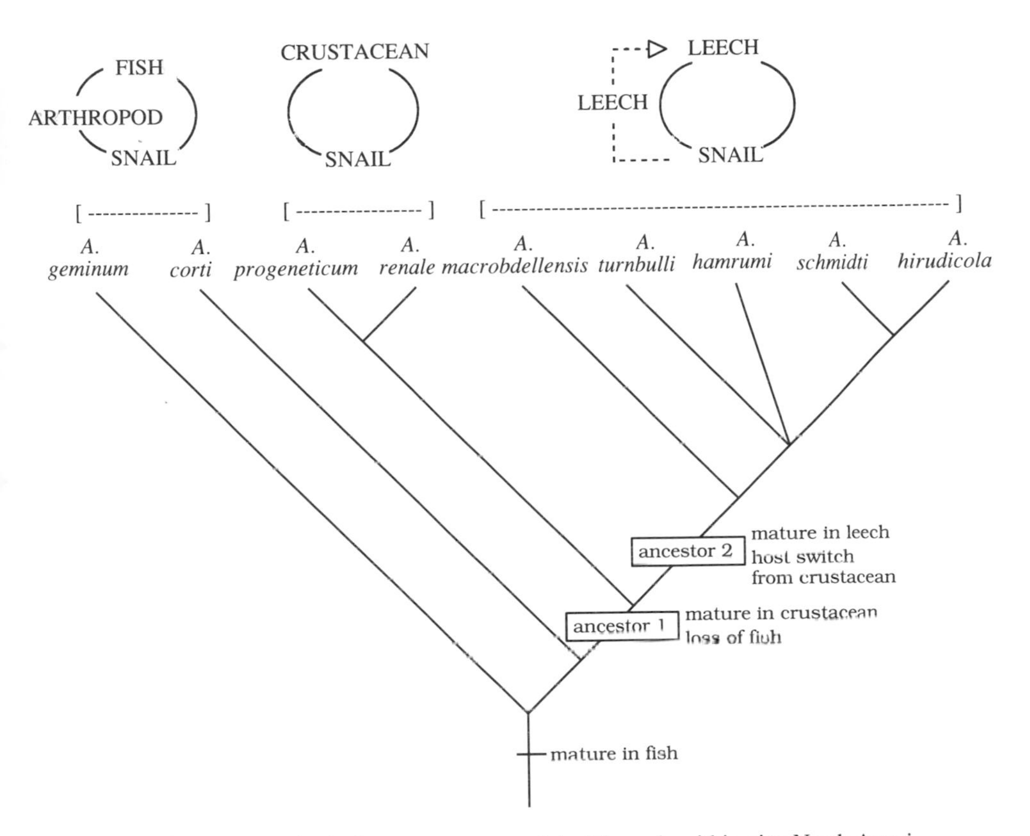

Figure 3.8. Changes in the host components of the life cycle within nine North American members of the flatworm genus *Alloglossidium*. Redrawn and modified from Carney and Brooks (1991).

Carney and Brooks (1991) investigated these hypotheses by reconstructing a phylogeny for *Alloglossidium* based upon morphological characters, then mapping life history characters onto that phylogenetic template (Fig. 3.8). Their results were unequivocal: the three-host life cycle with a fish definitive host is plesiomorphic for the group. This does not support Riggs and Ulmer's proposal that the species associated with leeches evolved before those inhabiting catfish. The relationships do support Font's ideas, with a slight modification. Both the lineage utilizing a crustacean as definitive host and the lineage utilizing leeches share a common life cycle trait, the lack of a fish definitive host. The most parsimonious argument for this shared novelty is that it originated in the common ancestor (ancestor 1) of both lineages. Three types of developmental change will result in the elimination of a terminal host: acceleration, post-displacement, or progenesis. At the moment we cannot chose among these alternatives because we do not have information about the rate or timing of development within *Alloglossidium*. Nevertheless, we can say that one of these developmental changes was responsible for the appearance of sexually mature individuals in what was primitively a second intermediate host.

Was that host a crustacean or a leech? The phylogenetic patterns cannot resolve this question because both options are equally parsimonious given the configuration of the tree. This is the point at which additional biological information is required! Freshwater crayfish have been identified as second intermediate hosts in *Alloglossidium geminum* and *A. corti;* therefore, it appears likely that the ancestral second intermediate turned definitive host was a crustacean. Given this, the presence of a leech definitive host represents a case of host switching by "ancestor 2." That ancestor arrived on its new host carrying the developmental change that was responsible for its truncated life cycle.

The primary purpose of this section has been to document the degree of phylogenetic constraints on parasite (in this case, digenean) life cycles. Although the precise documentation followed herein results from relatively recent phylogenetic systematic studies, there are notable precursors among parasitologists. We cannot think of a better summary for this discussion than the following words:

> Each species [of digenean] seems to play the game of parasitic life within the broad rules laid down by, and for, its family; but it appears to have developed within this code its own deviations from what we poor humans assume to be normal and regular for the family. [LaRue 1951]

MYTH VI: PARASITES ARE PARADIGMS OF ADAPTIVE RADIATION

> It has not been generally realized that the most extraordinary adaptive radiations on earth have been among parasitic organisms.
>
> —Price 1980

Price (1980) began his argument about the nature of adaptive evolution in parasites with the assumption that "parasite niches" are part of the environment (by which he meant the host). He then attempted to evaluate the potential for future adaptive radiations in parasites by assessing the number of unoccupied niches (i.e., the number of hosts and parts of hosts that were parasite free). This turns out to represent a substantial amount of the "environment," so Price concluded that adaptive radiation had been extensive in the past (there are a large number of parasite species) and that the potential for future adaptive radiations was high (there is a large amount of unoccupied niche space). He also concluded that both the large number of occupied and the large number of unoccupied host niches could be explained by host specificity. Based on this conclusion, he hypothesized that parasite populations and communities exist in nonequilibrium (specifically, below-equilibrium) conditions with respect to their host environments because changes in the host environment occurred faster than the parasites could adapt to them.

There are two questions that need to be asked at this point. First, if the host environment is changing faster than the parasites can adapt to it, why do we think that parasites are highly adaptive (i.e., why postulate that adaptive radiations have been more extensive among parasites than among free-living organisms)? It would seem that this scenario only works if hosts evolve more rapidly than parasites, in which case we would postulate that parasites are, if anything, adaptively conservative relative to their hosts, something postulated by the early parasite evolutionary biologists (see Chapter 1). This intrusion into Price's theory is made all the more troublesome by the emerging recognition that pronounced host specificity may be a constraint on, rather than a driving force of, parasite speciation (see Chapter 2).

The second question concerns the limitations on parasites' ability to adapt to changes in the host environment. If the changes in the environment create the potential for adaptive radiation, how can they also be the reason for observed constraints on adaptive radiations? In order to answer this, we must return to an idea discussed in Chapter 1, the question of host influences on parasite evolution. The belief that parasites are black boxes programmed by the environment (their host) is an orthogenetic assumption that has become part of the evolutionary folklore of parasitology. As we have demonstrated throughout this book, however, parasites do not exist as so many interchangeable beans in a bag. Like all other organisms, they have properties that might affect the rate, manner, and extent of their evolutionary responses to environmental stimuli.

What might these properties be? Darwin (1872) envisioned evolution as being the result of an interaction between rules governing inherited properties of organisms and the effect of the environment on those systems:

> The characters which naturalists consider as showing true affinity between any two or more species, are those which have been inherited from a common parent, all true classification being genealogical.

and

> Slight modifications, which in any way favoured the individuals of any species, by better adapting them to their altered conditions, would tend to be preserved; and natural selection would have free scope for the work of improvement.

Our position, therefore, is that parasite evolution is the result of interactions between the environment and populations tempered by a background of substantial inherited constraints. Since both parasites and hosts have evolutionary tendencies and capabilities, parasite evolution will be historically correlated in some way with host evolution, but will not necessarily be causally connected with it. Because of this reasoning, we advocate treating the niche as a property of the organism, not the environment.

To demonstrate this line of reasoning more clearly, we need to consider the concept of "genetic space" (Layzer 1978, 1980; Brooks and Wiley 1988; Brooks et al. 1989b; Brooks and McLennan 1990). Simply stated, genetic space refers to the realm of all possible genotypes. It can be enlarged (expanded) by processes such as mutation and sexual reproduction. At any given time, not all portions of genetic space are occupied, for a variety of reasons. Many of those reasons have to do with the evolutionary history of each species, which determines the degree to which and the way in which mutations integrate with the (greater) portions of the genome that have not changed. This history, in turn, ultimately determines the range of genotypes/phenotypes available to environmental selection. Mechanically, environmental selection acts to decrease the proportion of genetic space occupied, eliminating those organisms that are genetically and developmentally competent but poorly suited to current environments. The constraints on the occupation of genetic space that stem from the evolutionary history of the species are called phylogenetic constraints (Brooks and Wiley 1988); those stemming from the effects of the environment are called environmental constraints (Brooks and Wiley 1988). Interestingly, identifying a "constraint" as environmental or phylogenetic is a relative process, since the environmental constraints of the past can become phylogenetic constraints in the future.

How does the concept of evolutionary constraint integrate with the concept of parasite niches? To answer this question, we have to shift our perspective about "niche" away from the environment to the organism. From this angle, we begin to ask questions about the origins of characters that allow an organism to utilize portions of the environment that have not been utilized before, or to use the environment in novel ways. In other words, *evolution results in organisms that carve environmental space into niche space, not in an environment that is inherently organized as niche space and which molds organisms to fit it.* Phylogenetic constraints act to slow the rate at which new niches evolve, and this is a source of nonequilibrium conditions for parasites. Environmental selection will also be important because the niches of some parasites will not be allowed in some environments. Thus, environmental selection and the evolution of new hosts and new kinds of hosts will act in concert with phylogenetic constraints to maintain, or even intensify, nonequilibrium conditions. Price (1980) considered the occurrence of parasites in nonequilibrium conditions to be an *ecological* concept, because he considered the environment (primarily the host) to be the source of effects that created the nonequilibrium conditions. When one considers the potential role that phylogenetic constraints and the inherent evolutionary tendencies of the parasite have played in the evolution and diversification of parasitic groups, this becomes an *evolutionary* concept as well.

Futuyma (1986) defined adaptive radiation as "a term used to describe diversification into different ecological niches by species derived from a common ancestor." This concept has played an important role in evolutionary biology, as an explanation for differences in species richness among groups. Such differences are postulated to result from unusually high speciation rates in the more speciose group. Some authors have suggested that there should be an adaptive explanation for all speciation events (Stanley 1979; Stanley et al. 1981). Simpson (1953) believed that adaptive radiations resulted from diversification accelerated by ecological opportunity, such as dispersal into new territory (see the discussion of peripheral isolates allopatric speciation in Chapter 2), extinction of competitors, or adoption of a new way of life (i.e., an adaptive change in ecology or behavior). Other factors, including the adoption of a specialist foraging mode (Eldredge 1976; Eldredge and Cracraft 1980; Price 1980; Vrba 1980, 1984a,b; Cracraft 1984; Novacek 1984; Mitter et al. 1988), sexual selection and population structure (Spieth 1974; Wilson et al. 1975; Carson and Kaneshiro 1976; Ringo 1977; Templeton 1979; Gilinsky 1981; West-Eberhard 1983; Barton and Charlesworth 1984; Carson and Templeton 1984), or the origin of key ecological innovations in an ancestral species (Cracraft 1982b; Mishler and Churchill 1984; Brooks et al. 1985a), have also been postulated to have a positive effect on speciation rates.

Discussions of adaptive radiations tend to focus on either the "adaptive" or the "radiation" aspect of the term. One research group is fascinated by the observation that some sister groups are highly asymmetrical with respect to species richness. The disproportionate representation of one taxon within an assemblage is widespread throughout nature. For example, passerines dominate all other orders of birds, and teleost fishes dominate the actinopterygians. Some authors have argued that the perception of dominance is nothing more than a taxonomic artifact (Mayr 1969; Raikow 1986, 1988). Others have argued that such patterns can be explained by stochastic speciation and extinction processes (Raup et al. 1973; Raup and Gould 1974; Gould et al. 1977; Raup 1984). Flessa and Levinton (1975 in Dial and Marzluff 1989: 28) disagreed, concluding that "deterministic processes so pervade the evolution of diversity that stochastic processes must be regarded as secondary in importance" (see also Stanley et al. 1981; Fitzpatrick 1988; Vermeij 1988; Ehrlich and Wilson 1991). Dial and Marzluff (1989) compared patterns of species richness within 85 clades to the patterns predicted by five null models (Poisson, the simulation of Raup et al. 1973, the simulation of Anderson and Anderson 1975, simultaneous broken-stick, and canonical lognormal distribution). They discovered that nearly all of the groups were dominated to a significantly greater extent by one taxon than predicted by any of the null models regardless of the classification scheme used to delineate the clades. On average, the species-rich group tended to account for 37% (traditional, or evolutionary taxonomic, classification) to 39% (phylogenetic systematic, or cladistic, classification) of the subunits within each assemblage. The generality of this result across a wide biological spectrum, from plants through insects to mammals, prompted their conclusion that "overdominance of an assemblage by one unit is a common and *nonrandom* feature of distributions of taxonomic diversity."

Comparison of sister groups, one of which is species rich and the other of which is species poor, is a necessary starting point because sister groups are, by definition, of equal age. This comparison provides the most objective means possible for making certain we are dealing with evolutionarily "equivalent" units when assessing species richness and its causes for particular groups (see Mayden 1986 for an extended discussion). In these studies, an implicit or explicit assumption is made that *adaptive radiation equals speciation;* therefore, discussions revolve around differences in the amount or the rate of speciation in each of two sister clades. Since speciation and adaptation are not necessarily causally connected (Brooks and McLennan 1991), focusing our attention on speciation may tell us very little about adaptation.

The second research group is interested in uncovering the adaptive correlates of evolutionary diversification by searching for "key innovations": synapomorphic traits of a species-rich group that (1) are not found in its species-poor sister group, (2) are common to all the members of the species-rich clade, and (3) enhance speciation rates either directly or indirectly by aiding the survival and spread of species lineages so that the clade exists for a long time over a wide geographic area. A key innovation was traditionally considered to be any novel feature that characterized a clade (i.e., any synapomorphy) and is correlated with its adaptive radiation (Mayr 1960; Liem 1973). Liem (1973), for example, suggested that the extensive diversification of cichlid fishes in the African Rift lakes was due to the origin of a lower pharyngeal jaw suspended in a muscular sling in their common ancestor. Lauder (1981; see also Liem and Wake 1985 and Stiassny and Jensen 1987) raised several cautionary notes about this concept. First, if the species-rich clade is characterized by more than one autapomorphy, there is no a priori way to determine which of those traits might be the key innovation (perhaps even a combination of traits could be *the* innovation). This problem is even more complex because there is no theoretical reason to expect that the innovation originated with the ancestor of the species-rich clade. It may have appeared in one of the sister species of the original ancestor, or the clade may be comprised of several species-rich subunits, each of which may be characterized by a different "key innovation."

Second, simply asserting that every apomorphy is an adaptation, and therefore potentially responsible for an observed "adaptive" radiation, is of limited value at best and circular at worst. The question then is how to define the term "adaptive" in a manner that allows a hypothesis of adaptation to be falsified. Since the diversification of fixed characters can only be identified from a macroevolutionary perspective, the concept of adaptation must incorporate phylogenetic information, as well as information about the functional superiority of the putative adaptive character. Therefore, we will follow the position presented by Baum and Larson (1991) that an adaptive character is any derived state that confers a performance advantage on its possessor (see also Arnold 1983; Greene 1986; Coddington 1988) in comparison to its plesiomorphic state. Coddington (1988) extended the discussion, adding the following general criteria to the list of ways to recognize an adaptive radiation: the appearance of homoplasy correlated with functional change on a phylogenetic tree; the appearance of predicted homoplasy correlated with predicted functional changes; and the appearance of particular structural change correlated with particular functional change, regardless of homoplasy. And finally Lauder and Liem (1989) suggested an experimental approach to testing hypotheses of adaptive radiation. Their approach examines patterns of structural diversification throughout particular clades, thus equating adaptive radiation more with degree and extent of structural diversification than with speciation rates per se. They emphasize the importance of having a causal model that predicts what the relationship should be between possession of an innovation and the pattern of structural diversification in a clade, in order to recognize particular synapomorphies as key innovations.

Third, there is rarely strong evidence about the manner in which the innovative trait affects speciation rates in the clade. The null mode for speciation is geographic or vicariant allopatric speciation, in which speciation rates are determined by rates of geological change leading to physical fragmentation of ancestral species populations (Cracraft 1985). Vicariant speciation is sometimes termed a "nonadaptive" mode of speciation, because specific adaptations are not required for speciation to occur. Key innovations can influence speciation rates directly if they increase the likelihood that species within the clade will participate in other, relatively rapid, "adaptive" modes of speciation, such as peripheral isolates allopatric, parapatric, and sympatric speciation. Vicariance biogeographic methods are a useful way to study this interaction of adaptation and speciation because episodes of "adaptive" speciation are highlighted against a background of vicariance.

Key innovations can also influence speciation rates indirectly by decreasing the chance that a species will go extinct before it has a chance to be affected by vicariant speciation (Cracraft 1982a). Larson et al. (1981), for example, proposed a species selection argument in which key innovations gave the descendant species in the clade an advantage over competitors. Such lineages would be expected to survive longer and extend over a larger geographic range than other lineages, thus increasing the likelihood of vicariant speciation, which would result in increased species richness. This argument raises the question of whether there need be anything adaptive about simply surviving long enough to participate in many episodes of vicariant speciation. In general, the probability of bifurcation should increase with increased species longevity; however, the key innovation need not play a role in initiating those bifurcations. For example, long-lived species residing in areas subject to repeated vicariant episodes (hot spots: Cracraft 1982a) may show more bifurcations than equally long-lived species inhabiting more stable environments. The problem arises out of the assumption that *adaptation equals speciation*. If, as we discussed previously, the two processes are decoupled, then we must incorporate both processes into our definition of a key adaptation. In this book, then, we will define a key innovation as any apomorphic character that can be demonstrated to have adaptive value relative to its plesiomorphic antecedent, and can be demonstrated to play a direct role in initiating speciation (and thus increasing speciation rates). Given this definition, characters that simply promote the longevity of a species so that its chances of participating in vicariant speciation are increased are not key innovations.

Dial and Marzluff (1989) suggested that life history traits might play an important role in speciation rates. This perspective is not new to parasitologists. One of the worst memories an undergraduate may have is the seemingly never-ending procession of life cycle patterns paraded across the blackboards of parasitology classes. Parasitologists' preoccupation with life cycles stems from the fact that they (life cycles, not parasitologists) are so predictable that it is possible to design schemes for interrupting infections in disease-causing species. This predictability has also stirred the interest of evolutionary biologists, who have searched among life cycle patterns for evidence of key innovations leading to the adaptive radiation of parasite groups (e.g., Seurat 1920;

Baer 1950; Chabaud 1954, 1955, 1965a, 1982; Anderson 1957, 1958, 1982, 1984; Osche 1958; Combes 1972; Freeman 1973; Cable 1974; Odening 1974a,b; Euzet and Combes 1980; Price 1980; Moore 1981; Sprent 1982; Durette-Desset 1985).

Rogers (1962) suggested that adaptive radiation of parasite groups was marked by the evolution of key adaptations that increase the chances of transmitting offspring from generation to generation. These included mechanisms for (1) producing more eggs, (2) more efficient means of fertilization (adult reproductive behavior), (3) parthenogenesis and asexual reproduction, (4) more efficient protection of infective agents against unfavorable features of the environment, (5) changes in life cycle patterns, and (6) changes in the behavior of infective stages. We used this perspective to establish six categories of potential key innovations in parasite groups: adult morphology pertaining to reproductive activities (A_1); adult morphology not pertaining to reproductive activities (predominantly traits having to do with maintaining location within a host and feeding (A_2); modification of ontogenetic stages (O_1); origin of new ontogenetic stages (O_2); loss of old ontogenetic stages (O_3); and ecological and behavioral characteristics (E). Using these categories, we have (1) compared three major groups of parasites: the Sporozoea, the major groups of the Nematoda, and the Cercomeria; (2) performed comparisons of sister groups within the Cercomeria; and (3) examined particular groups of cercomerians in some detail, searching for the synapomorphies that (a) are correlated with species richness, (b) confer a fitness advantage when compared with the ancestral condition, and (c) may have been associated with an increase in adaptive modes of speciation in the species-rich clade.

Comparisons among Three Major Parasite Groups

The Sporozoea, Nematoda, and Cercomeria are not sister groups. Given that we are advocating the examination of adaptations, and adaptive radiations, within a phylogenetic framework, why are we comparing these distantly related groups? The answer to this question is simple: because we are searching for some generalizations about parasite adaptive radiations. Since "parasites" are not a monophyletic group, we have to compare distantly related clades. However, if "parasites as ecological entities" represents some sort of "real" evolutionary group, then we would expect to find some common convergent trends.

As one can see from Tables 3.9–3.11, it is difficult to extract any generalities about "parasite" evolution from the evolutionary diversification of our six character classes. Given that we had such a difficult time defining "parasite" in the first place, this is hardly an unexpected result. In general, we can say that: (1) Most of the character diversification occurs in adult morphology, with

Table 3.9
Potential correlates of adaptive radiation in major sporozoean groups

Taxon	Adult reproductive	Adult nonreproductive	Development (modification of stages)	Development (origin of stages)	Development (loss of stages)	Ecological
Lankesteria + Schizocystis	0	1	0	0	0	0
Lankesteria	0	0	1	0	0	0
Schizocystis	0	0	0	0	1	0
Group 2	2	0	1	0	0	0
Karyolysus	1	0	0	0	1	0
Haemogregarina/Cyrilia + Babesiosoma + Theileria/Babesia	2	0	1	0	1	0
Babesiosoma + Theileria/Babesia	1	0	1	0	0	0
Theileria/Babesia	0	5	0	0	0	0
Group 3	3	0	0	1	0	0
Coelotropha	0	0	0	0	1	0
Plasmodium	2	2	1	0	1	0
Aggregata + rest	2	0	0	0	0	0
Schellackia + rest	0	3	0	0	0	0
Schellackia	0	1	0	0	1	0
Sarcocystis + Eimeria + Cryptosporidium	0	1	0	0	0	0
Sarcocystis + Eimeria	0	0	1	0	0	0
Cryptosporidium	1	1	0	0	0	0

Note: See Appendix Figure 1. The total number of character changes (apomorphies) associated with the origin of each group is listed according to character category.

ontogenetic and ecological changes conservative by comparison. This result parallels a general evolutionary trend that has been observed in a variety of free-living groups; it is not unique to parasites. (2) On any given branch there are never changes in more than four catagories. It would appear, then, that the characters within these six categories are not acting as cohesive evolutionary units. The mean number of changes in a category per branch is 1.8 for sporozoans, 1.9 for nematodes, and 2.9 for cercomerians. At the moment, these mean

Table 3.10
Correlates of adaptive radiation in major parasitic nematode groups

Taxon	Adult reproductive	Adult nonreproductive	Development (modification of stages)	Development (origin of stages)	Development (loss of stages)	Ecological
Nematoda	4	4	0	1	0	0
Enoplimorpha	0	2	0	0	0	0
Enoplia	0	1	0	0	0	1
Dorylaimia	0	3	0	0	0	0
Muspiceida	1	1	0	0	0	0
Mermithida + Trichurida + Dioctophymida	0	1	0	0	0	0
Mermithida	0	1	1	0	0	2
Trichurida + Dioctophymida	2	0	0	0	0	0
Dioctophymatida	2	0	0	0	0	0
Chromadorimorpha	0	2	0	0	0	0
Monhysteria	1	0	0	0	0	0
Chromadoria	0	1	0	0	0	0
Phasmidia	1	4	0	0	0	0
Tylenchia	0	1	0	0	0	3
Rhabditia	1	0	0	0	0	1
Oxyurida	5	0	1	0	0	2 + 1
Strongylida	1	0	1	0	0	0 + 1
Rhabdiasida	1	0	0	1	0	0 + 1
Strongyloidina	1	0	0	0	0	0 + 1
Drilonematida	0	1	0	0	0	1
Ascaridida	0	1	0	0	0	0 + 1
Spirurida	0	2	0	0	0	1 + 1
Rhigonematida	1	1	0	0	0	1

Note: See Appendix Figure 74. The total number of character changes (apomorphies) associated with the origin of each group is listed according to character category. (The notation "+1" indicates the taxa that parasitize vertebrates; there may have been a single shift to vertebrates, or there may have been multiple shifts, but until the relationships among the major groups of vertebrate parasitic nematodes are better resolved, we will not know.)

Table 3.11
Correlates of adaptive radiation in major cercomerian platyhelminth groups

Taxon	Adult reproductive	Adult nonreproductive	Development (modification of stages)	Development (origin of stages)	Development (loss of stages)	Ecological
Cercomeria	4	7	1	0	0	2
Temnocephalidea	0	2	0	0	0	0
Neodermata	4	2	5	0	0	0
Udonellidea	0	2	0	0	0	1
Cercomeridea	2	2	0	0	0	2
Trematoda	5	3	1	0	0	1
Aspidobothrea	2	4	0	0	0	0
Digenea	2	0	0	16	0	0
Cercomeromorphae	0	2	2	0	0	0
Monogenea	6	4	0	5	0	2
Cestodaria	6	7	4	0	0	0
Gyrocotylidea	4	8	1	0	0	0
Cestoidea	2	2	4	0	0	0
Amphilinidea	5	2	0	0	0	1
Eucestoda	1	3	8	2	0	0

Note: See Appendix Figure 3. The total number of character changes (apomorphies) associated with the origin of each group is listed according to character category.

values are just interesting numbers because they reflect, in part, gaps in our knowledge for a variety of groups. (3) The metazoans do not lose developmental stages. Sporozoans, on the other hand, show six losses, including a putative loss of merogony, an asexual phase (which may be an artifact of our incomplete knowledge: J. Barta, S. Desser, and M. Siddall, personal communication).

Overall, then, there really is no such thing as a generalized parasite biology with respect to questions of adaptive radiation, only diverse biologies exhibited by the species we call parasites. This means that we must do a group-by-group analysis of adaptive radiations in order to uncover the key innovations and speciation processes involved in the radiations of each clade. At the moment, the only group for which we have a substantial data base for comparisons is the Cercomeridea.

Comparisons between Cercomerian Sister Groups

Setting the Stage: Life Cycle Patterns within the Cercomeridea

The association between vertebrates and cercomerideans appears to have begun a long time ago, possibly very shortly after the first (agnathan) vertebrates evolved, but almost certainly early in the evolution of the placoderms. The relationships depicted in Figure 3.9 suggest that the stem diversification of the Trematoda, the Monogenea, and the Cestodaria (gyrocotylideans, amphilinideans, and eucestodes) occurred in association with placoderm groups prior to the divergence of chondrichthyans from the rest of the gnathostomous vertebrates (actinopterygians, sarcopterygians, and tetrapods). Subsequent to that early diversification, the divergences of the aspidobothreans from the digeneans, of the gyrocotylideans from the cestoideans (amphilinideans plus eucestodes), and possibly of various monogenean groups are all correlated with the divergence of the chondrichthyans from the ostracoderm ancestor giving rise to the rest of the gnathostomous vertebrates. Hence, there is evidence of

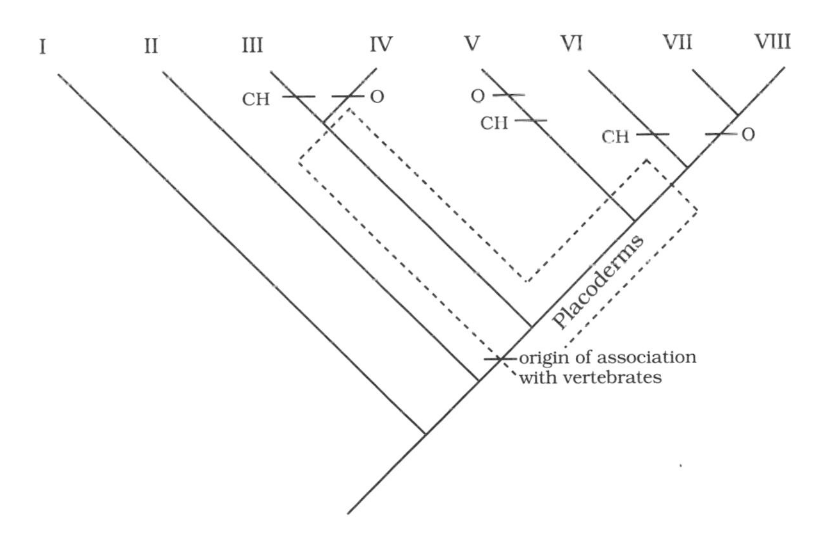

Figure 3.9. Phylogenetic patterns of diversification in vertebrate host groups inhabited by stem groups of major groups of cercomerian platyhelminthes. I = Temnocephala; II = Udonellidea; III = Aspidobothrea; IV = Digenea; V = Monogenea; VI = Gyrocotylidea; VII = Amphilinidea; VIII = Eucestoda. Host abbreviations are: CH = chondrichthyans; O = ostracoderms (to indicate the ancestor of all noncondrichthyan gnathostomous vertebrates). Note that the association between cercomerians and vertebrates occurred early in vertebrate evolution, with extensive radiation by the flatworms in the placoderms. Redrawn and modified from Brooks (1989b).

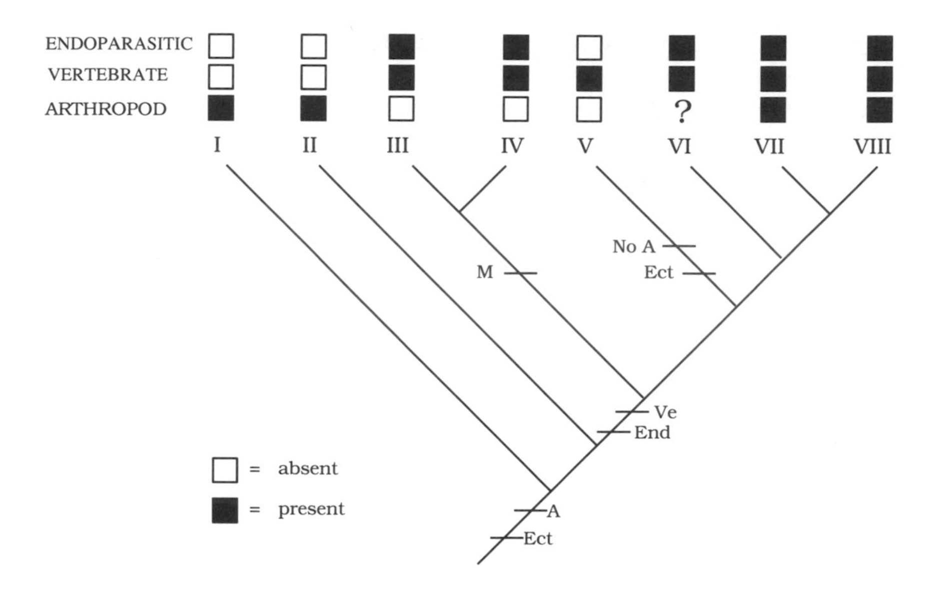

Figure 3.10. Phylogenetic patterns of life cycle diversification among major groups of cercomerians. I = Temnocephala; II = Udonellidea; III = Aspidobothrea; IV = Digenea; V = Monogenea; VI = Gyrocotylidea; VII = Amphilinidea; VIII = Eucestoda. Boxes above taxa indicate the distribution of traits for three components of life cycle patterns. Slash marks on tree, and accompanying abbreviations, summarize the data presented in the boxes phylogenetically. A = arthropod host acquired (primitive one-host ectoparasitic life cycle); Ve = vertebrate host acquired (primitive two-host endoparasitic life cycle); M = molluscan host acquired in exchange for the arthropod (derived two-host life cycle); No A = arthropod host lost (derived one-host life cycle); Ect = ectoparasitic adult; End = endoparasitic adult. From Brooks and McLennan (1991).

long-standing conservative association with particular vertebrate host groups in addition to the conservatism in diversification of life cycle patterns.

Within the Cercomeridea, the simplest life cycle is displayed by the Monogenea. The majority of adult monogeneans live on the exterior of their vertebrate hosts (a small number live in the urinary bladder and tract or in the stomach) and transmit their offspring directly to another vertebrate, where they mature and begin the cycle again. It has often been considered axiomatic among parasitologists that simple life cycles are more primitive than complex life cycles. If this is true, then the one-host/direct transmission pattern displayed by monogeneans should be the plesiomorphic pattern among flatworms parasitizing vertebrates. Given this expectation, we have to postulate that the addition of intermediate hosts and the appearance of endoparasitic modes of life have been independently derived by the digeneans and the tapeworms.

Figure 3.10 suggests an alternative interpretation. To begin with, the plesiomorphic life cycle pattern for all the parasitic flatworms appears to be one in which an arthropod is used as the only host by an ectoparasitic species (see the Temnocephalidea and Udonellidea in Figure 3.10). The pattern became more complicated in the ancestor of the cercomerideans as a vertebrate host was added and the adult parasites became endoparasitic (this pattern presumably arose as a result of predation by early vertebrates on crustaceans). At this level, then, the basal life cycle pattern involves an arthropod intermediate host plus a vertebrate final host, with the adult parasite living endoparasitically in the vertebrate. The current information on life cycle patterns in the Cestodaria (Gyrocotylidea + Amphilinidea + Eucestoda) suggests that most of them have *retained* this primitive life cycle pattern. The trematodes display one variation on this central life cycle theme; in their ancestor, a molluscan host was substituted for an arthropod host. Now where do the monogeneans fit? Figure 3.10 suggests that the monogeneans have a highly derived life cycle pattern, in

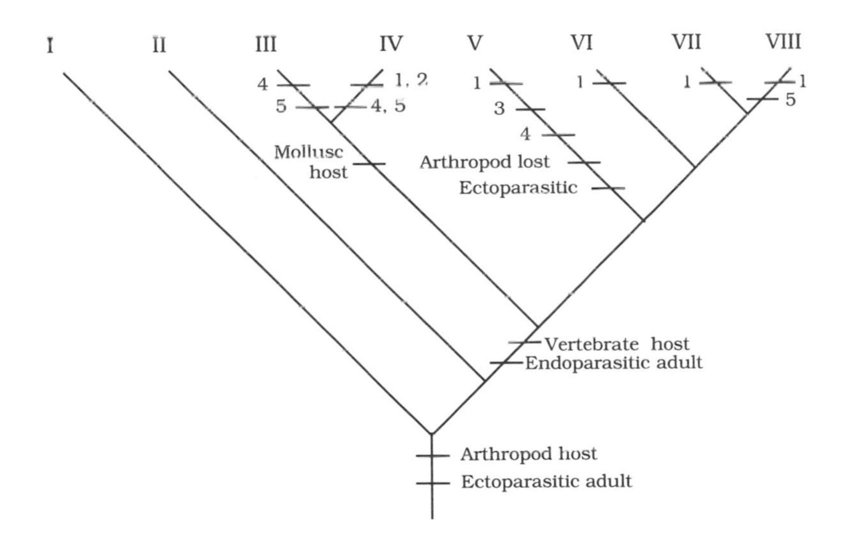

Figure 3.11. Summary of phylogenetic correlates of ontogenetic and ecological diversification among the major groups of cercomerians. I = Temnocephala; II = Udonellidea; III = Aspidobothrea; IV = Digenea; V = Monogenea; VI = Gyrocotylidea; VII = Amphilinidea; VIII = Eucestoda. Numbers refer to categories of developmental characters: 1 = unique larval or juvenile forms; 2 = asexual proliferation of larval stages; 3 = polyembryony; 4 = peramorphic heterochrony; 5 = paedomorphic heterochrony (for a summary of the actual characters for each group see Brooks 1989). Note phylogenetic constraints on both ontogenetic and ecological change, and ecological conservatism relative to developmental diversification. Redrawn and modified from Brooks (1989b).

which both the arthropod intermediate host and the endoparasitic life style have been lost. Virtually every discussion of the evolution of life cycle patterns in parasitic flatworms has assumed that the life cycle pattern displayed by monogeneans is primitively simple. The discovery that these flatworms display a *secondarily* simplified life cycle pattern challenges us, yet again, to rethink long-established assumptions and scenarios about parasite evolution.

Figure 3.11 depicts the covariation of developmental and ecological correlates of phylogeny for the major cercomerian groups, and clearly shows the conservative nature of the ecological diversification relative to the developmental diversification organized by phylogenetic constraints. The figure also delineates the developmental and ecological changes that are correlated with the origin of each cercomeridean lineage. Could any of these changes be key adaptations? In order to answer this question we must compare sister groups within the Cercomeridea, searching for two pieces of evidence. First, we need to identify sister groups that show differential species richness. In essence, we are reaching into all the possible sister group pairs within the Cercomeridea and extracting the groups that have experienced a radiation. Second, we need to identify the character changes that are correlated with species richness (i.e., are found in the species-rich group and not in its sister group). This second step allows us to reach into all the groups that have experienced a radiation and extract the ones that have also experienced an adaptive radiation.

Correlates of Species Richness: Comparisons of Sister Groups

Dial and Marzluff (1989) examined the distribution of species within 85 different groups of organisms. They discovered that, on average, 36% of the species in each group were concentrated within one clade. The three groups displaying the derived species number within the Platyhelminthes greatly exceed this median value when compared with their sister groups: the digeneans (flukes) comprise 99.1% of the known trematode species, the monogeneans comprise 66.5% of the cercomeridean species, and the eucestodes (tapeworms) comprise 99.8% of the cestodarian species. It appears, then, that these groups represent cases of radiation within the Cercomeria. Given this conclusion, is there any evidence that these radiations are adaptive?

Table 3.12 summarizes the distribution of apomorphies among the character categories for the sister groups within the Cercomeria that demonstrate differential species richness. There is no significant difference between species-rich groups and their sister groups in the origin of adult reproductive and nonreproductive characters; the patterns depicted for each sister group pairing represent a unique mosaic of conservative and derived characters. We will begin our search for key innovations within the character categories that exhibit apo-

Table 3.12
Identifying correlates of species richness in major parasitic platyhelminth groups

	Taxon					
Character category	Digenea (5000+ spp.)	Aspidobothrea (50+ spp.)	Monogenea (5000+ spp.)	Cestodaria (2518+ spp.)	Eucestoda (2500+ spp.)	Amphilinidea (8 spp.)
Adult reproductive	2	1	6	6	3	5
Adult nonreproductive	0	4	5	7	2	2
Development (modification)	13	0	4	4	8	0
Development (origin of stages)	3	0	1	0	2	0
Development (loss of stages)	0	0	0	0	0	0
Ecological	0	0	2	0	0	1

Note: The total number of character changes (apomorphies) associated with the origin of each group is listed according to character category. For example, the Digenea are distinguished by the presence of 18 autapomorphies: 3 origins of new developmental stages, 13 modifications of developmental stages and 2 changes in adult reproductive characters.

morphic changes in the species-rich groups and not in the species-poor groups. For example, apomorphies based upon modifications of larval characters typify the Digenea and Eucestoda. Changes in larval characters are absent from their respective sister groups. The monogeneans, on the other hand, are characterized by changes in ecological characters that are not matched in their cestodarian sister group. As mentioned earlier, however, there is no a priori way to determine which, if any, of the apomorphies displayed by each species-rich group is *the* innovation, so it is possible that the key innovation will reside in a character category that does not differ in the absolute number of synapomorphic changes demonstrated by a species-rich group and its sister group.

The Roots of Evolutionary Radiation among the Cercomerideans: The Species-Rich Groups

Monogenea The most obvious difference between the species-rich Monogenea and their sister group, the Cestodaria, is the appearance of two novel ecological traits in the monogeneans (Table 3.12). Monogeneans have lost the arthropod intermediate host that is plesiomorphic for the Cercomeridea, resulting in a direct life cycle using a vertebrate host, and have returned to the ancestral condition of ectoparasitism. A role for ectoparasitism and a direct life cycle as key innovations, *on their own,* can be ruled out because the species-poor groups Temnocephalidea and Udonellidea possess these characters. Temnocephalideans and udonellideans, however, display the characters in the plesiomorphic context of an arthropod host, whereas the monogeneans display the characters in the apomorphic context of a vertebrate host. Kearn (1986) discussed studies demonstrating that the transmission of monogeneans, either via the free-swimming oncomiracidium or by direct transfer from host to host, was enhanced by various aspects of the host's behavior (e.g., concentrations of host populations during spawning or during parental care). It appears, then, that if there is a fitness advantage conferred by these two ecological changes, it occurs because the monogeneans inhabit a derived host type relative to other parasitic platyhelminths having direct (single-host) life cycles.

The putative key innovation displayed by the monogeneans is thus a suite of traits, involving a complex interaction among three characters: "vertebrate host" (plesiomorphic for the monogeneans), "ectoparasitism" (apomorphic for the monogeneans) and "direct life cycle" (apomorphic for the monogeneans). There is currently little information available about the adaptive value, if any, of ectoparasitism over endoparasitism, so its potential role as a "key adaptation" or as part of a "key adaptive complex" cannot be evaluated. The direct life cycles in monogeneans result from the deletion of the arthropod intermediate host. Three different heterochronic changes (acceleration, postdisplacement, or progenesis) can result in precocial sexual maturation leading to deletion of an intermediate host. Thus, the characterization "direct life cycle" for the monogeneans may better be considered a description of the ecological effect of precocious sexual maturation. If so, it is the "change in developmental rate," not the "loss of the arthropod host with subsequent reduction to a direct life cycle" that should be considered the putative key innovation. Heterochronic changes in maturation rate originating in the ancestor of the monogeneans continued to play a role in character change throughout the evolutionary diversification of the group, leading ultimately to the appearance of a variety of autoinfective strategies, including reinfection of a host by a nonswimming oncomiracidium, ovoviviparity, and viviparity.

Studies have demonstrated that autoinfection has a positive effect upon the reproductive success of the bearer (see Tinsley 1983; Kearn 1986 and references therein). First, delaying hatching until embryos are well developed decreases the hazards of development in the external environment. This hazard is a substantial one. Wilson et al. (1982) estimated that of 100,000 *Fasciola hepatica* eggs deposited in a pasture from the middle of winter to the end of summer, only 17 would hatch. Second, remaining on the same host (resource patch) decreases the riskiness involved in parasite transfer involving a free-swimming but short-lived larva (Llewellyn 1968, 1981). Tinsley (1982, 1983) compared the reproductive success of two species of monogeneans, *Polystoma integerrimum* and *Pseudodiplorchis americanus*. *P. integerrimum* displays the ancestral monogenean life cycle "deposit eggs in the environment, ciliated oncomiracidia hatch and swim to vertebrate host." *P. americanus* displays the derived strategy of ovoviviparity. Tinsley compared a number of variables, including details of the life cycle, number of eggs and fully developed larvae produced, metabolic demands of reproduction, and prevalence and intensity of each species in its host. He concluded that, assuming no mortality, a single adult *P. americanus* could potentially produce 3.9×10^9 adults after four years, compared with a production of 2.5×10^3 adults for *P. integerrimum*. An enormous increase in fecundity is obviously associated with ovoviviparity when compared to the plesiomorphic condition. We can conclude, therefore, that the origin of a change in development in the context of a derived (vertebrate) definitive host is adaptive within the monogeneans.

Is the association between the monogenean radiation and the developmental change more than just a spurious correlation? From a theoretical perspective, the answer to this is yes, because the subsequent changes in life history characteristics have a direct effect on deme structure. Parasite demes have a tendency to be ephemeral. In the vast majority of cases, they must be reassembled each generation by the relatively random immigration of larval and/or juvenile stages, usually from a much larger gene pool (the parasite species population occurring in multiple hosts) than that represented by the members of the original deme. Parasites that autoinfect their individual hosts and inhabit long-lived hosts can produce more than one generation on or in the same host organism. This reduces the ephemerality of the deme structure, increasing the potential that differences appearing within a deme could be maintained by inbreeding. Tinsley (1983: 174) hypothesized that such inbreeding could "promote rapid increase in gene frequency and potential fixing of beneficial variation, and enable tracking of host and other environmental variations." The monogeneans are thus characterized by the appearance of an adaptive reproductive strategy that increases the opportunity for sympatric speciation to occur. In addition, the ability to produce a viable deme from one "pregnant" individual increases the likelihood that peripheral isolates speciation by host switching will occur.

Have adaptive modes of speciation occurred frequently in the monogeneans? Based on an analysis of 66 documented cases of vertebrate speciation,

Lynch (1989) suggested that 21% of the speciation events were due to adaptive modes of speciation (peripheral isolates allopatric speciation and sympatric speciation). Phylogenetic analyses of speciation in monogeneans have revealed that, depending upon the group, the combined frequency of peripheral isolates speciation and sympatric speciation ranged from 22% to 100%. The median value of 56% is much higher than the value reported in Lynch's study, supporting the conclusion that monogeneans exhibit a high rate of adaptively driven speciation. The hypothesis that the evolution of direct life cycles and autoinfection influences the rates of adaptively driven speciation is supported by an additional piece of evidence. Within the monogeneans, one of the most species-rich groups, the gyrodactylids, displays the extreme condition of this developmental trend, viviparity.

Overall, then, the monogeneans display the key adaptive complex "change in developmental rate of sexual maturity leading to the appearance of a direct life cycle and a variety of autoinfective strategies within the context of a derived definitive host." This change has a positive effect on the fitness of its bearers, and enhances the potential for sympatric speciation and speciation by peripheral isolation (host switching) to occur. Not surprisingly, adaptively driven modes of speciation occur more often in the monogeneans than in free-living groups. This, we believe, is a strong example of an adaptive radiation. Of course, the story does not stop here. Tinsley (1983) discussed a variety of novel characters that further enhanced the adaptive advantage of delayed development (e.g., uterine elongation). We cannot assess the putative adaptive values of these characters, nor can we examine their effects, if any, on speciation rates until detailed phylogenies are available for the monogenean families. Nevertheless, this discussion highlights the important and often overlooked fact that "key innovations," be they characters or complexes of characters, are themselves subject to evolutionary change. It is therefore important to trace the diversification of the innovation, in terms of both changes in the character itself and changes in other characters that may enhance or decrease the adaptive value of the innovation.

Eucestoda Unlike monogeneans, eucestodes originated within the plesiomorphic ecological context for the Cercomeridea (i.e., arthropod intermediate host, vertebrate definitive host). The major difference between the tapeworms and the amphilinids lies in modifications to larval stages (Table 3.12). Although most of these modifications appear relatively minor on an intuitive level (see Raikow 1986 for a similar discussion of passerine evolution), there is not enough information available to test hypotheses about their adaptive significance (Fitzpatrick 1988). The major correlate of the eucestode radiation for which any experimental data have been collected is the proliferation of adult reproductive structures (proglottids). The increase in sexual reproductive output for the tapeworms is dramatic. The average daily reproductive output of 100 eggs per worm in the Monogenea, and even the exceptional output of 24,000 eggs per worm in some of the larger digeneans (Tinsley 1983), pale by comparison to the 720,000 eggs expelled per day by the human beef tapeworm, *Taenia saginata* (Crompton and Joyner 1980). The lifetime fecundity of a tapeworm is much greater than the lifetime fecundity of an amphilinid (sister group bearing the plesiomorphic condition "lack of proglottids"), so the appearance of this new reproductive strategy can be assigned an adaptive status.

The proliferation of dispersal stages created by the new strategy increases the likelihood that a larva will find a suitable host, decreasing the chance that a given population will go extinct (Moore 1981 and references therein). This, in turn, has an indirect effect on speciation because it allows a species to persist long enough to encounter a variety of speciation-causing factors. These factors, however, need not fall into the domain of adaptively driven speciation; it is equally likely that they will have something to do with the long-term survival and geographic spread of species, increasing the likelihood of vicariant speciation. So, although the persistence of an individual's genealogical lineage is adaptive from that individual's perspective, it does not necessarily lead to an enhanced potential for adaptive speciation modes.

Proliferation of dispersal stages should have a more direct effect on speciation by increasing the probability that peripheral isolates allopatric speciation via host switching will occur. This hypothesis can be tested by comparing the frequency of host switching between the Eucestoda and its sister group, the Amphilinidea. At the moment, only one study of speciation within a phylogenetic framework is available for the tapeworms, so the hypothesis cannot be tested rigorously. Nevertheless, the data are tantalizing; 13% host switching in the amphilinids (2/15 speciation events: Bandoni and Brooks 1987b) versus 40% in the tapeworm group (6/15 speciation events: Hoberg 1986). Preliminary studies with other tapeworm groups also indicate that host switching has been widespread throughout the group (Moore and Brooks 1987). However, it is difficult to determine whether these results are more reflective of rampant host switching or of poorly resolved parasite phylogenies because the analysis produced multiple trees with a low consistency index.

Circumstantial support for the direct effect of an increase in dispersal stages on speciation rates is also found by investigating groups within the Eucestoda that have secondarily lost their segmentation, reverting to the plesiomorphic state "one set of reproductive organs per adult." One such group is the Caryophyllidea, monozoic tapeworms inhabiting the intestines of freshwater fish (predominantly ostariophysans) and most closely related to the species-rich diphyllobothriid pseudophyllideans (Brooks et al. 1991). Caryophyllideans are widespread, parasitize a species-rich host group, and are presumably quite old (Mackiewicz 1981); however, they have not enjoyed high speciation rates (approximately 111 known species). Mackiewicz (1981) attributed this to their low reproductive potential relative to polyzoic cestodes, and characterized them as an evolutionary dead end. Another example is found in the order Aporidea, parasites in the intestines of ducks and swans that are most closely related to the species-rich cyclophyllideans. Although not closely related to caryophyllideans, aporideans share two common traits with those tapeworms: low species number (fewer than six species in the order) and the monozoic life style.

In summary, the appearance of segmentation in the ancestor of the Eucestoda is associated with increased fecundity (and thus presumably also increased potential for speciation by host switching), and an increase in the frequency of that speciation mode compared to its frequency in the amphilinideans and in free-living groups. Within the Eucestoda, the ability to produce replicated sets of reproductive organs has been independently lost at least twice in the Caryophyllidea and the Aporidea. Both of these groups are species poor relative to their sister groups. In order to provide stronger support for the hypothesis that the character "polyzoic body" is a key innovation, we need better estimates of the role that peripheral isolates speciation has played in tapeworm diversification. If it has been widespread throughout the group, or if it is always associated with species-rich clades within the Eucestoda (e.g., the two most species-rich groups within the Eucestoda, the Tetraphyllidea and the Cyclophyllidea, originated following a host switch: see Brooks et al. 1991), then we can assign the new reproductive strategy the role of a key innovation. At the moment, we can only conclude that the hypothesis is interesting, but requires substantial input from speciation studies within a phylogenetic framework before it can be corroborated or discarded.

Digenea Like the eucestodes, the digeneans display an apomorphic increase in the number of dispersing larvae. Unlike the tapeworms, in which the larvae are products of sexual amplification, digeneans exhibit a strategy that is characterized by the origin of a completely new dispersal stage, the cercaria, and the origin of at least one, and generally two, phases of asexual amplification of larval stages between the miracidium and cercaria (the sporocyst and redia). Consequently, a single miracidium can produce 1000 or more infective cercariae. This is another example of a putative "key innovation" that is really an "innovation complex," the origin of a new free-swimming dispersal stage and of asexual reproductive phases intercalated in ontogeny.

The cercarial stage has also been perceived as an adaptation for enhancing the probability of infection, in this case not by amplifying the numbers of infective larvae but by dispersing those larvae spatially to the greatest extent possible, thereby increasing the likelihood of contacting a suitable host. We can evaluate that hypothesis by examining relative species richness among groups within the Digenea whose cercariae are not free-swimming. Some of those groups are species poor (Heronimidae, Troglotrematidae, perhaps the Ptychogonimidae), some exhibit moderate species richness (Gymnophallidae, Brachylaimidae), and others are quite species rich (the Opecoelidae + Zoogonidae + Lissorchiidae, which comprise the superfamily Opecoeloidea, and the Dicrocoeliidae). If the cercarial stage has adaptive significance in the evolution radiation of the Digenea, therefore, it is not spatial dispersion by swimming.

Alternatively, it has been postulated that the significance of the cercarial stage is tied to the evolutionary diversification of cercarial behaviors that opened up a broad spectrum of definitive hosts to infection by digeneans. The evolutionary diversification of cercarial behaviors cannot be assigned an adaptive status without a demonstration that each apomorphic change in cercarial behavior (e.g., from encysting on vegetation to encysting in a fish) conferred a fitness advantage to its bearers over organisms displaying the plesiomorphic behavior (Coddington 1988). We expect that this will be difficult to demonstrate because changes in cercarial behavior often result in the infection of new types of second intermediate hosts, which are eaten by new types of vertebrate definitive hosts. Although the origin of these novel behaviors could theoretically decrease competition for the plesiomorphic host resource, this would represent adaptive change only if the plesiomorphic host was a limiting resource. Given that the plesiomorphic cercarial behavior is encystment on vegetation, on exoskeletons of mollucs and arthropods, and in the open, it is difficult for us to conceive that the plesiomorphic resource was ever limiting.

The origin of asexual amplification is clearly an adaptation because it increases the reproductive output of each adult. This amplification should always have an indirect effect on speciation because, as discussed for the Eucestoda, it decreases the probability of population extinction, thereby increasing the chances for geographic spread and subsequent vicariance. It may also have a more direct effect on speciation by increasing the probability of peripheral isolates speciation by host switching. As with the eucestodes, however, we do not have enough information on the actual role that adaptive modes of speciation have played in the diversification of the Digenea, so we cannot yet determine whether their species richness is due to longevity or to accelerated speciation rates.

In summary, then, although the digeneans have experienced an evolutionary radiation, there is little concrete evidence that the radiation has been adaptive. Asexual amplification should increase the chances for speciation by host switching. If this has been the case, then the digeneans should show a significantly higher frequency of peripheral isolates speciation than the aspidobothreans, their sister group. Asexual amplification has arisen independently in another parasite group, the species-rich Taeniidae within the Eucestoda. Unfortunately, since many taeniids are not capable of asexual propagation (Moore 1981), and since there is no robust estimate of phylogenetic relationships within the group, we cannot determine whether asexual propagation is correlated with the taeniids' species richness. This will prove an interesting area for future research.

Comparisons among Species-Rich Groups: Evidence from Convergent Characters

One of the major problems encountered in studies of adaptive radiations is that, in many cases, such radiations may represent historically unique events. From a statistical perspective, a strong test of an adaptive radiation hypothesis is provided by the discovery of a convergent, adaptive trait correlated with species richness in more than one group (Coddington 1988). Although nature is rarely so obliging, one example exists in the parasitic platyhelminths.

Three species-rich groups, the Cyclophyllidea within the Eucestoda (representing virtually all of the tapeworms inhabiting terrestrial vertebrates and the majority of tapeworms occurring in birds and mammals) and the Opisthorchiformes and the Plagiorchiformes (the most species-rich group) within the Digenea, share the derived condition "larvae remain in the egg (extended intrauterine development) until ingested by the first intermediate host (delayed hatching)." This character has arisen independently in all of these lineages. Once again, we have evidence of a change in development, in this case an increase in the duration of intrauterine development, associated with the radiation of a group. These developmental changes have allowed the Cyclophyllidea to exploit a previously untapped resource—terrestrial arthropods and vertebrates. If we consider the movement into this new habitat as the "selective regime" (sensu Baum and Larson 1991), then the retention of the infective larval stage within the eggs until ingestion by the intermediate host provides a substantial performance benefit over the plesiomorphic condition of hatching in water and swimming in search of that host.

Several theories about the adaptive value of this strategy in the digeneans have been proposed, including avoiding the hazards of development in the external environment, promoting invasion of the intermediate host at the earliest opportunity, and providing protection for the embryo during the passage through the digestive system of the definitive host and through any unsuitable host that eats the eggs (Tinsley 1983). We now have three independent tests of the hypothesized adaptive value of this developmental change and of its involvement in the radiation of the groups that possess it. Indeed, when we also consider the basic biology of metacercarial stages in digeneans, the plerocercoid (and derivative) stage in eucestodes, dauerlarvae in nematodes, and cystacanths in acanthocephalans, this change would seem to offer a graphic example that "an important adaptation to parasitism is the suspension of development in the infective stage" (Rogers 1962). These conclusions will be strengthened by the demonstration that adaptively driven modes of speciation are widespread within these three groups.

The Roots of Evolutionary Radiation among the Cercomerideans: The Species-Poor Groups

We now turn to the three cercomeridean taxa that are the species-poor relatives of the Digenea, Monogenea, and Eucestoda. Simpson (1944) was among the first modern evolutionary biologists to consider general explanations for groups of unusually low species numbers. He considered all such groups *relicts* of one form or another and postulated that different processes could produce different kinds of relictual groups. We will be concerned with two major types of relicts. *Phylogenetic relicts* are "living fossils," members of groups that have existed for a long time without speciating very much. Such low speciation rates could result from phylogenetic or developmental constraints on phenotypic diversification, and/or from unusually pronounced ecological specialization (i.e., ecological constraints owing to the effects of strong, long-term stabilizing selection from the specialized habitat). *Numerical relicts,* by contrast, are the surviving members of once more species-rich groups whose ranks have been depleted by extinction.

Brooks and Bandoni (1988) suggested that a combination of phylogenetic, biogeographical, and ecological information could be used to distinguish between phylogenetic and numerical relicts among parasites. Establishing a group's relictual status first requires methods for determining that the group is old enough to be highly diverse. A number of methods are available for estimating the ages of clades, including molecular clock criteria, paleonotological data, and biogeographical analysis (see Chapter 2). Second, it must be established that the group is in fact unusually depauperate. This can be established by comparing sister groups (Mayden 1986).

Brooks and Bandoni further suggested that phylogenetic relicts should be ecologically conservative, whereas numerical relicts should be ecologically diverse. For example, Crocodilians (Crocodilia) are the sister group of the species-rich clade the birds (Aves). Living crocodilians, numbering about 22 species, inhabit a variety of estuarine to freshwater habitats throughout the tropical and subtropical regions of the world. They prey on a wide variety of vertebrates and some invertebrates. The fossil record indicates that crocodilians were once a species-rich group, including many fully marine species; in addition, the earliest known crocodilian fossils suggest a terrestrial origin for the group. Hence, the current diversity of crocodilians represents only a fraction of the species number and ecological diversity once encompassed by the group, so we consider crocodilians to be numerical relicts. As we discussed in Chapter 2, the helminth parasite fauna of crocodilians also appears to comprise numerical relicts.

Now consider the ratfish (holocephalans), the sister group of sharks, skates, and stingrays (elasmobranchs). There are 25 species of ratfish, compared with approximately 625 species of elasmobranchs. Ratfish occur worldwide in mid- to deep-water marine habitats, and forage on benthic invertebrates. The fossil record indicates that ratfish have been in existence for a considerable period of time, but have never been highly diverse. Both the fossil evidence and the ecological homogeneity of contemporaneous species suggest that ratfish are phylogenetic relicts.

The species-poor cercomerideans are all characterized by conservative developmental and ecological components of their life cycles. Consequently, we would consider them all to be phylogenetic relicts. There are, however, at least two types of phylogenetic relicts in parasite-host systems. The first type involves cases in which one of the associates becomes highly diverse while the other does not. In the second type, neither of the associated groups ever becomes very diverse.

Aspidobothrea Aspidobothreans are an example of a group that did not become as diverse as its host group. Phylogenetic analysis (Brooks et al. 1989a) and comparison with the phylogenetic relationships of their vertebrate hosts indicate that aspidobothreans have been around at least as long as the common ancestor of chondrichthyans and the rest of the gnathostome vertebrates. Both aspidobothreans and digeneans share an ancestral life cycle pattern involving a molluscan and a vertebrate host. One obvious difference between the groups is species richness; there are fewer than 50 described species of aspidobothreans and more than 5000 species of digeneans. Why this disparity?

In the aspidobothreans, larvae hatch from eggs and develop directly into juveniles in the molluscan host, are then ingested by a molluscan-eating vertebrate, and develop to the adult state. Hence, each embryo can potentially give rise to only a single adult. As we discussed earlier, digeneans are characterized by a series of complex developmental stages in the molluscan host, at least one (and usually two) of which produce a large number of cloned larvae or juveniles (depending on the species and the stage). It would appear, then, that the life cycle characters possessed by aspidobothreans have allowed them to persist over a very long period of time, but have not created the potential for increasing speciation rates. In this regard, they are similar to acanthocephalans, another group of helminth parasites with an apparently long history and relatively low diversity.

Amphilinidea and Gyrocotylidea Like the aspidobothreans, amphilinideans do not inhabit hosts that are themselves phylogenetic relicts; therefore, they are an example of a parasite group that has not become as diverse as its host group. Amphilinideans display an ecological synapomorphy. Rather than inhabiting the intestine (plesiomorphic for the Cercomeridea), all species occur as adults in the body cavity of their (primarily freshwater actinopterygian) hosts. Perhaps living in the body cavity of fishes has kept amphilinidean diversity low, since amphilinideans are relatively large and thus restricted to relatively large hosts. Additionally, amphilinidean eggs cannot reach the outside environment from the body cavity of all fishes, further restricting their range of suitable hosts.

The gyrocotylideans, on the other hand, are prime candidates for the second type of phylogenetic relict. They are ecologically conservative, being restricted as adults entirely to the spiral intestines of chimaeroid fishes, and are found in association with hosts that are themselves phylogenetic relicts. Like their hosts, the gyrocotylideans are less species rich than their sister group (in this case the amphilinideans plus the tapeworms). This finding suggests the possibility that the interaction between developmental conservatism and a conservative and highly specialized ecology limited diversity.

Overview of Cercomerian Radiation

The concept of adaptive radiation has come to mean many different things. To some researchers, it has been virtually synonymous with speciation. To others, it involves an association between overall diversification and adaptive changes in ecological and behavioral characters, as well as a high degree of homoplasious phenotypic change. And to still other researchers, speciation has dropped out of the picture and the term has become associated with character diversification. Recognizing that this diversification of definitions has weakened our ability to test the concept, several authors have begun to build a more rigorous framework for defining and testing hypotheses of adaptive radiation (Lauder 1981; Mayden 1986; Liem and Wake 1985; Stiassny and Jensen 1987;

Coddington 1988; Lauder and Liem 1989; Baum and Larson 1991; Brooks and McLennan 1991). We suggest that, for each putative case of an adaptive radiation, it must be demonstrated that (1) the group in question contains more species than its sister group, (2) species richness is a derived characteristic within the larger clade, (3) a character present in the more species-rich group enhances the potential that adaptively driven speciation (i.e., sympatric speciation or peripheral isolates speciation via host switching) will occur, and (4) adaptively driven speciation modes played the dominant role in the speciation of the more species-rich group. This latter point is extremely important because, if speciation has been driven primarily by vicariance, then we only have evidence for a "radiation," i.e., that whatever innovations have occurred in the ancestry of the group, the result has been only the persistence of the group long enough to participate in additional episodes of vicariance. In such cases, the presence of adaptive characters is not causally involved with any increase in speciation rates even though it is correlated with the presence of an evolutionary radiation.

Parasitic organisms are thought to be paradigm examples of adaptive evolution (Price 1980), and thus, by extension, good model systems for phylogenetic studies of adaptive radiations. Among parasites, the eight major groups of parasitic platyhelminths (Cercomeria) are well suited for such studies because their phylogenetic relationships are supported by an extensive data base, and because they exhibit a rich mosaic of evolutionary diversification in reproductive, developmental, and ecological characteristics. Our analysis of these groups has demonstrated that a strong postulate of adaptive radiation can be advanced only for the Monogenea, which satisfy all four of the criteria listed previously. The Digenea and the Eucestoda satisfy the first three criteria, but we do not at present have enough information about the frequency of adaptive modes of speciation to differentiate a radiation from an adaptive radiation for these groups. The remaining five groups are species poor. Given the paucity of comparable phylogenetic analyses for the free-living sister groups of the Cercomeria, or indeed for free-living groups in general, it is difficult to assess whether the cercomerians support Price's (1980) contention that parasites exhibit high rates of adaptive radiation. As a point of departure for future discussions, we note that the groups analyzed by Dial and Marzluff (1989) were generally dominated by one species-rich taxon. In the case of the Cercomeria, the group is characterized by three separate species-rich clades, indicating that these parasitic flatworms have at least undergone extensive radiation.

Species richness is associated with different phenomena in the Monogenea, Digenea, and Eucestoda, including apomorphic changes in the developmental rate of maturation, the production or amplification of dispersing larval or juvenile stages, and the amplification of sexual reproductive output, respectively. Species poverty in the other groups appears to be associated with ecological specialization and/or specialization on relictual hosts. Although these characteristics have allowed the groups to persist over a very long period of time, they have not resulted in increased rates of any speciation modes, including vicariant speciation. It is far too simplistic to suggest that species-poor groups have not speciated because they did not evolve the synapomorphies of the species-rich clades. It is more interesting to search for explanations for species poverty by examining characteristics of the groups themselves, because synapomorphies need not explain only species richness. For example, the Amphilinidea exhibit an ecological apomorphy that has restricted the range of suitable hosts.

So, parasite-host systems have produced a, if not the, paradigm-defining example of adaptive radiation, have highlighted the complexities underlying attempts to tease apart the evolutionary influences of speciation and adaptation, and have turned our attention towards questions about species stasis over time. This wealth of data reinforces our belief parasite-host systems are ideal models for studying evolutionary questions, and that both parasitologists and parasites should be at the forefront of evolutionary biology.

SUMMARY

Our consideration of adaptation and adaptive radiations in parasitic organisms has produced the following generalizations:

1. The degree of host specificity shows no macroevolutionary regularities—one cannot estimate the degree of phylogenetic congruence between host and parasite phylogenies, or the length of time hosts and parasites have been associated, from observations of host specificity.
2. Parasites are not simple and degenerate—this concept was derived from an improper use of comparative biology, in which the parasites and their hosts were compared, rather than the parasites and their free-living sister groups.
3. Parasites are not unusually adaptively plastic in their morphology—if anything, parasites seem unusually constrained in their morphological evolution.
4. Parasite ontogenies evolve as coherent units, and the degree of adaptive response by each stage is constrained by common evolu-

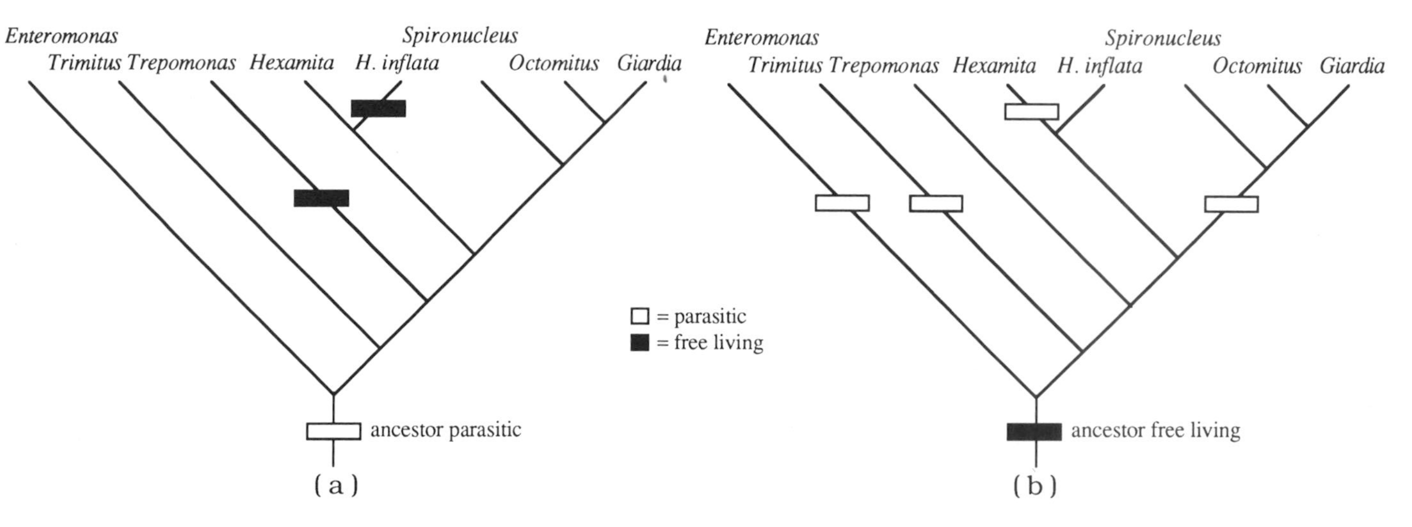

Figure 3.12. The reversibility of parasitism. Parasitic versus free-living modes of life are optimized onto the phylogenetic tree for the diplomonad flagellates. (a) The most parsimonious optimization for the data: free living has evolved twice from parasitism. (b) Maintaining the myth of irreversibility requires postulating that parasitism has arisen four times in these protists.

tionary history—larval and adult morphology is congruent phylogenetically.

5. The evolution of life cycles is the key element in the phylogenetic diversification and adaptive radiation of parasites—the evolution of life cycle patterns shows a rich mosaic of diversification in reproductive, developmental, and ecological characteristics in a strongly phylogenetic context.
6. The correlates of the evolutionary radiation of parasite groups appear to be primarily ontogenetic innovations, with changes in adult reproductive structures second, and ecological components of life cycles third—the evolution of changes in the biology of the parasites dictates the changes in life cycle patterns, including patterns of host utilization, not the reverse.
7. Species richness is correlated with different phenomena in different groups of parasites—these phenomena include changes in ecological components of life cycles, the production or amplification of dispersing larval or juvenile stages, and the amplification of sexual reproductive output.
8. Species poverty in parasite groups is correlated primarily with a "failure to evolve" and not with regressive evolution.

These generalizations tell us that the body of myths, metaphors, and misconceptions about parasite evolution that is entrenched in our textbooks is contradicted by empirical evidence made available by modern methods of phylogenetic analysis. These same data also tell us that parasite evolutionary biology is the same as the evolutionary biology of free-living organisms. Taken together, we conclude that parasite evolutionary biologists have an excellent opportunity for future research; their study organisms are excellent model systems for studying a wide variety of contemporary general evolutionary questions and no one has yet asked those questions widely within parasitology. We close this chapter by examining one last myth.

If, as our data indicate, ecological evolution in parasitic species is highly constrained phylogenetically, and if parasitism is an ecological phenomenon, one widespread belief in parasitology might still be true, even though all the other myths lie in ruins about our feet. This concept is that once the parasitic mode of existence has been adopted by a lineage, only parasitic descendant species will be produced. Alas, even this myth appears not to be inviolate (Siddall et al. in press). Figure 3.12 depicts the phylogenetic tree for the diplomonad flagellates (Siddall et al. 1992; Appendix) indicating which taxa are endoparasitic and which are free living. If we assume that the adoption of the parasitic mode of life is irreversible, we must assume that parasitism has arisen a minimum of four times in the eight genera considered. If we drop the "myth of irreversibility," however, we find a much simpler explanation for the same data: the common ancestor of the group was parasitic, and *Hexamita inflata* and members of *Trepomonas* have reverted to the free-living mode of existence.

4

EVOLUTIONARY BIOLOGY

A Unifying Theme for Parasitology in the Twenty-First Century

> All I have done is ask a great many questions, some of which I hope can be answered by my parasitological friends either at this conference or at a later occasion.
>
> —Mayr 1957

Over the centuries, biologists have managed to accumulate a large amount of morphological, developmental, and ecological information about parasitic organisms, information that is currently being expanded by the inclusion of molecular data. We hope that the preceding chapters have established that biologists can extract valuable information about the patterns and processes underlying the evolution of parasites in particular, and of biological evolution in general, from that data base. The study of parasite evolutionary biology should thus have a bright future, not only because so much is yet to be investigated, but also because of the potential for forging connections with other research programs. In this chapter, we will summarize the current data base and discuss some new areas of research that may benefit from the inclusion of parasites.

SIX GENERALIZATIONS FROM THE CURRENT DATA BASE

Physics decrees that we can never reverse the flow of time. Evolution decrees that the march of time be captured and encapsulated within biological entities. So, although we cannot go back in time, we can reconstruct its passage by studying the messages carried by organisms. Decoding this "always bilingual and usually garbled" (Manter 1966) information has allowed us to catch some interesting glimpses of the forces shaping parasite evolution. Until we have more phylogenetic analyses, however, these glimpses can provide us with only the following tentative, but nonetheless tantalizing, generalizations about parasite evolution.

1. Parasite phylogenetic relationships can be reconstructed based upon characters of the parasites themselves. As we indicated in Chapter 3, the belief that parasites are black boxes programmed by the environment (their hosts) is an orthogenetic legacy that has become part of parasitologists' evolutionary folklore. There is ample evidence, however, that robust phylogenetic hypotheses can be constructed using characters of the parasites, without incorporating information about host identity or relationship, and without invoking particular theories of evolutionary change. This means that, like all other organisms, parasites have properties that might affect the rate, manner, and extent of their evolutionary responses to environmental stimuli. These properties, in turn, are subject to the same constraining and diversifying influences that drive the evolution of free-living organisms. This recognition solves the problem of circularity inherent in the research programs of both the host and geographic schools of coevolution, and exorcises the last vestiges of orthogenesis from our methods for constructing parasite phylogenetic hypotheses.

2. Allopatric speciation has been the predominant speciation mode among parasites. Price (1980) proposed that the evolution of new parasite-host relationships occurred through colonization of new hosts, and interpreted this to be an example of sympatric speciation because *the hosts had to overlap geographically in order for the switch to occur.* This perspective, based upon distribution patterns of the host, changes when we view the speciation process from the perspective of the organism that is actually speciating. Sympatric speciation can only be identified by finding sympatric sister species of parasites (i.e., sister species in the same host). The sympatry of different host species is important because of its involvement with host switching (peripheral isolates allopatric speciation), not sympatric speciation.

As discussed in Chapter 2, although parasite populations are *maintained* by colonization of hosts, parasite-host relationships can *originate* in ways other than by colonization. For example, new parasite-host associations can arise when an ancestral parasite species inhabiting an ancestral host population is subdivided geographically with the host population (vicariant allopatric speciation), or when character changes occur within a population that result in reproductive isolation of the new parasite genotype (descendant) from the ancestral population without involving any changes around or to the host (sympatric speciation). Both of these examples have one thing in common: the new parasite-host association did not originate as a result of colonization because both the host and parasite are maintaining the same intimate association throughout the speciation process. We must, therefore, have a more comprehensive view about the origin of such associations. That view is based on the idea that parasites have their own evolutionary tendencies and fates, and thus may exhibit host and geographic distribution patterns that can be determined, evaluated, and explained without reference to a priori theories about host evolution or the influence of hosts on parasite evolution.

When we examine parasite speciation within a macroevolutionary framework we find that, based upon the evidence collected to date, vicariant speciation and peripheral isolates speciation via host switching have played the dominant roles in the evolution of most parasite groups. Just as with free-living systems, sympatric speciation runs a distant third overall, but may be more important within certain groups like the monogeneans and pinworms.

3. Parasites are neither unusually simplified nor unusually adaptively plastic. The patterns of morphological evolution do not support the view that parasite evolution has been characterized by widespread loss of traits, indicating that the parasite has given up much of its evolutionary independence to its host, or widespread homoplasy, indicating that parasites are so simplified that their options for morphological innovation are limited evolutionarily. In other words, parasites are not the degenerate, overspecialized, host-dependent creatures on their way to self-imposed extinction envisioned by the proponents of orthogenesis. They are instead successful, innovative creatures, some of which have persisted for a long time on this planet. Because of this, phylogenetic relationships can be reconstructed based upon characters of the parasites themselves. And, because of that, we can study hypotheses about the influences of host and geographical factors on parasite evolution without resorting to circular reasoning.

4. Host specificity is a property of parasite adaptation, not parasite speciation. The macroevolutionary trends do not support the belief that the degree of host specificity is causally connected with parasite speciation. Rather, it appears that host specificity plays a role in an organism's ability to adapt to changes in its resource base. Parasitological discussions of host specificity parallel debates within ecology in general about the influence of "specialist" versus "generalist" foraging modes on speciation rates among free-living organisms (e.g., Raikow 1986; Fitzpatrick 1988; Kochmer and Wagner 1988; Maurer 1989; and discussion in Brooks and McLennan 1991). The discovery that there are no general patterns of host specificity correlated with patterns of speciation in parasitic groups supports the hypothesis that speciation and adaptation are always phylogenetically correlated, but neither is causally dependent upon the other; a conclusion that was also reached based upon studies of free-living organisms (see examples in Brooks and McLennan 1991).

5. Life cycles are a combination of innovative and conservative characters organized into coherent and predictable patterns by phylogenetic history. Life cycles are not assembled de novo. Like other parts of an organism's phenotype, they are assembled in an evolutionarily coherent sequence that generally corresponds to the underlying genealogical relationships of their bearers. Because of this, we expect characters representing all parts of the life cycle to be useful in phylogenetic reconstructions. Phylogenetic analyses have confirmed that expectation. Congruent phylogenetic trees have been produced using both larval and adult characters, and both ontogenetic and ecological components of life cycle patterns.

The ecological components of life cycles are more conservative than the developmental component. For example, ecology does not appear to influence directly the evolution of development. Alterations in development, on the other hand, can have a direct effect on ecology. Although ontogenetic characters are more flexible in terms of innovation and, in some groups, homoplasy, there is little evidence of their secondary loss. This is particularly emphasized within the metazoan parasite groups, in which developmental stages are never lost once they have appeared, although the ecological context of their appearance may be altered and the duration of their appearance may be lengthened or shortened. In other words, this component of life cycles is not so adaptively plastic that stages can "come and go" in response to environmental pressures.

6. Parasites in general have not experienced unusually high degrees of adaptive radiation. Parasites are excellent model systems for studying both the "adaptive" and the "radiation" parts of the adaptive radiation concept. Within the parasitic flatworms, the monogeneans appear to have undergone an adaptive radiation, whereas the digeneans and the eucestodes appear to have experienced a radiation that may, or may not, have been adaptive. It is important to realize, however, that each species-rich group is balanced by a relatively species-poor sister group, so it is inaccurate to speak of "parasites" in general as having experienced high levels of adaptive radiations. An assessment of the relative extent of parasite adaptive radiations cannot be made until we have comparable data bases for free-living groups. At the moment, we can say that the monogeneans, digeneans, and eucestodes, not unlike ostariophysan and percomorph fish and passerine birds, provide a wealth of information about radiations, adaptive or not.

This information, in turn, supports the hypothesis that such radiations appear to be primarily a function of changes in life cycle components. For example, four putative key innovations were identified during our examination of the data base for the parasitic flatworms: (1) the evolution of a direct life cycle (a developmental change, possibly due to acceleration, with an ecological outcome), (2) the appearance of additional larval stages (a developmental change), (3) the appearance of asexual amplification of larval stages (a developmental change), and (4) the appearance of sexual amplification of reproductive output (again, a developmental change involving the repetitious production of segments). The success of these "key innovations" is based upon an interaction between the environment and populations tempered by a background of substantial inherited constraints. Since both parasites and hosts have evolutionary tendencies and capabilities, parasite evolution will be historically correlated in some way with host evolution, but will not necessarily be causally connected with it. Adaptive radiations thus result from active interaction between parasite and environmental (host) characteristics, rather than just from evolutionarily passive parasite responses to host characteristics.

These six generalizations are not written in stone. Like any conclusions in science, they are only tentative. We believe that the data base for the parasitic flatworms is one of the largest available for any group of organisms, be they free-living or parasitic. It has been assembled in little more than a decade by a relatively small number of workers, so what could a larger and equally determined group contribute to a new century of parasite evolutionary biology? More importantly, although the data base is extensive, it is still far from complete or even exhaustive. Because of this, all we can say is that, at the moment, the preceding generalizations are supported by the evidence; however, that support may change as we collect more data and perform more phylogenetic analyses for a variety of parasite groups. Nevertheless, we believe one powerful generalization that springs from the current data base that will not change with the accumulation of more information: *the evolutionary biology of parasites is the same as the evolutionary biology of nonparasites.*

We have already demonstrated that parasites are amenable to studies of phylogenetic systematics, speciation, the relationship among ontogeny, reproduction, and ecology during the evolution of life cycles, adaptive radiations, and historical biogeography. In closing, we would like to discuss two additional research programs in evolutionary biology that have benefited from an interaction with parasitologists. These programs are embedded within two cornerstones of evolutionary biology, natural selection and sexual selection.

NATURAL SELECTION AND PARASITE-HOST INTERACTIONS: STUDIES OF COEVOLUTIONARY DYNAMICS

The parasitological perspective on coevolution differs considerably from the nonparasitological perspective on the subject. Nonparasitologists began studying coevolutionary associations intensely only after the publication by Ehrlich and Raven (1964) discussing butterfly–host plant interactions. By contrast, parasitological evidence of long-term (co)evolutionary associations between parasites and hosts was assembled beginning a century ago, but such studies had foundered by 1940 (see Chapter 1). Thus, modern theories concerning coevolutionary processes have been developed almost entirely by nonparasitologists and have been derived from a population biological rather than phylogenetic perspective. The lack of communication between parasitologists and nonparasitologists interested in coevolution has had a significant impact. The few recent attempts by nonparasitologists to incorporate phylogenetic information into coevolutionary studies (e.g., Mitter et al. 1991; Farrell et al. 1992; Miller 1992; Stein 1992) have been methodologically outdated relative to the types of studies we discussed in Chapter 2. Not to be outdone, parasitologists have, until recently (e.g., Roza 1989; Behnke and Barnard 1990) failed to make use of the dynamic models of coevolution developed by nonparasitologists (see Futuyma and Slatkin 1983). As a consequence, we have little understanding of the similarities and differences between, for example, parasite-host and insect-plant systems at either the microevolutionary or macroevolutionary level.

One necessary step toward increased mutual appreciation is the articulation of phylogenetic patterns expected to result from the three major classes of coevolutionary dynamics postulated by nonparasitologists (see also Brooks and McLennan 1991).

Allopatric Cospeciation

Allopatric cospeciation (Brooks 1979b) or the "California Model" (Brooks and McLennan 1991) is based on the assumption that parasites and hosts are simply sharing space and energy. As the null model for studies of coevolution, it predicts *congruence between parasite and host phylogenies* based solely upon simultaneous allopatric speciation in parasite and host lineages (i.e., vicariance events). Like any null model, *support for the hypothesis of cospeciation offers relatively weak explanatory power.* For example, discovering that a particular set of associations has resulted from allopatric cospeciation eliminates coevolutionary models based on host switching, but does not allow us to distinguish the effects of a historical correlation from the effects of some mutual interaction that maintains or promotes the association and its diversification. And even if we assume the latter, the delineation of cospeciation patterns by themselves, no matter how detailed, does not allow us to differentiate among a variety of mechanisms by which the associated species might be causally, rather than casually, intertwined.

For example, Smiley (1978) presented a version of the cospeciation model based on the assumption that there is a general evolutionary tendency toward specialization, in this case specialization in host choice, by parasite species. Interestingly, this model is not substantially different from Szidat's (1940) proposal that host specificity was the inherent developmental end product (*Altersresistenz*) of an orthogenetic drive toward specialization that increased with both the age of the individual host and the age of the parasite-host associations. This model postulates that, as an evolving parasite lineage becomes progressively more specialized, it will be associated with fewer and fewer hosts, and cospeciation events will increase accordingly. In this case, the degree of cospeciation is the result of an evolutionary trend inherent in the parasite species and not of mutual adaptive modifications between host and parasite. As we have demonstrated in this book, however, there is little empirical support for such a model among parasites.

Alternatively, the association may be maintained by the presence of certain host resources that are of adaptive value to a parasite species. If an associate species requires a specific resource and/or a resource with a restricted distribution among host groups, then it is possible that the phylogenetic association between the host group and the associated group is due to the distribution of the resource. Consider a hypothetical ancestral fish species X characterized by trait q_0 that serves as a cue for members of tapeworm species A (Fig. 4.1). The tapeworm's reliance on q_0 is denoted by its possession of the coevolved trait q_1. Now let us follow two evolutionary scenarios involving speciation in the fish lineage producing descendant species Y and Z, as well as speciation in the tapeworm lineage producing descendant species B and C.

The evolutionary association between species Y and B, on the one hand, and between Z and C, on the other, represents an instance of cospeciation. Although both scenarios produce congruent parasite and host phylogenies, the processes underlying this congruence are somewhat different in each case. In the first scenario, q_0 and q_1 are phylogenetically conservative, so the associations are maintained, at least in part, by the common coadapted trait complex q_0:q_1. We have evidence here for "adaptively constrained" coevolution because *speciation in both lineages has occurred without changes in the causal basis for the primitive parasite-host interaction.* In the second scenario, the cospeciation of the fish and

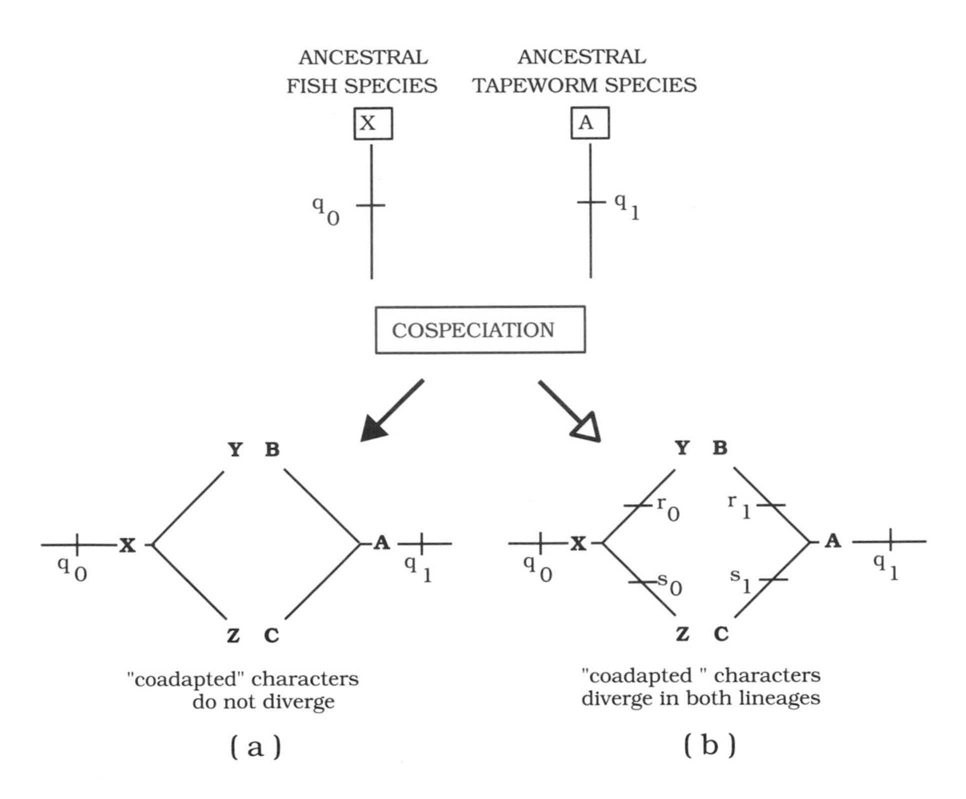

Figure 4.1. Examining the potential for coadaptation within the framework of cospeciation. In both of the scenarios, preliminary phylogenetic analysis has uncovered complete congruence between host and parasite phylogenies. (a) Host (X to Y + Z) and parasite (A to B + C) speciate simultaneously; descendant species retain the plesiomorphic coadapted trait complex (q_0 in the host; q_1 in the parasite). (b) Host and parasite speciate simultaneously, and diversification in the coadapted trait complex (q_0 to $r_0 + s_0$ in the host; q_1 to $r_1 + s_1$ in the parasite) is congruent with the speciation events. Redrawn and modified from Brooks and McLennan (1991).

tapeworm clades is matched by the diversification of the adaptively significant trait complex. Species Y and B now interact through the coadapted characters r0:r1, while the association between Z and C is maintained by traits s0:s1. This scenario, representing adaptively driven coevolution, provides ecological evidence that the observed cospeciation patterns are more than casual historical associations.

Cospeciation and Rates of Anagenesis

Parasitologists have discovered the exceedingly interesting fact that many parasites are strictly host specific and that a group of related species or genera is often restricted to related host species. The evident interpretation is that the

> association between host and parasite goes back into geological history, the evolution of the parasite being as closely correlated with that of the host as if it were an organ of the host. If, as seems to be true in some cases, the parasite is more conservative than the host (that is, its evolution is retarded as compared to that of the host) then the parasite will show phylogenetic relationships where they have become obscured in the vertebrate hosts. This hypothesis . . . raises a number of fundamental questions in evolutionary biology, particularly in cases of exceptions to this rule.
>
> —Mayr 1957

Hafner and Nadler (1988, 1990) approached the problem highlighted in the quotation above from a molecular biological perspective. They suggested that congruent parasite and host phylogenies can be produced in three different ways: (1) *synchronous cospeciation,* in which phylogenetic congruence is produced by synchronous speciation and is accompanied by equivalent degrees of evolutionary divergence in host and parasite lineages; and two forms of *delayed cospeciation,* in which (2) parasite speciation lags behind host speciation, so that parasite evolutionary divergence is less than host divergence or (3) host speciation lags behind parasite speciation, so that host evolutionary divergence is less than parasite divergence. Delayed cospeciation is not a form of resource tracking (*phylogenetic tracking* of Brooks and McLennan 1991), since it is postulated that all evolutionary products of the host and parasite lineages remain associated with each other throughout evolutionary history. That is, there is no host speciation, followed later by speciation of parasites via sequential colonization of hosts.

Hafner and Nadler's proposals tap into the recurring problem of determining rates of anagenesis and cladogenesis during parasite-host evolution. Recall that the host school supported the position that the historical development (anagenesis) and splitting (cladogenesis) of hosts is paralleled by the development and splitting of their parasites (Farenholz's Rule: Eichler 1942), whereas the geographic school advocated that parasites evolved more slowly than their hosts (Manter's Rule #1). Unfortunately, all parasitological rules, including Manter's rule, confound anagenetic and cladogenetic effects (Chapter 1), making them difficult to interpret in light of contemporary systematic and evolutionary approaches, which treat the two phenomena as historically correlated but not necessarily causally linked. So, in order to answer questions about relative rates of parasite-host evolution, we must first be able to distinguish anagenesis from cladogenesis. Hafner and Nadler were the first researchers to make this attempt.

They argued that, from the parasite's perspective, there are three types of anagenetic rates (equal to, greater than, or less than anagenetic rates for hosts) and three types of cladogenetic rates (equal to, greater than, or less than rates for their hosts). In all, nine different coevolutionary interactions are formed by combining rates of host and parasite anagenesis and cladogenesis. In order to differentiate among these possibilities, Hafner and Nadler made three assumptions about the processes underlying parasite-host coevolution. The first assumption was that a weak molecular clock—one that is relatively constant within host and parasite lineages, but need not be the same between the lineages—is operating in both groups. Given this assumption, they proposed that noting differences in the slope and *y* intercept of linear regressions performed on host and parasite genetic distances would provide a method for delineating the type of cospeciation mode affecting a particular parasite-host system.

Their second assumption rested on the widespread view that "parasites evolve more slowly than their hosts." As we discussed in the preceding paragraph, this "rule" was formulated without the benefit of information from developments in both modern evolutionary and modern systematic biology. Before it can be used as the foundation for the development of any theory about how hosts and parasites coevolve, therefore, it must be subjected to the bright light of empirical examination. Such examination requires that data be collected independently of any assumption about rates or mechanisms of evolutionary change—in other words, by comparing host and parasite phylogenies reconstructed using characters subjected to phylogenetic systematic analysis. The requirement for independence precludes the use of any data that invoke a hypothetical evolutionary mechanism (like the molecular clock) as part of the analysis.

Hafner and Nadler's third assumption is that there will still be phylogenetic congruence between parasites and hosts even when host speciation and parasite speciation are decoupled if there has been no host switching (delayed cospeciation). In order for this to occur, the group that speciates second must do so before the group that speciated first speciates again. If this does not happen, however, then the relationship between host and parasite phylogenetic patterns can become increasingly complex, depending upon the speciation/extinction rates in the more quickly speciating group. For example, the larger the number of speciation events experienced by one group before the second group speciates, the more possible ways there are for the second group to speciate, only one of which will result in identical parasite and host phylogenetic patterns. Allowing only bifurcating trees, if the host group speciates twice before the parasite group speciates once, there are 3 different possible outcomes for two parasite speciation events, one of which is congruent with the host phylogeny; if the host group speciates 3 times before the parasite group speciates, there are

15 different possible outcomes; and if the host group speciates four times, there are 105 different possible outcomes (Felsenstein 1978). Because of this, Hafner and Nadler must rely on parasite-host systems in which parasite speciation may be delayed relative to host speciation, but always occurs before the hosts speciate a second time.

Given the initial premise that parasite speciation and host speciation are decoupled (as required by delayed cospeciation), is it valid to assume that parasite speciation will generally occur before the hosts speciate a second time? Delayed cospeciation requires that gene flow in one group (group A) be maintained after gene flow in the other group (group B) has been lost. This might happen if group A did not speciate while group B underwent sympatric speciation, eventually resulting in two sympatrically localized sister species. There is a protocol for using phylogenetic, biogeographic, and population biological information to study speciation modes (Brooks and McLennan 1991 and references therein). This type of analysis provides an essential line of evidence for distinguishing synchronous cospeciation from delayed cospeciation because it allows us to identify parasite and host pairs that have experienced a common vicariance event. Since the initiation of speciation occurs at the same time within the pairs (i.e., begins with the vicariance event), we can examine the relative degree of genetic divergence for each parasite and its host in order to determine whether they diverged at the same rate following fragmentation of the ancestral species populations. It is also possible, using these protocols, to identify sympatric sister species for either the host or the parasite group, an approach that provides support for the possibility of delayed cospeciation of hosts or parasites.

Resource Tracking

Resource tracking models, which can also be referred to as *colonization* models (Ronquist and Nylin 1990), are based on the concept that hosts represent patches of necessary resources that parasites have tracked through evolutionary time (Kethley and Johnston 1975). Three pieces of information are required in order to examine thoroughly the explanatory power of this model: a phylogeny for the hosts, a phylogeny for the parasites, and an explicit description of the resource. Depending on the phylogenetic distribution of the resource, one of three general macroevolutionary outcomes will be produced.

The *sequential colonization model* (Jermy 1976, 1984), originally designed to explain insect-plant coevolution, proposes that the diversification of phytophagous insects took place after the radiation of their host plants. The insects are hypothesized to have colonized new host plants many times during their evolution. In each case the colonization was the result of the evolution of insects responding to a particular biotic resource that already existed in at least one plant species. That resource, in turn, is postulated to have been either *plesiomorphically* or *convergently* widespread, so the predicted macroevolutionary pattern for these two outcomes is that *parasite and host phylogenies will show no congruence.*

From the perspective of resource tracking models, the explanation for *congruence between portions of host and associate phylogenies* (the third outcome) is that the parasites are specialized on a host resource that is restricted to the host clade and has evolved in a manner congruent with the phylogenetic diversification of the host clade. This model differs from sequential colonization because it proposes that the resources are distributed in an apomorphic, rather than a plesiomorphic or convergent, fashion. It differs from allopatric cospeciation by assuming that parasites and hosts do not speciate simultaneously (association by descent); rather, the hosts evolve first and then parasites colonize them. The evolutionary sequence of events is as follows: a new host species evolves, characterized, in part, by an evolutionarily modified form of the required resource. This new species of host is then colonized by individuals from the parasite species associated with the ancestral host. Some of these individuals adapt to the new form of the resource, eventually producing, in their turn, a new species of parasite. And so the cycle continues.

In order for this evolutionary scenario to occur, the ancestral host species *must* persist after the speciation event that produces descendants bearing the modified resource; otherwise, the ancestral parasite species would have no resource base to support the population through the colonization phase. Ancestral species can persist following sympatric and peripheral isolates allopatric speciation, so, in order to distinguish phylogenetic tracking from allopatric cospeciation, we require an understanding of the manner in which the host group speciated, a method for distinguishing persistent ancestors on a phylogenetic tree, and a detailed mapping of the host resource and parasite tracking characters (if present) on the appropriate phylogenetic trees.

Evolutionary Arms Race

> The evolution of the parasite would take place in step with adaptations of the host to counter the unfavorable effects of the parasite so that a steady-state condition between the demands of the parasite and the reactions of the host would be maintained. This balance between host and . . . parasite would also be influenced strongly by the relation of the host to factors other than the parasite in its environment.
>
> —Rogers 1962

This is what nonparasitologists view as the classical coevolution model (Mode 1958; Ehrlich and Raven 1964; Feeny 1976; Berenbaum 1983), sometimes termed the *exclusion* model (Ronquist and Nylin 1990), and originally proposed for insect-plant systems. It may be summarized as follows: phytophagous insects reduce the fitness of their hosts. Plants that, by chance, acquire traits (defense mechanisms) that make them unpalatable to these insects will increase their fitness relative to their undefended brethren, and the new defense mechanism will spread throughout the plant population (new host species). However, eventually some mutant insects will, in their turn, overcome the new defense mechanism and be able to feed on the previously protected plant group. If this confers a fitness advantage on the mutants (e.g., through reduction of inter- or intraspecific competition for food) the counterdefense mechanism will spread throughout the insect population (new associate species). This new species of insect will be able to specialize on the previously protected plant group, and the cycle will begin anew.

The primary assumption in arms race models of coevolution is that coevolving ecological associations are maintained by mutual adaptive responses. For example, it is possible that during the course of evolution novel traits arise that protect the host from the effects of the parasite. It is also possible that traits countering such defense mechanisms may evolve in the parasite lineage. The macroevolutionary patterns that result depend upon the time scale on which the adaptive responses occur. First, in systems for which the defense and counterdefense traits arise on a microevolutionary scale, we would expect *fully congruent parasite and host phylogenies, with appropriate defense and counterdefense traits appearing at the same point in the common phylogeny.* In such cases, the coevolutionary arms race would not affect the patterns of macroevolutionary associations between parasite and host clades, leading us to expect phylogenetic congruence. This type of evolutionary arms race can be differentiated from allopatric cospeciation by the presence in both lineages of cooriginating, mutually adaptive traits.

Second, evolutionary arms race models generally assume that, in many cases, the time scale on which the defense and counterdefense traits originate in response to reciprocal selection pressure is longer than the time between speciation events. Given this assumption, we might expect to find macroevolutionary patterns in which *the parasite group is missing from most members of the host clade characterized by possession of the defense trait.*

A third possible macroevolutionary pattern results when one or more relatively plesiomorphic members of a host clade are colonized by more recently derived members of the parasite group bearing the counterdefense trait. In this case, *parasite and host phylogenies will demonstrate some degree of incongruence and we would expect to find evidence that some associates have back-colonized hosts* in the clade diagnosed by the presence of the defense trait.

And finally, Ehrlich and Raven (1964) postulated that coevolutionary patterns similar to ones expected for the sequential colonization resource tracking model would result when host shifts by insects with a counterdefense trait occurred between plants that had *convergently* evolved similar secondary metabolites in response to insect attack. In both cases, there is a departure from phylogenetic congruence between parasites and hosts; however, the resource tracking model requires only that the parasites be opportunistic. The host resource can be widespread owing to either plesiomorphic occurrence or convergence, but its evolutionary patterns are not affected by the presence or absence of the parasites. In contrast, the mechanism by which the host resource evolves in the Ehrlich and Raven case requires a high degree of convergent mutual modification on the part of the host and parasite groups. Differentiation between the two models requires information about the evolutionary elaboration of putative defense and counterdefense traits.

There is considerable overlap in the macroevolutionary patterns associated with each coevolutionary model. For example, one outcome of a resource tracking dynamic produces patterns of parasite and host phylogenetic congruence that look like allopatric cospeciation patterns, while another outcome produces patterns that resemble the results of a coevolutionary arms race. Outcomes of the arms race model range from strict cospeciation patterns to widespread host switching patterns. Thus, the macroevolutionary effects of coevolutionary processes tend to blur the distinctions among the models. In order to distinguish among these modes, then, we need information about host and parasite phylogenies (speciation) and information about the characters involved in the parasite-host association (resource type, distribution, host defense, parasite counterdefense traits).

SEXUAL SELECTION AND PARASITE-HOST INTERACTIONS: THE HAMILTON-ZUK HYPOTHESIS

The influence of female choice on the evolution of male epigamic characters has been a controversial issue ever since Darwin proposed his theory of intersexual selection (Darwin 1871). That controversy, originally surrounding the existence of female choice, now centers upon the mechanisms underlying such choice. Fisher (1930) proposed that the development of the male character and female preference for the character would "advance together" with ever-increasing speed until the process was checked by "severe counterselection" against males bearing the exaggerated trait. At the very heart of the Fisherian hypothesis lies the assumption that the relationship between female choice and the preferred male character is an arbitrary one in terms of male viability. That is, females who choose males with bigger tails, brighter colors, or more vigorous displays get nothing out of the interaction other than the production of male offspring with the preferred character (see, e.g., O'Donald 1962, 1967; Lande 1981; Kirkpatrick 1982, 1986). The generality of the Fisherian process was challenged by proponents of the "good genes" school (e.g., Trivers 1972; Zahavi 1975, 1977; Borgia 1979; Andersson 1982, 1986; Kodric-Brown and Brown 1984), who argued that there is an association between expression of the epigamic character and male viability. Because of this association, females who choose mates with bigger tails, brighter colors, or more vigorous displays also get mates with "better genes," and thus produce offspring that possess the preferred character and are more vigorous than offspring of nonpreferred males.

The chief objection to the good genes hypothesis comes from population genetics. If female choice of genetically "superior" males is a strong enough force to influence the evolutionary elaboration of the character advertising this superiority, how could genetic variability, the materials of selection, be maintained in the population (Maynard Smith 1978; Borgia 1979; Taylor and Williams 1982; Kirkpatrick 1986)? Hamilton (1982) and Hamilton and Zuk (1982) attempted to answer this objection by exploring the relationship among parasite fitness, host fitness, and the degree of host sexual dimorphism. Their proposal, based upon studies of parasite population structure, is the product of three assumptions. First, parasites and their hosts are involved in a microevolutionary arms race. That is, the time lag between changes in characters involved in parasite offense and changes in characters involved in host defense produces cycles of coadaptation. This temporally varying selection pressure maintains a significant level of additive genetic variability for viability in the host population (Hamilton 1982; Eshel and Aiken 1983; Eshel and Hamilton 1984; Anderson and May 1985). Second, parasites adversely affect the health (viability) of their hosts. And third, the condition of male epigamic characters is an accurate and direct reflection of the bearer's health (a "revealing handicap": Maynard Smith 1985). Given these assumptions, the interaction between the oscillating force of natural selection on host viability and female choice for a male character that accurately communicates underlying viability results in a form of directional intersexual selection that does not exhaust genetic variability in the population (but see Kirkpatrick 1986; Pomiankowski 1987a,b)

This theory is fundamentally important to studies of sexual selection. To date, however, it has been developed and is being tested with very little input from parasitologists. In some ways this is surprising, given the role played by parasitologists in the development of modern coevolutionary theory. In other respects, it is not so surprising. Parasitologists, in general, have neither corrected the simplistic assumptions about the relationships between parasites and their hosts that often underlie "parasites and sexual selection" studies, nor conducted research of their own into this fascinating area. In any case, this situation must change if the Hamilton-Zuk proposal is to move from an interesting idea to a robust evolutionary theory with well-defined and testable predictions.

What exactly can parasitologists offer students of sexual selection? There are at least two answers to this question. First, we can offer our expertise on the three aspects of parasite-host interactions that are of interest to students of sexual selection: the effects of parasites on the health of their hosts, host resistance, and parasite prevalence and intensity. Since parasitologists have been investigating these questions for many years, both the conceptual framework and the empirical data base are well developed in each area. This book is primarily concerned with the macroevolutionary patterns of parasite-host evolution, so we will not present a discussion of the outcome of these microevolutionary problems (for a detailed discussion and references see McLennan and Brooks 1991). Suffice it to say that such research can only benefit by the inclusion of a parasitologist in the group, and parasitology can only benefit by its inclusion in mainstream evolutionary theory.

Second, parasitologists can offer their expertise in the area of coevolution and its implications for the macroevolutionary components of the Hamilton-Zuk theory. The macroevolutionary manifestations of the theory are difficult to examine because the prediction is obscurely worded: "animals that show more strongly developed epigamic characters should be subject to a *wider variety* of parasites" (Hamilton and Zuk 1982, our italics). Confirmation of this prediction is provided by uncovering a strong correlation between some measure of parasite load and the degree of epigamic development across host taxa. However, in searching for this correlation, parasite species are often combined to produce one "infestation" value regardless of the differences in pathology that may be caused by different species of parasites, and prevalence, incidence, and intensity data are used interchangeably with each other and with more nebulous terms applied to the distribution of more than one parasite species (parasite load or burden). Thus it is initially unclear whether Hamilton and Zuk's prediction of a "wider variety" of parasites refers to more parasites absolutely, or to more species of parasites. Authors have criticized the first interpretation, arguing that the model could also make the opposite prediction: "species with bright males should be the least parasitized if female selection for male brightness has lowered the *level of infection* in the population" (Borgia 1986, our italics) and "if however, *parasite load* is reduced as a consequence of female choice for resistant mates, showiness might become negatively correlated with parasite load" (Read and Harvey 1989, our italics). This leaves us with the second interpretation: sexually dimorphic species should host more species of parasites than their monomorphic relatives. Let us investigate this prediction in more detail.

The Hamilton-Zuk hypothesis was originally formulated to explain the evolution of epigamic characters, not to explain the macroevolutionary distribution patterns of parasites among their hosts. However, incorporation of these distribution patterns into our picture of the interplay between parasites/hosts and sexual selection uncovers two important weaknesses in the macroevolutionary prediction as it is currently stated. The first problem is a methodological one. If we uncover a correlation between the number of parasite species inhabiting a host and degree of epigamic development in the host, we cannot accept that correlation as evidence of causality without determining that the factors involved in the observed asymmetrical distribution of parasite species could not, in turn, be involved in the evolution of the sexually selected character (Harvey and Partridge 1982). For example, based upon a correlation between parasite load and the degree of male showiness in passerine birds, Hamilton and Zuk concluded that the evolution of showiness was being driven by the interaction between a parasite-host arms race and female choice for a male character that accurately reflected the outcome of that race. Although this is one possible explanation for the observed correlation, it is not the only one. Kirkpatrick and Suthers (1988) discovered that passerine species with ground feeding habits were significantly more likely to be infected with blood parasites in general, and *Leucocytozoon* spp. in particular, than passerine species feeding above the ground. Bennett et al. (1978) and Kirkpatrick and Smith (1988) reported that the prevalence of blood parasites in colonially nesting birds was significantly higher than in noncolonial nesters. And finally, various authors have suggested that susceptibility to infection by insect-borne blood parasites may be influenced by the height of the nest location preferred by a given avian species (Greiner et al. 1975; but see Kirkpatrick and Suthers 1988). So, in this system, the number of parasite species associated with a host is strongly correlated with the foraging, nesting, and social habits of the birds, any one of which might, in turn, be associated with the evolution of sexual dimorphism. If this is the case, then the observed correlation between parasite load and sexual dimorphism is an artifact of the causal interaction between ecological factor *z* and parasite load and ecological factor *z* and sexual dimorphism.

The second problem is more fundamental and is inherent in the structure of the macroevolutionary prediction as it is currently formulated. Hamilton and Zuk extrapolated from the observation that parasite species are not equally distributed among host species to the conclusion that some hosts were more susceptible to parasite attack than other hosts. However, equating "susceptible" with "less capable of fending off parasitic invasion" creates a paradox: within a species, preferred males are more resistant to parasitism than nonpreferred males, but species containing these preferred individuals are less resistant to parasitism than species without this type of choice dynamic. The results of parasitological research can help us resolve this problem, not by revealing some magical answer but by demonstrating that the paradox does not really exist. Macroevolutionary patterns of parasite distribution are determined by a complex interaction among a variety of factors, only one of which may be changes in species-specific host "resistance." For example, the distribution of helminth parasites in frogs is determined, in part, by both parasite biology and host biology. The infective stages of the majority of helminth species are generally transmitted to the primary host in the water and, as adults, some anurans (e.g., toads and their relatives) are terrestrial or fossorial, while other anurans (e.g., leopard frogs, green frogs, bullfrogs) are aquatic. Because of the constraints imposed by the biology of the hosts and parasites we would expect the aquatic ranids to host more helminths than terrestrial bufonids. Not surprisingly, studies by parasitologists have all shown a strong positive relationship between the number of species of parasitic helminths infecting anurans and the aquatic habits of the anurans (Brandt 1936; Prokopic and Krivanec 1975; Brooks 1976). In this system, then, there will be differential distribution of parasite species regardless of the "susceptibility" of the hosts. Because the distribution of parasites among potential host species is more complex than just an extrapolation of population-level host resistance, there is no a priori reason to believe that a sexually dimorphic species will always be associated with more parasite species than its monomorphic relative.

We can approach this problem from another direction by restating the Hamilton-Zuk prediction in the following manner. The existence of female choice in species A does not *cause* species A to have more parasites than species B; however, the fact that species A has more parasites than species B may *cause* the evolution of epigamic characters via female choice in species A. For example, consider a hypothetical group of birds comprised of two sister groups, the ABCidae and DEFidae (Fig. 4.2). All the members of the ABCidae are sexually monomorphic and host five species of parasites. All the members of the DEFidae are sexually dimorphic and also host five species of parasites. Experimental investigations have demonstrated that females in the DEFidae prefer males with brighter feathers, and these males, in turn, have lower numbers of parasites than their nonpreferred conspecifics. The microevolutionary study confirms that a Hamilton-Zuk mechanism is operating in the An-

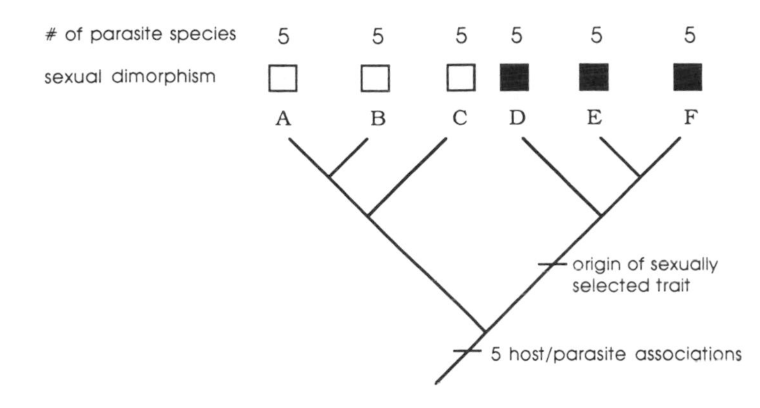

Figure 4.2. Distribution of parasites and sexual dimorphism among a clade of hypothetical birds. A + B + C = ABCidae; D + E + F = DEFidae. White boxes indicate that sexual dimorphism is absent; black boxes indicate that sexual dimorphism is present. From McLennan and Brooks (1991).

nidae. The macroevolutionary analysis, however, does not confirm this result because it contradicts the predicted relationship between degree of sexual showiness and number of parasite species. Is there a way out of this conflict?

Let us return to our initial formulation of the macroevolutionary prediction and refine it. The existence of female choice in species A does not cause species A to have more parasites than species B; however, the fact that species A has more parasites than species B may cause the evolution of epigamic characters via female choice in species A *once that trait appears in the population.* The presence of parasites does not cause the sexually selected trait to appear in the population. If it did, evolution would be Lamarckian, not Darwinian. Depending upon the details of the genetic/developmental system of the host, once the trait has appeared, the presence of parasites may influence its evolutionary success and elaboration. Given this time lag, there is no reason to believe that a relationship should exist between number of parasite species and degree of epigamic development. In the preceding example, the presence of an ancestral parasite-host association was sufficient to promote the spread of an epigamic character when it appeared. We can also make a case for sexually dimorphic species hosting fewer parasites than their monomorphic relatives. For example, the interaction between the dimorphic species and their parasites may be a more pathological one. On the other hand, as depicted in Figure 4.3, the monomorphic hosts may have continued to pick up parasite species, while the dimorphic hosts remained associated with parasites that had been originally involved in the parasite-host sexual selection dynamic. A case for sexually

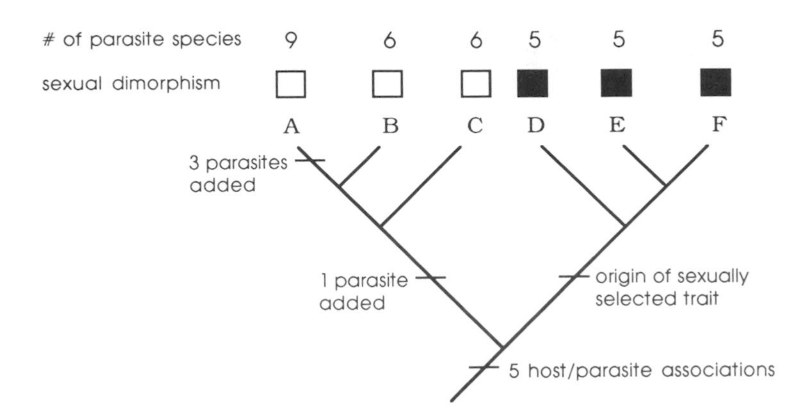

Figure 4.3. Association between sexual dimorphism and number of parasites. In this scenario, members of the ABCidae have continued to pick up parasites, while members of the DEFidae have retained the ancestral parasite-host associations. From McLennan and Brooks (1991).

selected species having more parasites than their relatives in some systems can also be made. Overall, then, the Hamilton-Zuk hypothesis does not make a prediction about the relationship between numbers of parasite species and degree of epigamic development in the hosts, aside from the trivial prediction that at some time in their history the dimorphic species must have been associated with parasites. It is important to emphasize here that this conclusion does not invalidate the hypothesis; it only forces us to rethink the macroevolutionary prediction.This suggests that the macroevolutionary outcome of the Hamilton-Zuk mechanism is qualitatively different from, rather than an extrapolation of, the microevolutionary dynamic based upon parasite numbers.

The Hamilton-Zuk hypothesis is concerned with mechanisms of evolutionary change, so it should make predictions about the evolutionary origins of, and associations among, characters through time. In this case the characters are male epigamic traits and parasite-host associations. Unfortunately, there are very few species-level phylogenies available for parasite groups, and of these, none are for parasites that have interested students of the Hamilton-Zuk hypothesis. For the sake of illustration, then, let us consider some wondrous fishes, the Gasterosteidae or sticklebacks, and their associations with a hypothetical group of tapeworms, monogeneans, and nematodes. The phylogenetic relationships of these parasites, numbered for phylogenetic analysis, are depicted in Figure 4.4. BPA of these data produces the host cladogram shown in Figure 4.5.

According to this analysis the sticklebacks' association with tapeworms (characters 1–11) originated with the ancestor of the fishes (association by

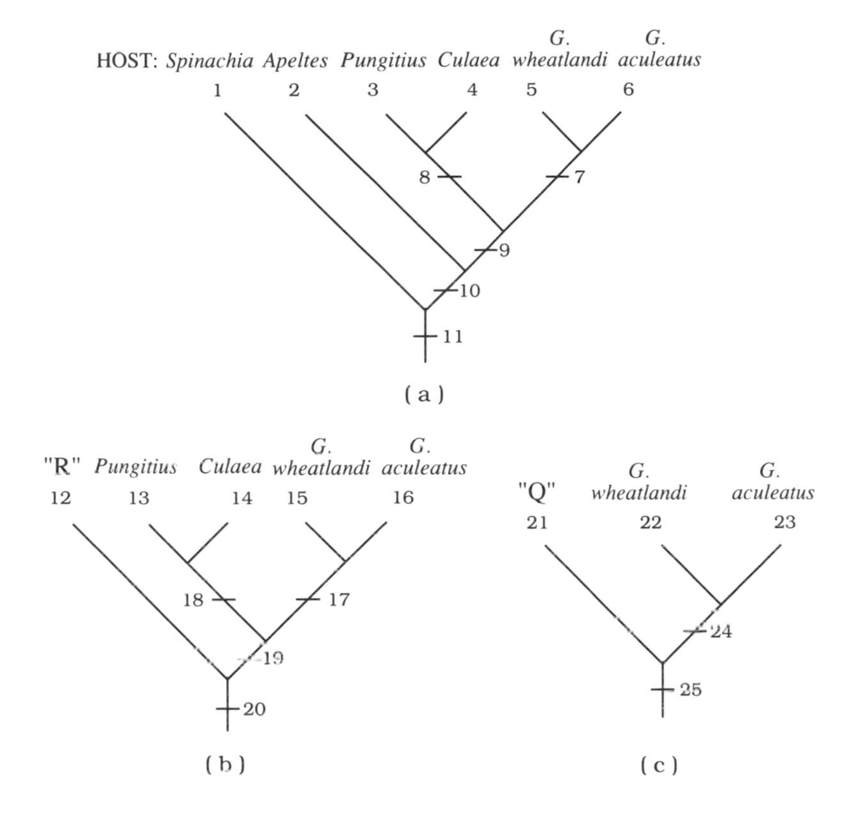

Figure 4.4. Phylogenetic trees for three hypothetical, but monophyletic, groups of parasites. (a) Tapeworms living in the upper intestine of gasterosteid fishes. (b) Monogeneans living on the body of the monophyletic group R and some members of the Gasterosteidae. (c) Nematodes living in the intestine of the monophyletic group Q and some members of the Gasterosteidae. Hosts: R = a monophyletic group of theoretical trout species; Q = a monophyletic group of theoretical blenny species. *G.* = *Gasterosteus*. Redrawn and modified from McLennan and Brooks (1991).

descent), and speciated along with their host group (cospeciation). The monogeneans (characters 12–20) are present because of a host switch (association by colonization) by ancestor 20 from the monophyletic group R to the ancestor of the *Pungitius* + *Culaea* + *Gasterosteus* clade. Once the host switch had occurred the monogenean ancestor speciated, producing descendant 19, and the remainder of the parasite phylogeny is congruent with the gasterosteid speciation pattern (cospeciation). Finally, the presence of the nematodes (characters 21–25) also resulted from a host switch, in this case from the monophyletic group Q to the ancestor of *Gasterosteus wheatlandi* and *G. aculeatus*. When

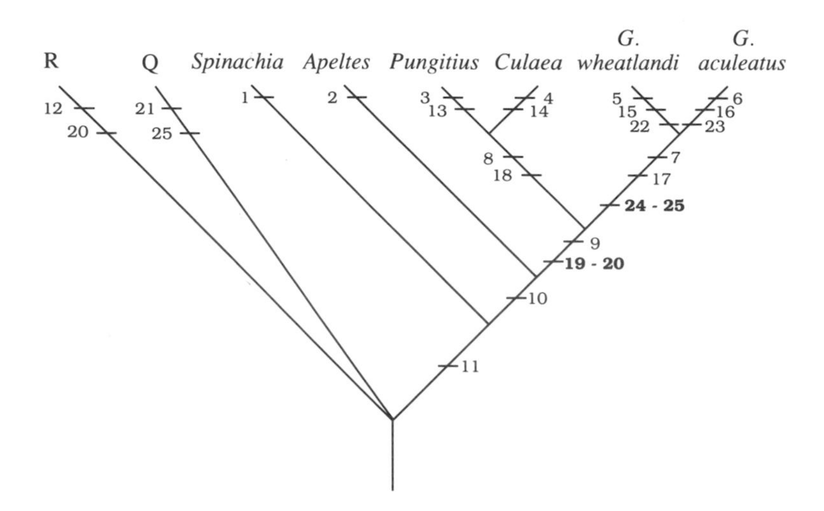

Figure 4.5. Phylogenetic tree for the hosts reconstructed from the parasite data. Numbers refer to the three parasite phylogenies depicted in Fig. 4.4: tapeworms (characters 1–11), monogeneans (characters 12–20), and nematodes (characters 21–25). Boldface numbers indicate instances of presumptive host switching. Hosts: R = a monophyletic group of theoretical trout species; Q = a monophyletic group of theoretical blenny species. *G.* = *Gasterosteus*. Redrawn and modified from McLennan and Brooks (1991).

that ancestor speciated, so did the ancestral nematode, producing descendants 22 and 23 in their respective hosts. Thus, each of the three parasite groups began its association with gasterosteids at a different time. Now consider the origin of the three stickleback-parasite associations in relation to the origin of the sexually dimorphic character "whole body nuptial coloration" as shown in Figure 4.6.

Three types of macroevolutionary associations are delineated in this figure. First, the parasite-host association appears before the origin of the sexually selected character (tapeworms and sticklebacks). In this scenario, the host and parasite have shared a long history of cospeciation. Although researchers are just beginning to ask questions about the relationship between parasite virulence and the age of the parasite-host association, the working assumption is that tightly cospeciated groups will fall at the low end of Anderson and May's (1982) virulence scale. Thus, by the time the sexually dimorphic character appeared in the ancestor of the *Pungitius* + *Culaea* + *Gasterosteus* clade, the tapeworm associated with this ancestor would be hypothesized to have only a minor effect on host health. Past this point, the parasite-host interaction would continue to move toward minimal virulence, so that, by the time *Pungitius, Culaea,* and *Gasterosteus* appeared, the dynamic among host, parasite, male

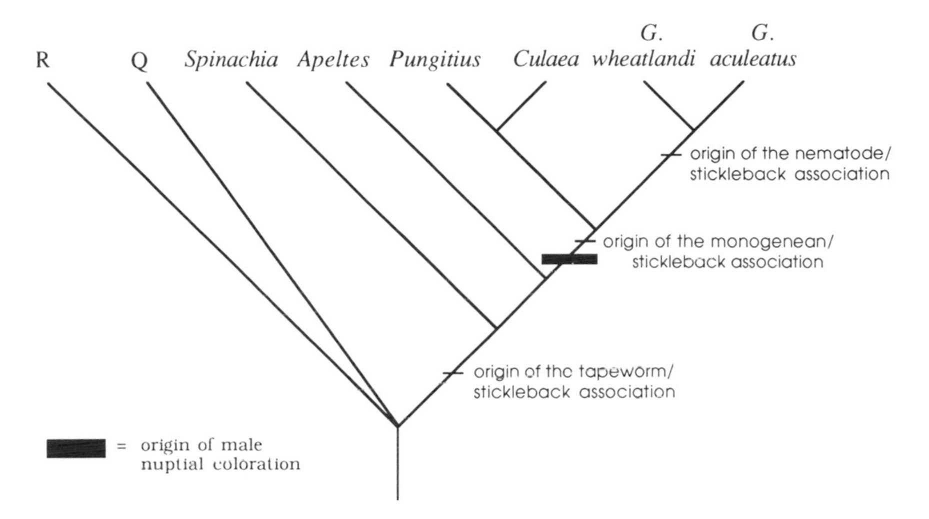

Figure 4.6. Origins of parasite-host associations and male nuptial coloration in Gasterosteidae. Redrawn and modified from McLennan and Brooks (1991).

color, and female choice would be further weakened, rendering it less likely that a female choosing males based upon color would also be choosing mates demonstrating a superior ability to resist invasion/damage from the appropriate tapeworm species. If these assumptions about parasite pathology and the age of the parasite-host association are correct, then this type of pattern provides, at best, weak support for the Hamilton-Zuk hypothesis. Second, the parasite-host association originated in the same species in which the sexually dimorphic trait arose (monogeneans and sticklebacks). In this scenario, the appearance of a new parasite species in a host species (by a host switch) is accompanied by the appearance of the sexually selected character in the host. Since host switches are presumed to be accompanied by more virulent adaptive/coadaptive cycling between the parasite and its new host (Stock and Holmes 1987), both the macroevolutionary pattern and the biological interactions assumed to underlie such a pattern provide strong support for the Hamilton-Zuk hypothesis. And third, the parasite-host association originates after the sexually selected character (nematodes/sticklebacks). This pattern refutes the hypothesis that the interaction between the parasite-host dynamic and female choice for a male character reflecting that dynamic is driving the *initial* elaboration of the male character. Although this conclusion tells us nothing about the mechanism of sexual selection, it does not rule out the possibility that this interaction is positively reinforcing this mechanism once it occurs.

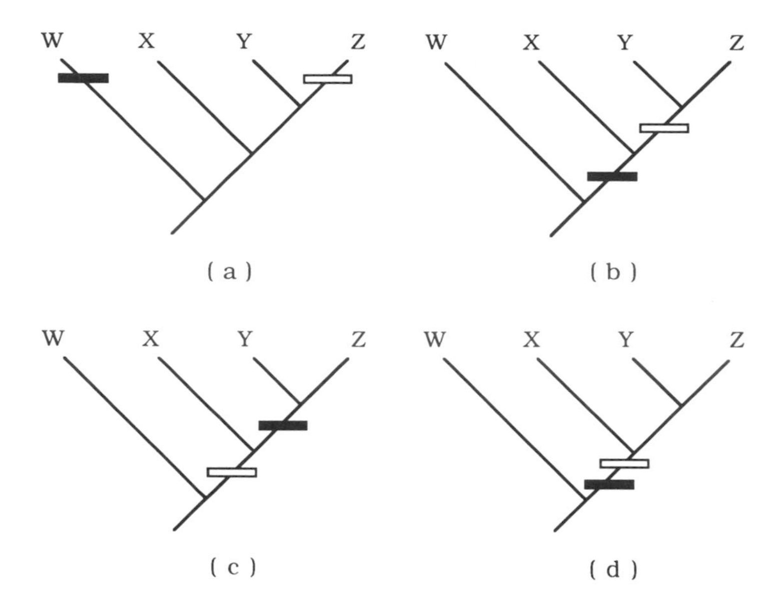

Figure 4.7. Examining phylogenetic relationships between two different characters. Black boxes denote the origin of the parasite-host association. White boxes denote the origin of the sexually selected trait. (a) The parasite-host association and the sexually selected trait arose independently in different species; the traits neither co-originate nor co-occur. (b) The parasite-host association arose before the sexually dimorphic trait; the traits do not co-originate but they do co-occur in some members of the group. (c) The sexually selected trait arose before the parasite-host association; the traits do not co-originate but they do co-occur in some members of the group. (d) Both characters originated in the same ancestor; the traits co-originate and co-occur. Patterns (a) and (c) refute the Hamilton-Zuk hypothesis, pattern (b) provides weak support, and pattern (d) provides strong support for the hypothesis. From McLennan and Brooks (1991).

Overall, then, there are four potential macroevolutionary patterns, two of which support (Fig. 4.7b and 4.7d) and two of which refute (Fig. 4.7a and 4.7c) the hypothesis that there is an evolutionary interaction between the presence of parasites and the development of sexual dimorphism in the study group. Because the sequence of character origin cannot be determined without a phylogeny, a macroevolutionary analysis is a critical first step in distinguishing between systems that will, and those that will not, provide a strong test of the Hamilton-Zuk hypothesis. For example, suppose a researcher decided to investigate a host species because it was "sexually dimorphic" and "had parasites." If this species was like Y and Z in Figure 4.7c, this investigator would either (1) uncover a significant association between parasite burden and degree of male epigamic development and incorrectly conclude that the Hamilton-Zuk mechanism was involved in the initial elaboration of the male character, or (2) uncover no association and correctly conclude that the hypothesis was not supported for this group. However, if phylogenies had been available prior to the experiments (and this is currently a large "if"), the researcher could have determined that the epigamic trait originated prior to the parasite-host association, rejected the hypothesis on macroevolutionary grounds, and saved the time and expense of running the experiment.

The strongest macroevolutionary evidence for the Hamilton-Zuk hypothesis comes from cases in which a host switch coincides with the origin of the sexually selected trait. The proposal that host switches are an important component of the hypothesis complies with the assumption that the interaction between the host and parasite represents a coevolutionary arms race dynamic. However, as discussed in the previous section, there are two general classes of coevolutionary models that address the influence of host switching on parasite-host interactions, only one of which involves a specific cycling of mutual modification, or reciprocal adaptation, of the population ecology or population genetics of the ecologically associated species. Any given episode of association by colonization could be due to an arms race dynamic if phylogenetic patterns of host switching are correlated with the origins of particular defense and counterdefense traits, or such episodes could be due to a resource tracking dynamic if the required resource exhibits a plesiomorphic or convergent distribution among hosts. It is important to distinguish between the two types of host switching because they carry very different implications for the theoretical foundations of the hypothesis. If a macroevolutionary analysis demonstrates that the host switch correlated with the appearance of the sexually selected trait fits a resource tracking model, then we cannot assume a priori that there is a cycle of adaptation/coadaptation occurring in the parasite-host association. On its own, then, this type of pattern does not provide support for the Hamilton-Zuk hypothesis. If, on the other hand, the host switch represents an instance of the evolutionary arms race, then we have strong macroevolutionary support for the existence of the tight coadaptive interaction between parasite and host required by the hypothesis.

This is still not the whole story though. It is important to realize that one macroevolutionary outcome of an arms race dynamic is the production of parasite-free hosts, i.e., hosts that "won" the arms race. The discovery of parasite-free but sexually dimorphic host species would not appear, at first glance, to support the Hamilton-Zuk hypothesis. However, consider the situation depicted in Figure 4.8. In this scenario, the appearance of a new parasite counterdefense character is associated with a host shift by the new parasite species from the old host (species A) to a new host (ancestor z). This host

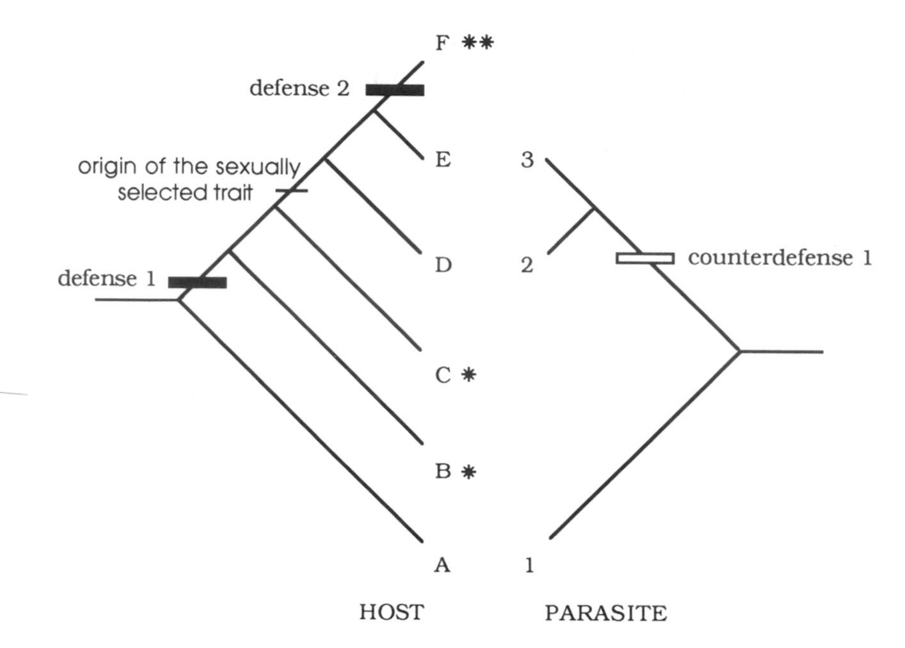

Figure 4.8. One possible outcome of an evolutionary arms race scenario. * = Parasite absent because this host developed defense 1 trait, and was never back-colonized by the parasite species bearing the counterdefense 1 trait; ** = parasite absent because this host developed defense 2 trait and there are no extant parasite species bearing the appropriate counterdefense. From McLennan and Brooks (1991).

switch, in turn, is associated with the appearance and elaboration of a sexually selected character in ancestor z. At this point in the phylogeny we have strong macroevolutionary evidence supporting the Hamilton-Zuk hypothesis. The interaction between a parasite-host arms race dynamic and female choice for a male character reflecting that dynamic is transmitted from ancestor z to three of its descendants, species D and the ancestor of species E + F, and species E, maintaining sexual dimorphism in those species. A novel defense character appears in species F, allowing the host to win the arms race, for the time being, with the parasite. This outcome, in turn, should eventually disrupt the parasite-driven sexual selection dynamic as less information about male quality is incorporated in the male signal. However, the time required for this decay will depend upon a variety of factors, including the initial strength of the interaction (i.e., costs versus benefits of choice for both males and females), the degree of development of the male character, and the genetic/developmental constraints upon changes in the male character. Since this process will be very system dependent, we must consider two alternate explanations for the existence of parasite-free, sexually dimorphic species F: (1) something other than the Hamilton-Zuk dynamic is maintaining sexual dimorphism in this species or (2) insufficient time has passed for us to see the predicted decay in the relationship between the intensity of the male signal and male quality.

At the moment, we have very little information about the difference, if any, in the relationship between degree of pathology and (1) the age of the parasite-host relationship and (2) the type of host switch. For example, the assertion that high pathogenicity is evidence of recent and still imperfect parasite-host relationships (Hegner 1926; Chandler 1940) does not seem unreasonable. Thus, in the invasion of a new host by a prospective parasite, selection would act against the more pathogenic individuals among the parasite species and the less resistant individuals among the host species. This process would lead to a more stable parasite-host relationship. The condition in which a high pathogenicity indicates a newly established parasite might be expected if parasitism arose directly (Baer 1952) and did not form a more benign association. This may be true for some species but there is also evidence that suggests that parasitism may have arisen through intermediate conditions like commensalism (Caullery 1952). As Ball (1943) has pointed out, if this hypothesis is true the view that high pathogenicity indicates a newly established parasite is less plausible. In fact, examples that indicate that high pathogenicity can occur in both newly established and old parasites are known (Ball 1943). It seems, then, that generalizations about the evolution of parasites based on pathogenicity should be avoided until more data are available (Rogers 1962). This information, however, is crucial to the development of a strong coevolutionary theory, and to the understanding of how a particular coevolutionary interaction may interact with the other evolutionary forces operating on that system. Information from macroevolutionary studies (see McLennan and Brooks 1991), integrated with information about the biological basis for parasite-host relationships at the microevolutionary level (for reviews see Møller 1990 and Clayton 1991) will allow us to investigate the Hamilton-Zuk hypothesis within a more robust evolutionary framework. This theory represents more than just an attempt by the good genes school to discover a mechanism for their evolutionary perspective. It also represents a unique opportunity for parasitologists to make an impact on modern evolutionary ideas, and finally to lay to rest the widespread misconception that parasites are nothing more than dismal, degenerate, disease carriers.

OVERTURE

Our interest in studying parasite evolutionary biology, and our hope and optimism for the future of such studies, rest on five perspectives:

1. It is the biology of the parasite and not the biology of the host that is fundamentally important in understanding parasite evolution.
2. Almost all of the traditional generalizations about parasite evolution are myths, metaphors, and misconceptions unsupported by data, and based on a lack of a robust analytical method for reconstructing phylogeny, improper comparisons between parasites and their hosts rather than between parasites and their free-living sister groups, and assumptions that host biology determines parasite biology.
3. It is a good thing that those generalizations are contradicted by data, because they represent archaic views that reinforce the notion that parasites and parasitologists have nothing substantive to offer evolutionary biology.
4. The data suggest that parasite evolutionary biology is just like the evolutionary biology of free-living species, so parasitologists have an opportunity to expand their own discipline, and to do it in such a way that we can make direct comparisons between the findings in parasite evolutionary biology and in general evolutionary biology.
5. Because parasite evolutionary biology does not differ from the evolutionary biology of non-parasites, we will find that parasite systems are excellent models for a variety of general evolutionary studies, and may become the systems of choice for some.

In closing we would like to offer some ideas about research areas to which parasites and parasitologists can make valuable contributions:

1. The information presented in the appendix represents one of the most extensive phylogenetically analyzed morphological data bases available. Consequently, it could play a strong role in ongoing efforts to evaluate the phylogenetic information in new types of data, as well as the efficacy of new methods for analyzing such information. This is particularly relevant to systematic studies using molecular data, an area that holds great promise for expanding our horizons in evolutionary biology but which is experiencing growing pains in the area of phylogenetic analysis (see Hillis and Moritz 1990).
2. Parasites and their hosts are outstanding systems for studies of speciation modes. They are especially suited to studies of peripheral isolates allopatric speciation, because hosts represent highly

discrete evolutionary real estate and host switching is not subject to the caveats about postspeciation dispersal that plague free-living organisms. Once more extensive phylogenetic studies have been performed, some groups (such as the monogeneans) may also be outstanding systems for the study of sympatric speciation.

3. Nonparasitologists should be impressed by the extent to which the individual components of reproductive biology, development, and ecology, as well as their complex interactions, can be highlighted and examined in parasite-host systems. Parasitologists interested in this aspect of evolutionary biology should find an interested audience for their work. Phylogenetic analysis also allows us to examine phylogenetically associated changes in reproductive and nonreproductive, male and female, characters. Based upon this examination, we can ask a variety of questions, such as: What are the the costs and benefits of different reproductive strategies? Do male and female characters covary in either their origin or their loss (digeneans, monogeneans)? What is the relationship between reproductive conservatism and reproductive flexibility in male or female characters (digeneans, monogeneans)? What is the relationship between sexual reproduction and the appearance of character novelty (eucestodes)? If asexual reproduction is good, and sex is better, is sex plus asexual reproduction the best (digeneans)? Finally, the array of biological information known about parasites, coupled with their obvious evolutionary success, makes them excellent models for studies of adaptive radiations.
4. Price (1986) suggested:

> One major advantage of parasite communities over others is that the habitat they live in, the host, has such a well defined structure. . . . The host microcosm is replicated through time and space much more so than habitats for most other organisms. Therefore, the study of comparative community structure is very powerful.

This is an excellent observation; unfortunately there is a major drawback to implementing comparative studies of community structure. There is currently little overlap between parasite groups for which we have extensive community ecological information and groups for which we have extensive phylogenetic information. In addition, there is much misunderstanding on the part of some parasite ecologists about the use of phylogenetic methods and the possible forms of phylogenetic components in community structure, which has led to unproductive and inappropriate polarization of perspectives (e.g., Bush et al. 1990). Ironically, parasitological studies do not lag behind the study of free-living community ecology in this regard. There are, however, indications that this perspective is changing, as a variety of researchers begin to recognize that communities are mosaics of species that evolved elsewhere and dispersed into the area (colonizers) and species that evolved in situ (residents: see discussion in Brooks and McLennan 1991, in press a; in press b; see also contributions in Mayden 1992 and Ricklefs and Schluter in press). Because parasite communities are so well defined and so easily studied (relatively speaking), parasitologists have an opportunity to lead the way in breaking new ground in the study of community evolution.

5. Parasites track broadly through ecosystems, telling us a myriad of interesting things about hosts. For example, parasites can tell us (a) what hosts eat (and what eats them), (b) how much time hosts spend in different types of microhabitats (recall, e.g., that even though *Terrapene carolina* is a terrestrial turtle, it does manage to acquire *Telorchis robustus,* which means it must be eating tadpoles, all of which, in North America, are aquatic), (c) whether hosts are picking up parasites via host switching (and if so, which hosts might be competing with each other), (d) whether or not any of the hosts harbor parasites that are likely to cause disease problems, (e) whether the host changes diet during its lifetime, and (f) which hosts are residents and which are colonizers in the community. Because such a wide range of information can be gleaned with relatively little effort, parasites should be highly useful in biodiversity studies.

We can maximize this information if we begin to think of parasites as biodiversity probes *par excellence,* as libraries of natural and of earth history, rather than as infinitely adaptive, degenerative agents of disease who have abdicated their ability for self-determination in return for an evolutionary sugar daddy cum host. Parasites are developmentally and ecologically complex organisms, subject to, and constrained by, the same rules that govern the evolution of all biological systems. Because of this situation, we need to reexamine the merits of classical research

> based upon the assumption that their evolution is somehow special. The results of such an examination will not force us to abandon decades of research. They will, however, encourage us to throw in our hand with a wider realm of evolutionary biologists, because, as Ernst Mayr (1957) recognized nearly half a century ago, the study of parasites "is not only valuable for the parasitologist, but is also a potential gold-mine for the evolutionist and general biologist."

Parasitologists were at the forefront of the evolutionary debates of the early twentieth century. Parasitologists should be at the forefront of evolutionary biology at the beginning of the twenty-first century. The use of parasitological data in general studies of natural and sexual selection is increasing (see, e.g., the review by Rennie 1992), and large phylogenetic data bases for parasites are accumulating, such as the one presented herein for the parasitic platyhelminths, with its implications for studies of macroevolution. Parasites are still an enigma. But because the pioneering work of five generations of parasite evolutionary biologists can now be integrated within a phylogenetic framework, they need no longer carry an evolutionary stigma.

APPENDIX: THE PHYLOGENETIC DATA BASE

This appendix contains the data base from which we derive our discussions of, and assertions about, parasite evolution. We summarize the current phylogenetic data base for parasitic helminths and protists. The number of studies published each year is increasing, so we cannot claim to have included all of them; in addition, we have not included some studies because we could not reproduce the published results. The reasons for this situation varied: so many equally parsimonious trees that we had no confidence in any minimal groupings (e.g., Brooks 1978b; Moore and Brooks 1987); incomplete data for critical characters and character complexes (Baldwin and Schouest 1990); only a fraction of the species representing the ingroup were studied (e.g., Brooks 1978b; Moore and Brooks 1987; Guegan and Agnese 1991); no outgroups were used (e.g., Hugot 1988; Sey 1991); or a number of the taxa have yet to be described (e.g., Gardner 1991).

We concentrate on helminths and protists for several reasons. First, they are the groups studied by the majority of biologists who call themselves parasitologists. Systematists studying parasitic arthropods, for example, traditionally have stronger ties either with entomology (insects and chelicerates; see Kim 1985 and references therein for a summary of the data base and evolutionary implications for arthropods parasitizing mammals) or with crustacean biology (parasitic crustaceans; e.g., Cressey et al. 1983; Ho and Do 1985; Deets 1987; Benz and Deets 1988; Deets and Ho 1988; Dojiri and Deets 1988; Ho 1988). Second, by assembling the phylogenetic data base for these groups in one place we hope to show how much progress has been made and indicate the scope of the work remaining to be done. And finally, this data base is large enough for us to begin asking some general macroevolutionary questions, including questions about both the geographic and host contexts of parasite speciation (Chapter 2), about parasite adaptation and adaptive radiation (Chapter 3), and about the relationship between macroevolutionary (phylogenetically based) and microevolutionary (population-based) studies using host-parasite systems (Chapter 4).

REPRESENTING THE DATA BASE

Phylogenetic information can be stored and presented in the form of *phylogenetic trees, classifications,* and *diagnoses of taxa.* One of the goals of phylogenetics is to achieve consistency among these modes of data storage and presentation.

Phylogenetic trees are explicit hypotheses of genealogical relationships among taxa.These hypotheses, in turn, serve as the templates for evolutionary explanations about the origins and patterns of diversification of traits, groups, and associations.

Classifications serve as convenient cataloging and information storage and retrieval systems. The phylogenetic approach is based upon Darwin's recognition that evolution is a genealogical process (i.e., natural groups are monophyletic groups). Given this, all taxa depicted in classifications must be monophyletic in order to reflect underlying genealogy. One by-product of this evolutionary criterion is that phylogenetic trees and their classifications are consistent with one another, i.e., one can reconstruct a phylogenetic tree from its classification and vice versa. Phylogenetic classifications can be *exhaustive* or *annotated.* Exhaustive classifications recognize every genealogical division within a group, ensuring that the classification is completely congruent with the phylogenetic tree. This approach, however, stretches the Linnaean system of taxonomic ranks to, and perhaps beyond, its limits. Some authors have suggested that we abandon the Linnaean system in favor of an alternative form of indexing. Others have argued that this step would dissociate earlier research efforts from current ones, and would deprive us of a mechanism for reconstructing the history of ideas about relationships. In general, we will try to preserve as much of the traditional nomenclatorial structure as possible, without violating the condition that this structure must be built with monophyletic blocks.

Wiley et al. (1991) presented three rules that phylogeneticists use to achieve these goals:

> *Rule 1.* Only monophyletic groups will be formally classified.
>
> *Rule 2.* All classifications will be logically consistent with the phylogenetic hypothesis accepted by the investigator.
>
> *Rule 3.* Each classification must be capable of expressing the sister group relationships among the taxa classified.

Biological diversity is so great that a number of variations are possible even within the strictures of these three rules. Consequently, at least nine "conventions" have been proposed to help maintain clarity and consistency (Wiley et al. 1991). The following three conventions are the main ones:

> *Convention 1*. The Linnaean system of ranks will be used.
>
> *Convention 2*. Minimum taxonomic decisions will be made to construct a classification or to modify an existing one (maintain as much continuity with previous nomenclature as possible).
>
> *Convention 3*. Taxa forming an asymmetrical part of a phylogenetic tree may be placed at the same rank and sequenced in their order of branching (the "sequencing" convention). When such a list is encountered the sequence of the list follows the sequence of the branching.

Two additional conventions are used to indicate sources of uncertainty in classifications:

> *Convention 4*. Taxa whose relationships are polytomous (more than two branches at a single node on the phylogenetic tree) will be denoted *sedis mutabilis* (changeable status) and assigned the same rank.
>
> *Convention 5*. Monophyletic taxa of uncertain relationships will be denoted *incertae sedis* (uncertain status) at a level in the phylogenetic hierarchy where their relationships are known with some certainty. This convention is used to identify any grouping for which there are neither synapomorphies (evolutionarily derived homologies indicating genealogical relationships) supporting its monophyly nor synapomorphies linking some members of the group with other groups. This treatment highlights groups that require further research in order to establish or refute their monophyletic status.

The remaining four conventions listed by Wiley et al. (1991) pertain to the manner in which collective, fossil, stem, and hybrid taxa might be indicated. We have not used those conventions in this book.

Diagnoses accompanying phylogenetic classifications provide the evidence supporting each genealogical grouping proposed in the phylogenetic tree. Because of this, diagnoses cannot be the traditional exhaustive lists of all the characteristics displayed by all species included within a group. They are, instead, lists of the synapomorphies that diagnose each group, i.e., those traits that are postulated by phylogenetic systematic analysis to have originated in the stem species of the group. This characteristic draws our attention to the importance of basing diagnoses on the characters that bind a monophyletic group together. Phylogenetic diagnoses also provide hypotheses about the patterns of origin and diversification of various traits. *It is important to recognize that characters may evolve into different states as they are passed from ancestral to descendant species; therefore, there is no a priori reason to believe that all members of a monophyletic group will possess the same state for every character that delineates the group as a whole.* For example, the vertebrate group Tetrapoda is diagnosed by the possession of four appendages even though some members of the group, such as caecilians, legless lizards, and snakes, lack limbs. The larger group to which those limbless taxa belong is diagnosed phylogenetically by the presence of four limbs because the common ancestor of those taxa was the species in which four limbs first appeared. Some assertions about "mistakes" in phylogenetic diagnoses stem from the belief that a character deemed diagnostic at one level cannot be one that has evolved further within the group.

If diagnoses are attached to exhaustive classifications, each synapomorphy will appear only once. If diagnoses are attached to annotated classifications, there may be some redundancy in the listing of synapomorphies. Note, however, that understanding the sequencing convention and phylogenetic methods will still allow you to reconstruct the phylogenetic tree and to place all the synapomorphies at their hypothesized point of origin.

There are two major ways to indicate which traits support each branch on a phylogenetic tree, each having strengths and weaknesses. In one approach, each transformation series has a number, and each state within the transformation series another number. For example, the apomorphic state for binary character 1 in a theoretical analysis would be listed as "1-1" or "1(1)," or perhaps just as "1." For binary character 2, the notation would be "2-1," "2(1)," or "2." Each time a character state appears as a putative apomorphy on a phylogenetic tree thus labeled, it is represented by the same number. Evolutionary reversals to the plesiomorphic condition would be indicated by codes such as "1-0" or "1(0)," or sometimes by "1" accompanied by a special symbol indicating the reversal. Codings for linear multistate transformation series are unambiguous using this notation (i.e., we can assume that state 3-2 is derived from state 3-1), but confusion can result when two apomorphic states are each derived from the plesiomorphic condition (i.e., 3-1 and 3-2 are derived independently from 3-0). This approach allows a specialist to reconstruct a data matrix readily for any given tree, but the reader cannot gain a quick indication of the relative numbers and distribution of apomorphies and homoplasies. In addition, this numerical notation on the tree can become confusing for a complex analysis, and even more so when multiple data sets for different groups are being assembled (i.e., transformation series "6" for one group may not be transformation series "6" for another). Identification of the traits corresponding to the numbers is obtained by referring to phylogenetic diagnoses for the taxa, or to discussions of the transformation series.

In the second approach to labeling phylogenetic trees, each apomorphic trait (character state) is given a different number. The total number of apomorphic traits for each group is reflected in the number accompanying the appropriate branch in the accompanying figure. Each homoplasious trait is indicated by a special symbol on the tree, giving the reader a quick indication of the number and distribution of apomorphies and homoplasies. Like the first type of notation, the identities of each trait, and an indication of which are homoplasious, is obtained from phylogenetic diagnoses accompanying the trees. This second type of notation makes it difficult to reconstruct a data matrix directly from a tree, focusing attention on the characters rather than on numerical codes. It does, however, decrease the potential for confusion when many data sets are considered.

In this appendix, we summarize many different data sets, while in this book, we wish to focus attention on the characters that indicate patterns of phylogenetic relationships and evolutionary diversification among parasite groups. Consequently, we have used the second form of notation for the phylogenetic trees, in conjunction with annotated classifications and cladistic diagnoses. The number of character changes on each branch of each tree is indicated by a number range, and the number of homoplasious changes among those character changes is indicated by the number of asterisks accompanying the range. In the diagnoses each trait is accompanied by the number to which it corresponds on the tree, and an asterisk if that trait is homoplasious on the tree.

THE DATA BASE

Protista

Phylum Apicomplexa Levine 1970 (Fig. 1)

References: Barta (1989); Siddall and Desser (1991).

Group 1 *incertae sedis*

Diagnosis: With characters of the class.

Selenidioides* Levine 1971 *incertae sedis

Diagnosis: With characters of the class.

***Lankesteria* Mingazi 1891**

Diagnosis: Extracellular merogony (1); trophozoites modified into two "body" regions (2).

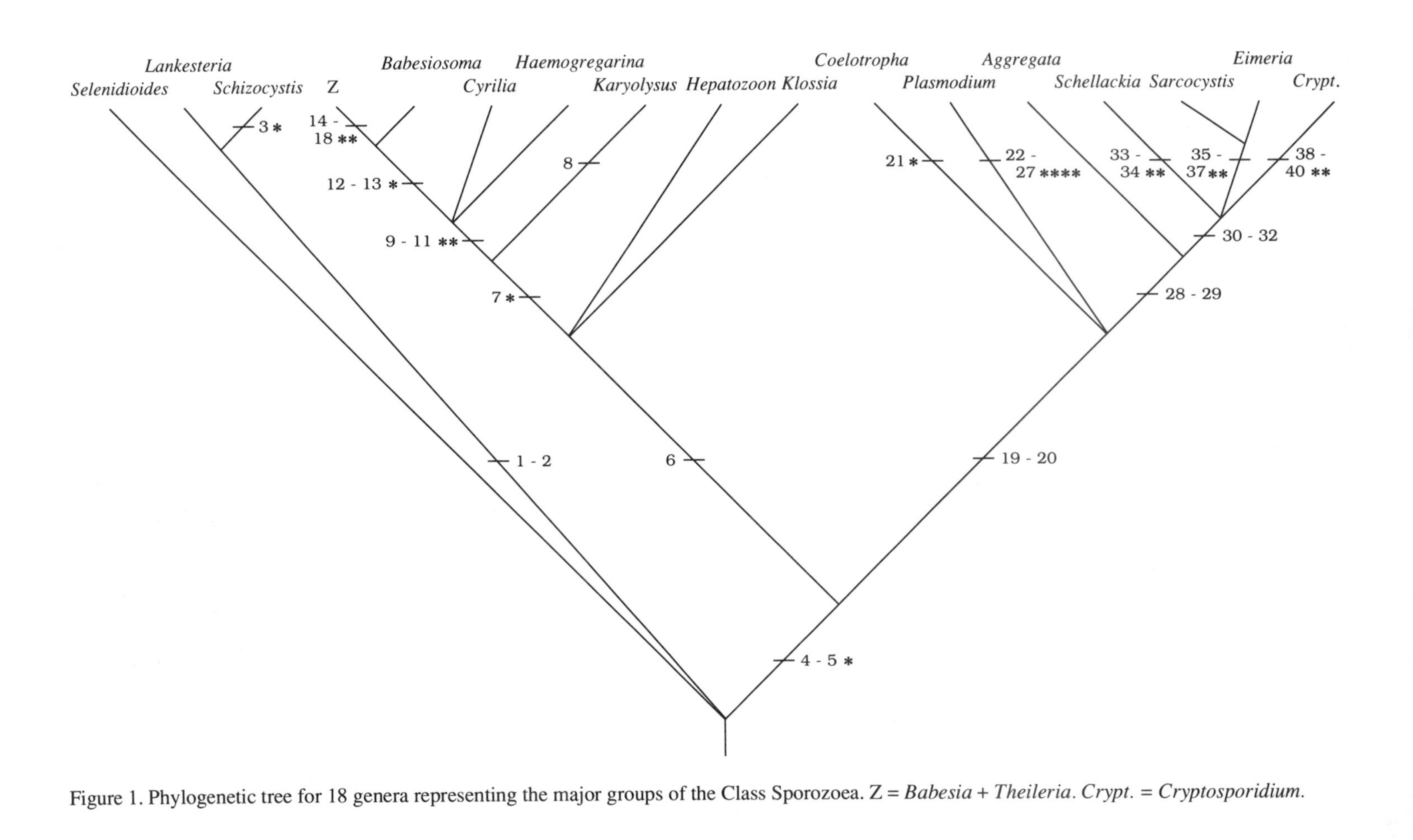

Figure 1. Phylogenetic tree for 18 genera representing the major groups of the Class Sporozoea. Z = *Babesia* + *Theileria*. *Crypt.* = *Cryptosporidium*.

***Schizocystis* Leger and Duboscq 1908**

Diagnosis: Trophozoites modified into two "body" regions (2); no merogony *(3).

Group 2

Diagnosis: No gametocyst (4); anisogamous gamonts *(5); four microgametes formed from each microgamont (6).

Hepatozoon* Miller 1908/*Klossia* Schneider 1875 *incertae sedis

Diagnosis: With characters of group 2.

***Karyolysus* Labbe 1894**

Diagnosis: No persistent cysts *(7); two microgametes formed from each microgamont (8).

Haemogregarina* Danilewsky 1885/*Cyrilia* Lainson 1981 *incertae sedis

Diagnosis: No persistent cysts *(7); intracellular sporogony (9); aflagellate microgametes *(10); isogamous gametes *(11).

Babesiosoma* Jakowska and Nigrelli 1956 *incertae sedis

Diagnosis: No persistent cysts *(7); intracellular sporogony (9); aflagellate microgametes *(10); isogamous gametes *(11); one microgamete formed from each microgamont (12); motile zygote *(13).

***Theileria* Bettencourt, Franca, and Borges 1907/*Babesia* Starcovici 1893**

Diagnosis: No persistent cysts *(7); intracellular sporogony (9); aflagellate microgametes *(10); isogamous gametes *(11); one microgamete formed from each microgamont (12); motile zygote *(13); conoids present in some nongametes *(14); mitochondria acristate (15); mitosis acentriolar (16); amylopectin granules lacking *(17); strahlenkorper present (18).

Group 3

Diagnosis: No gametocyst (4); anisogamous gamonts *(5); no syzygy (19); two or more flagella on microgametes (20).

Coelotropha* Hennere 1963 *sedis mutabilis

Diagnosis: No merogony *(21).

Plasmodium* Marchiafava and Celli 1885 *sedis mutabilis

Diagnosis: Conoids present in some nongametes *(22); amylopectin granules lacking *(23); eight microgametes formed from each microgamont (24); motile zygote *(25); no persistent cysts *(26); microgamete flagella internal (27).

Aggregata* Frenzel 1885 *incertae sedis

Diagnosis: Gametogenesis intracellular (28); flagella in microgametes anterior (29).

Schellackia* Reichenow 1919 *sedis mutabilis

Diagnosis: Gametogenesis intracellular (28); flagella in microgametes anterior (29); crystalloid bodies lacking (30); polar ring complex lacking (31); perforatorium present (32); paranuclear bodies present *(33); no persistent cysts *(34).

Sarcocystis* Lankester 1882 plus *Eimeria* Schneider 1875 *sedis mutabilis

Diagnosis: Gametogenesis intracellular (28); flagella in microgametes anterior (29); crystalloid bodies lacking (30); polar ring complex lacking (31); perforatorium present (32); endodyogeny (35); paranuclear bodies present *(36); wall-forming bodies present *(37).

Cryptosporidium* Tyzzer 1907 *sedis mutabilis

Diagnosis: Gametogenesis intracellular (28); crystalloid bodies lacking (30); polar ring complex lacking (31); perforatorium present (32); no micropores (38); wall-forming bodies present *(39); aflagellate microgametes *(40).

Phylum Sarcomastigophora

Order Diplomonadida (Wenyon 1926) Brugerolle 1975 (Fig. 2)

Reference: Siddall et al. (1992).

Diagnosis: No finlike projections on flagella (1); shallow kinetosomal pockets in nucleus (2); proximal portion of supranuclear fibers microtubular around nucleus/nuclei (3); recurrent flagellum longer than cell (4); infranuclear fiber present (5).

Enteromonas* da Fonseca 1915 *incertae sedis

Diagnosis: With characters of the Diplomonadida.

Trimitus* Alexeieff 1910 *incertae sedis

Diagnosis: Secondary direct microtubules lacking (6); microtubule organizing center of direct fiber simple at KR (recurrent kinetosome) (7); direct fiber not recurved below cytostome (8); doublet plus singlet microtubules associated with distal portion of direct fiber (9).

***Trepomonas* Dujardin 1838**

Diagnosis: Secondary direct microtubules lacking (6); microtubule organizing center of direct fiber simple at KR (7); direct fiber not recurved below cytostome (8); doublet plus singlet microtubules associated with distal portion of direct fiber (9); binary axial symmetry (10); only number 1 flagellum exiting anterior to kinetosomes (11); nucleus pyriform (12); external fibrillar lamella lateral/flange (13); kinetid posterior to center of nucleus (14); recurrent flagellum shorter than cell *(15).

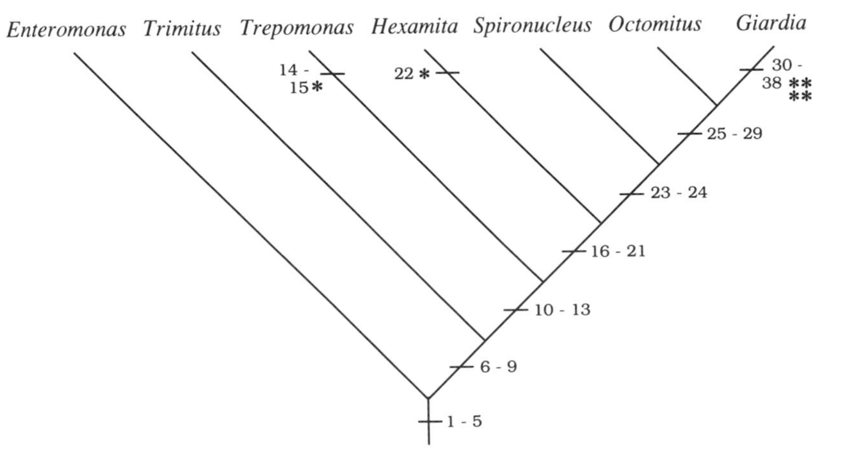

Figure 2. Phylogenetic tree for six genera representing the Diplomonadida.

***Hexamita* Dujardin 1838**

Diagnosis: Secondary direct microtubules lacking (6); direct fiber not recurved below cytostome (8); binary axial symmetry (10); external fibrillar lamella lateral/flange (13); endoplasmic reticulum (ergastoplasm) compact/recurrent (16); cytostomal opening posterior (17); no flagella exiting anterior to kinetosomes (18); cytopharynx a proximal pocket (19); microtubule organizing center of direct fiber indistinct (20); microtubules associated with distal portion of direct fiber (21); nucleus round or oval *(22).

***Spironucleus* Lavier 1936**

Diagnosis: Secondary direct microtubules lacking (6); direct fiber not recurved below cytostome (8); binary axial symmetry (10); nucleus pyriform (12); external fibrillar lamella lateral/flange (13); endoplasmic reticulum (ergastoplasm) compact/recurrent (16); cytostomal opening posterior (17); no flagella exiting anterior to kinetosomes (18); cytopharynx a proximal pocket (19); microtubule organizing center of direct fiber indistinct (20); microtubules associated with distal portion of direct fiber (21); recurrent flagellum medial to posterior of nucleus (23); kinetosomal pocket in nucleus deep (24).

Octomitus* Prowazek 1904 *incertae sedis

Diagnosis: Secondary direct microtubules lacking (6); direct fiber not recurved below cytostome (8); binary axial symmetry (10); nucleus pyriform (12); endoplasmic reticulum (ergastoplasm) compact/recurrent (16); cytostomal opening posterior (17); no flagella exiting anterior to kinetosomes (18); microtubule organizing center of direct fiber indistinct (20); microtubules associated with distal portion of direct fiber (21); recurrent flagellum medial to posterior of nucleus (23); kinetosomal pocket in nucleus deep (24); kinetosomes at K1 and K2 opposing in medial plane of cell (25); axoneme from kinetosome KR intracytoplasmic (26); kinetosome on medial surface of nuclei (27); cytopharynx lacking (28); external fibrillar lamella lacking (29).

***Giardia* Kunstler 1882**

Diagnosis: Secondary direct microtubules lacking (6); direct fiber not recurved below cytostome (8); binary axial symmetry (10); cytostomal opening posterior (17); no flagella exiting anterior to kinetosomes (18); microtubule organizing center of direct fiber indistinct (20); kinetosomes at K1 and K2 opposing in medial plane of cell (25); axoneme from kinetosome KR intracytoplasmic (26); kinetosome on medial surface of nuclei (27); cytopharynx lacking (28); external fibrillar lamella lacking (29); finlike projections on some flagella *(30); kinetosomal pocket in nucleus lacking *(31); endoplasmic reticulum (ergastoplasm) scattered (32); proximal portion of supranuclear fibers microtubular between nuclei (33); nucleus round or oval *(34); infranuclear fiber lacking *(35); triplet microtubules associated with distal portion of direct fiber (36); flagellum number 1 crosses over sagittal plane of cell (37); two rows of microtubules in supranuclear fiber (38).

Metazoa

Phylum Platyhelminthes

Subphylum Rhabdocoela *sensu* Ehlers 1984

Infraphylum Typhloplanoida *sensu* Ehlers 1984

Sub-infraphylum Doliopharyngophora *sensu* Ehlers 1984

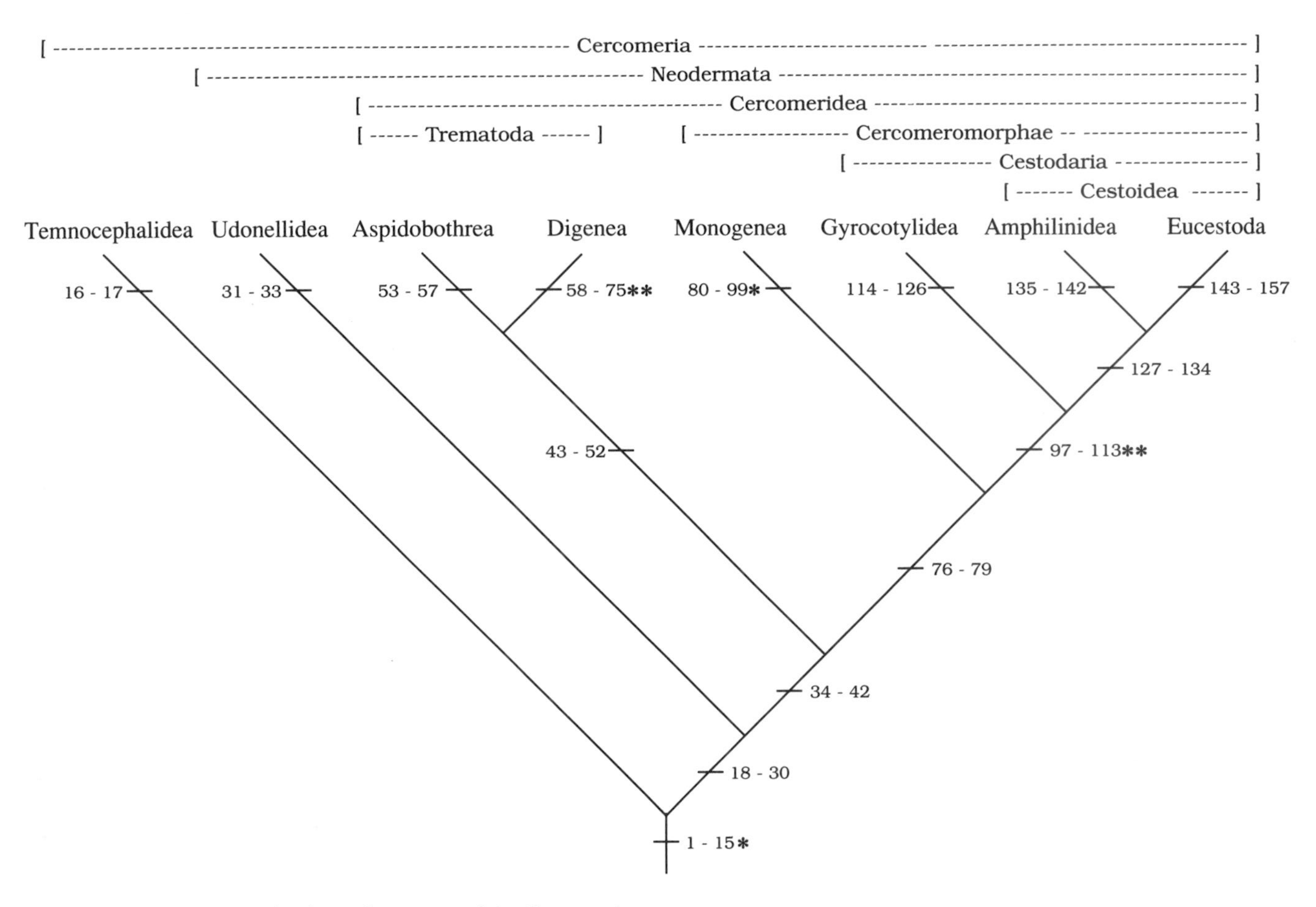

Figure 3. Phylogenetic tree for the major groups of the Cercomeria.

Superclass Cercomeria Brooks 1982 (Fig. 3)

References: Brooks (1982, 1989a,b); Ehlers (1984, 1985a,b, 1986); Wirth (1984); Brooks et al. (1985a,b, 1989); Xylander (1986, 1987a–d, 1988, 1989, 1990); Bandoni and Brooks (1987a,b); Justine (1991a); Boeger and Kritsky (in press).

Diagnosis (doliiform pharynx and reduction of the dual-gland adhesive system indicate membership in Doliopharyngophora): Rhabdocoelous platyhelminths lacking a vagina (1); single ovary (2); paired testes (3); paired lateral excretory vesicles (4); doliiform pharynx (5); saccate gut (6); copulatory stylet (7); without locomotory cilia in adults (8); posterior adhesive organ formed by an expansion of the parenchyma into, minimally, an external pad (9); terminal or subterminal mouth (10); reduction of the dual-gland adhesive system (11); amphistomous juvenile (12); one-host life cycle with arthropod host (13); ectoparasitic *(14); genital aperture midline (15).

Sub-superclass Temnocephalidea Benham 1901

Diagnosis: Cercomeria with cephalic tentacles (16); peripheral layer of microtubules in spermatozoa spirally arranged (17).

Sub-superclass Neodermata Ehlers 1984

Diagnosis: Genital pores in anterior half of body (18); single ventrolateral vagina connecting with oviduct (19); Mehlis's gland (20), vitellaria in adults lateral and follicular (21); no dictyosomes or endoplasmic reticulum in larval epidermis (22); cytoplasmic granules lacking in spermatozoa (23); larval epidermis shed at end of larval stage (24); protonephridia with two-cell weir (25); syncytial postlarval neodermis (26); cilia of larval epidermis with only one rostrally directed rootlet (27); epithelial sensory cells with electron microscopically dense collars (28); epidermal cells in larvae separated by neodermis material (29); copulatory stylet lacking (30).

Class Udonellidea Ivanov 1952

Diagnosis: Secondary protonephridial system of canals and pores (31); giant paranephrocytes (32); arthropod host itself is parasitic on vertebrate (33).

Class Cercomeridea Brooks, O'Grady, and Glen 1985

Diagnosis: Male genital pore and uterus proximate (34); oral sucker present (35); uterus with lateral coiling (36); adult intestine bifurcate (37); spermiogenesis with a proximal-distal fusion, resulting in dorsal and ventral microtubules present in the principal region of the spermatozoon (38); two-host life cycle involving an arthropod and a vertebrate (39); endoparasitic (40); muscular cirrus (41); oviduct straight, intercecal (42).

Subclass Trematoda Rudolphi 1808

Diagnosis: Dorsal vagina a Laurer's canal (43); posterior adhesive organ a sucker (44); male genitalia in adults consisting of cirrus sac, pars prostatica, and internal seminal vesicle (45); male genital pore opening into genital atrium independent of uterine opening (46); operculate eggs usually longer than 50 µm (47); adults with pharynx near oral sucker (48); lamellated walls in protonephridia (49); cercomer shifted to ventral surface (50); single medial excretory vesicle opening posterodorsally (51); two-host life cycle involving a molluscan and a vertebrate (52).

Infraclass Aspidobothrea Burmeister 1856

Diagnosis: Vaginal opening lost (53); specialized microvilli and microtubules in neodermis (54); posterior suckers fused anteriorly (55); hypertrophy and linear subdivision of posterior sucker by transverse septa (56); atrophy of oral sucker (57).

Infraclass Digenea Van Beneden 1858

Diagnosis: First larval stage a miracidium (58); miracidium hatches from egg and swims to snail host (59); miracidum with single pair of flame cells (60); saclike sporocyst stage ("mother sporocyst") in snail host follows miracidium (61); cercaria stage developing in snail follows mother sporocyst (62); cercariae with simple tails (63); cercariae amphistomous (64); cercarial excretory system anepitheliocystid (65); cercarial excretory ducts stenostomatous (66); cercariae with secondary dorsal excretory pore (67); cercariae with primary excretory pore at posterior end of tail (68); cercariae remain in sporocyst until snail host is ingested (69); cercarial intestine bifurcate (70); uteri in adults passing postovarian, then anteriorly to just postbifurcal (71); gut development paedomorphic (does not appear until redial or cercarial stage) (72); tiers of epidermal cells in miracidium (73); no evidence of endoderm in embryos *(74); vitellogenic cells with only one kind of electron-dense vesiculated inclusions *(75).

Subclass Cercomeromorphae Bychowsky 1937

Diagnosis: Posterior adhesive organ armed with hooks, called a cercomer (76); cerebral commissures doubled (77); posterior nervous system commissures doubled (78); 16 hooks on cercomer in larvae (79).

Infraclass Monogenea Van Beneden 1858

Diagnosis: Three rows of ciliary epidermal bands in oncomiracidium larva (one at each end, one in middle) (80); four rhabdomeric eye spots (81); one-host life cycle involving a vertebrate (loss of arthropod host) (82); ectoparasitic *(83); single testis (84); testes postovarian (85); sperm microtubules lying along entire cell periphery (86); cirrus spinose (87); ovate cirrus (88); single egg filament (89); no anchors on oncomiracidium (90); 16 marginal hooks on oncomiracidium (91); haptor disk-shaped (92); 16 marginal hooks in adult (93); dactylogyrid hook shape (94); anchors present in at least one stage of development (95); one pair of ventral anchors (96).

Infraclass Cestodaria Monticelli 1891

Diagnosis: Osmoregulatory system becomes reticulate in late ontogeny (97); intestine lacking (98); posterior body invagination present (99); cercomer paedomorphic, reduced in size, and at least partially invaginated (100); male genital pore not proximate to uterine opening (101); oral sucker/pharynx complex vestigial (102); ovary follicular (103); ovary bilobed (104); testes multiple, in two lateral bands (105); ten equal-sized hooks on cercomer in larvae (106); larval epidermis syncytial (107); vitelloducts syncytial (108); neodermis does not protrude to surface between epidermal cells (109); no desmosomes in the passage of the first excretory canal cells (110); no evidence of endoderm in embryos *(111); vitellogenic cells with only one kind of electron-dense vesiculated inclusions *(112); inner longitudinal muscle layer well developed (113).

Cohort Gyrocotylidea Poche 1926

Diagnosis: Rosette at posterior end of body (114); funnel connecting with rosette short (115); funnel narrow (116); anterolateral genital notch present (117); body margins crenulate (118); body spines small over most of body, large at pharyngeal level (119); large body spines long and narrow (120); testes extending posteriorly only to level of metraterm (121); vitellaria encircling entire body, extending along entire body length (122); no nuclei in larval epidermis (123); no multiciliary nervous receptors (124); no extensions of neodermis into intercellular space between epidermis and basal lamina (125); copulatory papilla present (126).

Cohort Cestoidea Rudolphi 1808

Diagnosis: Male genital pore and vagina proximate (127); cercomer totally invaginated during ontogeny (128); excretory system opens posteriorly in later ontogeny (129); hooks on larval cercomer in two size classes (six large and four small) (130); protonephridial ducts lined with microvilli (131); subepidermal ciliary receptors with true photoreceptor functions lacking in larvae (132); protonephridia in larvae in posterior end of body (133); genital pores marginal (134).

Subcohort Amphilinidea Poche 1922

Diagnosis: Uterine pore and genital pores not proximate (135); male pore at posterior end (136); vaginal pore at posterior end (137); tegument of adults with irregular ridges and depressions (138); uterus N-shaped (139); uterine pore proximal to vestigial pharynx (140); inner longitudinal muscle layer weakly developed (141); adults parasitic in body cavity (142).

Subcohort Eucestoda Southwell 1930

Diagnosis: Body of adults polyzoic (143); cercomer lost during ontogeny (144); six hooks on larval cercomer (145); excretory system reticulate in early ontogeny (146); medullary portion of proglottids restricted (147); hexacanth embryo hatches from egg, is ingested in water (148); second larval stage a procercoid (149); third larval stage a plerocercoid (150); protein embedments in epidermis of hexacanth (151); tegument covered with microtriches (152); sperm lacking mitochondria (153); cerebral development paedomorphic, none seen in larvae (154); "polylecithal" eggs (a large component of vitelline material forming a true shell that is quinone tanned) (155); one embryonic membrane formed by the embryo (with the consequent lack of an embryophore) (156); hexacanths with unicellular protonephridium (157).

Molecular Data

Baverstock et al. (1991) examined the phylogenetic relationships among the major cercomeridean groups, using fragments of 18S rRNA. They expressed preference for a tree that did not unambiguously support the monophyly of the Eucestoda, that suggested that the Monogenea was paraphyletic, that portrayed the Gyrocotylidea as the sister group of one of the two monogenean groups, and that placed the Temnocephalidea in a polytomy with a species of polyclad turbellarian.

Although it is most important to note that there is strong agreement in the results using morphological and rRNA data, it is also important to investigate the possible sources of the disagreement. To do this, we must first attempt to determine which, if either, of the data bases is more robust. Phylogenetic analysis of the rRNA data produced three equally parsimonious trees, compared with a single tree for the morphological data (in addition, an exhaustive search of all 135,135 possible trees for the morphological data demonstrated that the second most parsimonious tree is five steps longer than the most parsimonious one). Baverstock et al. (1991) did not report the three equally parsimonious trees, but presented a strict consensus tree based on them, which differed from their preferred tree only in placing the polyclad turbellarian as the sister species of the temnocephalidean. The consistency index (CI) for the rRNA data is 66% (tree length of 638 steps based on 387 bases), whereas for the morphological data the consistency index is 98.1% (tree length of 157 based on 154 morphological characters). Baverstock et al. (1991) reported that placing the Gyrocotylidea as the sister group of the Amphilinidea plus Eucestoda, as supported by the morphological data, required an additional 52 homoplasious changes in DNA bases, which would decrease the CI 9.9% from 66% to 56.1%. Similarly, placing the Gyrocotylidea as the sister group of the Monogenea requires an additional 17 homoplasious changes in morphological (anatomical and ultrastructural) characters (characters 97–113 mentioned earlier), which would decrease the CI 9.6% from 98.1% to 88.5%.

Baverstock et al. (1991) performed 20 iterations of a bootstrapping analysis in an effort to determine the relative robustness of some of the groupings supported by the RNA data. They found that (1) 60% of the time the two species of eucestodes in the analysis grouped together, supporting the monophyly of the Eucestoda, (2) 100% of the time the Eucestoda were grouped as the sister group of the Amphilinidea, (3) 55% of the time the Gyrocotylidea plus one monogenean taxon grouped as the sister group of the Eucestoda + Amphilinidea, (4) 100% of the time the Gyrocotylidea and that monogenean taxon grouped as sister taxa, (5) 75% of the time the second monogenean taxon grouped as the sister group of the Eucestoda, Amphilinidea, Gyrocotylidea, and first monogenean taxon, (6) 55% of the time the digenean grouped as the sister group of the Cercomeromorphae taxa, (7) 50% of the time the Temnocephalidea grouped as the sister group of the Digenea + Cercomeromorphae, (8) 50% of the time the polyclad turbellarian grouped as the sister group of the Cercomeria taxa, and (9) 85% of the time the triclad turbellarian grouped as the sister group of all the rest. Multiple 100-iteration bootstrapping runs for the morphological data, using 50%, 80%, 95%, and 98% confidence intervals, resulted in (1) 100% support for the Eucestoda + Amphilinidea (Cestoidea), the Cestoidea + Gyrocotylidea (Cestodaria), the Digenea + Aspidobothrea (Trematoda), the Trematoda + Cercomeromorphae (Cercomeridea), the Cercomeridea + Udonellidea (Neodermata), and the Neodermata + Temnocephalidea (Cercomeria) groupings; (2) 95–98.5% support for the Monogenea + Cestodaria (Cercomeromorphae); and (3) 1.5–5% support for two alternative groupings, (a) the Monogenea as the sister group of the Trematoda + Cestodaria, and (b) the Monogenea, the Aspidobothrea, and the Digenea + Cestodaria as sister groups in an unresolved trichotomy. In all cases the Monogenea and the Eucestoda were supported as monophyletic taxa, and in no case was there any support for placing the Gyrocotylidea as the sister group of any of the Monogenea.

The above analyses suggest that the morphological data base is more robust than the current rRNA data base with respect to phylogenetic reconstructions. There may be several reasons for this. First, no representative of the Udonellidea or Aspidobothrea was included in the analysis by Baverstock et al. (1991). Second, no rhabdocoel turbellarian, the group considered to be the sister group of the Cercomeria, was used as an outgroup. Third, only a single species each of temnocephalidean, digenean, gyrocotylidean, and amphilinidean; two species of monogenean; and two species of eucestode were sampled for RNA, whereas the morphological data base is augmented by extensive background information for a large number of species within each taxon. Finally, it may be that the cercomerians are such an old group that the phylogenetic "signal" carried by the relatively simple RNA alphabet of four letters (bases) has been obscured by independent evolution and differential evolutionary rates in some of the lineages (Hillis and Moritz 1990). In this context it is pertinent to note that even among the outgroups, the Baverstock et al. analysis reversed the relationships of the triclad and polyclad turbellarians relative to the phylogenetic analysis of the phylum Platyhelminthes (Ehlers 1984, 1985a,b, 1986).

Cercomeria: Additional Resolution

Infraclass Aspidobothrea Burmeister 1856 (Fig. 4)

References: Gibson (1987); Brooks et al. (1989a).

Order Aspidobothriiformes Monticelli 1892

Diagnosis: With characters of the infraclass.

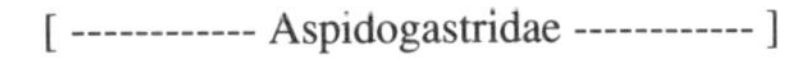

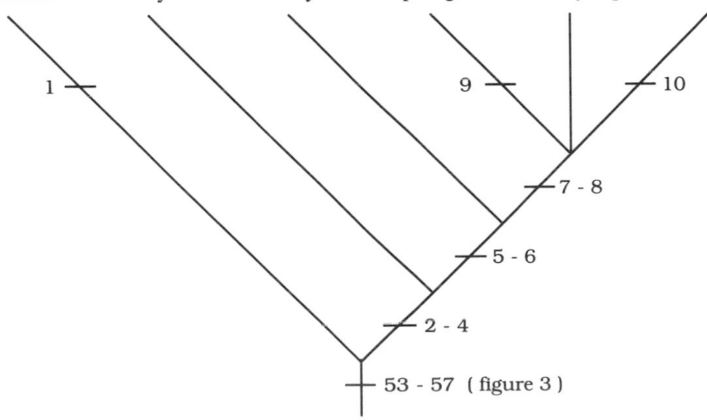

Figure 4. Phylogenetic tree for the families and subfamilies of the Aspidobothrea.

Family Rugogastridae Schell 1973

Diagnosis: Aspidobothriiforms with multiple testes (1).

Family Stichocotylidae Faust and Tang 1936

Diagnosis: Aspidobothriiforms with saccate intestines (2); septate oviducts (3); septum dividing body from disk (4).

Family Multicalycidae Gibson and Chinabut 1984 *incertae sedis*

Diagnosis: Aspidobothriiforms with saccate intestines (2); septate oviducts (3); septum dividing body from disk (4); single testis (5); marginal glands (6).

Family Aspidogastridae Poche 1907

Diagnosis: Aspidobothriiforms with saccate intestines (2); septate oviducts (3); septum dividing body from disk (4); single testis (5); marginal glands (6); loculi present on posterior sucker disk face (7); lateral growth and vertical subdivision of sucker face by septa (8).

Subfamily Cotylaspidinae Chauhan 1954 *sedis mutabilis, incertae sedis*

Diagnosis: With characters of the family.

Subfamily Aspidogastrinae Poche 1907 *sedis mutabilis*

Diagnosis: Aspidogastrids with longitudinal septum on posterior sucker disk face (9).

Subfamily Rohdellinae Gibson and Chinabut 1984 *sedis mutabilis*

Diagnosis: Aspidogastrids with hermaphroditic ducts (10).

Infraclass Digenea Van Beneden 1858 (Fig. 5): Ordinal Relationships

References: Brooks et al. (1985b); Gibson (1987); Cribb (1988); Shoop (1988); Brooks et al. (1989a).

Order Heronimiformes Odening 1961

Diagnosis: Digenea with symmetrically branched sporocysts (1); eggs hatching in utero (2); ventral sucker degenerating in adults (3).

Order Paramphistomiformes Szidat 1936

Diagnosis: Digenea with bioculate cercariae (4); redial stage with appendages *(5); cercariae leaving snail and encysting in the open ("on" something, which may be animal, vegetable, or mineral) (6); pharynx in adults at junction of esophagus and cecal bifurcation (7); some apharyngeate cercariae (8); cycloid excretory system in cercariae (9).

Order Echinostomiformes LaRue 1957

Diagnosis: Digenea with redial stage with appendages *(5); cercariae leave snail host and encyst in the open (6); primary excretory pore in anterior half of cercarial tail (10); ventral sucker in cercariae midventral (11); secondary excretory pore terminal (12); acetabulum in adult midventral (13); adult body spinose (14); no eye spots in cercariae (15); rediae with collars (16); uterus extending from ovary to preacetabular (17).

Order Haploporiformes Brooks, O'Grady, and Glen 1985

Diagnosis: Digenea with bioculate cercariae (4); cercariae leaving snail and encysting in the open (6); primary excretory pore in anterior half of cercarial tail (10); ventral sucker in cercariae midventral (11); secondary excretory pore terminal (12); acetabulum in adult midventral (13); adult body spinose (14); rediae without appendages (18);

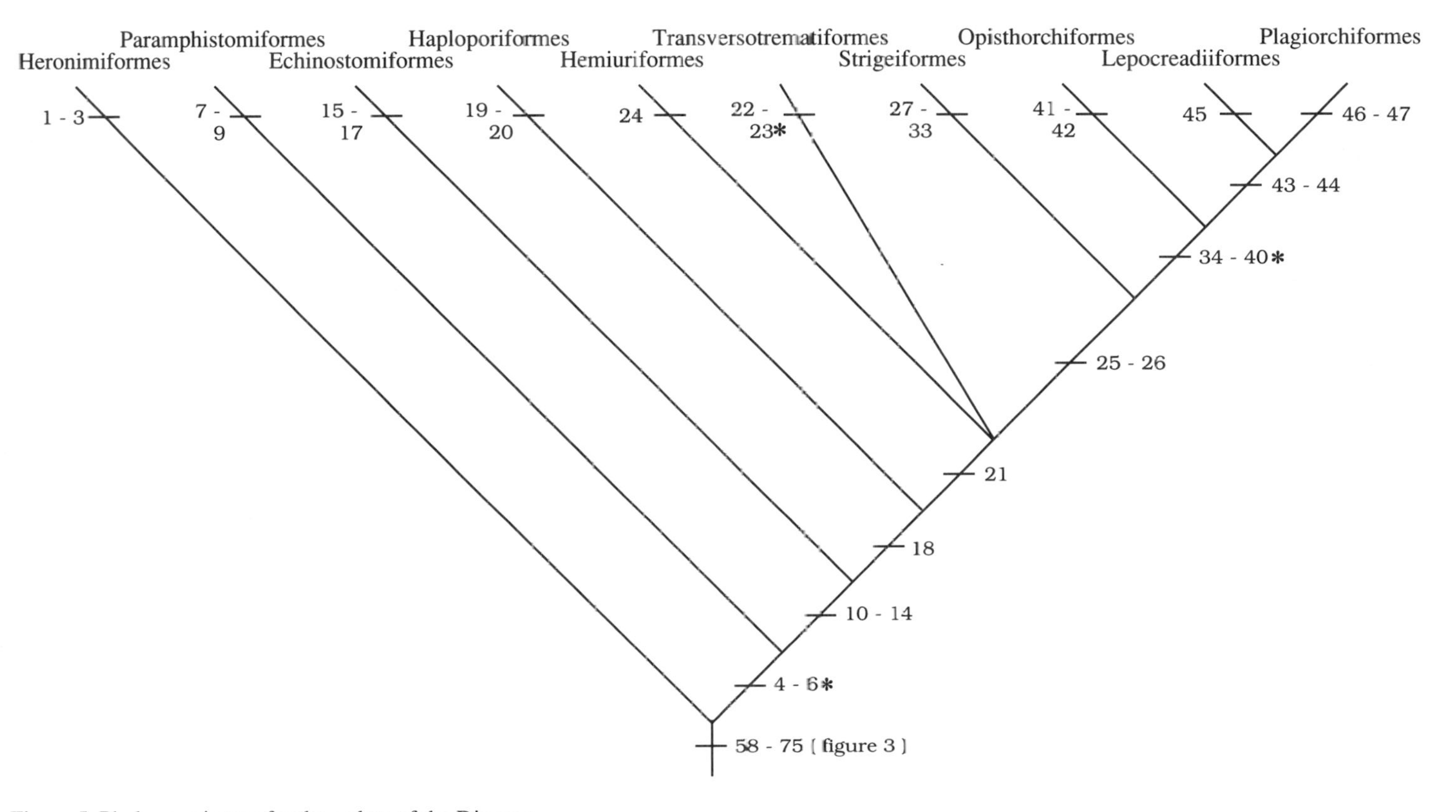

Figure 5. Phylogenetic tree for the orders of the Digenea.

hermaphroditic duct present (19); uterus extending from ovary anteriorly to halfway between bifurcation and pharynx (20).

Order Transversotrematiformes n. ord. ***sedis mutabilis incertae sedis***

Diagnosis: Digenea with bioculate cercariae (4); primary excretory pore in anterior half of cercarial tail (10); furcocercous cercariae (21); body transversely elongate (22); rediae with appendages *(23).

Order Hemiuriformes Travassos et al. 1969 ***sedis mutabilis***

Diagnosis: Digenea with bioculate cercariae (4); primary excretory pore in anterior half of cercarial tail (10); ventral sucker in cercariae midventral (11); secondary excretory pore terminal (12); acetabulum in adult midventral (13); adult body spinose (14); rediae without appendages (18); furcocercous cercariae (21); cystophorous (sensu lato) cercaria (24).

Order Strigeiformes LaRue 1926

Diagnosis: Digenea with bioculate cercariae (4); acetabulum in adult midventral (13); adult body spinose (14); rediae without appendages (18); cercariae encyst "in" second intermediate host (25); mesostomatous excretory system (26); two pairs of flame cells in miracidium (27); no secondary excretory pore in cercariae (28); brevifurcate cercariae (29); primary excretory pores at tips of furcae in cercarial tail (30); ovary between testes (31); genital pore mid-hindbody (32); uterus extending anteriorly from ovary to near acetabulum then posteriorly to genital pore (33).

Order Opisthorchiformes LaRue 1957

Diagnosis: Digenea with bioculate cercariae (4); secondary excretory pore in cercariae terminal (12); acetabulum in adult midventral (13); adult body spinose (14); rediae without appendages (18); cercariae encyst "in" second intermediate host (25); mesostomatous excretory system (26); cercarial tail not furcate *(34); cercarial excretory bladder lined with epithelium (35); oviduct seminal receptacle present (36); primary excretory vesicle in cercariae extending a short distance into tail (37); dorsoventral finfold present on cercarial tail (38); eggs small, generally less than 40 μm (39); eggs ingested and hatch in molluscan host (40); no cirrus sac (41); no cirrus (42).

Order Lepocreadiiformes Brooks, O'Grady, and Glen 1985

Diagnosis: Digenea with bioculate cercariae (4); secondary excretory pore in cercariae terminal (12); acetabulum in adult midventral (13); adult body spinose (14); rediae without appendages (18); cercariae encyst "in" second intermediate host (25); mesostomatous excretory system (26); cercarial tail not furcate *(34); cercarial excretory bladder lined with epithelium (35); oviduct seminal receptacle present (36); primary excretory vesicle in cercariae extending a short distance into tail (37); dorsoventral finfold present on cercarial tail (38); eggs small, generally less than 40 μm (39); eggs ingested and hatch in molluscan host (40); primary excretory pore not extending into tail of cercariae (43); dorsoventral finfold in cercariae small (44); lateral setae on cercarial tail (45).

Order Plagiorchiformes LaRue 1957

Diagnosis: Digenea with secondary excretory pore in cercariae terminal (12); acetabulum in adult midventral (13); adult body spinose (14); rediae without appendages (18);

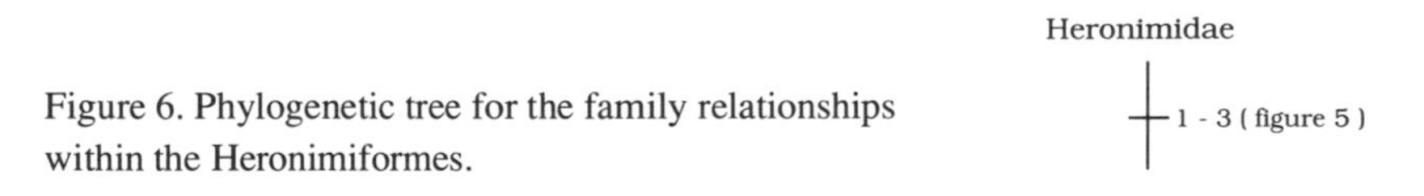

Figure 6. Phylogenetic tree for the family relationships within the Heronimiformes.

cercariae encyst "in" second intermediate host (25); mesostomatous excretory system (26); cercarial tail not furcate *(34); cercarial excretory bladder lined with epithelium (35); oviduct seminal receptacle present (36); primary excretory vesicle in cercariae extending a short distance into tail (37); dorsoventral finfold present on cercarial tail (38); eggs small, generally less than 40 µm (39); eggs ingested and hatch in molluscan host (40); primary excretory pore not extending into tail of cercariae (43); dorsoventral finfold in cercariae small (44); xiphidiocercariae (with anterior stylet) (46); no external seminal vesicle (47).

Infraclass Digenea Van Beneden 1858: Family relationships within orders

References: Brooks et al. (1985b); Lotz (1986); Gibson (1987); Shoop (1988); Brooks et al. (1989a).

Order Heronimiformes Odening 1961 (Fig. 6)

Family Heronimidae Ward 1917

Diagnosis: With characters of the order.

Order Paramphistomiformes Szidat 1936 (Fig. 7)

Family Gyliauchenidae Ozaki 1933

Diagnosis: Paramphistomiforms with short ceca in adults (1).

Family Paramphistomidae Fischoeder 1901 *incertae sedis*

Diagnosis: Paramphistomiforms with posttesticular ovary (2); some apharyngeate adults (3); uterus in adults of some species proceeding directly anteriorly from ovary to immediately postbifurcal (4).

Superfamily Pronocephaloidea Looss 1899

Diagnosis: Paramphistomiforms with posttesticular ovary (2); apharyngeate adults (3); uterus in adults proceeding directly anteriorly from ovary to immediately postbifurcal (4); dorsoterminal secondary excretory pore in cercariae (5); no ciliated patches in cercarial excretory ducts (6); primary excretory pore midtail in cercariae (7); posterior adhesive organs in cercariae (8); no acetabulum in cercariae (9); no acetabulum in adults (10).

Family Microscaphidiidae Looss 1900 *incertae sedis*

Diagnosis: Pronocephaloids with no cirrus sac (11).

Deuterobaris* Group *incertae sedis

Diagnosis: Pronocephaloids with no cirrus sac (11); ventral glands (12).

Family Pronocephalidae Looss 1899

Diagnosis: Pronocephaloids with ventral glands (12); testes in adults symmetrical, at posterior end of body (13); trioculate cercariae (14); esophageal bulb lacking (15); head collar in adults (16).

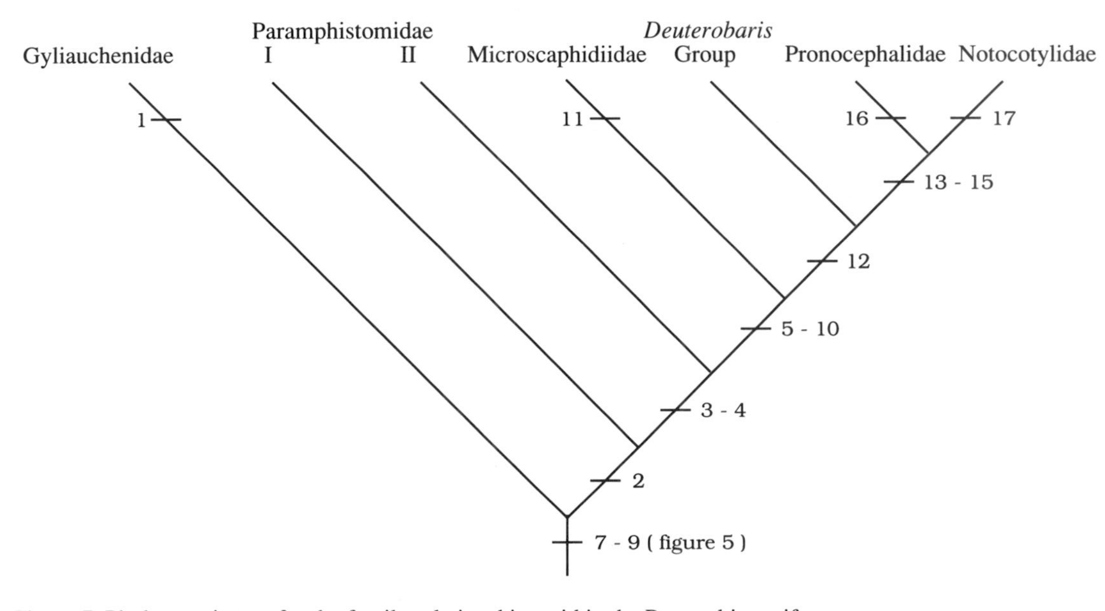

Figure 7. Phylogenetic tree for the family relationships within the Paramphistomiformes.

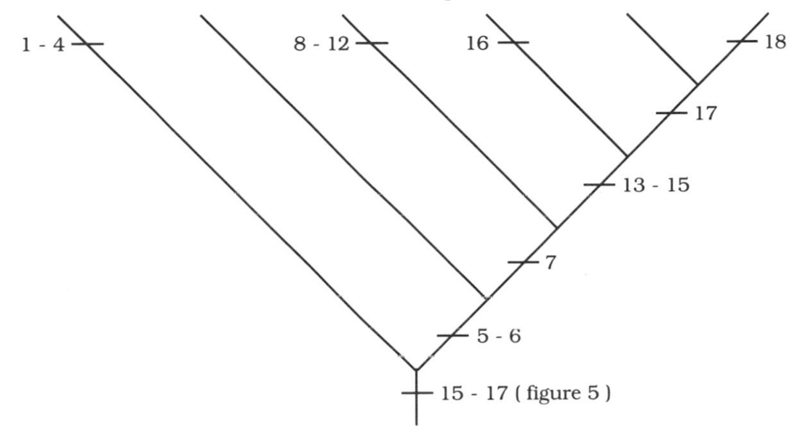

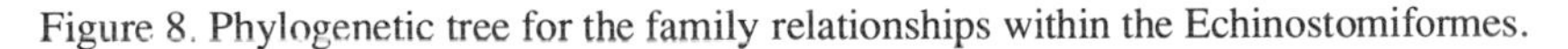
Figure 8. Phylogenetic tree for the family relationships within the Echinostomiformes.

Family Notocotylidae Luhe 1909

Diagnosis: Pronocephaloids with ventral glands (12); testes in adults symmetrical, at posterior end of body (13); trioculate cercariae (14); esophageal bulb lacking (15); eggs less than 50 μm long (17).

Order Echinostomiformes LaRue 1957 (Fig. 8)

Superfamily Cyclocoeloidea Stossich 1902

Diagnosis: Echinostomiforms with acetabulum degenerating in adults (usually lost) (1); uterine opening at pharyngeal level (2); cercariae with cyclocoel (3); adults with cyclocoel (4).

Family Cyclocoelidae Stossich 1902

Diagnosis: With characters of the superfamily.

Superfamily Psilostomoidea Looss 1900

Diagnosis: Echinostomiforms with primary excretory pore in anterior one third of cercarial tail (5); no external seminal vesicle (6).

Family Psilostomidae Looss 1900

Diagnosis: With characters of the superfamily.

Superfamily Fascioloidea Railliet 1895

Diagnosis: Echinostomiforms with primary excretory pore in anterior one third of cercarial tail (5); no external seminal vesicle (6); ventral sucker in anterior one-third of adult body (7); unipartite uncoiled internal seminal vesicle (8); extensive vitellaria throughout body (9); dendritic testes (10); dendritic ovaries (11); dendritic ceca (12).

Family Fasciolidae Railliet 1895

Diagnosis: With characters of the superfamily.

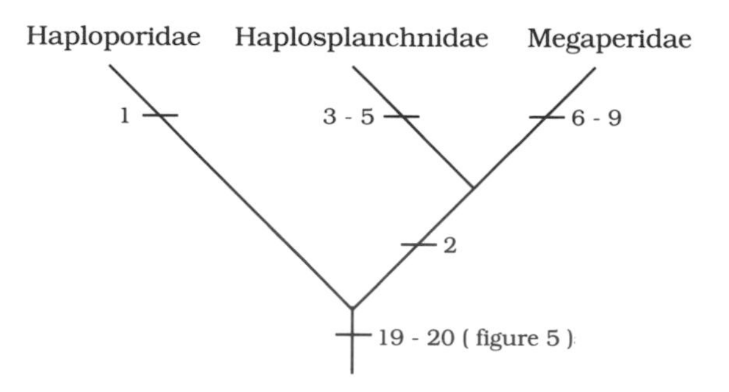

Figure 9. Phylogenetic tree for the family relationships within the Haploporiformes.

Superfamily Echinostomoidea Looss 1899

Diagnosis: Echinostomiforms with primary excretory pore in anterior one-third of cercarial tail (5); no external seminal vesicle (6); ventral sucker in anterior one-third of adult body (7); bipartite internal seminal vesicle (13); head collar in adults (14); spinose head collar in cercariae (15).

Family Philophthalmidae Looss 1899

Diagnosis: Echinostomoids with megalurus cercariae (16).

Family Echinostomidae Looss 1899

Diagnosis: Echinostomoids with armed head collar in adults (lost in some) (17).

Family Rhopaliasidae Looss 1899

Diagnosis: Echinostomoids with proboscides (18).

Order Haploporiformes Brooks, O'Grady, and Glen 1985 (Fig. 9)

Family Haploporidae Looss 1902

Diagnosis: Haploporiforms in which some adults have a single testis *(1).

Family Haplosplanchnidae Poche 1926

Diagnosis: Haploporiforms with digitiform processes on cercarial tail (2); single testis in adult *(3); single cecum (4); no redial stage (5).

Family Megaperidae Manter 1934

Diagnosis: Haploporiforms with membrane linking projections on cercarial tail (6); dorsoventral finfold on cercarial tail (7); preacetabular testes (8); uterus extending from ovary to immediately preacetabular (9).

Order Transversotrematiformes *sedis mutabilis* (Fig. 10)

Family Transversotrematidae Yamaguti 1954

Diagnosis: With characters of the order.

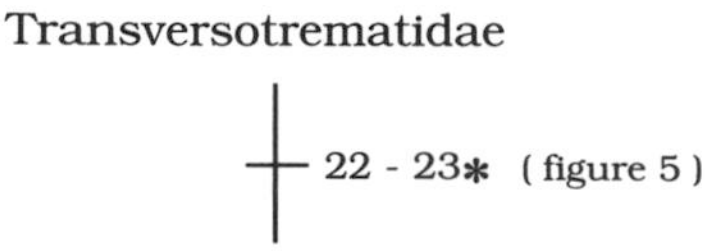

Transversotrematidae

22 - 23* (figure 5)

Figure 10. Phylogenetic tree for the family relationships within the Transversotremati-formes.

Order Hemiuriformes Travassos et al. 1969 (Fig. 11)

Superfamily Bivesiculoidea Yamaguti 1934

Diagnosis: Hemiuriforms with furcocerocus rediae (1); no acetabulum (2); single testis with two vasa efferentia (3); small, ovoid, flattened bodies (4); oculate miracidia (5); V-shaped excretory vesicle (6).

Family Bivesiculidae Yamaguti 1934

Included Group: Tetrodemidae Yamaguti 1971.

Diagnosis: With characters of the superfamily.

Superfamily Hemiuroidea Looss 1899

Diagnosis: Hemiuriforms with miracidia with condensed cilia on surface (7); genital cone present (8); hermaphroditic duct present (9); adults lacking body spines (10); anoculate cercariae (11); postovarian vitellaria (12); arms of excretory ducts usually united anteriorly (13).

Family Ptychogonimidae Dollfus 1937

Diagnosis: Hemiuroids with miracidia with "condensed cilia" on outer surface (7); genital cone present (8); hermaphroditic duct present (9); adults lacking body spines (10); anoculate cercariae (11); postovarian vitellaria (12); arms of excretory ducts usually united anteriorly (13); genital sac absent *(14). Cercariae may be microcercous.

Family Azygiidae Luhe 1909

Diagnosis: Hemiuroids with miracidia with "condensed cilia" on outer surface (7); genital cone present (8); hermaphroditic duct present (9); adults lacking body spines (10); anoculate cercariae (11); postovarian vitellaria (12); arms of excretory ducts usually united anteriorly (13); ovary posttesticular *(15); no interceding uterine coils (16); testes immediately postacetabular (17); genital sac a prostatic sac (18).

Family Hirudinellidae Dollfus 1932 *sedis mutabilis*

Diagnosis: Hemiuroids with miracidia with "condensed cilia" on outer surface (7); genital cone present (8); adults lacking body spines (10); anoculate cercariae (11); postovarian vitellaria (12); arms of excretory ducts usually united anteriorly (13); ovary posttesticular *(15); no interceding uterine coils (16); testes immediately postacetabular (17); condensed cilia in the form of spines (19); genital pore between cecal bifurcation and mouth (20); cystophorous cercariae (sensu stricto) (21); compact vitellaria (22); drusenmagen (prececal sacs of Manter 1970; stomach of Yamaguti 1971) (23); hermaphroditic duct lost *(24); excretory arms leave excretory vesicle dorsoventrally *(25).

Family Bathycotylidae Dollfus 1932 *sedis mutabilis*

Diagnosis: Hemiuroids with miracidia with "condensed cilia" on outer surface (7); hermaphroditic duct present (9); adults lacking body spines (10); anoculate cercariae (11); postovarian vitellaria (12); arms of excretory ducts usually united anteriorly (13); ovary posttesticular *(15); no interceding uterine coils (17); condensed cilia in the form of spines (19); genital pore between cecal bifurcation and mouth (20); cystophorous cercariae (sensu stricto) (21); compact vitellaria (22); drusenmagen (prececal sacs of Manter 1970; stomach of Yamaguti 1971) (23); genital cone absent (26); genital sac absent *(27); intertesticular ovary (28).

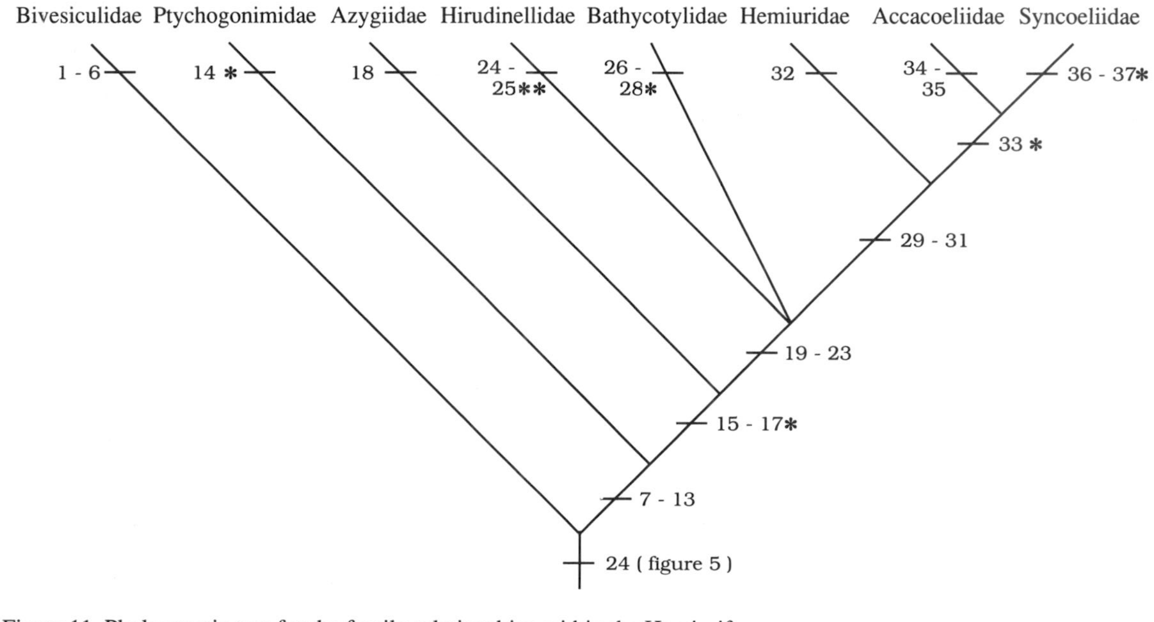

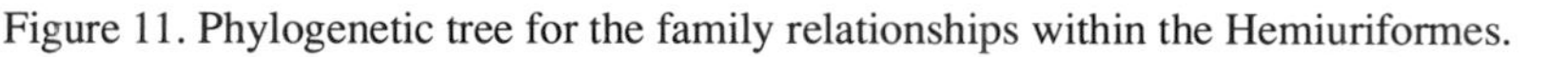
Figure 11. Phylogenetic tree for the family relationships within the Hemiuriformes.

Family Hemiuridae Looss 1899

Diagnosis: Hemiuroids with miracidia with "condensed cilia" on outer surface (7); genital cone present (8); adults lacking body spines (10); anoculate cercariae (11); postovarian vitellaria (12); arms of excretory ducts usually united anteriorly (13); ovary posttesticular *(15); testes immediately postacetabular (17); condensed cilia in the form of spines (19); genital pore between cecal bifurcation and mouth (20); cystophorous cercariae (sensu stricto) (21); compact vitellaria (22); drusenmagen (prececal sacs of Manter 1970; stomach of Yamaguti 1971) (23); ovary with interceding uterine coils (29); long hermaphroditic duct (30); external seminal vesicle only (31); Laurer's canal modified into a seminal receptacle (up to and including a Juel's organ: Gibson and Bray 1979) (32).

Family Accacoeliidae Odhner 1911

Diagnosis: Hemiuroids with miracidia with "condensed cilia" on outer surface (7); genital cone present (8); hermaphroditic duct present (9); adults lacking body spines (10); anoculate cercariae (11); postovarian vitellaria (12); arms of excretory ducts usually united anteriorly (13); ovary posttesticular *(15); testes immediately postacetabular (17); condensed cilia in the form of spines (19); genital pore between cecal bifurcation and mouth (20); cystophorous cercariae (sensu stricto) (21); compact vitellaria (22); ovary with interceding uterine coils (29); long hermaphroditic duct (30); external seminal vesicle only (31); excretory arms oriented dorsoventrally *(33); drusenmagen with anterior diverticula (34); pedunculate acetabulum (35). Cercariae possibly diplocercous.

Family Syncoeliidae Looss 1899

Diagnosis: Hemiuroids with miracidia with "condensed cilia" on outer surface (7); genital cone present (8); hermaphroditic duct present (9); adults lacking body spines (10); anoculate cercariae (11); postovarian vitellaria (12); arms of excretory ducts usually united anteriorly (13); ovary posttesticular *(15); testes immediately postacetabular (17); condensed cilia in the form of spines (19); genital pore between cecal bifurcation and mouth (20); cystophorous cercariae (sensu stricto) (21); compact vitellaria (22); ovary with interceding uterine coils (29); long hermaphroditic duct (30); external seminal vesicle only (31); excretory arms oriented dorsoventrally *(33); sinuous ceca (36); drusenmagen absent *(37).

Order Strigeiformes LaRue 1926 (Fig. 12)

Superfamily Clinostomoidea Luhe 1901

Diagnosis: Strigeiforms with body finfold on cercariae (1); retractable head collar in adults (2).

Family Clinostomidae Luhe 1901

Diagnosis: With characters of the superfamily.

Superfamily Schistosomatoidea Stiles and Hassall 1898

Diagnosis: Strigeiforms with paedomorphic rediae ("daughter sporocysts") (3); apharyngeate cercariae (4); apharyngeate adults (5); no metacercarial stage (6).

Family Sanguinicolidae Graff 1907

Diagnosis: Schistosomatoids with X- or H-shaped ceca in adults (7); no oral sucker in adults (8); no oral sucker in cercariae (9); no acetabulum in adults *(10); no acetabulum in cercariae *(11): body finfold on cercariae (12); no cirrus sac *(13); no cirrus *(14).

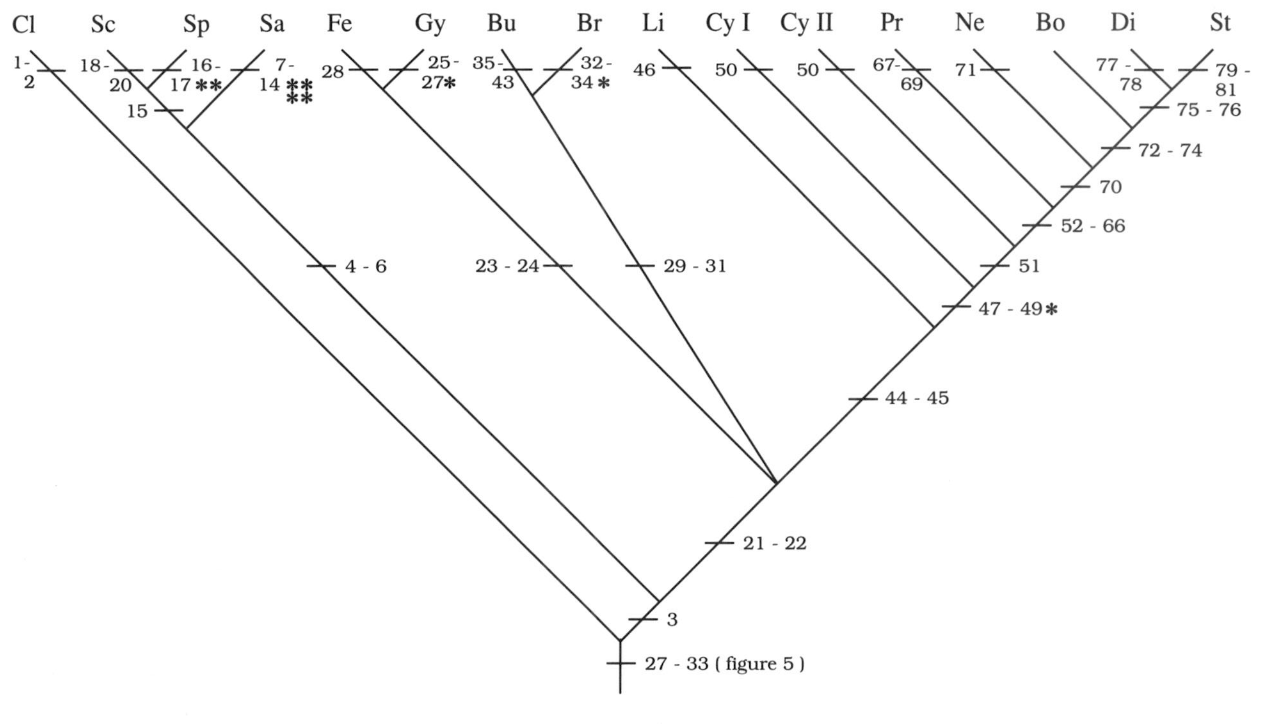

Figure 12. Phylogenetic tree for the family relationships within the Strigeiformes. Cl = Clinostomidae; Sc = Schistosomatidae; Sp = Spirorchidae; Sa = Sanguinicolidae; Fe = Fellodistomidae; Gy = Gymnophallidae; Bu = Bucephalidae; Br = Brachylaemidae; Li = Liolopidae; Cy = Cyathocotylidae (probably paraphyletic, hence Cy I and Cy II); Pr = Proterodiplostomidae; Ne = Neodiplostomidae; Bo = Bolbophoridae; Di = Diplostomidae; St = Strigeidae.

Family Spirorchidae Stunkard 1921

Diagnosis: Schistosomatoids with two types of penetration glands in cercariae (15); acetabulum lacking in some adults *(16); acetabulum lacking in some cercariae *(17).

Family Schistosomatidae Stiles and Hassall 1898

Diagnosis: Schistosomatoids with two types of penetration glands in cercariae (15); postacetabular uterine opening (18); adults dioecious (19); gynecophoric canal or chamber present (20).

Superfamily Gymnophalloidea Odhner 1905 *sedis mutabilis*

Diagnosis: Strigeiforms with paedomorphic rediae ("daughter sporocysts") (3); longifurcate cercariae (21); spinose cirrus in adults (22); setiferous cercarial tail (23); large excretory vesicle in cercariae (vesicle extends into caudal region) (24).

Family Gymnophallidae Odhner 1905

Diagnosis: Gymnophalloids with no cirrus sac *(25); dichotoma or cercariaeum cercariae (26); germinal sacs reported in some descriptions (27).

Family Fellodistomidae Nicoll 1909

Diagnosis: Gymnophalloids with aspinose tegument (28).

Superfamily Brachylaemoidea Joyeux and Foley 1930 *sedis mutabilis*

Diagnosis: Strigeiforms with paedomorphic rediae ("daughter sporocysts") (3); longifurcate cercariae (21); spinose cirrus (22); subterminal genital pore (29); branched rediae ("branched sporocysts") (30); miracidia with ciliated bars (31).

Family Brachylaemidae Joyeux and Foley 1930

Diagnosis: Brachylaemoids with cercariaeum or obscuromicrocercous cercariae (32); stenostomate excretory system (33); testes in adults tandem or oblique at posterior end of the body (34).

Family Bucephalidae Poche 1907

Diagnosis: Brachylaemoids with relatively compact preovarian vitellaria (35); acetabulum at anterior end of cercarial body (36); acetabulum at anterior end of adult body (37); mouth midventral in cercariae (38); mouth midventral in adults (39); cercarial intestine saccate (40); adult intestine saccate (41); genital pore terminal (42); longifurcate cercariae with short tail stem (gasterostome type cercariae) (43).

Superfamily Strigeoidea Railliet 1909 *sedis mutabilis*

Diagnosis: Strigeiforms with paedomorphic rediae ("daughter sporocysts") (3); longifurcate cercariae (21); spinose cirrus (22); paranephridial plexus in cercariae (44); paranephridial plexus in adults (45).

Family Liolopidae Odhner 1912

Diagnosis: Strigeoids with lateral genital pores (46).

Family Cyathocotylidae Muhling 1898 *incertae sedis*

Diagnosis: Strigeoids with no cirrus *(47); nonpapillose tribocytic organ (48); terminal genital pore (49); finfolds on cercarial tail furcae (50); ovary anterior to testes in some species (51).

Family Proterodiplostomidae Dubois 1936

Diagnosis: Strigeoids with no cirrus *(47); finfolds on cercarial tail furcae (50); primary excretory pores in middle of cercarial furcae (52); vitellaria extensive (53);

hermaphroditic duct present (54); adult excretory pore dorsal, subterminal (55); papillose tribocytic organs (56); vitelline follicles in fore- and hindbody (57); large paraprostate gland (58); without suckers in genital atrium (59); genital bursa lacking (60); no genital diverticula (61); genital atrium not inflated (62); dorsal, subterminal genital pores (63); no dorsal accessory genital suckers (64); no anterior pseudosuckers (65); Mehlis's gland intertesticular (66); paraprostate and male ejaculatory duct fused, uterus separate (67); paraprostate organ in adults (68); paraprostate organ in metacercariae (69).

Family Neodiplostomidae Shoop 1989

Diagnosis: Strigeoids with no cirrus *(47); finfolds on cercarial tail furcae (50); primary excretory pores in middle of cercarial furcae (52); vitellaria extensive (53); hermaphroditic duct present (54); adult excretory pore dorsal, subterminal (55); papillose tribocytic organs (56); vitelline follicles in fore- and hindbody (57); large paraprostate gland (58); without suckers in genital atrium (59); genital bursa lacking (60); no genital diverticula (61); genital atrium not inflated (62); dorsal, subterminal genital pores (63); no dorsal accessory genital suckers (64); no anterior pseudosuckers (65); Mehlis's gland intertesticular (66); lobate testes (70); paranephridial plexus comprising three longitudinal vessels with some anastomoses forming distinct transverse commissures (71).

Family Bolbophoridae Shoop 1989

Diagnosis: Strigeoids with no cirrus *(47); finfolds on cercarial tail furcae (50); primary excretory pores in middle of cercarial furcae (52); vitellaria extensive (53); hermaphroditic duct present (54); adult excretory pore dorsal, subterminal (55); papillose tribocytic organs (56); vitelline follicles in fore- and hindbody (57); large paraprostate gland (58); without suckers in genital atrium (59); genital bursa lacking (60); no genital diverticula (61); genital atrium not inflated (62); dorsal, subterminal genital pores (63); no dorsal accessory genital suckers (64); Mehlis's gland intertesticular (66); lobate testes (70); anterior pseudosuckers (72); paranephridial plexus comprising three longitudinal vessels with distinct transverse commissures reduced to three or fewer and some smaller transverse anastomoses (73); anterior pseudosuckers on metacercaria (74).

Family Diplostomidae Poirier 1886

Diagnosis: Strigeoids with no cirrus *(47); finfolds on cercarial tail furcae (50); primary excretory pores in middle of cercarial furcae (52); vitellaria extensive (53); hermaphroditic duct present (54); adult excretory pore dorsal, subterminal (55); papillose tribocytic organs (56); vitelline follicles in fore- and hindbody (57); large paraprostate gland (58); without suckers in genital atrium (59); genital bursa lacking (60); no genital diverticula (61); genital atrium not inflated (62); dorsal, subterminal genital pores (63); no dorsal accessory genital suckers (64); Mehlis's gland intertesticular (66); lobate testes (70); anterior pseudosuckers (72); paranephridial plexus comprising three longitudinal vessels, three or fewer transverse commissures, and no other anastomoses (75); anterior pseudosuckers on metacercaria (76); no metacercarial encystment (77); enclosed limebodies in metacercaria (78).

Family Strigeidae Railliet 1919

Diagnosis: Strigeoids with no cirrus *(47); finfolds on cercarial tail furcae (50); primary excretory pores in middle of cercarial furcae (52); vitellaria extensive (53);

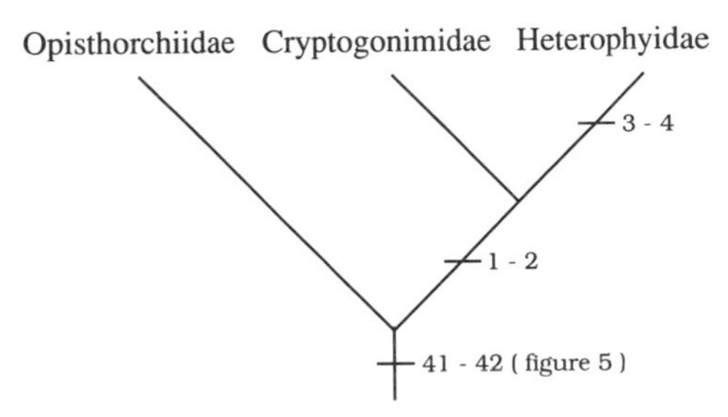

Figure 13. Phylogenetic tree for the family relationships within the Opisthorchiformes.

hermaphroditic duct present (54); adult excretory pore dorsal, subterminal (55); vitelline follicles in fore- and hindbody (57); large paraprostate gland (58); without suckers in genital atrium (59); genital bursa lacking (60); no genital diverticula (61); genital atrium not inflated (62); dorsal, subterminal genital pores (63); no dorsal accessory genital suckers (64); Mehlis's gland intertesticular (66); lobate testes (70); anterior pseudosuckers (72); paranephridial plexus comprising three longitudinal vessels, three or fewer transverse commissures, and no other anastomoses (75); anterior pseudosuckers on metacercaria (76); cup-shaped forebody (79); bilobate tribocytic organ (80); cup-shaped forebody in metacercaria (81).

Order Opisthorchiformes LaRue 1957 (Fig. 13)
Family Opisthorchiidae Looss 1899 *incertae sedis*
Diagnosis: With characters of the order.
Family Cryptogonimidae Ward 1917 *incertae sedis*
Diagnosis: Opisthorchiforms with ventrogenital sac (1); hermaphroditic duct *(2).
Family Heterophyidae Odhner 1914
Diagnosis: Opisthorchiforms with ventrogenital sac (1); hermaphroditic duct *(2); some cercariae without acetabulum *(3); some parapleurolophocercous (tails with lateral finfolds) cercariae (4).

Order Lepocreadiiformes Brooks, O'Grady, and Glen 1985 (Fig. 14)
Family Deropristidae Cable and Hunninen 1942 *incertae sedis*
Diagnosis: With characters of the order.

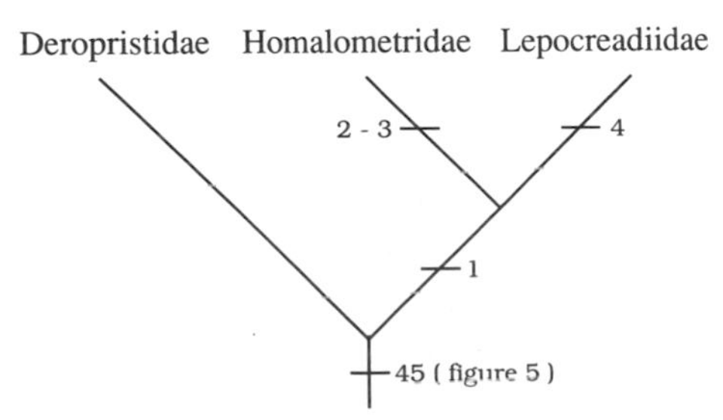

Figure 14. Phylogenetic tree for Lepocreadiiformes.

Family Homalometridae Cable and Hunninen 1942

Diagnosis: Lepocreadiiforms with preovarian uterus *(1); no cirrus sac *(2); hermaphroditic duct *(3).

Family Lepocreadiidae Odhner 1905

Diagnosis: Lepocreadiiforms with preovarian uterus *(1); trichocercous (proliferation of caudal setae) cercariae (4).

Order Plagiorchiformes LaRue 1957 (Fig. 15a,b)

Suborder Allocreadiata Brooks, O'Grady, and Glen 1985

Diagnosis: Plagiorchiforms without body spines (1).

Superfamily Allocreadioidea Stossich 1903

Diagnosis: With characters of the suborder.

Family Allocreadiidae Stossich 1903

Diagnosis: With characters of the superfamily.

Suborder Acanthocolpata Brooks, O'Grady and Glen 1985

Diagnosis: Plagiorchiforms with a large spinose cirrus (2); spinose lining of the metraterm and distal portion of the uterus (3); a long prepharynx (4); a subterminal to terminal oral sucker (5).

Superfamily Acanthocolpoidea Luhe 1906

Diagnosis: With characters of the suborder.

Family Acanthocolpidae Luhe 1906

Diagnosis: Acanthocolpoidea with spinose oral sucker (6).

Family Campulidae Odhner 1926

Diagnosis: Acanthocolpoidea with highly lobate or dendritic testes (7); triangular eggs (8).

Suborder Troglotrematata Brooks, O'Grady, and Glen 1985

Diagnosis: Plagiorchiforms with anoculate cercariae (9); uterus extending posteriorly from ovary, then anteriorly to immediately postbifurcal (10); seminal receptacle reduced or lacking *(11); chaeto- or sulcatomicrocercous cercariae (12); miracidium hatches and swims to molluscan host *(13).

Superfamily Troglotrematoidea Braun 1915

Diagnosis: With characters of the suborder.

Family Troglotrematidae Braun 1915

Diagnosis: With characters of the superfamily.

Suborder Renicolata LaRue 1957

Diagnosis: Plagiorchiforms with anoculate cercariae (9); paedomorphic rediae (daughter sporocysts) (14); rudimentary acetabulum (15); no cirrus sac *(16).

Superfamily Renicoloidea Dollfus 1939

Diagnosis: With characters of the suborder.

Family Renicolidae Dollfus 1939

Diagnosis: With characters of the superfamily.

Suborder Macroderoidiata suborder n. *incertae sedis*

Diagnosis: Plagiorchiforms with paedomorphic rediae (daughter sporocysts) (14); ornatae xiphidiocercariae (17).

Superfamily Macroderoidea *incertae sedis*

Diagnosis: With characters of the suborder.

Family Macroderoididae McMullen 1937

Diagnosis: With characters of the superfamily.

Suborder Plagiorchiata LaRue 1957 *incertae sedis*

Diagnosis: Plagiochiforms with paedomorphic rediae (daughter sporocysts) (14); armatae xiphidiocercariae (or modification) (18).

Infra-suborder Opecoelatea Brooks, Bandoni, Macdonald, and O'Grady 1989

Diagnosis: Plagiorchiates with short ceca *(19); transverse testes (20); transverse genital sacs (21); marginal or submarginal genital pores (22).

Superfamily Opecoeloidea Ozaki 1925

Diagnosis: Opecoelates with cotylomicrocercous cercariae *(23).

Family Opecoelidae Ozaki 1925

Diagnosis: Opecoeloids with preovarian uterus *(24).

Family Zoogonidae Odhner 1902

Diagnosis: Opecoeloids with cercariaeum cercariae *(25); vitelline follicles few in number or in single compact mass, postovarian (26); miracidia hatching and swimming to first host (27).

Family Lissorchiidae Poche 1926

Diagnosis: Opecoeloids with cercariaeum cercariae *(25).

Superfamily Microphalloidea Ward 1901

Diagnosis: Opecoelates with V-shaped excretory vesicles (28); small "daughter sporocysts" with few cercariae (29); extremely small adult bodies (30).

Family Microphallidae Ward 1901

Diagnosis: Microphalloids with prominent and highly modified terminal genitalia, including muscularized genital atrium (31); "ubiquita" cercariae (32).

Family Lecithodendriidae Odhner 1910

Diagnosis: Microphalloids with vitellaria bunched in "shoulder" region (33); "virgulate" cercariae (34).

Family Prosthogonimidae Luhe 1909

Diagnosis: Microphalloids with vitellaria bunched in "shoulder" region (33); genital pores near the oral sucker (35); secondarily elongate cirrus sacs *(36).

Infra-suborder Plagiorchiatea LaRue 1957 *incertae sedis*

Diagnosis: With characters of the suborder.

Superfamily Plagiorchioidea Ward 1917 *sedis mutabilis*

Diagnosis: With characters of the infra-suborder.

Family Plagiorchiidae Ward 1917

Diagnosis: With characters of the superfamily.

Superfamily Dicrocoelioidea Looss 1899 *sedis mutabilis*

Diagnosis: Plagiorchiateans with very simple male genitalia—very small cirrus sac, cirrus, and seminal vesicle (37).

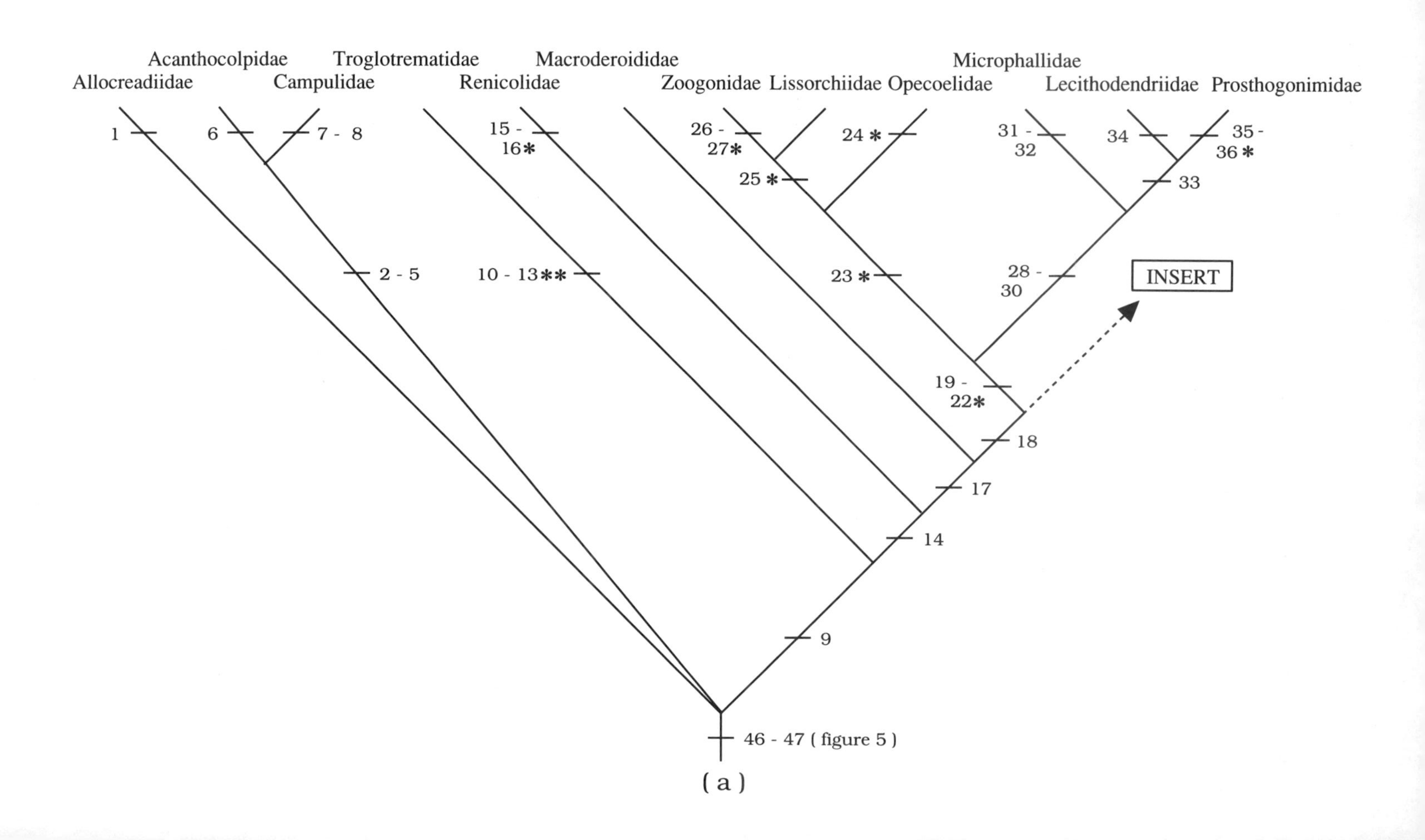
Acanthocolpidae
Troglotrematidae
Macroderoididae
Microphallidae
Allocreadiidae
Campulidae
Renicolidae
Zoogonidae
Lissorchiidae
Opecoelidae
Lecithodendriidae
Prosthogonimidae
1
6
7 - 8
15 - 16*
26 - 27*
24 *
31 - 32
34
35 - 36 *
25 *
33
2 - 5
10 - 13**
23 *
28 - 30
INSERT
19 - 22*
18
17
14
9
46 - 47 (figure 5)
(a)

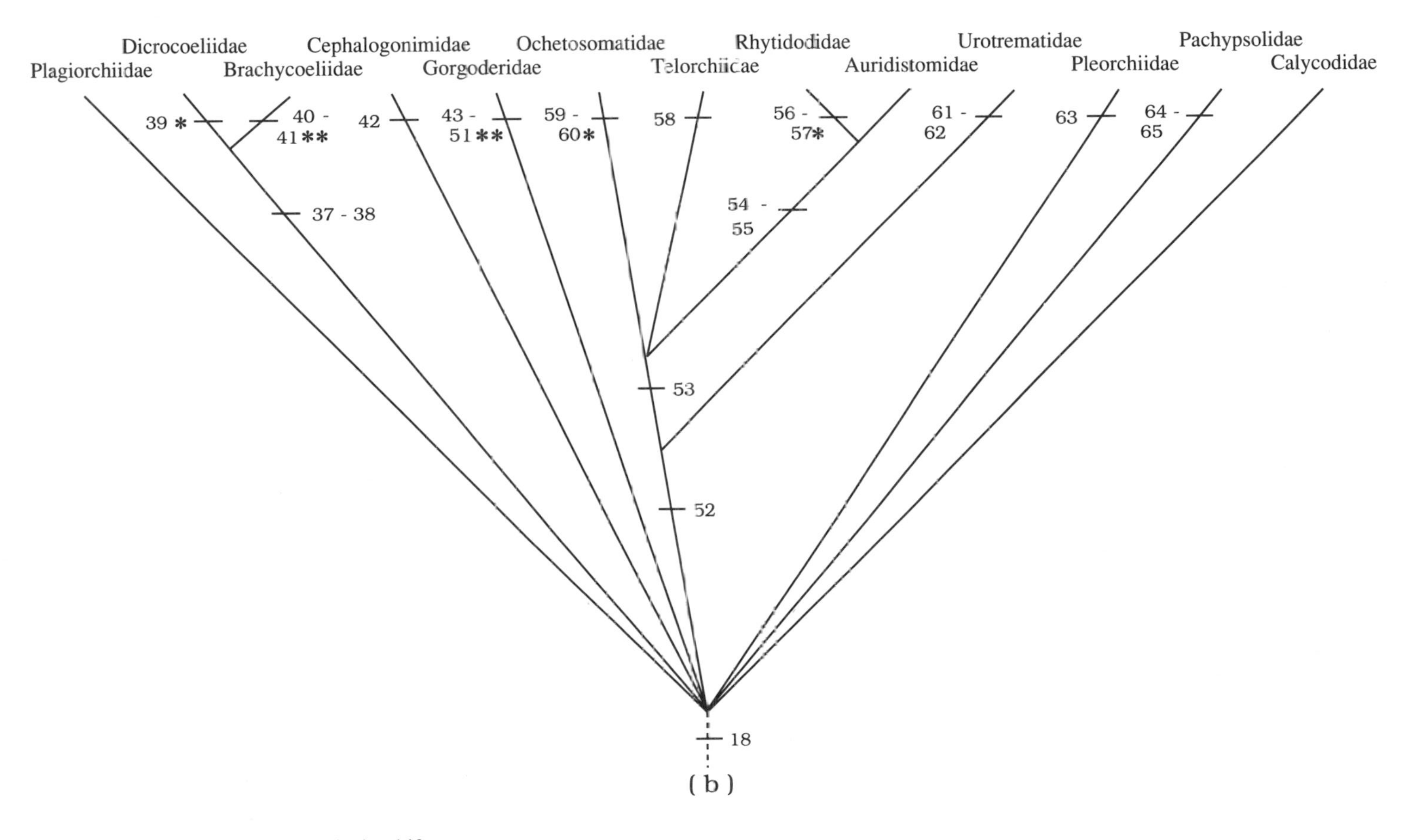

Figure 15. Phylogenetic tree for Plagiorchiformes.

Family Dicrocoeliidae Looss 1899

Diagnosis: Dicrocoelioids with posttesticular ovary (38); some cotylomicrocercous cercariae *(39).

Family Brachycoeliidae Looss 1899

Diagnosis: Dicrocoelioids with posttesticular ovary (38); short ceca *(40); cercariaeum cercariae *(41).

Family Cephalogonimidae Nicoll 1915

Diagnosis: Plagiorchiateans with genital pore dorsal to oral sucker (42).

Superfamily Gorgoderoidea Looss 1899 *sedis mutabilis*

Diagnosis: Plagiorchiateans with cystocercous cercariae (43); vitellaria compact (44); vitellaria pretesticular (45); vitellaria postacetabular (46); cirrus lacking (47); no cirrus sac *(48); seminal receptacle small or lacking *(49); body foliate in shape (50); extracecal uterine loops confined to hindbody (51).

Family Gorgoderidae Looss 1899

Diagnosis: With characters of the superfamily.

Superfamily Telorchioidea Looss 1899 *sedis mutabilis*

Diagnosis: Plagiorchiateans with ovary pretesticular and separated from the testes by extensive uterine folding (52); prominent glandular and muscular metraterm (53).

Family Auridistomidae Stunkard 1924

Diagnosis: Telorchioids with extensive vitellaria in hindbody (54); papillae around oral sucker (55).

Family Rhytidodidae Odhner 1926

Diagnosis: Telorchioids with extensive vitellaria in hindbody (54); papillae around oral sucker (55); preovarian uterus *(56); diverticulate anterior arms of excretory vesicle (57).

Family Telorchiidae Looss 1899

Diagnosis: Telorchioids with ovary pretesticular and separated from the testes by extensive uterine folding (52); at least one testis postuterine and near posterior end of body (58).

Family Ochetosomatidae Leao 1944

Diagnosis: Telorchioids with ceca extending only one-half to two-thirds body length posterior (59); testes at or near posterior ends of ceca (60).

Family Urotrematidae Poche 1926

Diagnosis: Telorchioids with ovary pretesticular and separated from testes by extensive uterine coiling (52); genital pores at posterior end of body (61); one testis near posterior end of body (62).

Family Pleorchiidae Poche 1926

Diagnosis: Plagiorchiateans with multiple testes (64).

Family Pachypsolidae Yamaguti 1958

Diagnosis: Plagiorchiateans with vitellarium in stellate groups (65); ceca with anterior diverticula (66).

Family Calycodidae Dollfus 1929

Diagnosis: With characters of the infra-suborder.

Infraclass Digenea Van Beneden 1858: Infrafamilial relationships

Family Hemiuridae Looss 1899 (Fig. 16)

References: Gibson and Bray (1979); Brooks et al. (1985b).

Subfamily Isoparorchiinae Travassos 1922

Diagnosis: Hemiurids with tubular ovaries (1).

Subfamily Sclerodistominae Odhner 1927 *incertae sedis*

Diagnosis: Hemiurids with no external opening of Laurer's canal (2); rudimentary Juel's organ (3).

Subfamily Didymozoinae Poche 1907

Diagnosis: Hemiurids with no external opening of Laurer's canal (2); fully developed Juel's organ (as a blind seminal receptacle) (4); without genital sac (5); tubular ovaries (6).

Subfamily Dictysarcinae Skrjabin and Guschanskaja 1955 *incertae sedis*

Diagnosis: Hemiurids with no external opening of Laurer's canal (2); fully developed Juel's organ (as a blind seminal receptacle) (4); lobed vitellaria (7); amuscular or poorly developed genital cone (8).

Subfamily Lecithasterinae Odhner 1905

Diagnosis: Hemiurids with no external opening of Laurer's canal (2); fully developed Juel's organ (as a blind seminal receptacle) (4); lobed vitellaria (7); saccate seminal vesicle (9); without genital cone (10).

Subfamily Derogeninae Nicoll 1910

Diagnosis: Hemiurids with no external opening of Laurer's canal (2); fully developed Juel's organ (as a blind seminal receptacle) (4); smooth vitellaria (11); amuscular or poorly developed genital cone (8); saccate seminal vesicle (9); internal seminal vesicle only (12).

Subfamily Hemiurinae Looss 1899 *incertae sedis*

Diagnosis: Hemiurids with no external opening of Laurer's canal (2); fully developed Juel's organ (as a blind seminal receptacle) (4); smooth vitellaria (11); amuscular or poorly developed genital cone (8); saccate seminal vesicle (9); tegumental plications present (13); ecsoma present (14).

Subfamily Bunocotylinae Dollfus 1950 *incertae sedis*

Diagnosis: Hemiurids with no external opening of Laurer's canal (2); fully developed Juel's organ (as a blind seminal receptacle) (4); amuscular or poorly developed genital cone (8); saccate seminal vesicle (9); smooth vitellaria (11); tegumental plications present (13); ecsoma present (14).

Family Syncoeliidae Looss 1899 (Fig. 17)

References: Gibson and Bray (1979); Brooks et al. (1985b, 1989).

Subfamily Sclerodistomoidinae Kamegai 1971 *incertae sedis*

Diagnosis: With characters of the family.

Subfamily Syncoeliinae Looss 1899

Diagnosis: Syncoeliids with lobed vitellaria (1); tubular ovary (2); multiple testes (3).

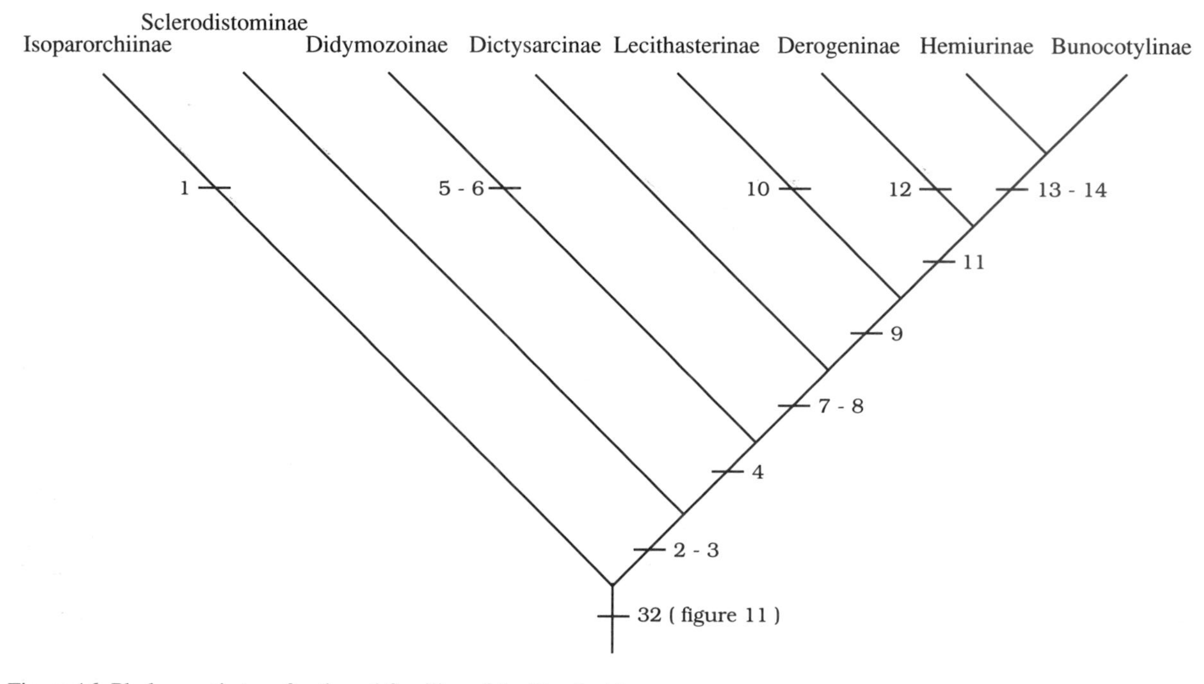

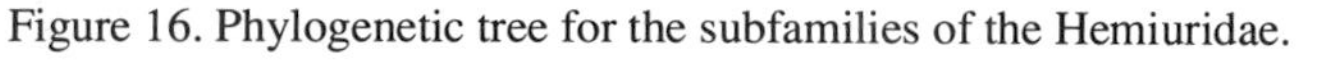
Figure 16. Phylogenetic tree for the subfamilies of the Hemiuridae.

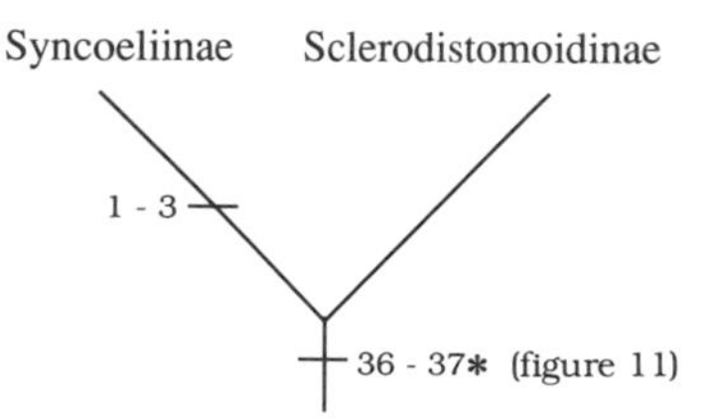

Figure 17. Phylogenetic tree for the subfamilies of the Syncoeliidae.

Family Spirorchidae Stunkard 1921

Subfamily Hapalotrematinae Poche 1926 *incertae sedis*

Diagnosis: With characters of the family.

***Hapalorhynchus* Stunkard 1922 (Fig. 18)**

Reference: Platt (1988).

Diagnosis: Hapaotrematines with dorsolateral genital pore (1).

***albertoi* Lamothe-Argumedo 1978**

Diagnosis: *Hapalorhynchus* with acetabulum narrower than oral sucker (2); esophageal diverticula (3).

***brooksi* Platt 1988**

Diagnosis: *Hapalorhynchus* with conspicuous *pars prostatica* (4); branched stem of excretory vesicle (5); outpocketings of the anterolateral portion of the ceca (6).

***stunkardi* Byrd 1939**

Diagnosis: *Hapalorhynchus* with conspicuous *pars prostatica* (4); no cirrus sac (7); vitellaria extending anteriorly to midesophagus (9); tricornuate eggs (10).

***reelfooti* Byrd 1939**

Diagnosis: *Hapalorhynchus* with conspicuous *pars prostatica* (4); no cirrus sac (7); oval eggs (8); vitellaria extending anteriorly to midesophagus (9).

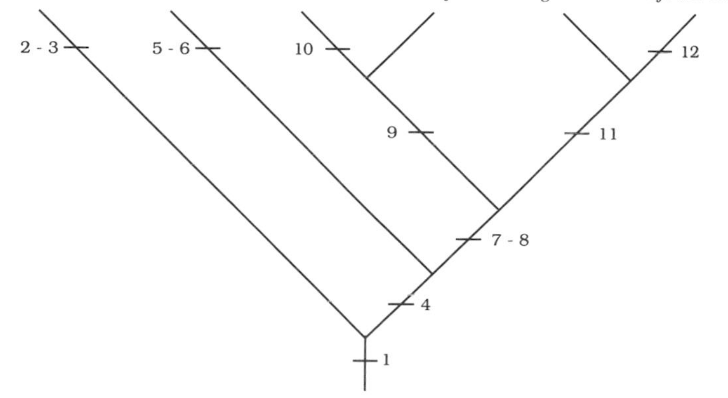

Figure 18. Phylogenetic tree for the species of *Hapalorhynchus*.

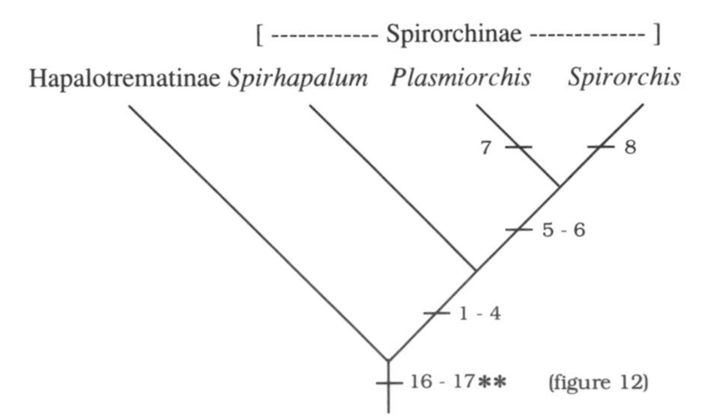

Figure 19. Phylogenetic tree for the genera of the Spirorchinae.

***gracilis* Stunkard 1922**

Diagnosis: *Hapalorhynchus* with conspicuous *pars prostatica* (4); no cirrus sac (7); with oval eggs (8); ceca terminating more than 20% of total body length from posterior end (11).

***foliorchis* Brooks and Mayes 1975**

Diagnosis: *Hapalorhynchus* with conspicuous *pars prostatica* (4); no cirrus sac (7); with oval eggs (8); ceca terminating more than 20% of total body length from posterior end (11); oval ovary (12).

Subfamily Spirorchinae Stunkard 1921 (Fig. 19)

Reference: Platt (1992).

Diagnosis: Spirorchids with lateral outpocketings surrounding esophagus near cecal bifurcation (1); plicate organ present (2); median esophageal pouch present (3); multiple anterior testes (4).

Spirhapalum* Ejsmont 1927 *incertae sedis

Diagnosis: With characters of the subfamily.

***Plasmiorchis* Mehra 1934**

Diagnosis: Testicular field linear (5); posterior testis lacking (6); ceca looping anteriorly before turning posteriad (7).

***Spirorchis* MacCallum 1919**

Diagnosis: Testicular field linear (5); posterior testis lacking (6); acetabulum lacking (8).

Family Fellodistomidae Nicoll 1909 (Fig. 20)

References: Bray (1988); Brooks et al. (1989a); Cribb and Bray (in press).

Subfamily Baccigerinae Yamaguti 1958

Diagnosis: Fellodistomids with no furcae on cercarial tail (1); tegumental spines *(2); posterodorsal opening of Laurer's canal (3); canalicular seminal receptacle (4).

Subfamily Lintoniinae Yamaguti 1970 *incertae sedis*

Diagnosis: Fellodistomids with no furcae on cercarial tail (2).

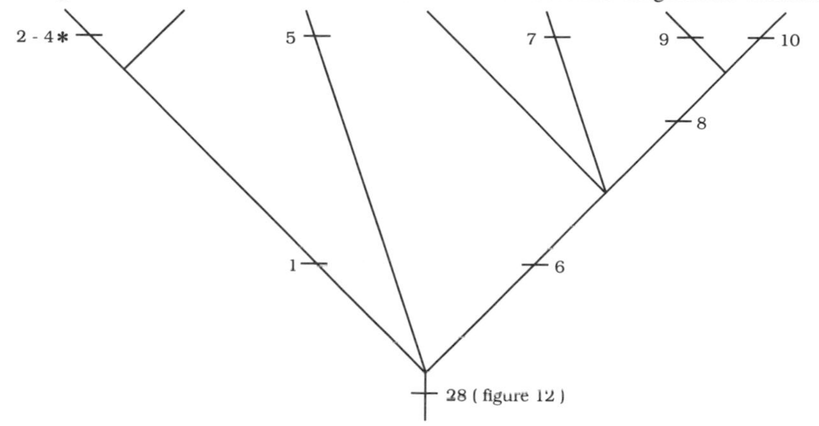

Figure 20. Phylogenetic tree for the subfamilies of the Fellodistomidae. Tan = Tandanicolinae.

Subfamily Tandanicolinae Johnston 1927

Diagnosis: Fellodistomids with accessory reproductive organ (5).

Subfamily Fellodistominae Nicoll 1909

Diagnosis: Fellodistomids without cercarial tail setae (6).

Subfamily Proctoecinae Skrjabin and Koval 1957

Diagnosis: Fellodistomids without cercarial tail setae (6); microcercous cercariae (7).

Subfamily Tergestiinae Skrjabin and Koval 1957

Diagnosis: Fellodistomids without cercarial tail setae (6); elongate pharynx (8); cephalic papillae and collarettes (9).

Subfamily Monascinae Dollfus 1947

Diagnosis: Fellodistomids without cercarial tail setae (6); elongate pharynx (8); single cecum (10).

Family Liolopidae Odhner 1912 (Fig. 21)

Reference: Brooks and Overstreet (1978).

Subfamily Liolopinae Odhner 1912 *sedis mutabilis, incertae sedis*

Diagnosis: With characters of the family.

***Liolope* Cohn 1902**

Diagnosis: With cirrus sac abutting acetabulum (1); no esophagus (2).

Subfamily Harmotrematinae Yamaguti 1933

Diagnosis: With vitellaria extending anterior to anterior margin of acetabulum (3).

***Moreauia* Johnston 1915**

Diagnosis: With laterally elongate body (4).

***Dracovermis* Brooks and Overstreet 1978**

Diagnosis: All gonads in posterior one-third of body (5).

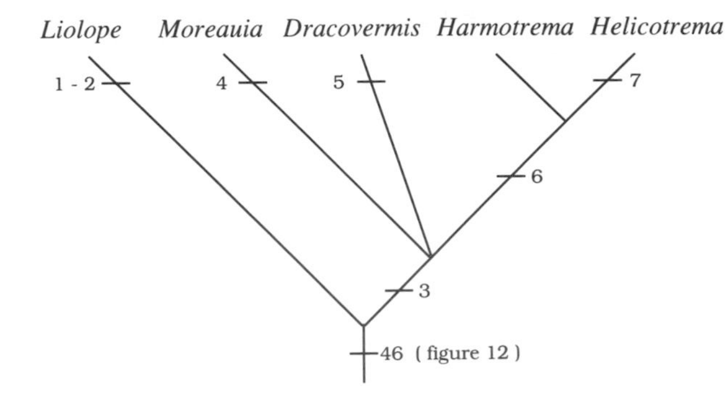

Figure 21. Phylogenetic tree for the genera of the Liolopidae.

Harmotrema* Nicoll 1914 *incertae sedis

Diagnosis: Body length more than four times body width (6).

***Helicotrema* Odhner 1912**

Diagnosis: Body length more than four times body width (6); tegumental spines lacking (7).

***Dracovermis* Brooks and Overstreet 1978 (Fig. 22)**

Reference: Brooks and Overstreet (1978).

***occidentalis* Brooks and Overstreet 1978**

Diagnosis: With vitellaria confluent posteriorly *(1); folded metraterm (2).

***nicolli* (Mehra 1936) Brooks and Overstreet 1978**

Diagnosis: With eggs 120 µm long by 80 µm wide (3); forebody greater than 40% total body length (4); ratio of oral sucker to acetabulum less than 1:2 (5).

***rudolphi* (Tubangui and Masilungen 1936) Brooks and Overstreet 1978**

Diagnosis: With vitellaria confluent posteriorly *(6); eggs 120 µm long by 80 µm wide (3).

***brayi* Brooks and Overstreet 1978**

Diagnosis: With forebody greater than 40% total body length (4); vitellaria confluent posteriorly *(6); sucker ratio 1:2 (7); cecal bifurcation 25% total body length from

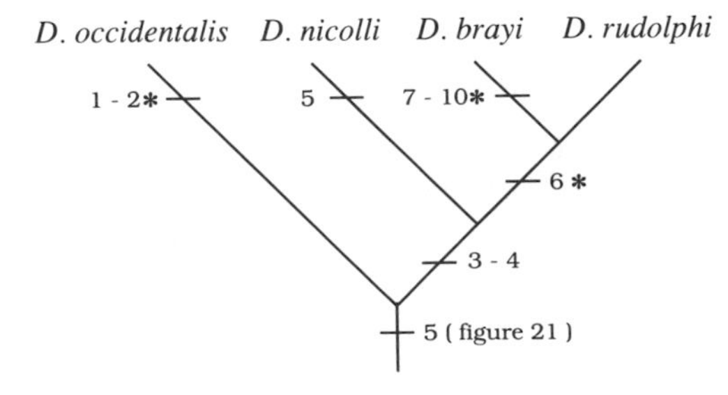

Figure 22. Phylogenetic tree for the species of *Dracovermis*.

anterior end (8); ratio of oral sucker to pharynx 1:1 (9); eggs 100 μm long by 60 μm wide *(10).

Family Proterodiplostomidae Dubois 1936 (Fig. 23)

References: Brooks (1979b); Brooks et al. (1992).

Subfamily Heterodiplostominae Dubois 1936 *incertae sedis, sedis mutabilis*

Diagnosis: With characters of the family. Parasites of snakes.

Ophiodiplostomum* Dubois 1936, *Heterodiplostomum* Dubois 1936, *Petalodiplostomum* Dubois 1936 *sedis mutabilis, incertae sedis

Diagnosis: As above.

Subfamily Proterodiplostominae Dubois 1936 *sedis mutabilis*

Diagnosis: Proterodiplostomes with papillose tribocytic organ (1); vitelline follicles confined to forebody (2). Parasites of crocodilians.

Tribe Pseudoneodiplostomini Dubois 1936 *sedis mutabilis*

Diagnosis: Proterodiplostomines with inflated genital atrium *(3).

***Neelydiplostomum* Gupta 1958**

Diagnosis: Pseudoneodiplostomini with terminal genital pores *(4).

***Pseudoneodiplostomum* Dubois 1936**

Diagnosis: Pseudoneodiplostomini with vitellaria in fore- and hindbody *(5).

Tribe Pseudocrocodilicolini Byrd and Reiber 1942 *sedis mutabilis*

Diagnosis: Proterodiplostomines with paraprostate and male ejaculatory duct fused, and uterus fused to that common duct forming a hermaphroditic duct (6).

Pseudocrocodilicola* Byrd and Reiber 1942 *sedis mutabilis, incertae sedis

Diagnosis: With characters of the tribe.

***Polycotyle* Willemoes-Suhm 1870**

Diagnosis: Pseudocrocodilicolini with very small paraprostate gland *(7); sucker in genital atrium (8); inflated genital atrium *(9); many dorsal accessory suckers (10); Mehlis's gland between ovary and anterior testis (11).

***Crocodilicola* Poche 1925**

Diagnosis: Pseudocrocodilini with paraprostate fused with uterine duct, and those fused with ejaculatory duct to form a hermaphroditic duct (12); very small paraprostate gland *(13).

***Archaeodiplostomum* Dubois 1936**

Diagnosis: Pseudocrocodilicolini with paraprostate fused with uterine duct, and those fused with ejaculatory duct to form a hermaphroditic duct (12); vitellaria in fore- and hindbody *(14).

Tribe Proterodiplostomini Dubois 1936 *sedis mutabilis*

Diagnosis: Proterodiplostomines with paraprostate not fused with male duct, opening separately into genital atrium (15).

***Mesodiplostomum* Dubois 1936**

Diagnosis: Proterodiplostomini with genital diverticula (16); vitellaria in fore- and hindbody *(17).

***Cystodiplostomum* Dubois 1936**

Diagnosis: Proterodiplostomini with very small paraprostate gland *(18); gonads close to forebody (19); gonads close to tribocytic organ (20); ventral hindbody constriction (21).

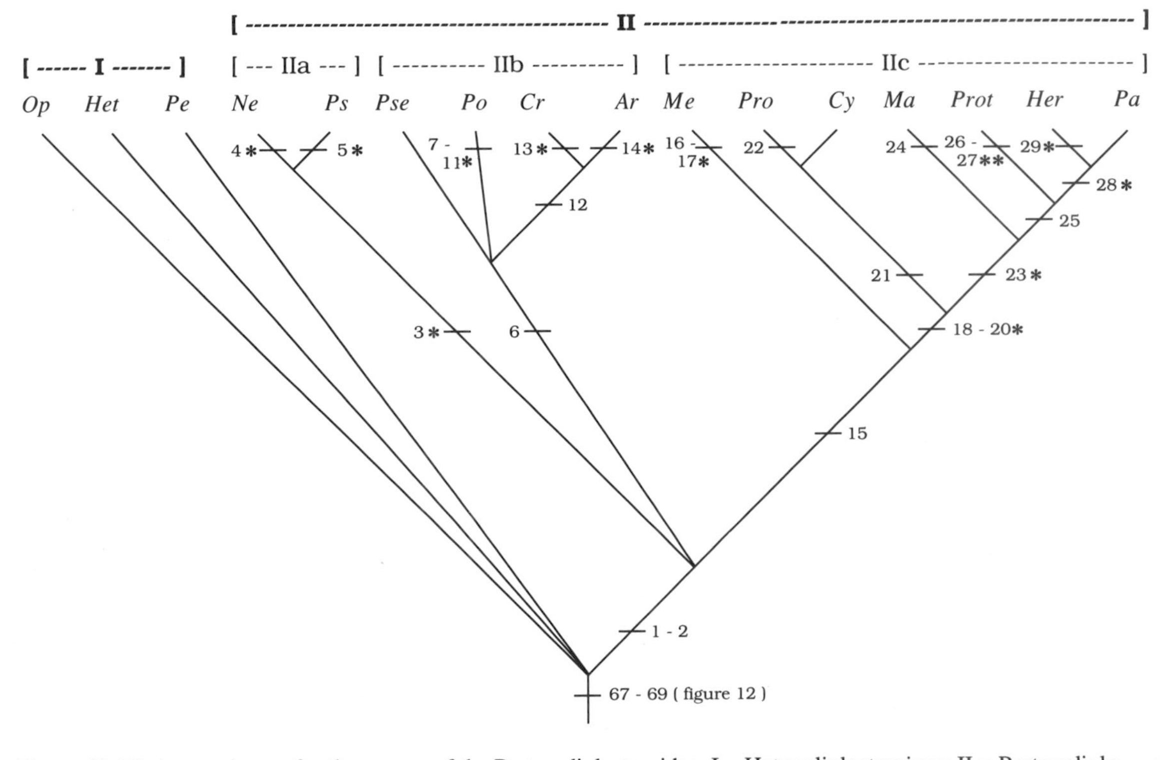

Figure 23. Phylogenetic tree for the genera of the Proterodiplostomidae. I = Heterodiplostominae; II = Proterodiplostominae; IIa = Pseudoneodiplostomini; IIb = Pseudocrocodilicolini; IIc = Proterodiplostomini. *Op* = *Ophiodiplostomum; Het* = *Heterodiplostomum; Pe* = *Petalodiplostomum; Ne* = *Neelydiplostomum; Ps* = *Pseudoneodiplostomum; Pse* = *Pseudocrocodilicola; Po* = *Polycotyle; Cr* = *Crocodilicola; Ar* = *Archaeodiplostomum; Me* = *Mesodiplostomum; Pro* = *Prolecithodiplostomum; Cy* = *Cystodiplostomum; Ma* = *Massoprostatum; Prot* = *Proterodiplostomum; Her* = *Herpetodiplostomum; Pa* = *Paradiplostomum.*

Prolecithodiplostomum **Dubois 1936**

Diagnosis: Proterodiplostomini with small paraprostate gland *(18); gonads close to forebody (19); gonads close to tribocytic organ (20); hindbody constriction around entire body (22).

Massoprostatum **Caballero 1947**

Diagnosis: Proterodiplostomini with small paraprostate gland *(18); gonads close to forebody (19); gonads close to tribocytic organ (20); terminal genital pore *(23); vitellaria confined to hindbody (24).

Proterodiplostomum **Dubois 1936**

Diagnosis: Proterodiplostomini with small paraprostate gland *(18); gonads close to forebody (19); gonads close to tribocytic organ (20); noneversible genital atrium (25); vitellaria in fore- and hindbody *(26); dorsal subterminal genital pore *(27).

Paradiplostomum **LaRue 1926**

Diagnosis: Proterodiplostomini with small paraprostate gland *(18); gonads close to forebody (19); gonads close to tribocytic organ (20); terminal genital pore *(23); noneversible genital atrium (25); inflated genital atrium *(28).

Herpetodiplostomum **Dubois 1936**

Diagnosis: Proterodiplostomini with gonads close to forebody (19); gonads close to tribocytic organ (20); terminal genital pore *(23); noneversible genital atrium (25); inflated genital atrium *(28); large paraprostate gland *(29).

Family Neodiplostomidae Shoop 1989 (Fig. 24)

Reference: Shoop (1989).

Subfamily Crassiphialinae Shoop 1989

Diagnosis: Neodiplostomids with bisegmented metacercarial body shape (1).

Conodiplostomum **Group**

Diagnosis: With characters of the subfamily.

Conodiplostomum **Dubois 1937**

Diagnosis: With characters of the group.

Posthodiplostomum **Group**

Diagnosis: Crassiphialines with genital prepuce (2).

Posthodiplostomum **Dubois 1936**

Diagnosis: With characters of the group.

Mesoophorodiplostomum **Dubois 1936**

Diagnosis: *Posthodiplostomum* group with intertesticular ovary (3).

Ornithodiplostomum **Dubois 1936,** ***Prolobodiplostomum*** **Baer 1959**

Diagnosis: *Posthodiplostomum* group with unsegmented adult body shape (4).

Lophosicyadiplostomum **Group**

Diagnosis: Crassiphialines with ringed oral sucker (5).

Lophosicyadiplostomum **Dubois 1936**

Diagnosis: With characters of the group.

Crassiphiala **Group**

Diagnosis: Crassiphialines with vitellaria confined to the hindbody (6).

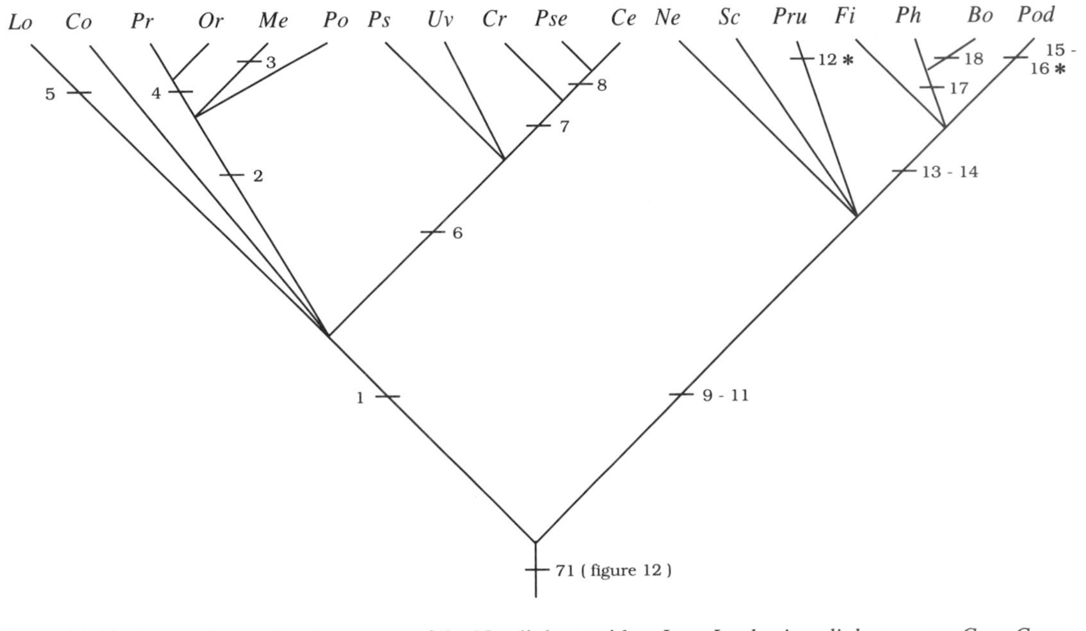

Figure 24. Phylogenetic tree for the genera of the Neodiplostomidae. *Lo* = *Lophosicyadiplostomum; Co* = *Conodiplostomum; Pr* = *Prolobodiplostomum; Or* = *Ornithodiplostomum; Me* = *Mesoophorodiplostomum; Po* = *Posthodiplostomum; Ps* = *Pseudodiplostomum; Uv* = *Uvulifer; Cr* = *Crassiphiala; Pse* = *Pseudocercocotyla; Ce* = *Cercocotyla; Ne* = *Neodiplostomum; Sc* = *Scolopacitrema; Pru* = *Prudhoella; Fi* = *Fibricola; Ph* = *Pharyngostomum; Bo* = *Bolbocephalodes; Pod* = *Podospathalium.*

***Uvulifer* Yamaguti 1934, *Pseudodiplostomum* Dubois 1936**

Diagnosis: With characters of the group.

***Crassiphiala* van Haitsma 1925**

Diagnosis: *Crassiphiala* group with vestigial (or absent) acetabulum (7).

***Pseudocercocotyla* Yamaguti 1971, *Cercocotyla* Yamaguti 1939**

Diagnosis: *Crassiphiala* group with vestigial (or absent) acetabulum (7); sucker in genital bursa (8).

Subfamily Neodiplostominae Shoop 1989

Diagnosis: Neodiplostomids with no genital cone (9); unencysted metacercaria (10); enclosed limebodies in metacercariae (11).

Neodiplostomum* Group *incertae sedis

Diagnosis: Neodiplostomines with lingual-shaped tribocytic organ *(12).

***Neodiplostomum* Railliet 1919, *Scolopacitrema* Sudarikov and Rykovsky 1959, *Prudhoella* Beverly-Burton 1960**

Diagnosis: With characters of the group.

***Pharyngostomum* Group**

Diagnosis: Neodiplostomines with vitellaria confined to the forebody (13); paranephridial plexus comprising three longitudinal vessels, three or fewer transverse commissures and no other anastomoses (14).

***Fibricola* Dubois 1932**

Diagnosis: With characters of the group.

***Podospathalium* Dubois 1932**

Diagnosis: *Pharyngostomum* group with fleshy forebody (15); lingual-shaped tribocytic organ *(16).

***Pharyngostomum* Ciurea 1922**

Diagnosis: *Pharyngostomum* group with symmetrical testes (17).

***Bolbocephalodes* Strand 1935**

Diagnosis: *Pharyngostomum* group with symmetrical testes (17); no oral sucker (18).

Family Bolbophoridae Shoop 1989

Reference: Shoop (1989).

Bolbophorus* Dubois 1935 *Posthodiplostomoides* Williams 1969, *Dolichorchis* Dubois 1961, *Adenodiplostomum* Dubois 1937 *incertae sedis, sedis mutabilis

Diagnosis: With characters of the family.

Family Diplostomidae Poirier 1886 (Fig. 25)

Reference: Shoop (1989).

***Subuvulifer* Group**

Diagnosis: With characters of the family.

***Subuvulifer* Dubois 1952, *Neoharvardia* Gupta 1963**

Diagnosis: As above.

***Neoalaria* Group**

Diagnosis: Diplostomids with unsegmented adult body shape (1).

***Neoalaria* Lal 1939, *Tylodelphys* Diesing 1850, *Glossodiplostomoides* Bhalerao 1942**

Diagnosis: With characters of the group.

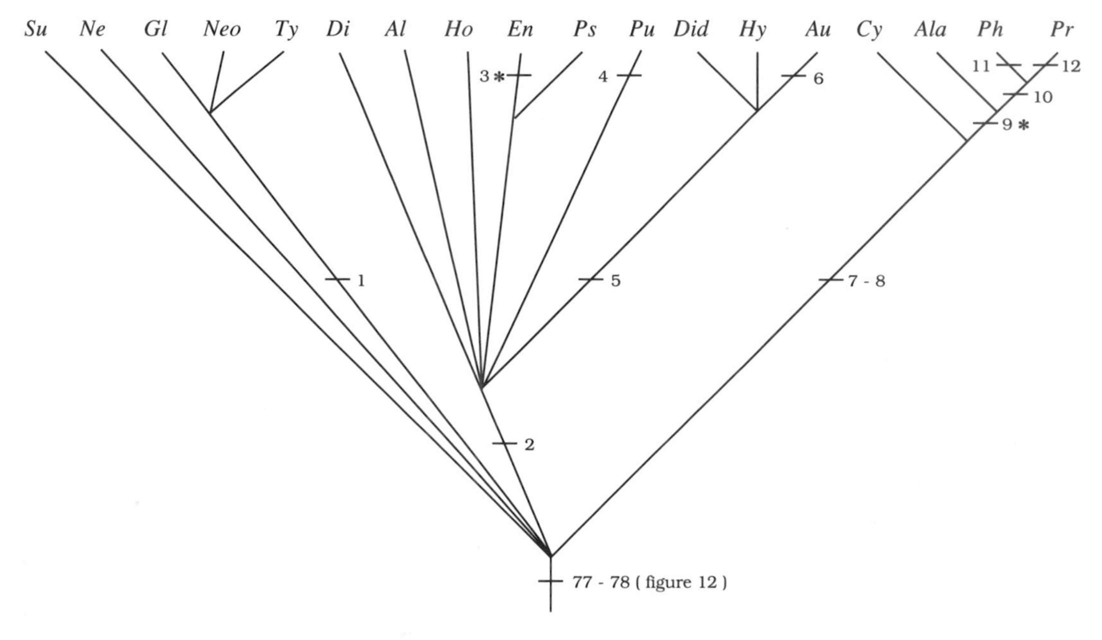

Figure 25. Phylogenetic tree for the genera of the Diplostomidae. *Su* = *Subuvulifer; Ne* = *Neoharvardia; Gl* = *Glossodiplostomoides; Neo* = *Neoalaria; Ty* = *Tylodelphys; Di* = *Diplostomum; Al* = *Allodiplostomum; Ho* = *Harvardia; En* = *Enhydrodiplostomum; Ps* = *Pseudoscolopacitrema; Pu* = *Pulvinifer; Did* = *Didelphodiplostomum; Hy* = *Hysteromorpha; Au* = *Austrodiplostomum; Cy* = *Cynodiplostomum; Ala* = *Alaria; Ph* = *Pharyngostomoides; Pr* = *Procyotrema*.

Subfamily Diplostominae Monticelli 1888

Diagnosis: Diplostomids with no genital cone (2).

Tribe *Diplostomini* Monticelli 1888

Diagnosis: With characters of the subfamily.

***Diplostomum* von Nordmann 1832, *Allodiplostomum* Yamaguti 1935, *Harvardia* Baer 1932**

Diagnosis: With characters of the tribe.

***Enhydrodiplostomum* Group**

Diagnosis: Diplostomines with lingual-shaped tribocytic organ *(3).

***Enhydrodiplostomum* Dubois 1944, *Pseudoscolopacitrema* Palmieri, Krishnasamy, and Sullivan 1979**

Diagnosis: With characters of the group.

***Pulvinifer* Group**

Diagnosis: Diplostomines with vitellaria confined to the forebody (4).

***Pulvinifer* Yamaguti 1953**

Diagnosis: With characters of the group.

***Austrodiplostomum* Group**

Diagnosis: Diplostomines with pyriform-shaped tribocytic organ (5).

***Didelphodiplostomum* Dubois 1944, *Hysteromorpha* Lutz 1931**

Diagnosis: With characters of the group.

***Austrodiplostomum* Szidat and Nani 1951**

Diagnosis: *Austrodiplostomum* group with vestigial or lacking acetabulum (6).

Subfamily Alariinae Hall and Wigdor 1918

Diagnosis: Diplostomids with vitellaria confined to the forebody (7); mesocercarial stage (8).

***Cynodiplostomum* Dubois 1936**

Diagnosis: With characters of the subfamily.

***Alaria* Group**

Diagnosis: Alariines with lingual-shaped tribocytic organ *(9).

***Alaria* Schrank 1788**

Diagnosis: With characters of the group.

***Pharyngostomoides* Harkema 1942**

Diagnosis: *Alaria* group with symmetrical testes (10); spherical ovaries (11).

***Procyotrema* Harkema and Miller 1959**

Diagnosis: *Alaria* group with symmetrical testes (10); elongate testes (12). Adults not in gastrointestinal tract.

Family Strigeidae Railliet 1919 (Fig. 26)

Reference: Shoop (1989).

Subfamily Strigeinae Railliet 1919

Diagnosis: Strigeids with mesocercaria stage (1).

***Strigea* Abildgaard 1790**

Diagnosis: With characters of the subfamily.

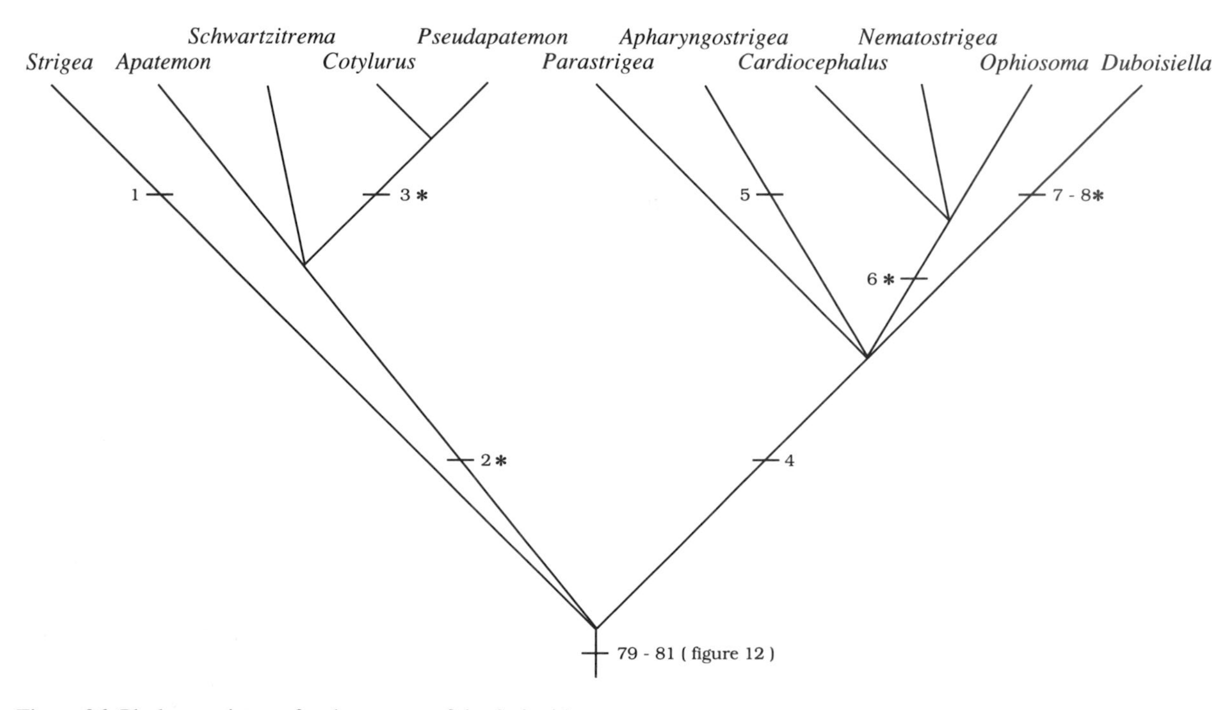

Figure 26. Phylogenetic tree for the genera of the Strigeidae.

Subfamily Cotylurinae Dubois 1936

Diagnosis: Strigeids with vitellaria confined to the hindbody *(2).

***Apatemon* Szidat 1928, *Schwartzitrema* Perez Viqueras 1941**

Diagnosis: With characters of the subfamily.

***Cotylurus* Szidat 1928, *Pseudapatemon* Dubois 1936**

Diagnosis: Cotylurines with no genital cone *(3).

***Cardiocephalus* Group**

Diagnosis: Strigeids with vestigial anterior pseudosuckers (4).

***Parastrigea* Szidat 1928**

Diagnosis: With characters of the group.

***Apharyngostrigea* Ciurea 1927**

Diagnosis: *Cardiocephalus* group with no pharynx (5).

***Cardiocephalus* Szidat 1928, *Nematostrigea* Sandground 1934, *Ophiosoma* Szidat 1928**

Diagnosis: *Cardiocephalus* group with vitellaria restricted to hindbody *(6).

***Duboisiella* Baer 1938**

Diagnosis: *Cardiocephalus* group with vitellaria restricted to forebody (7); genital cone absent *(8).

Family Cryptogonimidae Ward 1917

Subfamily Acanthostominae Poche 1926 (Fig. 27)

References: Brooks (1981a); Brooks and Caira (1982); Blair et al. (1988); Brooks and Holcman (in press).

Diagnosis: Terminal oral sucker armed with single row of spines (1); preacetabular pit (2); no ventrogenital pit (3); genital pore not in preacetabular pit (4); seminal vesicle coiled posteriorly (5); suckerlike gonotyl present (6).

***Timoniella* Rebecq 1960**

Diagnosis: Preovarian seminal receptacle (7).

***Gymnatrema* Morozov 1955**

Diagnosis: Some uterine loops lateral to testes but none posttesticular (8); vitelline follicles not extending anteriorly to posterior margin of seminal vesicle (10); vitelline

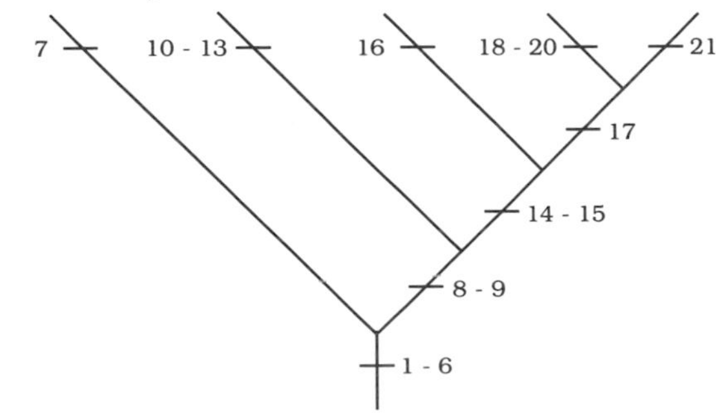

Figure 27. Phylogenetic tree for the genera of the Acanthostominae.

follicles confluent posttesticularly (11); one cecum atrophied (12); one cecum opening laterally and one cecum ending blindly (13).

***Proctocaecum* Baugh 1957**

Diagnosis: Some uterine loops lateral to testes but none posttesticular (8); ceca opening separately and laterally at even levels (9); excretory vesicle Y-shaped with short stem and constriction of arms in middle (14); eggs averaging more than 30 μm long (15); gonotyl large, solid, muscular (16).

***Caimanicola* Teixeira de Freitas and Lent 1938**

Diagnosis: Some uterine loops lateral to testes but none posttesticular (8); ceca opening separately and laterally at even levels (9); excretory vesicle Y-shaped with short stem and constriction of arms in middle (14); eggs averaging more than 30 μm long (15); gonotyl lacking (17); esophagus longer than pharynx (18); tegumental spines unusually robust in mid-forebody (19); maximum body length 7–16 mm (20).

***Acanthostomum* Looss 1899**

Diagnosis: Some uterine loops lateral to testes but none posttesticular (8); ceca opening separately and laterally at even levels (9); eggs averaging more than 30 μm long (15); gonotyl lacking (17); excretory vesicle with long stem and short arms (21).

***Timoniella* Rebecq 1960**

***Timoniella (Timoniella)* Rebecq 1960 (Fig. 28)**

Diagnosis: Vitelline follicles not extending anteriorly to posterior margin of seminal vesicle (1); length of body occupied by uterine loops more than 50% of total body length (2); seminal vesicle not coiled posteriorly (3); prepharynx shorter than pharynx

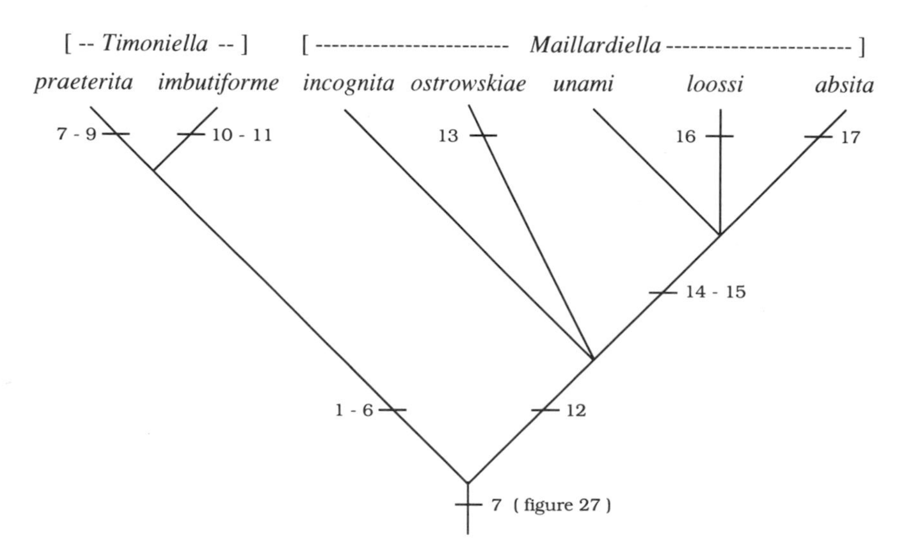

Figure 28. Phylogenetic tree for the species of *Timoniella*.

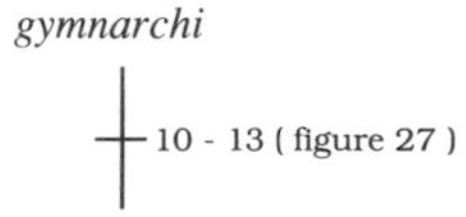

gymnarchi

10 - 13 (figure 27)

Figure 29. Phylogenetic tree for the species of *Gymnatrema.*

(4); ratio of oral sucker width to pharyngeal width 1:0.25–0.40 (5); ratio of body length to width averaging 7.5–15:1 (6).

***praeterita* (Looss 1901) Maillard 1974**

Diagnosis: Cyclocoel (7); forebody 10–20% of total body length (8); maximum body length 7–16 mm (9).

***imbutiforme* (Molin 1859) Brooks 1981**

Diagnosis: Ratio of oral sucker to acetabular width 1:0.8–1.3 (10); oral spines averaging 25–30 in number (11).

***Timoniella (Maillardiella)* Brooks and Holcman in press**

Diagnosis: No gonotyl (12).

***incognita* Brooks 1981**

Diagnosis: As above.

***ostrowskiae* Brooks and Holcman in press**

Diagnosis: Ceca opening separately at posterior end of body (13).

***unami* (Pelaez and Cruz 1953) Brooks 1981**

Diagnosis: Ceca opening into excretory vesicle (14); vitelline follicles extending posteriorly to middle of posterior testis (15).

***loossi* (Perez Vigueras 1957) Brooks 1981**

Diagnosis: Ceca opening into excretory vesicle (14); vitelline follicles extending posteriorly to middle of posterior testis (15); vitelline follicles confluent dorsally (16).

***absita* Blair, Brooks, Purdie, and Melville 1988**

Diagnosis: Ceca opening into excretory vesicle (14); vitelline follicles extending posteriorly to middle of posterior testis (15); constriction in seminal vesicle (17).

***Gymnatrema* Morozov 1955 (Fig. 29)**

***gymnarchi* (Dollfus 1950) Morozov 1955**

Diagnosis: With characters of the genus.

***Proctocaecum* Baugh 1957 (Fig. 30)**

***Proctocaecum (Proctocaecum)* Baugh 1957**

Diagnosis: Relative length of uterine loops less than 45% of total body length (1).

***gonotyl* (Dollfus 1950) Brooks 1981**

Diagnosis: As above.

***vicinum* (Odhner 1902) Brooks 1981**

Diagnosis: Ceca opening separately and laterally at uneven levels (2).

***coronarium* (Cobbold 1861) Brooks 1981**

Diagnosis: Ceca opening separately and laterally at uneven levels (2); vitelline follicles not extending anteriorly to posterior margin of seminal vesicle (3); one cecum atrophied (4); maximum oral spine length more than 100 µm *(5).

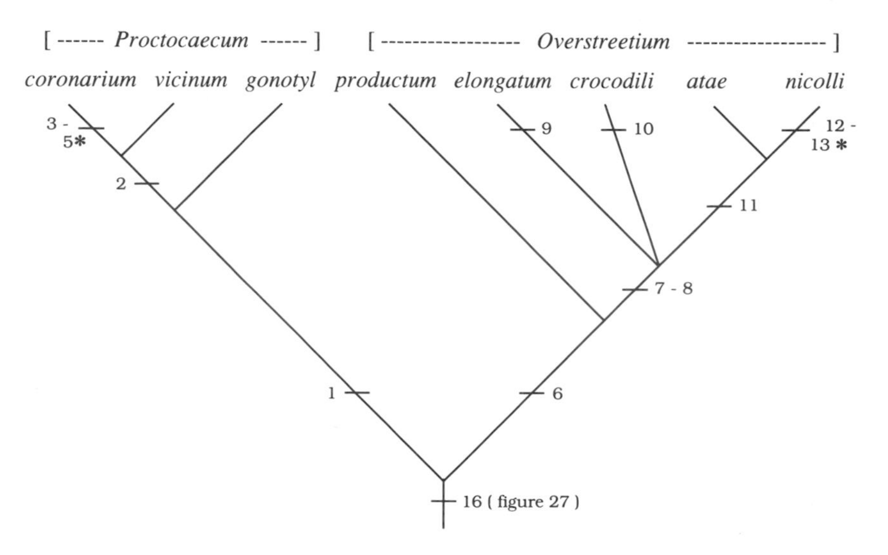

Figure 30. Phylogenetic tree for the species of *Proctocaecum*.

***Proctocaecum* (*Overstreetium*) Brooks and Holcman in press**

Diagnosis: Ratio of body length to width averaging 7.5–15:1 (6).

***productum* (Odhner 1902) Brooks 1981**

Diagnosis: As above.

***elongatum* (Tubangui and Masilungen 1936) Brooks 1981**

Diagnosis: Ceca opening separately at posterior end of body (7); maximum body length 7–16 mm (8); ratio of body length to width averaging more than 20:1 (9).

***crocodili* (Yamaguti 1954) Brooks 1981**

Diagnosis: Ceca opening separately at posterior end of body (7); maximum body length 7–16 mm (8); forebody less than 10% of total body length (10).

***atae* (Tubangui and Masilungen 1936) Brooks 1981**

Diagnosis: Ceca opening separately at posterior end of body (7); maximum body length 7–16 mm (8); ratio of oral sucker width to pharyngeal width 1:0.25–0.40 (11).

***nicolli* Brooks 1981**

Diagnosis: Ceca opening separately at posterior end of body (7); maximum body length 7–16 mm (8); ratio of oral sucker width to pharyngeal width 1:0.25–0.40 (11); vitelline follicles extending anteriorly to posterior margin of acetabulum (12); maximum oral spine length more than 100 µm *(13).

***Caimanicola* Teixeira de Freitas and Lent 1938 (Fig. 31)**

***pavida* (Brooks and Overstreet 1977) Brooks 1981**

Diagnosis: Oral spines averaging 25–30 in number (1).

***caballeroi* (Pelaez and Cruz 1953) Brooks 1981**

Diagnosis: Eggs averaging less than 30 µm long (2); maximum body length 2–6 mm (3).

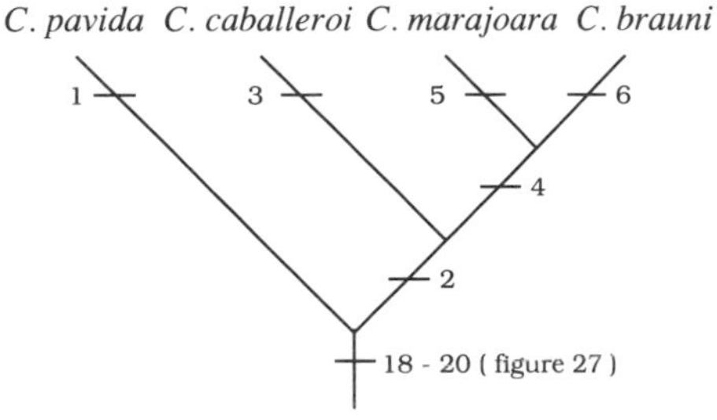

Figure 31. Phylogenetic tree for the species of *Caimanicola*.

***marajoara* Texeira de Freitas and Lent 1938**

Diagnosis: Eggs averaging less than 30 μm long (2); length of body occupied by uterine loops more than 50% of total body length (4); vitelline follicles not extending anteriorly to posterior margin of seminal vesicle (5).

***brauni* (Mane-Garzon and Gil 1961) Brooks 1981**

Diagnosis: Eggs averaging less than 30 μm long (2); length of body occupied by uterine loops more than 50% of total body length (4); ratio of oral sucker width to acetabular width 1:0.3–0.7 (6).

***Acanthostomum* Looss 1899 (Fig. 32)**

***Acanthostomum* (*Blairium*) Brooks and Holcman in press**

Diagnosis: Ceca opening separately at posterior end of body (1).

***scyphocephalum* Braun 1899**

Diagnosis: With characters of the subgenus.

***americanum* (Perez Vigueras 1957) Herber 1961**

Diagnosis: Length of body occupied by uterine loops more than 50% of total body length *(2).

***megacetabulum* Thatcher 1963**

Diagnosis: Length of body occupied by uterine loops more than 50% of total body length *(2); ratio of oral sucker to acetabular width 1:0.8–1.3 (3).

***gnerii* Szidat 1954**

Diagnosis: Testes oblique (4); vitelline follicles sparse (5).

***minimum* Stunkard 1938**

Diagnosis: Testes oblique (4); vitelline follicles sparse (5); one cecum atrophied *(6).

***astorquii* Watson 1976**

Diagnosis: Testes oblique (4); vitelline follicles sparse (5); one cecum atrophied *(6).

Acanthostomum* (*Gibsonium*) Brooks and Holcman in press *sedis mutabilis

Diagnosis: Vitelline follicles not extending anteriorly to posterior margin of seminal vesicle *(7); oral spines averaging less than 20 in number (8).

***absconditum* Looss 1901**

Diagnosis: As above.

***Acanthostomum* (*Acanthostomum*) Looss 1899**

Diagnosis: Oral spines averaging 25–30 in number (9); vitelline follicles extending anteriorly to posterior margin of seminal vesicle (10); eggs averaging less than

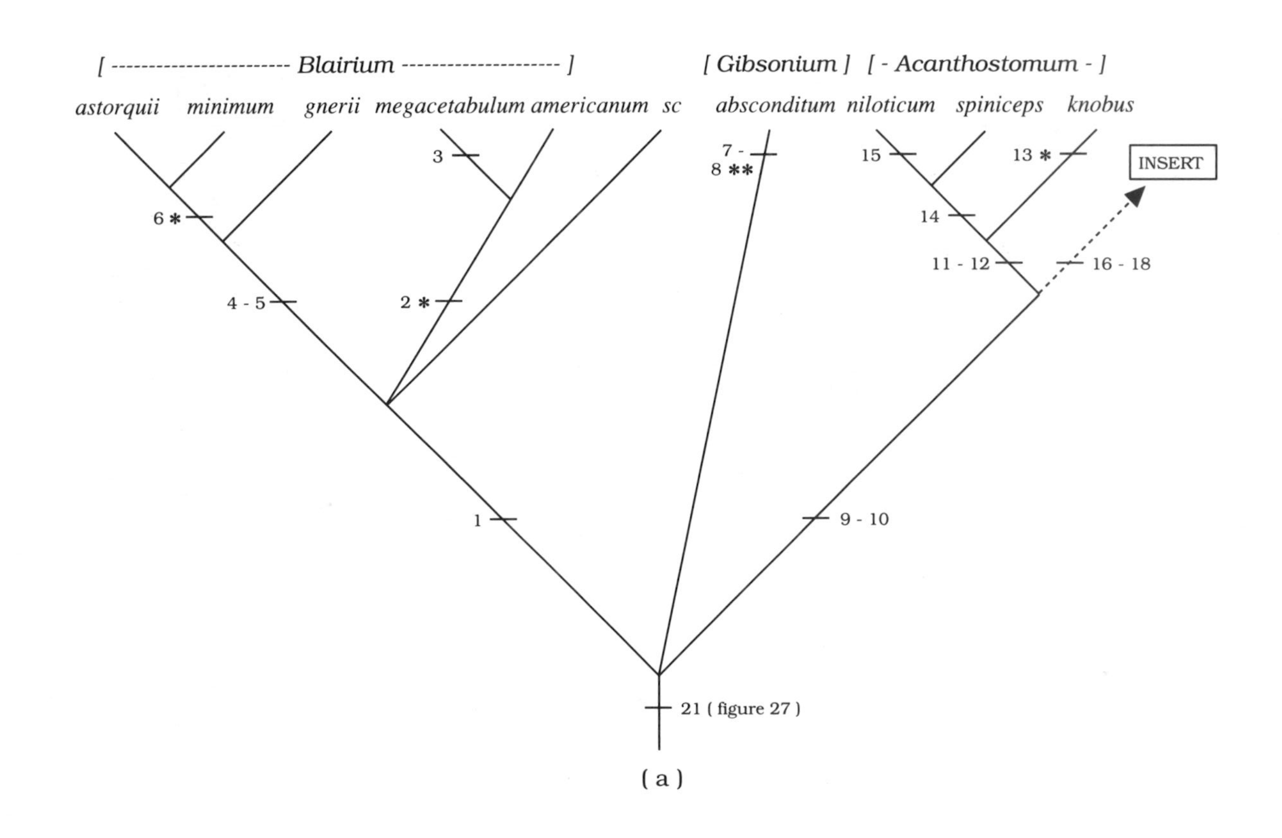

[------------------------ Blairium ---------------------]
[Gibsonium]
[- Acanthostomum -]
astorquii
minimum
gnerii
megacetabulum
americanum
sc
absconditum
niloticum
spiniceps
knobus
INSERT
3
7 -
8 **
15
13 *
6 *
14
11 - 12
16 - 18
4 - 5
2 *
1
9 - 10
21 (figure 27)

(a)

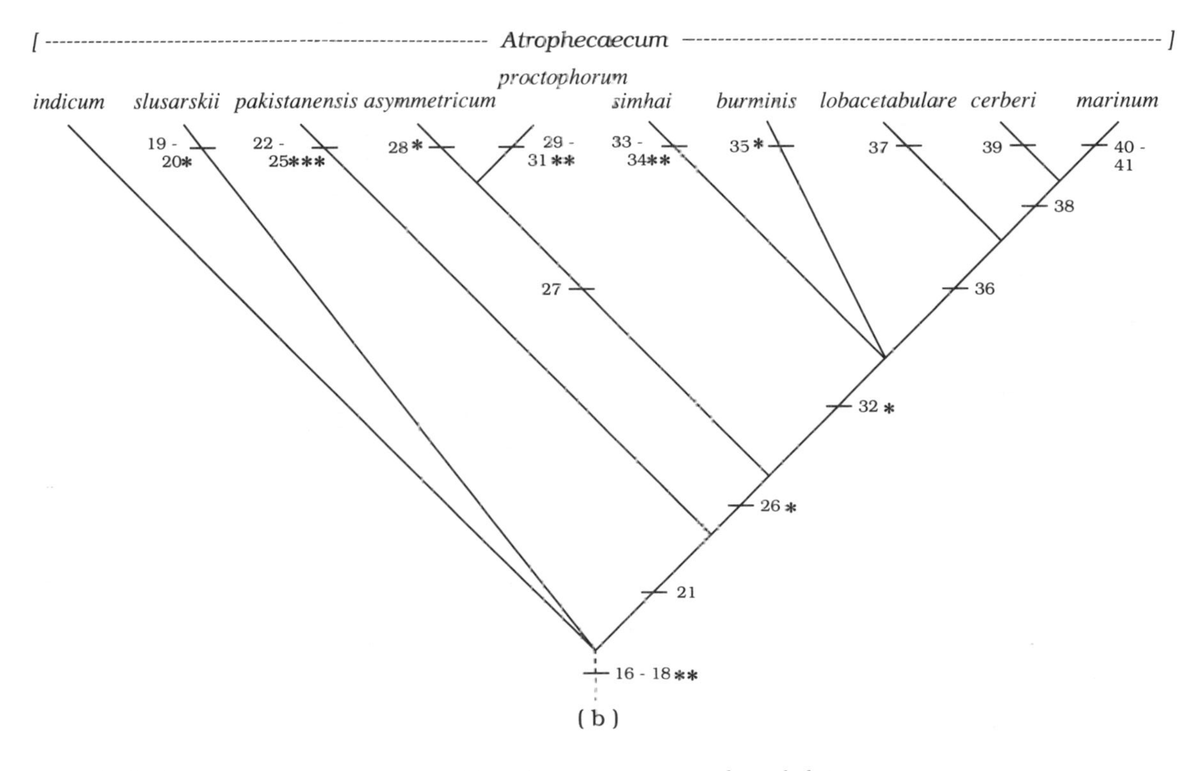

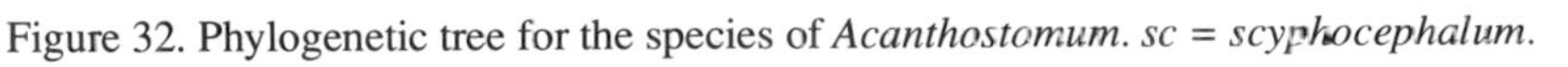
Figure 32. Phylogenetic tree for the species of *Acanthostomum*. *sc* = *scyphocephalum*.

30 µm long (11); cecal bifurcation approximately 10% of total body length pre-acetabular (12).

***knobus* Issa 1962**

Diagnosis: Ratio of body length to width averaging 7.5–15:1 *(13).

***spiniceps* Looss 1896**

Diagnosis: Ratio of oral sucker width to pharyngeal width 1:0.25–0.40 (14).

***niloticum* Issa 1962**

Diagnosis: Ratio of oral sucker width to pharyngeal width 1:0.25–0.40 (14); vitelline follicles extending anteriorly to posterior margin of acetabulum (15).

***Acanthostomum (Atrophecaecum)* Bhalerao 1940**

Diagnosis: Vitelline follicles not extending anteriorly to posterior margin of seminal vesicle *(7); oral spines averaging 25–30 in number (9); vitelline follicles extending anteriorly to posterior margin of seminal vesicle (10); vitelline follicles terminating preovarially (16); length of body occupied by uterine loops more than 50% of total body length *(17); forebody 10–20% of total body length *(18).

***indicum* Sinha 1942**

Diagnosis: As above.

***slusarskii* Kalyankar 1977**

Diagnosis: Ceca opening into excretory vesicle (19); maximum body length 7–16 mm *(20).

***pakistanensis* Coil and Kuntz 1960**

Diagnosis: Prepharynx shorter than pharynx (21); forebody 10–20% of total body length *(22); ratio of body length to width averaging 7.5–15:1 *(23); oral spines averaging 20–24 in number *(24); maximum oral spine length more than 100 µm (25).

***asymmetricum* (Simha 1958) Khalil 1963**

Diagnosis: Prepharynx shorter than pharynx (21); one cecum lost (27); maximum body length 7–16 mm *(28).

***proctophorum* (Dwivedi 1966) Yamaguti 1971**

Diagnosis: Prepharynx shorter than pharynx (21); one cecum lost (27); length of body occupied by uterine loops more than 50% of total body length *(29); oral spines averaging 20–24 in number *(30); vitelline follicles confluent preovarially (31).

***simhai* Khalil 1963**

Diagnosis: Prepharynx shorter than pharynx (21); one cecum atrophied *(26); vitelline follicles not extending anteriorly to posterior margin of seminal vesicle *(32); ratio of body length to width averaging 7.5–15:1 *(33); maximum body length 7–16 mm *(34).

***burminis* (Bhalerao 1926) Bhalerao 1940**

Diagnosis: Prepharynx shorter than pharynx (21); one cecum atrophied *(26); vitelline follicles not extending anteriorly to posterior margin of seminal vesicle *(32); length of body occupied by uterine loops more than 50% of total body length *(35).

***lobacetabulare* Brooks and Caira 1982**

Diagnosis: Prepharynx shorter than pharynx (21); one cecum atrophied *(26); vitelline follicles not extending anteriorly to posterior margin of seminal vesicle *(32); subterminal mouth (36); lobate acetabulum (37).

***cerberi* (Fischthal and Kuntz 1963) Brooks and Caira 1982**

Diagnosis: Prepharynx shorter than pharynx (21); one cecum atrophied *(26); vitelline follicles not extending anteriorly to posterior margin of seminal vesicle *(32); subterminal mouth (36); no oral spines (38); no esophagus (39).

***marinum* (Coil and Kuintz 1960) Brooks and Caira 1982**

Diagnosis: One cecum atrophied *(26); vitelline follicles not extending anteriorly to posterior margin of seminal vesicle *(32); subterminal mouth (36); no oral spines (38); no prepharynx (40); secondary group of vitelline follicles surrounding testes (41).

Family Allocreadiidae Stossich 1903

Subfamily Bunoderinae (Fig. 33)

References: Caira (1989); Brooks (1992b).

Diagnosis: Allocreadiids with at least one pair of ventral papillae associated with the oral sucker (1).

Paracreptotrematina* Amin and Myer 1982 *incertae sedis

Diagnosis: With characters of the subfamily.

***Bunoderella* Schell 1964**

Diagnosis: Bunoderines with one pair of dorsolateral papillae (2); four pairs of penetration glands in the cercaria (3).

***metteri* Schell 1964**

Diagnosis: With characters of the genus.

Bunodera* Railliet 1896 *incertae sedis

Diagnosis: Bunoderines with one pair of dorsolateral papillae (2); one pair of dorsomedial papillae (4).

***mediovitellata* Tsimbaliuk and Roitman 1966**

Diagnosis: *Bunodera* with vitellaria extending posteriorly only to mid-hindbody (5).

***sacculata* Van Cleave and Mueller 1932**

Diagnosis: *Bunodera* with vitellaria extending posteriorly only to mid-hindbody (5); short ceca (6); distal portion of uterus expanded and saclike in older individuals *(8).

***eucaliae* (Miller 1936) Miller 1940**

Diagnosis: *Bunodera* with vitellaria extending posteriorly only to mid-hindbody (5); short ceca (6); oral papillae very small (7); acetabulum larger than oral sucker (9); testes at or immediately posterior to cecal ends (10).

***inconstans* (Lasee, Font, and Sutherland 1988) Brooks 1992**

Diagnosis: *Bunodera* with vitellaria extending posteriorly only to mid-hindbody (5); short ceca (6); acetabulum larger than oral sucker (9); testes at or immediately posterior to cecal ends (10); oral papillae lacking (11).

***luciopercae* (Mueller 1776) Stiles and Hassall 1898**

Diagnosis: *Bunodera* with three pairs of penetration glands in cercariae (12); distal portion of uterus expanded and saclike in older individuals *(13).

***Crepidostomum* Braun 1900**

Diagnosis: Bunoderines with one pair of dorsolateral papillae (2); one pair of dorsomedial papillae (4); pretesticular uterus (14); flame cell formula 2[(2 + 2 + 2) + (2 + 2 + 2)] (15).

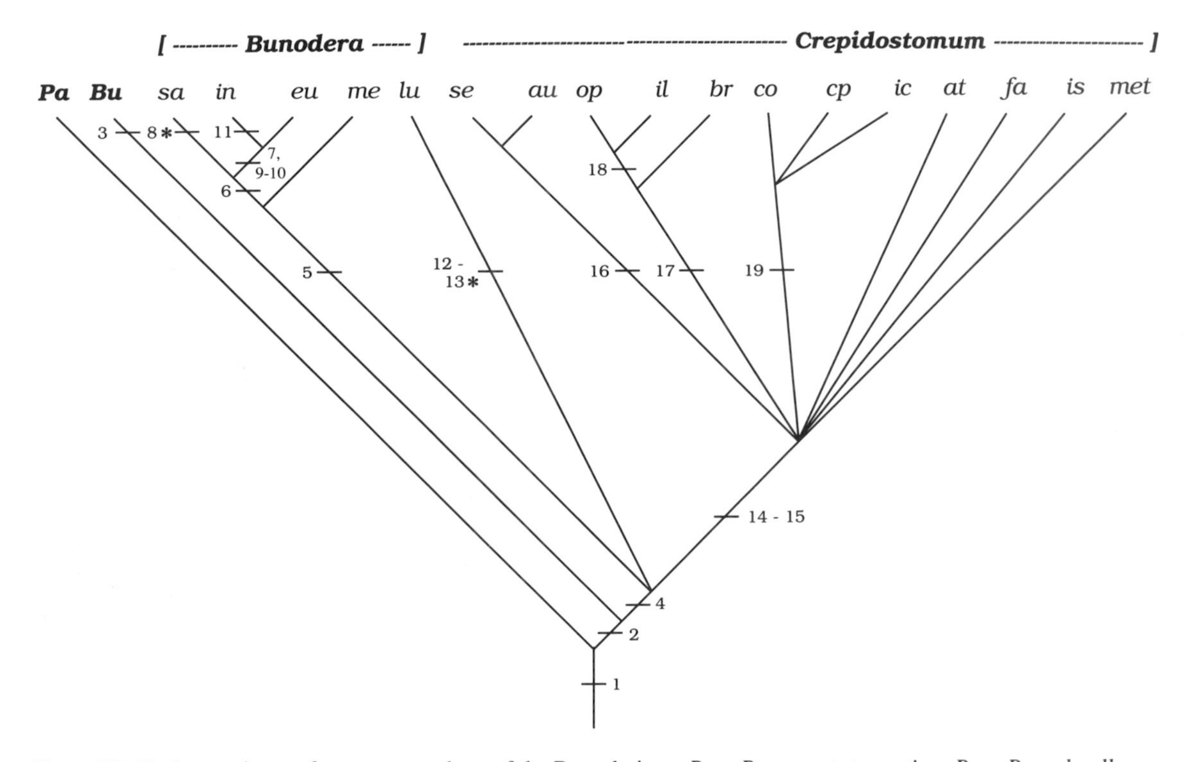

Figure 33. Phylogenetic tree for some members of the Bunoderinae. *Pa* = *Paracreptotrematina; Bu* = *Bunoderella; sa* = *Bunodera sacculata; in* = *B. inconstans; eu* = *B. eucaliae; me* = *B. mediovitellata; lu* = *B. luciopercae; se* = *Crepidostomum serpentinum; au* = *C. auriculatum; op* = *C. opeongoensis*; *il* = *C. illinoiense; br* = *C. brevivitellum; co* = *C. cornutum; cp* = *C. cooperi; ic* = *C. ictaluri; at* = *C. auritum; fa* = *C. farionis; is* = *C. isostomum; met* = *C. metoechus.*

***auritum* (MacCallum 1919) Hopkins 1934, *farionis* (Muller 1784) Nicoll 1909, *isostomum* Hopkins 1931, *metoechus* (Braun 1900) Braun 1900**

Diagnosis: With characters of *Crepidostomum.*

***serpentinum* Talbott and Hutton 1935, *auriculatum* (Wedl 1858) Pratt 1902**

Diagnosis: *Crepidostomum* with very large dorsomedial papillae (16).

***brevivitellum* Hopkins 1934**

Diagnosis: *Crepidostomum* with dorsolateral papillae broader than the dorsomedial papillae (17).

***opeongoensis* Caira 1985, *illinoiense* Faust 1918**

Diagnosis: *Crepidostomum* with dorsolateral papillae broader than the dorsomedial papillae (17); ventral papillae extending to, or posterior to, the posterior margin of the oral sucker (18).

***cooperi* Hopkins 1931, *cornutum* (Osborn 1903) Stafford 1904, *ictaluri* (Surber 1928) Van Cleave and Mueller 1934**

Diagnosis: *Crepidostomum* with anterior protrusion of acetabulum in cercariae smaller than the posterior protrusion (19).

Family Macroderoididae McMullen 1937

Subfamily Glyptheminthinae *incertae sedis*

Diagnosis: With characters of the family.

***Glypthelmins* Stafford 1905 *incertae sedis* (Fig. 34)**

References: Brooks (1977); O'Grady (1987).

Diagnosis: With characters of the subfamily.

***linguatula* (Rudolphi 1819) Travassos 1924**

Diagnosis: With characters of the genus.

***hepatica* (Lutz 1928) Yamaguti 1958**

Diagnosis: Ratio of oral sucker to ventral sucker 1:1 *(1); posterior extent of vitellaria no further posterior than one testis diameter posttesticular *(2).

***vitellinophilum* Dobbin 1958**

Diagnosis: Eggs 30–40 µm *(3); ratio of oral sucker to ventral sucker 1:0.5 (4).

***incurvatum* Nasir 1966**

Diagnosis: Eggs 30–40 µm *(3).

***proximus* Texeira de Freitas 1941**

Diagnosis: No extracecal uterine loops (5); tegumental spines extending from anterior end to mid-hindbody *(6).

***palmipedis* (Lutz 1928) Texeira de Freitas 1941**

Diagnosis: No extracecal uterine loops (5); tegumental spines extending from anterior end to mid-hindbody *(6); ratio of oral sucker to acetabulum 1:1 *(7).

***robustus* Brooks 1976**

Diagnosis: No extracecal uterine loops (5); tegumental spines extending from anterior end to mid-hindbody level *(6); ratio of oral sucker to ventral sucker 1:1 *(7); Laurer's canal arising distal to common vitelline duct (8); ratio of oral sucker to pharynx more than 1:0.7 (9).

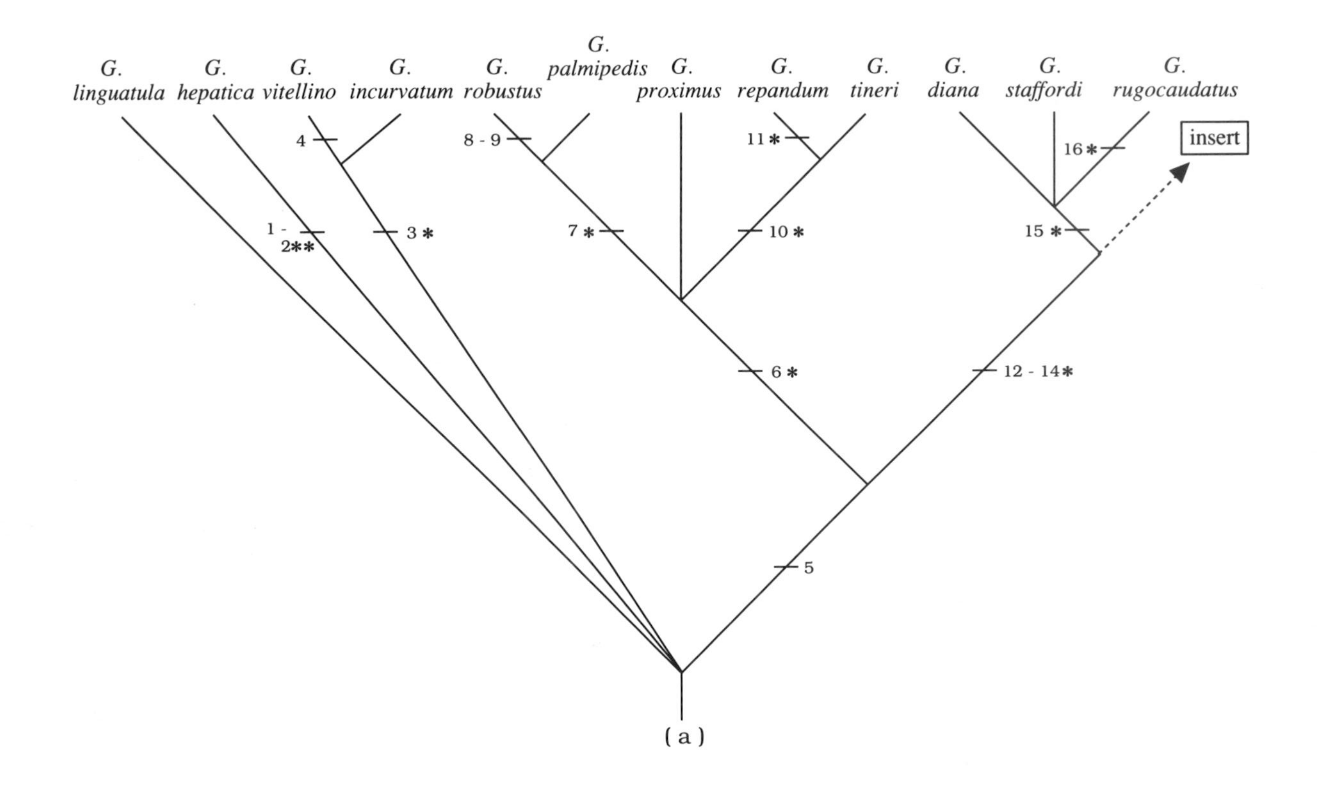
G. linguatula
G. hepatica
G. vitellino
G. incurvatum
G. robustus
G. palmipedis
G. proximus
G. repandum
G. tineri
G. diana
G. staffordi
G. rugocaudatus
insert
4
1 - 2**
3 *
8 - 9
7 *
11 *
10 *
6 *
16 *
15 *
12 - 14 *
5
(a)

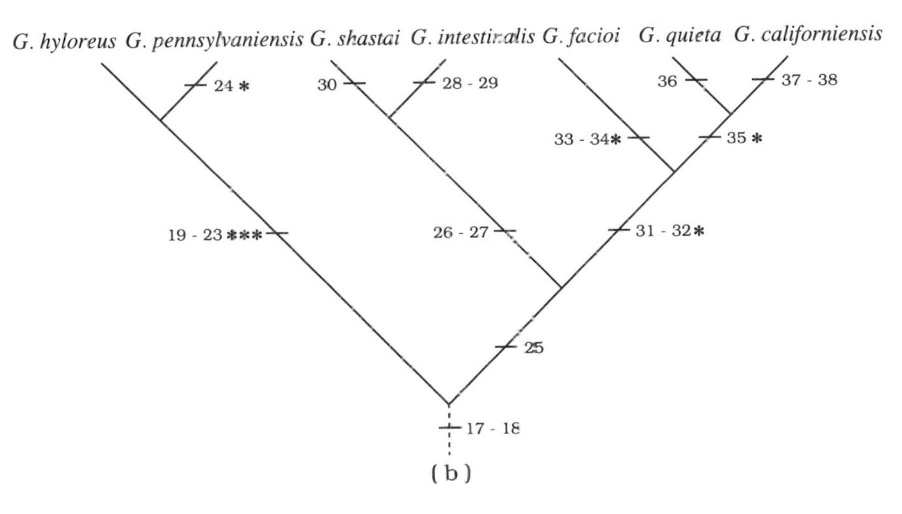

Figure 34. Phylogenetic tree for the species of *Glypthelmins*. *vitellino* = *vitellinophilum.*

***tineri* (Babero 1951) Brooks 1977**

Diagnosis: No extracecal uterine loops (5); tegumental spines extending from anterior end to mid-hindbody *(6); internal seminal vesicle bipartite, straight *(10).

***repandum* (Rudolphi 1819) Nasir and Diaz 1970**

Diagnosis: No extracecal uterine loops (5); tegumental spines extending from anterior end to mid-hindbody *(6); internal seminal vesicle bipartite, straight *(10); testes symmetrical *(11).

***staffordi* Tubangui 1928**

Diagnosis: No extracecal uterine loops (5); pretesticular uterine loops lacking (12); internal seminal vesicle bipartite, straight *(13); left-handed ovary (14); testes symmetrical *(15).

***rugocaudatus* (Yoshida 1916) Yahata 1934**

Diagnosis: No extracecal uterine loops (5); pretesticular loops lacking (12); internal seminal vesicle bipartite, straight *(13); left-handed ovary (14); testes symmetrical *(15); eggs 30–40 μm long *(16).

***diana* Belous in Skrjabin and Antipin 1959**

Diagnosis: No extracecal uterine loops (5); pretesticular loops lacking (12); internal seminal vesicle bipartite, straight *(13); left-handed ovary (14); testes symmetrical *(15).

***hyloreus* Martin 1969**

Diagnosis: Internal seminal vesicle bipartite, straight *(13); left-handed ovary (14); eggs more than 40 μm long (17); excretory vesicle bifurcating at or posterior to level of testes (18); pretesticular uterine loops present *(19); cercarial stylet lacking (20); oral sucker to ventral sucker ratio greater than 1:0.75 (21); testes symmetrical *(22); uterine loops extracecal *(23).

***pennsylvaniensis* Cheng 1961**

Diagnosis: Internal seminal vesicle bipartite, straight *(13); left-handed ovary (14); eggs more than 40 μm long (17); excretory vesicle bifrucating at or posterior to level of testes (18); pretesticular loops present *(19); cercarial stylet lacking (20); oral sucker to ventral sucker ratio greater than 1:0.75 (21); testes symmetrical *(22); uterine loops extracecal *(23); posterior extent of vitellaria no further posterior than one testis diameter posttesticular *(24).

intestinalis* (Lucker 1931) *comb. n.

Diagnosis: No extracecal uterine loops (5); pretesticular loops lacking (12); internal seminal vesicle bipartite, straight *(13); left-handed ovary (14); eggs more than 40 μm long (17); excretory vesicle bifurcating at or posterior to level of testes (18); dorsomedial confluence of posterior vitelline fields (25); cirrus sac more than half length of forebody (26); medial glands at level of pharynx and esophagus (27); anterior vitelline duct lacking, no anterior vitelline field (28); testes tandem (29).

***shastai* Ingles 1936**

Diagnosis: No extracecal uterine loops (5); pretesticular loops lacking (12); internal seminal vesicle bipartite, straight *(13); left-handed ovary (14); eggs more than 40 μm long (17); excretory vesicle bifurcating at or posterior to level of testes (18); dorsomedial confluence of anterior vitelline glands (25); cirrus sac more than half length of forebody (26); medial glands at level of pharynx and esophagus (27); ratio of oral sucker to pharynx less than 1:0.5 (30).

***facioi* Brenes Madrigal, Arroyo Sancho, Jimenez-Quiros, and Delgado Flores 1959**

Diagnosis: No extracecal uterine loops (5); pretesticular loops lacking (12); internal seminal vesicle bipartite, straight *(13); left-handed ovary (14); eggs more than 40 µm long (17); excretory vesicle bifurcating at or posterior to level of testes (18); dorsomedial confluence of anterior vitelline glands (25); tegumental scales (broad overlapping spines) (31); posterior extent of vitellaria no further posterior than one testis diameter posttesticular *(32); tegumental spines from anterior end to mid-hindbody *(33); penetration glands persist in adults (34).

***quieta* (Stafford 1900) Stafford 1905**

Diagnosis: No extracecal uterine loops (5); pretesticular loops lacking (12); internal seminal vesicle bipartite, straight *(13); left-handed ovary (14); eggs more than 40 µm long (17); excretory vesicle bifurcating at or posterior to level of testes (18); dorsomedial confluence of anterior vitelline glands (25); tegumental scales (broad overlapping spines) (31); posterior extent of vitellaria no further posterior than one testis diameter posttesticular *(32); testes symmetrical *(35); prominent pharyngeal glands present (36).

***californiensis* (Cort 1919) Miller 1930**

Diagnosis: No extracecal uterine loops (5); pretesticular loops lacking (12); internal seminal vesicle bipartite, straight *(13); left-handed ovary (14); eggs more than 40 µm long (17); excretory vesicle bifurcating at or posterior to level of testes (18); dorsomedial confluence of anterior vitelline glands (25); tegumental scales (broad overlapping spines) (31); testes symmetrical *(35); anterior vitelline duct present, anterior vitelline field extending posteriorly only to pharyngeal level (37); posterior extent of vitellaria no further posterior than testicular level (38).

Molecular Data

Rannala (1990) presented electrophoretic data covering 14 loci comparing *Glypthelmins californiensis* and *G. intestinalis,* demonstrating that their inferred degree of genetic similarity was similar to that found for other congeneric species, and supporting the inclusion of *Haplometrana intestinalis* in *Glypthelmins.*

Subfamily Macroderoidinae McMullen 1937 (Fig. 35)

Reference: Carney and Brooks (1991).

Diagnosis: Macroderoidids with simple uterine structure (1); cecal bifurcation 30% of total body length from anterior end (2); no seminal receptacle (3).

***Macroderoides* Pearse 1924**

Diagnosis: Macroderoidines with large spinose cirrus (4).

***Paramacroderoides* Venard 1941**

Diagnosis: Macroderoidines with large spinose cirrus (4); prominent spines on oral sucker (5).

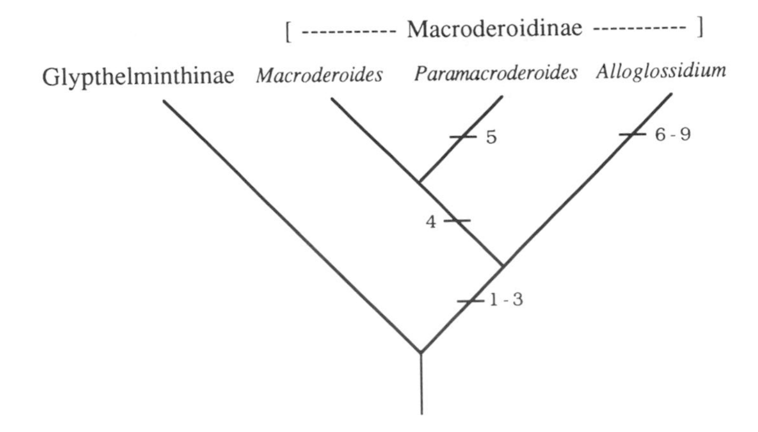

Figure 35. Phylogenetic tree for the genera of the Macroderoidinae.

***Alloglossidium* Simer 1929**

Reference: Carney and Brooks (1991).

Diagnosis: Macroderoidines with small cirrus (6); acetabulum smaller than oral sucker (7); cecal bifurcation averaging 20% of total body length from anterior end (8); vitellaria extending anteriorly to acetabular level (9).

***Alloglossidium* Simer 1929 (Fig. 36)**

Reference: Carney and Brooks (1991).

***geminum* (Lamont 1921) Van Cleave and Mueller 1932**

Diagnosis: With characters of the genus.

***corti* (Mueller 1930) Van Cleave and Mueller 1932**

Diagnosis: Vitellaria extend anteriorly to acetabular level (1).

***progeneticum* (Sullivan and Heard 1969) Font and Corkum 1975**

Diagnosis: Vitellaria extend anteriorly to acetabular level (1); anterior testis post-equatorial (2); ovarian diameter greater than testicular diameter *(3); distal portion of gravid uterus a distended sac (4); cecal bifurcation 10–15% of total body length from anterior end (5).

***renale* Font and Corkum 1975**

Diagnosis: Vitellaria extend anteriorly to acetabular level (1); anterior testis post-equatorial (2); ovarian diameter greater than testicular diameter *(3); distal portion of gravid uterus a distended sac (4); cecal bifurcation 10–15% of total body length from anterior end (5); body oval (6).

***macrobdellensis* Beckerdite and Corkum 1974**

Diagnosis: Vitellaria extend anteriorly to acetabular level (1); anterior testis post-equatorial (2); body filiform (7); large-sized tegumental spines (more than 12 μm long) (8); acetabulum and oral sucker equal in size (9); forebody 10–20% of total body length (10); postcecal length at least 20% of total body length (11); excretory vesicle extending anterior to anterior testis (12).

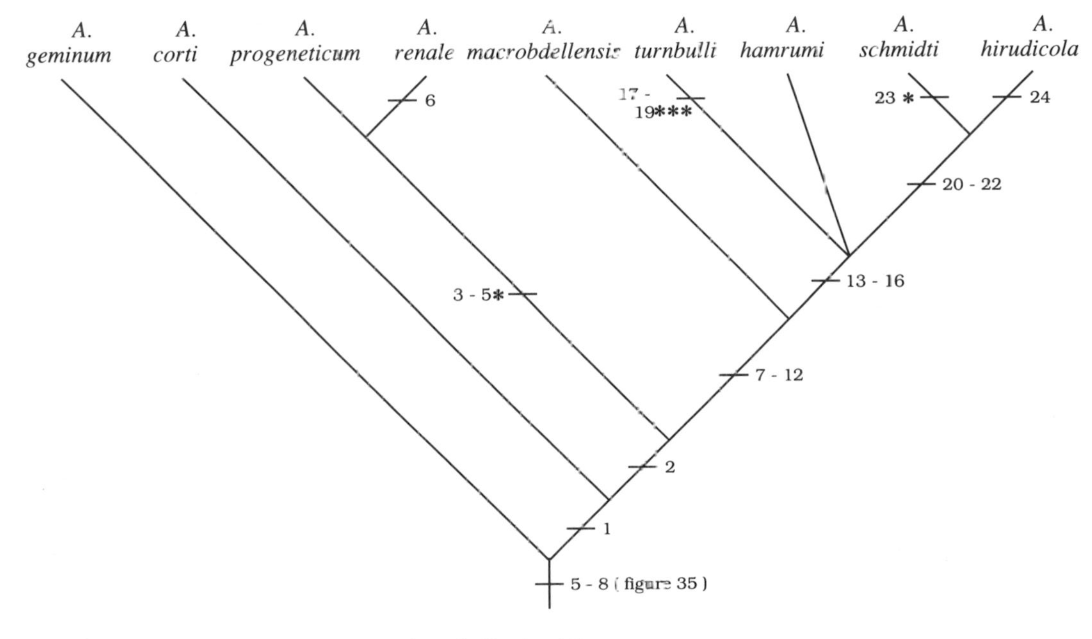

Figure 36. Phylogenetic tree for the species of *Alloglossidium*.

turnbulli **Neumann and Vande Vusse 1976**

Diagnosis: Body filiform (7); acetabulum and oral sucker equal in size (9); forebody 10–20% of total body length (10); postcecal length at least 20% of total body length (11); excretory vesicle extending anterior to anterior testis (12); uterine loops highly coiled (13); tegumental spines around oral sucker not notched or tapered (14); tegumental spines terminating at midbody (15); anterior testis preequatorial (16); ovarian diameter greater than testicular diameter *(17); medium-sized tegumental spines (4–7 μm × 4–5 μm) *(18); vitellaria do not extend anteriorly to acetabulum *(19).

hamrumi **Neumann and Vande Vusse 1976**

Diagnosis: Vitellaria extend anteriorly to acetabular level (1); body filiform (7); large-sized tegumental spines (more than 12 μm long) (8); acetabulum and oral sucker equal in size (9); forebody 10–20% of total body length (10); postcecal length at least 20% of total body length (11); excretory vesicle extending anterior to anterior testis (12); uterine loops highly coiled (13); tegumental spines around oral sucker not notched, not tapered (14); tegumental spines terminating at midbody (15); anterior testis preequatorial (16).

schmidti **Timmers 1979**

Diagnosis: Vitellaria extend anteriorly to acetabular level (1); acetabulum and oral sucker of equal size (9); forebody 10–20% of total body length (10); postcecal length at least 20% of total body length (11); excretory vesicle extending anterior to anterior testis (12); uterine loops highly coiled (13); tegumental spines around oral sucker not notched, not tapered (14); anterior testis preequatorial (16); tegumental spines terminating in region of oral sucker (20); body elongate *(21); minute tegumental spines (22); ovarian diameter greater than testicular diameter *(23).

hirudicola **Schmidt and Chaloupka 1969**

Diagnosis: Vitellaria extend anteriorly to acetabular level (1); forebody 10–20% of total body length (10); postcecal length at least 20% of total body length (11); excretory vesicle extending anterior to anterior testis (12); uterine loops highly coiled (13); tegumental spines around oral sucker not notched, not tapered (14); anterior testis preequatorial (16); tegumental spines terminating in region of oral sucker (20); body elongate *(21); minute tegumental spines (22); acetabulum larger than oral sucker (24).

Family Telorchidae Looss 1899 (Fig. 37)

Reference: Macdonald and Brooks (1989a).

Pseudotelorchis **Yamaguti 1971**

Diagnosis: Telorchiids with irregularly shaped ovaries (1); irregularly positioned testes (2).

Loefgrenia **Travassos 1920**

Diagnosis: Telorchiids with distance between ovary and acetabulum at least one ovarian diameter (3); without esophagus (4).

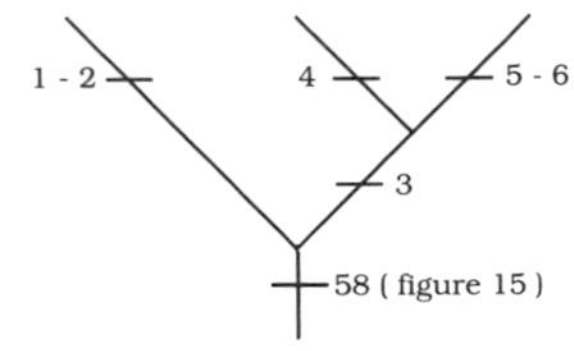

Figure 37. Phylogenetic tree for the genera of the Telorchidae.

***Telorchis* Luhe 1899**

Diagnosis: Telorchiids with distance between ovary and acetabulum at least one ovarian diameter (3); tandem testes (5); uterus composed of well-ordered ascending and descending transverse loops (6).

***Telorchis* Luhe 1899 endemic to North America (Fig. 38)**

Reference: Macdonald and Brooks (1989b).

***stunkardi* Chandler 1923**

Diagnosis: *Telorchis* with average ratio of body length to width less than 5:1 (1); vitelline follicles in lateral, clustered groups (2); average ratio of oral sucker to pharynx to acetabulum width 1:0.4:0.4 (3).

***sirenis* (Zeliff 1937) Wharton 1940**

Diagnosis: *Telorchis* with average ratio of body length to width less than 5:1 (1); vitelline follicles in lateral, clustered groups (2); average ratio of oral sucker to pharynx to acetabulum width 1:0.4:0.4 (3); body linguiform (4).

***corti* Stunkard 1915**

Diagnosis: *Telorchis* with average ratio of oral sucker to pharynx to acetabulum width 0.75:0.6:0.7 (5); internal seminal vesicle bipartite (6).

***dollfusi* (Stunkard and Franz 1977) Macdonald and Brooks 1989**

Diagnosis: *Telorchis* with average ratio of oral sucker to pharynx to acetabulum width 0.75:0.6:0.7 (5); internal seminal vesicle bipartite (6); oral sucker with pair of ventrolateral lappets (7); body length of ovigerous adults always greater than 4.8 mm *(8).

***auridistomi* (Byrd 1937) Wharton 1940**

Diagnosis: *Telorchis* with internal seminal vesicle bipartite (6); oral sucker with pair of dorsolateral lappets (9); average ratio of oral sucker to pharynx to acetabulum width 0.5:0.3:0.6 (10).

***angustus* (Stafford 1900) Barker and Covey 1911**

Diagnosis: *Telorchis* with average ratio of oral sucker to pharynx to acetabulum width 0.75:0.6:0.7 (5); internal seminal vesicle bipartite (6); oral sucker with pair of ventrolateral lappets (7); genital pore dorsolateral to left cecum (11); posterior end of cirrus sac at acetabular level (12).

***scabrae* Macdonald and Brooks 1987**

Diagnosis: *Telorchis* with average ratio of oral sucker to pharynx to acetabulum width 0.75:0.6:0.7 (5); body length of ovigerous adults always greater than 4.8 mm *(13); cirrus sac thick-walled and muscular (14); genital pore ventral to left cecum (15); posterior end of cirrus sac midway between acetabulum and ovary *(16).

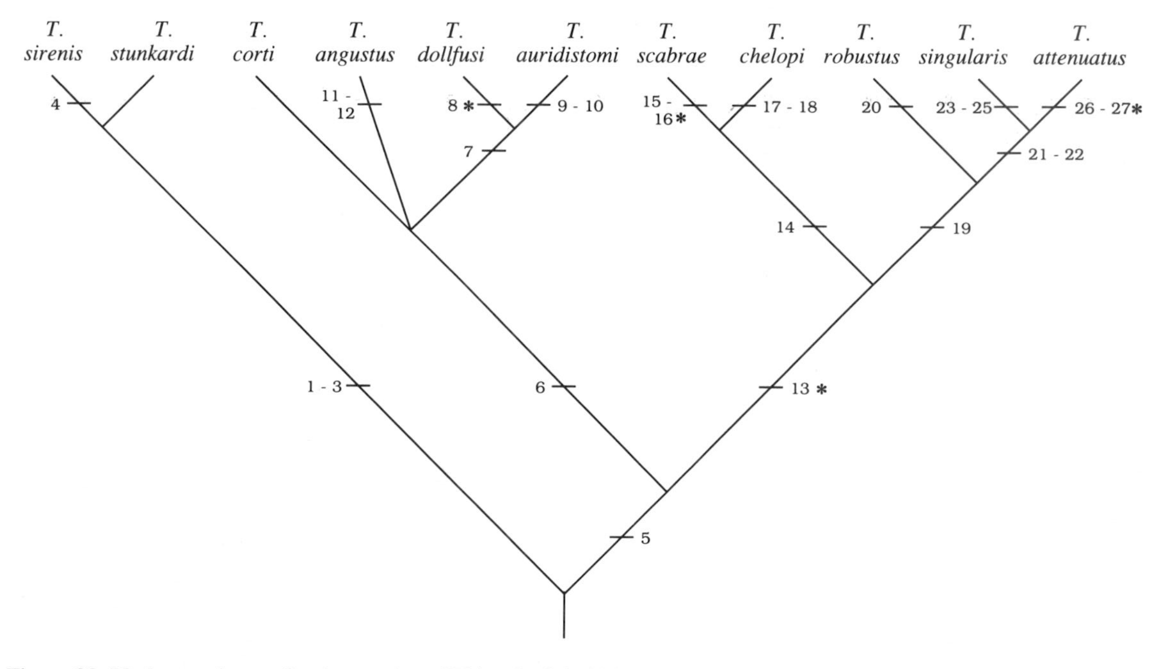

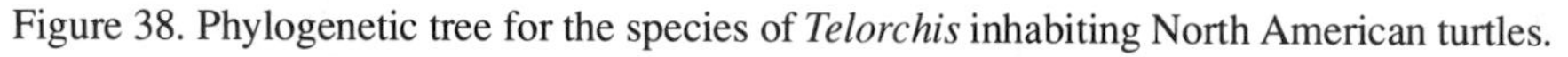
Figure 38. Phylogenetic tree for the species of *Telorchis* inhabiting North American turtles.

chelopi **MacCallum 1919**

Diagnosis: *Telorchis* with average ratio of oral sucker to pharynx to acetabulum width 0.75:0.6:0.7 (5); body length of ovigerous adults always greater than 4.8 mm *(13); cirrus sac thick-walled and muscular (14); cecal epithelium thickened at bifurcation (17); prominent pharyngeal glands present (18).

robustus **Goldberger 1911**

Diagnosis: *Telorchis* with average ratio of oral sucker to pharynx to acetabulum width 0.75:0.6:0.7 (5); body length of ovigerous adults always greater than 4.8 mm *(13); cirrus sac thick-walled and muscular (14); cecal bifurcation less than 6% of total body length from anterior end (19); esophagus lacking (20).

singularis **(Bennett 1935) Wharton 1940**

Diagnosis: *Telorchis* with body length of ovigerous adults always greater than 4.8 mm *(13); cirrus sac thick-walled and muscular (14); cecal bifurcation less than 6% of total body length from anterior end (19); body filiform (21); ovary closer to testes than to ventral sucker (22); genital atrium present (23); metraterm with muscular bulb or sphincter (24); average ratio of oral sucker to pharynx to acetabulum width 0.9:0.75:0.75 (25).

attenuatus **Goldberger 1911**

Diagnosis: *Telorchis* with average ratio of oral sucker to pharynx to acetabulum width 0.75:0.6:0.7 (5); body length of ovigerous adults always greater than 4.8 mm *(13); cirrus sac thick-walled and muscular (14); cecal bifurcation less than 6% of total body length from anterior end (19); body filiform (21); ovary closer to testes than to ventral sucker (22); average ratio of body length to width 16:1 (26); posterior end of cirrus sac midway between acetabulum and ovary *(27).

Family Gorgoderidae Looss 1899 (Fig. 39)

References: Brooks and Macdonald (1986).

Subfamily Gorgoderinae Looss 1899

Diagnosis: Gorgoderids with extracecal uterine loops in forebody (1); I-shaped excretory vesicles *(2).

Phyllodistomoides **Brooks 1977** ***incertae sedis***

Diagnosis: With characters of the subfamily.

Dendrorchis **Travassos 1926** ***incertae sedis***

Diagnosis: Gorgoderines with no pharynx (3).

Phyllodistomum **Braun 1899** ***sedis mutabilis, incertae sedis***

Diagnosis: Gorgoderines with no pharynx (3); compact, nonfollicular vitellaria (4); extracecal uterine loops confined to hindbody (5).

Gorgotrema **Dayal 1938** ***sedis mutabilis***

Diagnosis: Gorgoderines with no pharynx (3); compact, nonfollicular vitellaria (4); extracecal uterine loops confined to hindbody (5); compact follicular testes (6).

Xystretum **Linton 1910** ***sedis mutabilis***

Diagnosis: Gorgoderines with no pharynx (3); compact, nonfollicular vitellaria (4); extracecal uterine loops confined to hindbody (5); cyclocoel (7).

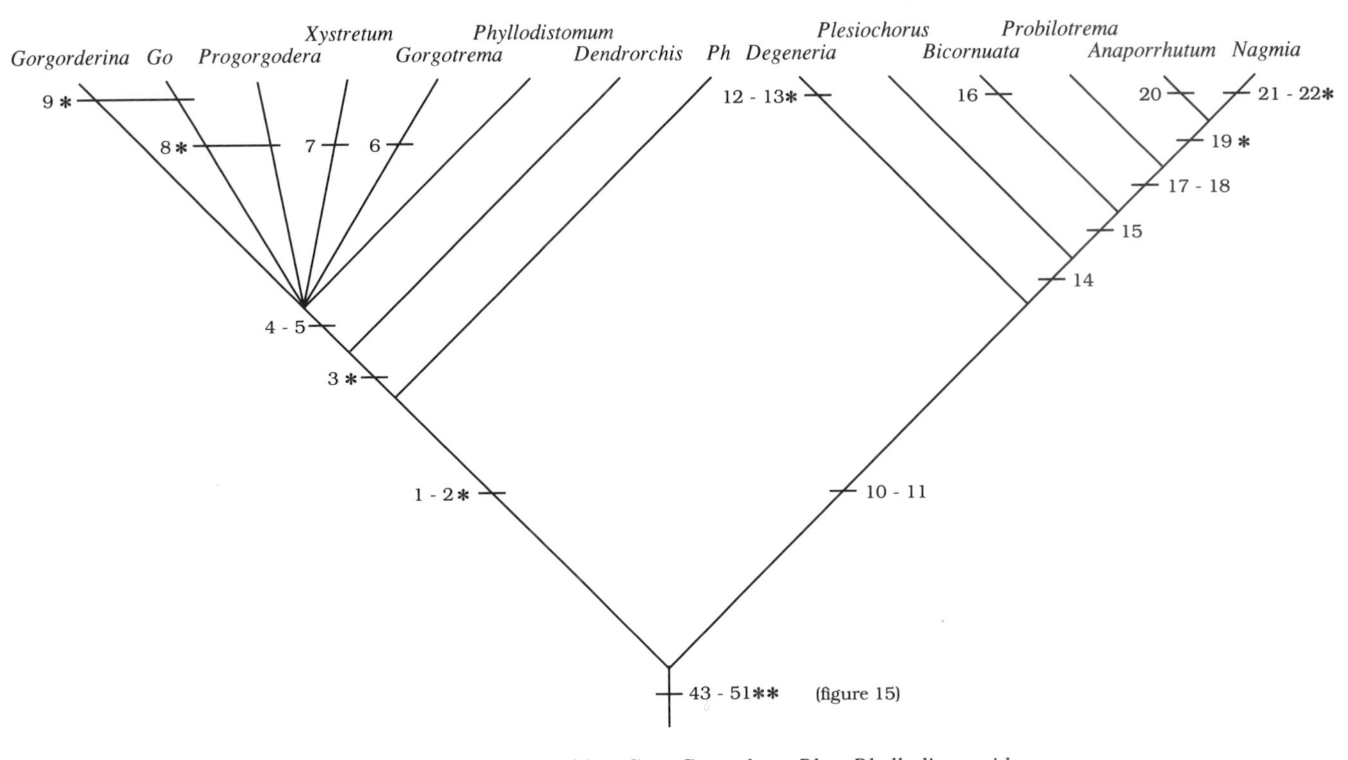

Figure 39. Phylogenetic tree for the genera of the Gorgoderidae. *Go* = *Gorgodera; Ph* = *Phyllodistomoides.*

Progorgodera* Brooks and Buckner 1976 *sedis mutabilis, incertae sedis

Diagnosis: Gorgoderines with no pharynx (3); compact, nonfollicular vitellaria (4); extracecal uterine loops confined to hindbody (5); multiple testes *(8).

Gorgodera* Looss 1899 *sedis mutabilis, incertae sedis

Diagnosis: Gorgoderines with no pharynx (3); compact, nonfollicular vitellaria (4); extracecal uterine loops confined to hindbody (5); multiple testes *(8); elongate body *(9).

Gorgoderina* Looss 1899 *sedis mutabilis, incertae sedis

Diagnosis: Gorgoderines with no pharynx (3); compact, nonfollicular vitellaria (4); extracecal uterine loops confined to hindbody (5); elongate body *(9).

Subfamily Anaporrhutinae Looss 1901

Diagnosis: Gorgoderids with cecal or extracecal testes (10); no Laurer's canal (11).

***Degeneria* Campbell 1977**

Diagnosis: Anaporrhutines with extracecal uterine loops extending into forebody (12); genital sac present *(13).

Plesiochorus* Looss 1901 *incertae sedis

Diagnosis: Anaporrhutines with extracecal vitellaria (14).

***Bicornuata* Pearse 1930**

Diagnosis: Anaporrhutines with extracecal vitellaria (14); uterine loops intercecal only (15); truncate body (16).

Probilotrema* Looss 1902 *incertae sedis

Diagnosis: Anaporrhutines with extracecal vitellaria (14); uterine loops intercecal only (15); vitellaria compact with diffuse follicles (17); testes comprised of dispersed follicles (18).

***Anaporrhutum* Ofenheim 1900**

Diagnosis: Anaporrhutines with extracecal vitellaria (14); uterine loops intercecal only (15); vitellaria compact with diffuse follicles (17); testes comprised of dispersed follicles (18); vitellaria intercecal only *(19); excretory vesicle H-shaped (20).

***Nagmia* Nagaty 1930**

Diagnosis: Anaporrhutines with uterine loops intercecal only (15); vitellaria compact with diffuse follicles (17); testes comprised of dispersed follicles (18); vitellaria intercecal only *(19); ceca diverticulate (21); excretory vesicles I-shaped *(22).

Family Zoogonidae Odhner 1902 (Fig. 40)

References: Bray (1986, 1987); Brooks (1990a).

***Lecithostaphylus* Group**

Diagnosis: Zoogonids with pedunculate acetabulum (1).

Genus *Lecithostaphylus* Odhner 1911

Diagnosis: With characters of the group.

Subfamily Lepidophyllinae Stossich 1903 *incertae sedis*

Diagnosis: Zoogonids with narrow ceca, extending posteriorly to testicular level (2).

Genus *Steganoderma* Stafford 1904 *sedis mutabilis*

Diagnosis: Lepidophyllines with seminal receptacle almost always empty (3).

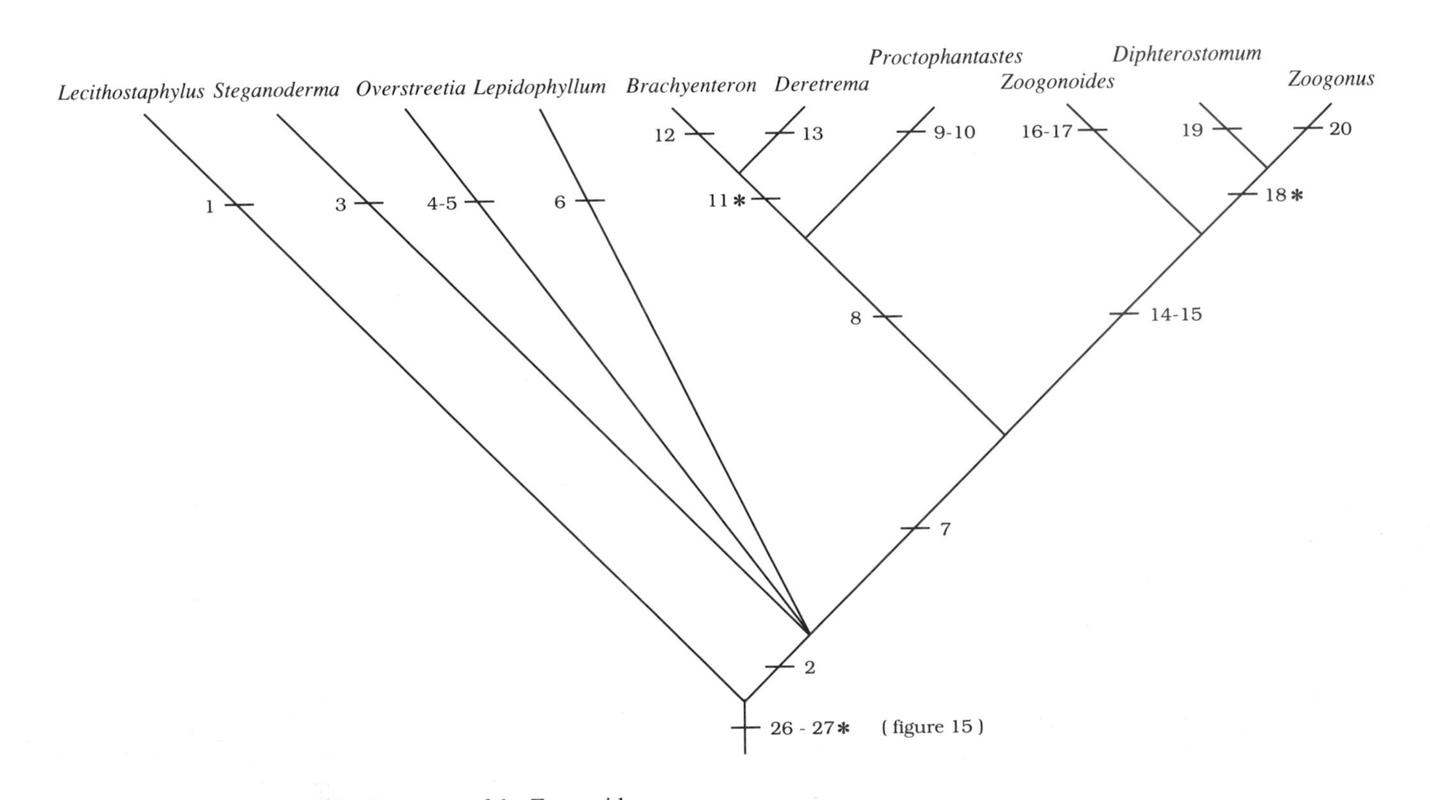

Figure 40. Phylogenetic tree for the genera of the Zoogonidae.

Genus *Overstreetia* Bray 1985 *sedis mutabilis*

Diagnosis: Lepidophyllines with enlarged spines in one or more rows around the oral sucker (4); tandem or subtandem testes (5).

Genus *Lepidophyllum* Stossich 1903 *sedis mutabilis*

Diagnosis: Lepidophyllines with deeply lobate testicular margins (6).

***Deretrema* Group**

Diagnosis: Zoogonids with narrow ceca, extending posteriorly to testicular level (2); bipartite seminal vesicle (7); vitellarium mainly in forebody, but overlapping acetabulum (8).

Genus *Proctophantastes* Odhner 1911

Diagnosis: Deretrematines with acetabulum divided equatorially by internal muscular ridge and external cleft (9); periatrial glands in the form of several small, claviform clusters of cells apparently feeding into the genital atrium (10).

Genus *Brachyenteron* Manter 1934

Diagnosis: Deretrematines with short, saccate ceca *(11); vitellarium lateral to acetabulum (12).

Genus *Deretrema* Linton 1910

Diagnosis: Deretrematines with short, saccate ceca *(11); elongate, coiled seminal vesicle (13).

Subfamily Zoogoninae Odhner 1902

Diagnosis: Zoogonids with narrow ceca extending posteriorly to testicular level (2); bipartite seminal vesicle (7); egg capsules not tanned, thin, and practically colorless (14); vitellarium a single median mass (15).

Genus *Zoogonoides* Odhner 1902

Diagnosis: Zoogonines with genital atrium sac (16); posttesticular ovary (17).

Genus *Diphterostomum* Stossich 1903

Diagnosis: Zoogonines with short, saccate ceca *(18); vitellarium a pair of small median masses (19).

Genus *Zoogonus* Odhner 1902

Diagnosis: Zoogonines with short, saccate ceca *(18); genital pore at acetabular level or in anterior portion of hindbody (20).

***Lecithostaphylus* Odhner 1911 (Fig. 41)**

***hemirhamphi* (Manter 1947) Yamaguti 1953, *parexocoeti* (Manter 1947) Yamaguti 1971, *depauperati* Yamaguti 1970**

Diagnosis: With characters of the genus.

***retroflexum* (Molin 1859) Odhner 1911, *nitens* (Linton 1898) Linton 1940**

Diagnosis: Lecithostaphylines with uterine loops half pre- and half postgonadal (1).

***Lepidophyllum* Stossich 1903 (Fig. 42)**

***atherinae* (Price 1934) comb. n., *macrophallos* (Szidat and Nani 1951) comb. n., *oviformis* (Szidat 1962) comb. n.**

Diagnosis: *Lepidophyllum* with lobate ovarian margins *(1).

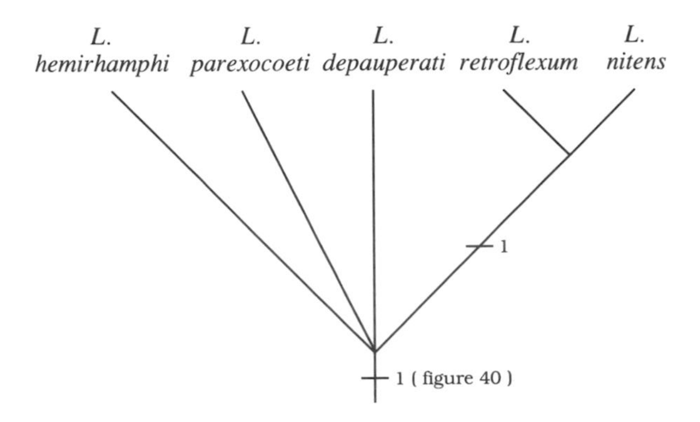

Figure 41. Phylogenetic tree for species of *Lecithostaphylus.*

***hispidum* (Yamaguti 1934) comb. n., *hirudinacea* (Zhukov 1957) comb. n.**

Diagnosis: *Lepidophyllum* with lobate ovarian margins *(1) body flattened, spatulate, or leaf-shaped *(2); long ceca, extending nearly to posterior end of body (3).

***steenstruppi* Odhner 1902, *appyi* Bray and Gibson 1988, *armatum* Zhukov 1957, *brachycladium* Zhukov 1957, *cameroni* Arai 1969, *pleuronectini* Zhukov 1957, *pyriforme* (Yamaguti 1934) Bray 1987, *schantaricum* Kulikov, Tsimbaliuk, and Kasatschenko 1968**

Diagnosis: *Lepidophyllum* with body flattened, spatulate, or leaf-shaped *(2).

***Proctophantastes* Odhner 1911 (Fig. 43)**

***abyssorum* Odhner 1911, *gillissi* (Overstreet and Pritchard 1977) Bray and Gibson 1988**

Diagnosis: With characters of the genus.

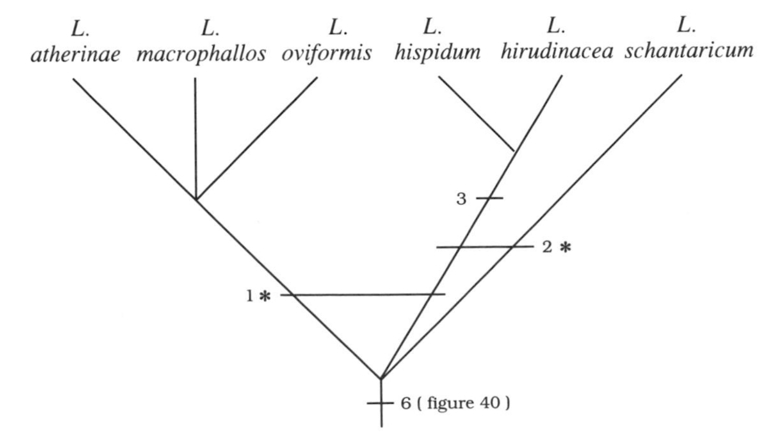

Figure 42. Phylogenetic tree for species of *Lepidophyllum. L. schantaricum* includes *L. steenstruppi, L. appyi, L. armatum, L. brachycladium, L. cameroni, L. pleuronectini,* and *L. pyriforme.*

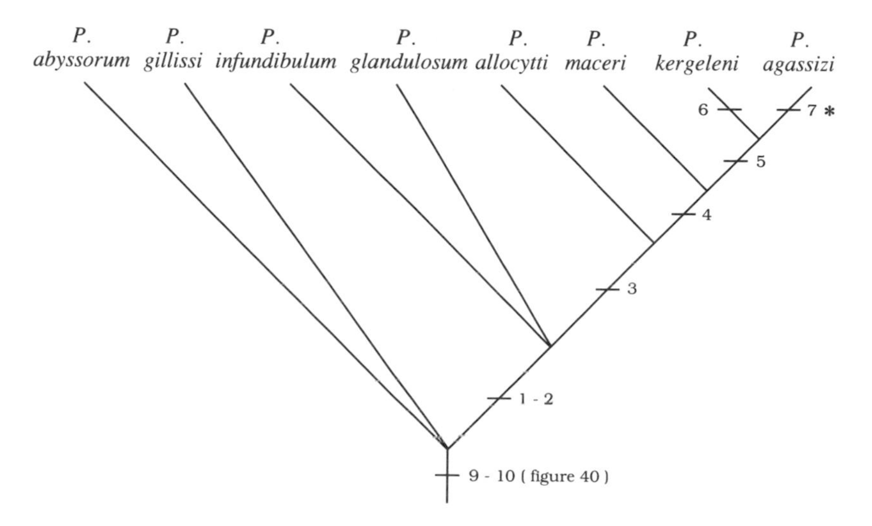

Figure 43. Phylogenetic tree for species of *Proctophantastes.*

infundibulum **Kamegai 1973,** ***glandulosum*** **(Byrd 1964) Yamaguti 1971**

Diagnosis: *Proctophantastes* with genital atrium large, muscular, and pocketed (1); periatrial glands a large undivided mass of cells surrounding the genital atrium and part of the cirrus sac, apparently delimited from parenchyma (2).

allocytti **Tkachuk 1979**

Diagnosis: *Proctophantastes* with genital atrium large, muscular, and pocketed (1); periatrial glands a large undivided mass of cells surrounding the genital atrium and part of the cirrus sac, apparently delimited from parenchyma (2); acetabulum with internal row of large papillae (3).

maceri **(Bray and Gibson 1988) comb. n.**

Diagnosis: *Proctophantastes* with genital atrium large, muscular, and pocketed (1); periatrial glands a large undivided mass of cells surrounding the genital atrium and part of the cirrus sac, apparently delimited from parenchyma (2); acetabulum with internal row of large papillae (3); tandem or subtandem testes (4).

kergeleni **(Parukhin and Lyadov 1979) comb. n.**

Diagnosis: *Proctophantastes* with genital atrium large, muscular, and pocketed (1); periatrial glands a large undivided mass of cells surrounding the genital atrium and part of the cirrus sac, apparently delimited from parenchyma (2); acetabulum with internal row of large papillae (3); tandem or subtandem testes (4); uterine loops mostly pre-gonadal (5); elongate, coiled seminal vesicle (6).

agassizi **(Campbell 1975) comb. n.**

Diagnosis: *Proctophantastes* with genital atrium large, muscular, and pocketed (1); periatrial glands a large undivided mass of cells surrounding the genital atrium and part of the cirrus sac, apparently delimited from parenchyma (2); acetabulum with internal row of large papillae (3); tandem or subtandem testes (4); uterine loops mostly pre-gonadal (5); saccate seminal vesicle *(7).

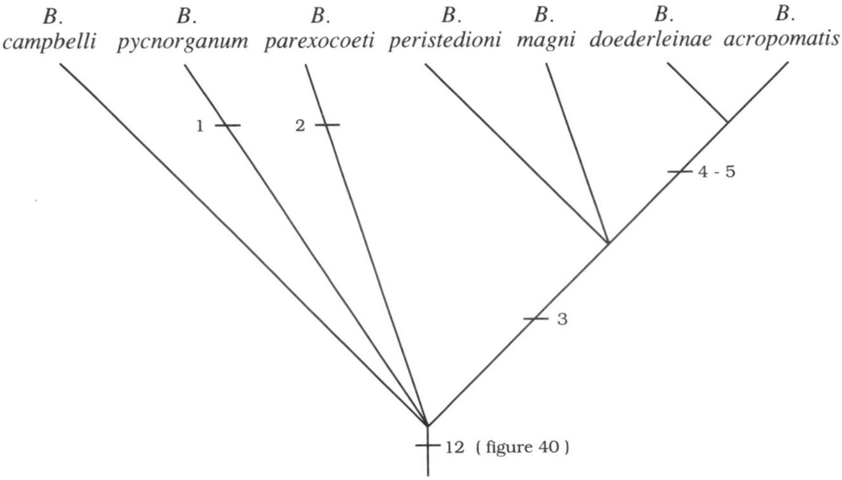

Figure 44. Phylogenetic tree for species of *Brachyenteron*. *magni* = *magnibursatum*.

***Brachyenteron* Manter 1934 (Fig. 44)**

***campbelli* Bray and Gibson 1988**

Diagnosis: With characters of the genus.

***pycnorganum* (Rees 1953) Bray 1987**

Diagnosis: *Brachyenteron* with tegumental pit on posterior lip of acetabulum (1).

***parexocoeti* Manter 1947**

Diagnosis: *Brachyenteron* with pedunculate acetabulum (2).

***peristedioni* Manter 1934, *magnibursatum* Gaevskaya and Rodiuk 1983**

Diagnosis: *Brachyenteron* with testes close to or lateral to acetabulum (3).

***doederleinae* Yamaguti 1938, *acropomatis* Yamaguti 1938**

Diagnosis: *Brachyenteron* with testes close to or lateral to acetabulum (3); saccate seminal vesicle (4); vitellarium entirely in forebody but posterior to cirrus sac on poral side (5).

***Deretrema* Linton 1910 (Fig. 45)**

***minutum* Manter 1954**

Diagnosis: With characters of the genus.

***spinosa* (Zubchenko 1978) comb. n.**

Diagnosis: *Deretrema* with posttesticular ovary *(1); vitellarium entirely in forebody but posterior to cirrus sac on poral side (2); testes entirely in forebody (3).

***bridgeri* (Bray and Gibson 1988) comb. n.**

Diagnosis: *Deretrema* with posttesticular ovary *(1); vitellarium entirely in forebody but posterior to cirrus sac on poral side (2); testes close to or lateral to acetabulum *(4).

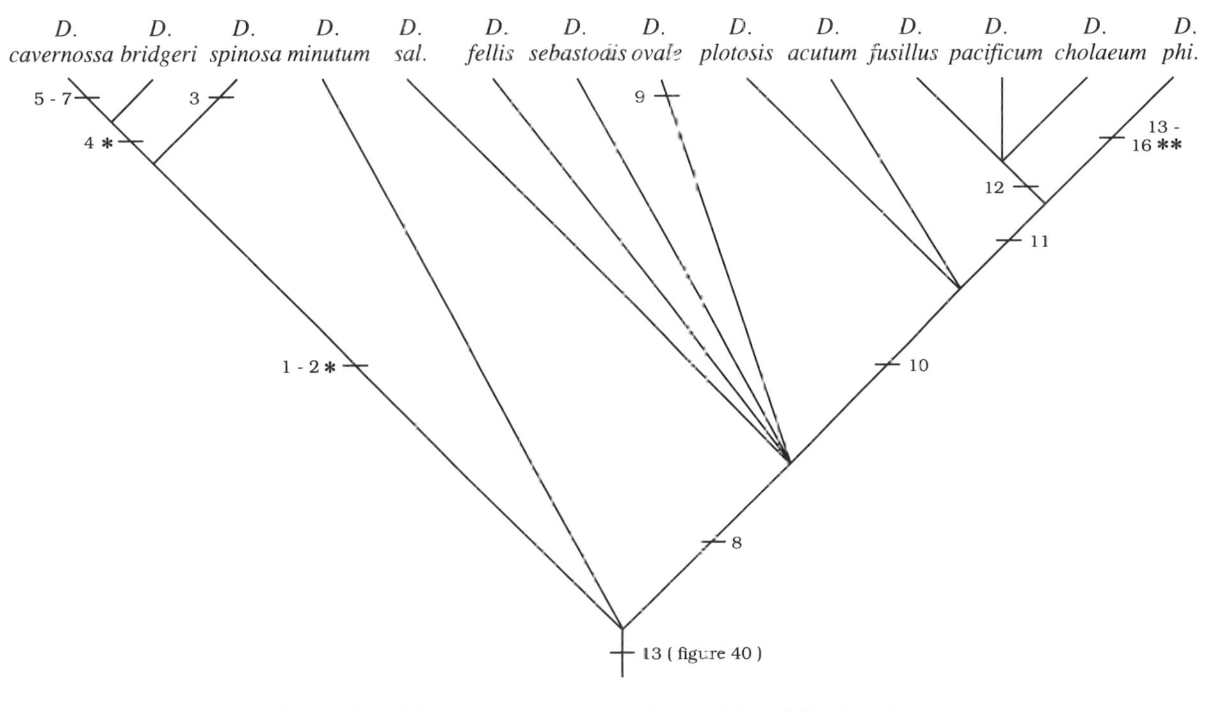

Figure 45. Phylogenetic tree for species of *Deretrema*. *sal* = *salmonicola*; *phi.* = *philippinensis*.

***cavernossa* (Overstreet and Pritchard 1977) comb. n.**

Diagnosis: *Deretrema* with posttesticular ovary *(1); testes close to or lateral to acetabulum *(4); tegumental pit on posterior lip of acetabulum (5); saccate seminal vesicle (6); vitellarium entirely in forebody and partly anterior to cirrus sac on poral side (7).

***salmonicola* (Dollfus 1951) comb. n., *fellis* (Yamaguti 1934) Yamaguti 1940, *sebastodis* (Yamaguti 1934) Yamaguti 1940**

Diagnosis: *Deretrema* with ceca narrow, reaching only to testicular level (8).

***ovale* Machida 1984**

Diagnosis: *Deretrema* with ceca narrow, reaching only to testicular level (8); ovary intertesticular (9).

***plotosi* Yamaguti 1940, *acutum* Pritchard 1963**

Diagnosis: *Deretrema* with ceca narrow, reaching only to testicular level (8); spination lining esophagus and/or pharynx (10).

***fusillus* Linton 1910, *pacificum* Yamaguti 1942, *cholaeum* McFarlane 1936**

Diagnosis: *Deretrema* with ceca narrow, reaching only to testicular level (8); spination lining esophagus and/or pharynx (10); uterine loops half pre- and half postgonadal (11); aspinose tegument (12).

***philippinensis* Beverly-Burton and Early 1982**

Diagnosis: *Deretrema* with spination lining esophagus and/or pharynx (10); uterine loops half pre- and half postgonadal (11); esophagus shorter, and wide or narrow (13); ceca extending postesticular but not near posterior end of body (14); testes close to or lateral to acetabulum *(15); posttesticular ovary *(16).

***Zoogonoides* Odhner 1902 (Fig. 46)**

***longicecus* (Siddiqi and Cable 1960) comb. n.**

Diagnosis: *Zoogonoides* with ceca extending posteriorly to near posterior end of body (1).

***malacanthi* (Siddiqi and Cable 1960) comb. n.**

Diagnosis: *Zoogonoides* with ceca extending posteriorly to near posterior end of body (1); elongate, coiled seminal vesicle (2).

***californicus* (Arai 1954) comb. n.**

Diagnosis: Zoogonoides with ceca extending posteriorly to near posterior end of body (1); prepharynx longer than pharynx (3); saccate seminal vesicle (4).

***viviparus* (Olsson 1868) Odhner 1902**

Diagnosis: *Zoogonoides* with intertesticular ovary (5).

***pyriformis* Pritchard 1963, *yamagutii* Kamegai 1973, *acanthogobii* Yamaguti 1938**

Diagnosis: *Zoogonoides* with testes close to or lateral to acetabulum (6).

***laevis* Linton 1940**

Diagnosis: *Zoogonoides* with testes close to or lateral to acetabulum (6); ventral lips of acetabulum prominently lamellar (7); genital atrium lacking (8); genital pore dextral only (9).

***subaequiporus* Odhner 1911, *ugui* (Shimazu 1974) comb n.**

Diagnosis: *Zoogonoides* with vitellarium a pair of small median masses (10).

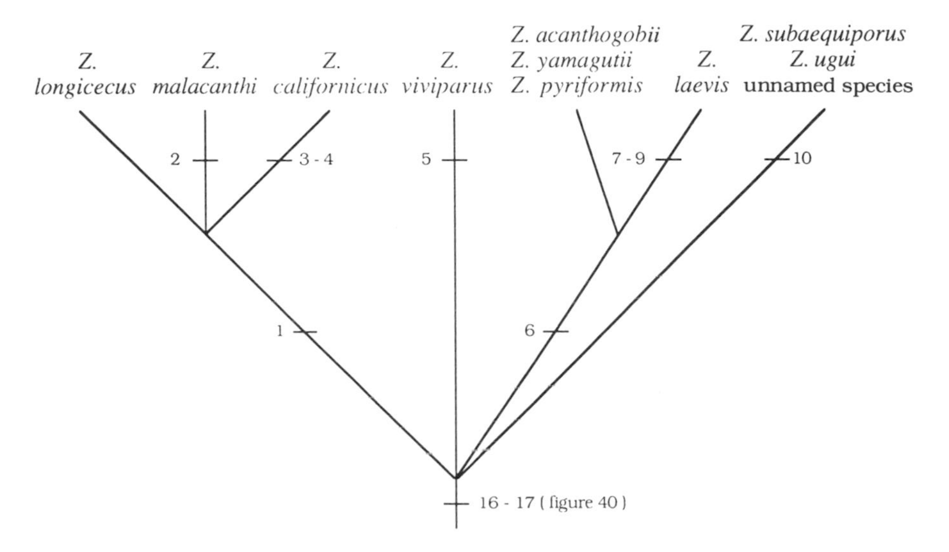

Figure 46. Phylogenetic tree for species of *Zoogonoides*.

***Diphterostomum* Stossich 1903 (Fig. 47)**

***vividum* (Nicoll 1912) Bray and Gibson 1988**

Diagnosis: With characters of the genus.

***spinosus* (Overstreet 1971) comb. n.**

Diagnosis: *Diphterostomum* with enlarged spines in one or more rows around oral sucker (1).

***albulae* Overstreet 1969**

Diagnosis: *Diphterostomum* with intertesticular ovary (2); testes close to or lateral to acetabulum (3).

***betencourti* (Monticelli 1893) Odhner 1911**

Diagnosis: *Diphterostomum* with intertesticular ovary (2); testes close to or lateral to acetabulum (3); esophagus shorter and wide or narrow (4).

***magnacetabulum* Yamaguti 1938**

Diagnosis: *Diphterostomum* with intertesticular ovary (2); ventral lips of acetabulum prominently lamellar (5).

***americanum* Manter 1947**

Diagnosis: *Diphterostomum* with intertesticular ovary (2); ventral lips of acetabulum prominently lamellar (5); esophagus lacking (6).

***indicum* Madhavi 1979**

Diagnosis: *Diphterostomum* with ventral lips of acetabulum prominently lamellar (5); ovary pretesticular (7); genital pore dextral or sinistral (8).

***brusinae* (Stossich 1888) Stossich 1903**

Diagnosis: *Diphterostomum* with ventral lips of acetabulum prominently lamellar (5); ovary pretesticular (7).

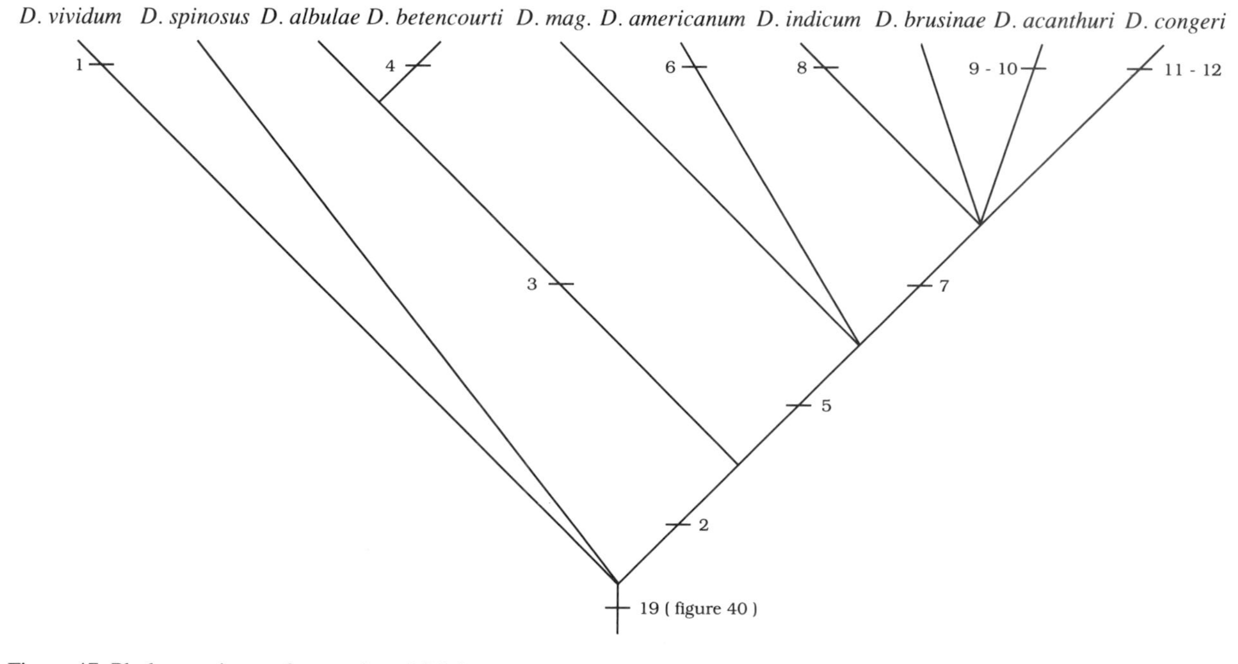

Figure 47. Phylogenetic tree for species of *Diphterostomum*. *D. mag.* = *D. magnacetabulum*.

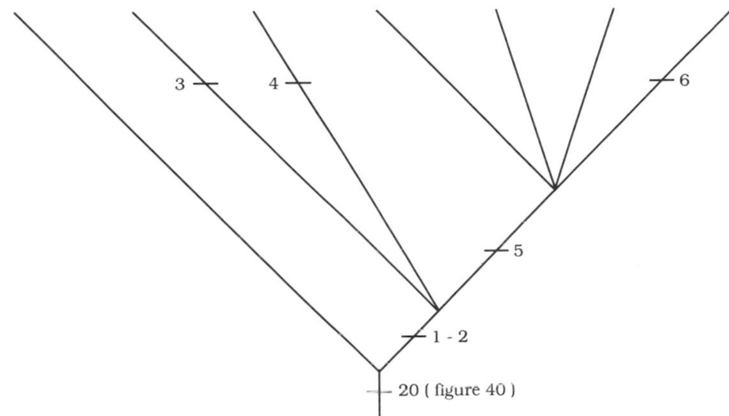

Figure 48. Phylogenetic tree for species of *Zoogonus*.

***acanthuri* (Pritchard 1963) comb. n.**

Diagnosis: *Diphterostomum* with ventral lips of acetabulum prominently lamellar (5); ovary pretesticular (7); vitellarium in single median mass (9); single cecum (10).

***congeri* (Manter 1954) comb. n.**

Diagnosis: *Diphterostomum* with ventral lips of acetabulum prominently lamellar (5); ovary pretesticular (7); vitellaria in single median field of follicles (11); vitellarium mainly in hindbody but overlapping acetabulum (12).

***Zoogonus* Odhner 1902 (Fig. 48)**

***israelense* (Fischthal 1980) comb. n.**

Diagnosis: With characters of the genus.

***mazuri* (Korotaeva 1975) Bray 1986**

Diagnosis: *Zoogonus* with cecal bifurcation dorsal or posterodorsal to acetabulum (1); posttesticular ovary (2); vitellarium a pair of small median masses (3).

***dextrocirrus* Aldrich 1961**

Diagnosis: *Zoogonus* with cecal bifurcation dorsal or posterodorsal to acetabulum (1); posttesticular ovary (2); genital pore dextral only (4).

***argentopi* Madhavi 1979, *lasius* (Leidy 1891) Stunkard 1940, *rubellus* (Olsson 1868) Odhner 1902**

Diagnosis: *Zoogonus* with cecal bifurcation dorsal or posterodorsal to acetabulum (1); posttesticular ovary (2); prepharynx longer than pharynx (5).

***pagrosomi* Yamaguti 1939**

Diagnosis: *Zoogonus* with cecal bifurcation dorsal or posterodorsal to acetabulum (1); posttesticular ovary (2); prepharynx longer than pharynx (5); genital pore in forebody (6).

Infraclass Monogenea Van Beneden 1858: Resolution to the familial level (Fig. 49)

References: Justine (1991b); Boeger and Kritsky (in press).

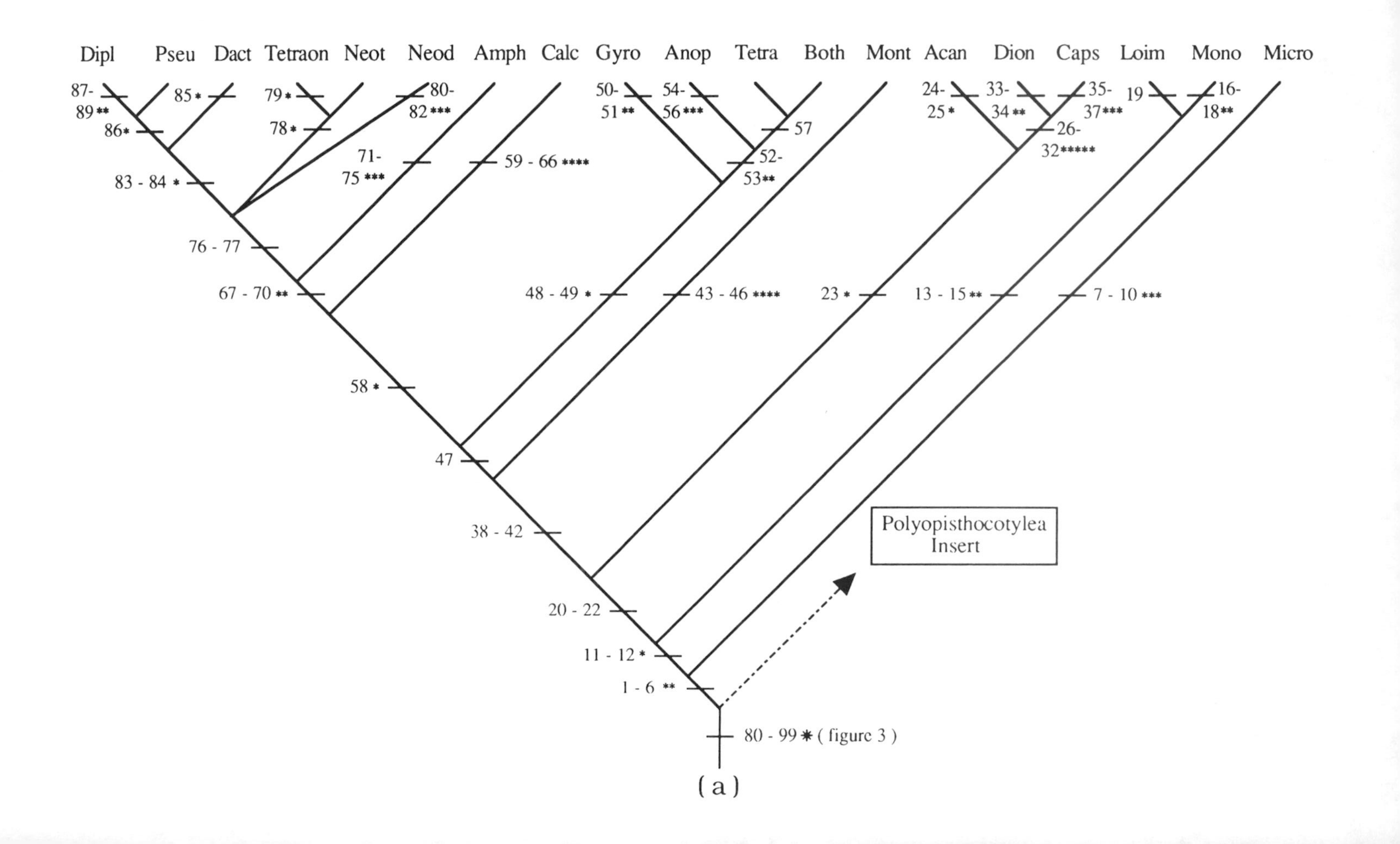
Dipl
Pseu
Dact
Tetraon
Neot
Neod
Amph
Calc
Gyro
Anop
Tetra
Both
Mont
Acan
Dion
Caps
Loim
Mono
Micro
87-89**
85*
86*
83 - 84 *
79*
78*
80-82***
71-75 ***
59 - 66 ****
76 - 77
67 - 70 **
58 *
50-51 **
54-56***
57
52-53**
48 - 49 *
43 - 46 ****
47
24-25*
33-34 **
35-37***
26-32*****
19
16-18**
23 *
13 - 15 **
7 - 10 ***
38 - 42
20 - 22
11 - 12 *
1 - 6 **
80 - 99 * (figure 3)
Polyopisthocotylea
Insert
(a)

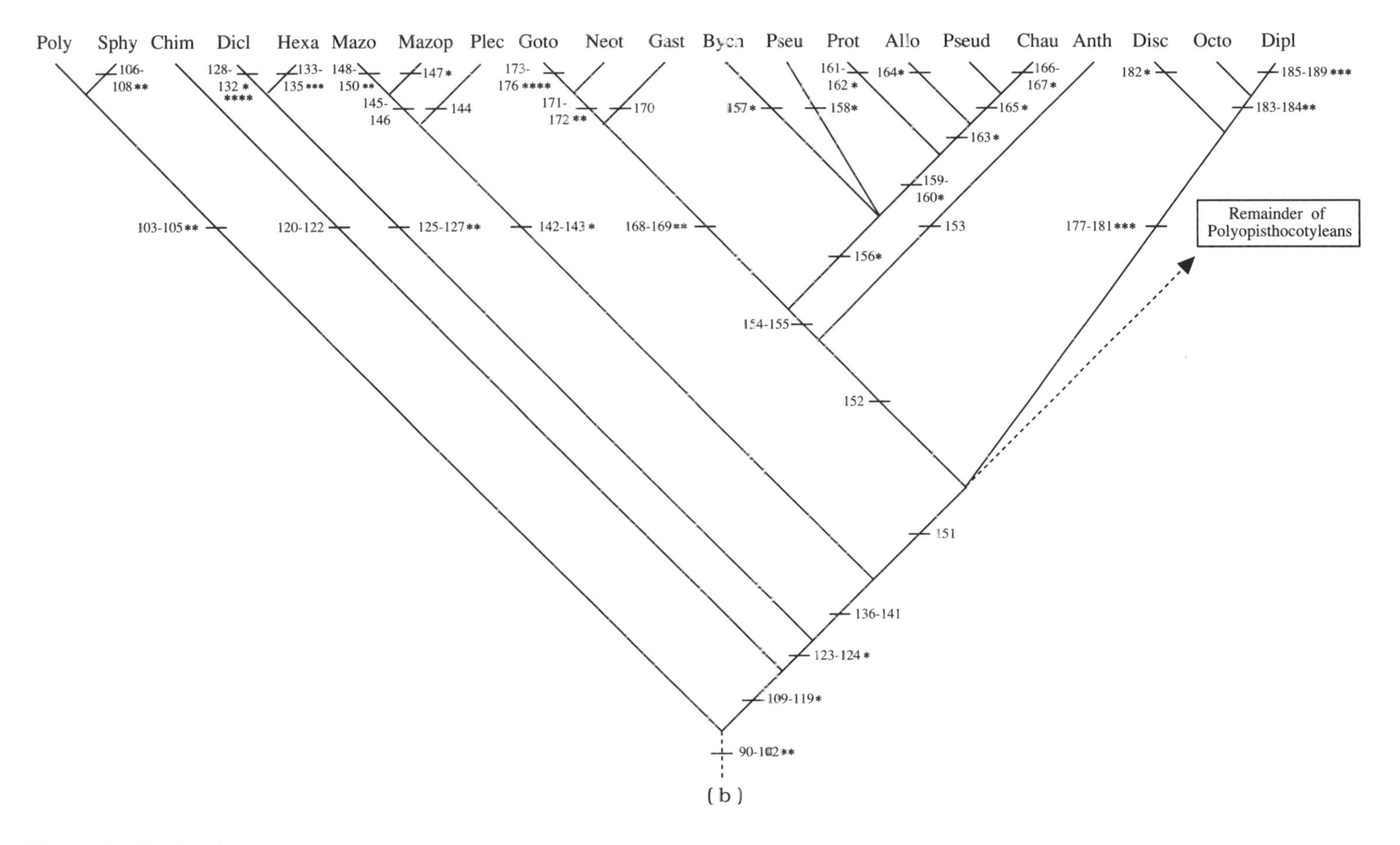

Figure 49—*Continued on next page.*

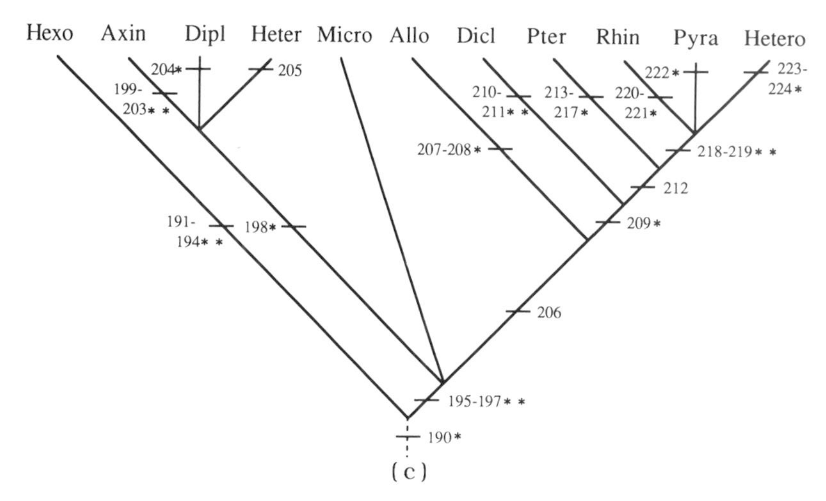

Figure 49. Phylogenetic tree for the families of the Monogenea. (a) Polyonchoinea: Dipl = Diplectanidae; Pseu = Pseudomurraytrematidae: Dact = Dactylogyridae; Tetraon = Tetraonchidae; Neot = Neotetraonchidae; Neod = Neodactylodiscidae; Amph = Amphibdellatidae; Calc = Calceostomatidae; Gyro = Gyrodactylidae; Anop = Anoplodiscidae; Tetra = Tetraonchoididae; Both = Bothitrematidae; Mont = Montschadskiellidae; Acan = Acanthocotylidae; Dion = Dionchidae; Caps = Capsalidae; Loim = Loimoidae; Mono = Monocotylidae; Micro = Microbothriidae. (b) Polyopisthocotylea: Poly = Polystomatidae; Sphy = Sphyranuridae; Chim = Chimaericolidae; Dicl = Diclybothriidae; Hexa = Hexabothriidae; Mazo = Mazocraeidae; Mazop = Mazoplectidae; Plec = Plectanocotylidae; Goto = Gotocotylidae; Neot = Neothoracocotylidae; Gast = Gastrocotylidae; Bych = Bychowskycotylidae; Pseu = Pseudodiclidophoridae; Prot = Protomicrocotylidae; Allo = Allodiscocotylidae; Pseud = Pseudomazocraeidae; Chau = Chauhaneidae; Anth = Anthocotylidae; Disc = Discocotylidae; Octo = Octomacridae; Dipl = Diplozooidae. (c) Polyopisthocotylea (continued): Hexo = Hexostomatidae; Axin = Axinidae; Dipl = Diplasiocotylidae; Heter = Heteraxinidae; Micro = Microcotylidae; Allo = Allopyragraphoridae; Dicl = Diclidophoridae; Pter = Pterinotrematidae; Rhin = Rhinecotylidae; Pyra = Pyragaphoridae; Hetero = Heteromicrocotylidae.

Cohort Polyonchoinea Bychowsky 1937 (Fig. 49a)

Diagnosis: Oral sucker lacking *(1); cirrus sclerotized (2); 14 marginal and 2 central hooks in oncomiracidium (3); 14 marginal and 2 central hooks in adult *(4); no intercentriolar body in spermatozoa (5); no striated roots in spermatozoa (6).

Order Microbothriiformes Lebedev 1988

Diagnosis: Gut diverticula present *(7); haptor a sclerotized groove (8); hooks lacking in adult *(9); anchors lacking in all stages of development *(10).

Family Microbothriidae Price 1936

Diagnosis: With characters of the order.

Order Monocotyliformes Lebedev 1988

Diagnosis: Sperm microtubules lying along one-quarter of cell periphery (11); mouth ventral *(12); 14 marginal hooks in oncomiracidium (13); 14 marginal hooks in adult *(14); distal region containing only the nucleus in mature spermatozoan *(15).

Family Monocotylidae Taschenberg 1879

Diagnosis: Oviduct looping right cecum *(16); eggs tetrahedric *(17); one axoneme plus one disappearing axoneme during spermiogenesis (18).

Family Loimoidae Price 1936

Diagnosis: One axoneme plus one altered axoneme during spermiogenesis (19).

Order Capsaliformes Lebedev 1988

Diagnosis: Mouth ventral *(12); no microtubules in spermatid zone of differentiation (20); cytoplasmic middle process and flagella fused from the beginning (21); no external ornamentation of cell membrane (22); genital aperture marginal *(23).

Family Acanthocotylidae Monticelli 1903

Diagnosis: Acanthocotylid hooks (24); anchors lacking in all stages of development *(25).

Family Dionchidae Johnston and Tiegs 1922

Diagnosis: Ceca confluent *(26); two testes (27); sclerotized portion of cirrus lacking *(28); cirrus elongate *(29); cirrus aspinose *(30); cirrus sac present *(31); beadlike giant mitochondrion present (32); vagina lacking *(33); 14 marginal hooks in adult *(34).

Family Capsalidae Baird 1853

Diagnosis: Ceca confluent *(26); two testes (27); sclerotized portion of cirrus lacking *(28); cirrus elongate *(29); cirrus aspinose *(30); cirrus sac present *(31); beadlike giant mitochondrion present (32); gut diverticula present *(35); eggs tetrahedric *(36); anchors present in oncomiracidium *(37).

Order Montschadskielliformes Lebedev 1988

Diagnosis: No microtubules in spermatid zone of differentiation (20); cytoplasmic middle process and flagella fused from the beginning (21); no external ornamentation of cell membrane (22); vas deferens looping around left cecum (38); accessory piece present (39); single axoneme during spermiogenesis (40); single centriole in spermatozoon (41); axonemal b microtubules incomplete during spermiogenesis (42); subterminal mouth *(43); gut diverticula present *(44); oviduct looping around right cecum *(45); 14 marginal hooks in adult *(46).

Family Montschadskiellidae Bychowsky, Korotajeva, and Gussev 1970

Diagnosis: With characters of the order.

Order Gyrodactyliformes Bychowsky 1937

Diagnosis: Mouth ventral *(12); no microtubules in spermatid zone of differentiation (20); cytoplasmic middle process and flagella fused from the beginning (21); no external ornamentation of cell membrane (22); vas deferens looping around left cecum (38); accessory piece present (39); single axoneme during spermiogenesis (40); single centriole in spermatozoon (41); axonemal b microtubules incomplete during spermiogenesis (42); two ventral bars (47); gyrodactylid hooks (48); 16 marginal hooks in adult (49).

Family Gyrodactylidae van Beneden and Hesse 1863

Diagnosis: Vagina lacking *(50); sclerotized portion of cirrus lacking *(51).

Family Anoplodiscidae Tagliani 1912

Diagnosis: Gut single *(52); eggs tetrahedric *(53); gut diverticula present *(54); hooks lacking in adult *(55); anchors lacking in all stages of development *(56).

Family Tetraonchoididae Bychowsky 1951

Diagnosis: Gut single *(52); eggs tetrahedric *(53); posterior pair of eyes fused (57).

Family Bothitrematidae Price 1936

Diagnosis: Gut single *(52); eggs tetrahedric *(53); posterior pair of eyes fused (57).

Order Dactylogyriformes Bychowsky 1937

Diagnosis: Mouth ventral *(12); no microtubules in spermatid zone of differentiation (20); cytoplasmic middle process and flagella fused from the beginning (21); no external ornamentation of cell membrane (22); vas deferens looping around left cecum (38); accessory piece present (39); single axoneme during spermiogenesis (40); single centriole in spermatozoon (41); axonemal b microtubules incomplete during spermiogenesis (42); two ventral bars (47); two ventral pairs of anchors *(58).

Suborder Calceostomatinea Gussev 1977

Diagnosis: Eggs tetrahedric *(59); anchor present in oncomiracidium (60); 12 marginal, 2 central hooks in oncomiracidium *(61); 12 marginal, 2 central hooks in adult (62); distal region containing only the nucleus in mature spermatozoon *(63); axoneme structure in mature spermatozoon noncircular *(64); lateral crest on mature spermatozoon (65); external microtubules associated with spermatid (66).

Family Calceostomatidae Parona and Perugia 1890

Diagnosis: With characters of the suborder.

Suborder Amphibdellatinea Boeger and Kritsky in press

Diagnosis: Centriole adjunct present (67); oviduct looping around right cecum *(68); haptor not disk-shaped and not a sclerotized groove *(69); one ventral and one dorsal pair of anchors (70); eggs tetrahedric *(71); one ventral bar (72); axoneme structure in mature spermatozoon noncircular *(73); axonemal b microtubules complete during spermiogensis *(74); anterior region of sperm nucleus coiled (75).

Family Amphibdellatidae Carus 1885

Diagnosis: With characters of the suborder.

Suborder Neodactyliscinea Boeger and Kritsky in press *sedis mutabilis*

Diagnosis: Centriole adjunct present (67); oviduct looping around right cecum *(68); haptor not disk-shaped and not a sclerotized groove *(69); eight marginal, two central, and four dorsal hooks in adult (76); two lateral vaginae *(80); two ventral pairs of anchors *(81); no bars *(82).

Family Neodactylodiscidae Kamegai 1973

Diagnosis: With characters of the suborder.

Suborder Tetraonchinea Bychowsky 1937 *sedis mutabilis*

Diagnosis: Centriole adjunct present (67); oviduct looping around right cecum *(68); haptor not disk-shaped and not a sclerotized groove *(69); one ventral and one dorsal pair of anchors (70); eight marginal, two central, and four dorsal hooks in adult (76); gut single *(78).

Family Neotetraonchidae Bravo-Hollis 1968 *incertae sedis*

Diagnosis: With characters of the suborder.

Family Tetraonchidae Monticelli 1903

Diagnosis: One ventral, two dorsal bars *(79).

Suborder Dactylogyrinea Bychowsky 1937 *sedis mutabilis*

Diagnosis: Centriole adjunct present (67); oviduct looping around right cecum *(68); haptor not disk-shaped and not a sclerotized groove *(69); one ventral, one dorsal pair of anchors (70); one dorsal, one ventral bar (77); 12 marginal, 2 central hooks in oncomiracidium *(83); ten marginal, two central, and four dorsal hooks in adult (84).

Family Dactylogyridae Bychowsky 1933

Diagnosis: Oviduct straight, intercecal *(85).

Family Pseudomurraytrematidae Kritsky, Mizelle, and Bilqees 1978

Diagnosis: One ventral, two dorsal bars *(86).

Family Diplectanidae Bychowsky 1957

Diagnosis: One ventral, two dorsal bars *(86); squamodisk present (87); distal region containing only the nucleus in mature spermatozoon *(88); axonemal b microtubules complete during spermiogenesis *(89).

Cohort Polyopisthocotylea Odhner 1912 (Fig. 49b, c)

Diagnosis: Gut diverticula present *(90); more than two testes (91); genitointestinal canal present (92); two lateral vaginae present *(93); no haptoral suckers in oncomiracidium (94); haptoral suckers present in adult (95); all pairs of haptoral suckers subequal (96); three pairs of haptoral suckers (97); two rows of haptoral suckers (98); number of suckers in each row at least approximately equal (99); angle between haptoral sucker rows 0° to less than 83° (100); haptoral sucker rows parallel or forming an angle up to 45° with body midline (101); layer of peripheral microtubules in spermatozoa not interrupted at axoneme level (102).

Subcohort Polystomatoinea Lebedev 1986

Diagnosis: Ceca confluent *(103); egg filaments lacking *(104); hooks associated with haptoral suckers in adult (105).

Order Polystomatiformes Lebedev 1988

Diagnosis: With characters of the subcohort.

Family Polystomatidae Gamble 1896

Diagnosis: With characters of the order.

Family Sphyranuridae Poche 1926

Diagnosis: Haptor not disk-shaped and not a sclerotized groove *(106); 14 marginal, 2 central hooks in adult *(107); one pair of haptoral suckers (108).

Subcohort Oligonchoinea Bychowsky 1937

Diagnosis: Two oral suckers (buccal organ) present (109); vaginal duct connecting with vitelline ducts (110); two ventral hooks in adult (111); crochet en fleau hooklike in oncomiracidium (112); crochet en fleau present in adult (113); hook present in haptoral sucker during development but lost in adult (114); four pairs of haptoral suckers *(115); most sclerites elongate (116); midsclerite of suckers terminating in a hook (117); lateral sclerites subequal to midsclerite (118); lateral sclerites rod-shaped (119).

Order Chimaericoliformes Bychowsky 1957

Diagnosis: Buccal organ comprising nonmuscular pouches of prepharynx (120); lobate ovary (121); one pair of lateral sclerites (122).

Family Chimaericolidae Brinkmann 1942

Diagnosis: With characters of the order.

Order Diclybothriiformes Bychowsky 1957

Diagnosis: Ovary elongate, U-shaped *(123); ten marginal hooks in oncomiracidium (124); oral sucker lacking *(125); ceca confluent *(126); suckers present in haptoral appendix (127).

Family Diclybothriidae Price 1936

Diagnosis: Mouth ventral *(128); cirrus elongate *(129); cirrus sac present *(130); egg filaments lacking *(131); anchors present in oncomiracidium *(132).

Family Hexabothriidae Price 1942

Diagnosis: Hooks lacking in adult *(133); one oral sucker present *(134); crochet en fleau lacking in adult *(135).

Order Mazocraeiformes Bychowsky 1937

Diagnosis: Ten marginal hooks in oncomiracidium (124); one pair of eyespots fused (136); buccal organ present as muscular suckers (137); ovary elongate, inverted, U-shaped (138); two egg filaments (139); flared or truncate midsclerite of sucker (140); two pairs of lateral sclerites (141).

Suborder Mazocraeinea Bychowsky 1957

Diagnosis: Posterior midsclerite platelike (142); two pairs of posterolateral sclerites *(143).

Family Plectanocotylidae Monticelli 1903

Diagnosis: Plectanocotyloid crochet en fleau in oncomiracidium (144).

Family Mazoplectidae Mamaev and Slipchenko 1975

Diagnosis: Anterolateral sclerite fused distally (145); distal posterolateral sclerite fused distally (146); vagina lacking *(147).

Family Mazocraeidae Price 1936

Diagnosis: Anterolateral sclerite fused distally (145); distal posterolateral sclerite fused distally (146); elongate U-shaped ovary (148); two dorsal vaginae *(149); anterior midsclerite occurring as two separate pieces (150).

Suborder Gastrocotylinea Lebedev 1972 *sedis mutabilis*

Diagnosis: Microcotylid crochet en fleau in oncomiracidium (151); accessory sclerite parallel to midsclerite (152).

Superfamily Anthocotyloidea Price 1943

Diagnosis: Anterior pair of haptoral suckers larger than other pairs (153).

Family Anthocotylidae Price 1936

Diagnosis: With characters of superfamily.

Superfamily Protomicrocotyloidea Johnston and Tiegs 1922

Diagnosis: Gastrocotylid crochet en fleau in oncomiracidium (154); accessory sclerite perpendicular to midsclerite (155); cirrus sac present *(156).

Family Bychowskycotylidae Lebedev 1969

Diagnosis: Numerous pairs of haptoral suckers *(157).

Family Pseudodiclidophoridae Yamaguti 1965

Diagnosis: Ceca confluent *(158).

Family Protomicrocotylidae Johnston and Tiegs 1922

Diagnosis: Testes preovarian (159); single ventrolateral vagina (160); haptoral appendix expanded laterally (161); cirrus sac lacking *(162).

Family Allodiscocotylidae Tripathi 1959

Diagnosis: Testes preovarian (159); single ventrolateral vagina (160); cirrus elongate *(163); crochet en fleau lacking in adult *(164).

Family Pseudomazocraeidae Lebedev 1972

Diagnosis: Testes preovarian (159); single ventrolateral vagina (160); cirrus elongate *(163); cirrus aspinose *(165).

Family Chauhaneidae Euzet and Trilles 1960

Diagnosis: Testes preovarian (159); cirrus elongate *(163); cirrus aspinose *(165); genital aperture marginal *(166); one midventral vagina (167).

Superfamily Gastrocotyloidea Price 1943

Diagnosis: Gastrocotylid crochet en fleau in oncomiracidium (154); accessory sclerite perpendicular to midsclerite (155); one middorsal vagina *(168); numerous pairs of haptoral suckers *(169).

Family Gastrocotylidae Price 1943

Diagnosis: Single row of haptoral suckers (170).

Family Neothoracocotylidae Lebedev 1969

Diagnosis: Hooks lacking in adult *(171); crochet en fleau lacking in adult *(172).

Family Gotocotylidae Yamaguti 1963

Diagnosis: Hooks lacking in adult *(171); crochet en fleau lacking in adult *(172); elongate cirrus *(173); cirrus sac present *(174); two dorsal vaginae *(175); lateral sclerite flattened bilaterally *(176).

Suborder Discocotyliinea Bychowsky 1957

Diagnosis: Microcotylid crochet en fleau in oncomiracidium (151); cirrus aspinose *(177); haptoral suckers present in oncomiracidium (178); six marginal hooks in oncomiracidium (179); hooks lacking in adult *(180); anchors absent in all stages of development *(181).

Family Discocotylidae Price 1936

Diagnosis: Egg filaments lacking *(182).

Family Octomacridae Yamaguti 1963

Diagnosis: Vagina lacking *(183); single testis *(184).

Family Diplozooidae Tripathi 1959

Diagnosis: Vagina lacking *(183); single testis *(184); gut single *(185); permanent copula (186); one egg filament *(187); four marginal hooks in oncomiracidium (188); two pairs of posterolateral sclerites *(189).

Suborder Hexostomatinea Boeger and Kritsky in press

Diagnosis: Microcotylid crochet en fleau in oncomiracidium (151); one middorsal vagina *(190); two pairs of eyespots *(191); cirrus aspinose *(192); two spined plates surrounding vaginal aperture (193); sclerites short, modified (194).

Family Hexostomatidae Price 1936

Diagnosis: With characters of suborder.

Suborder Microcotylinea Lebedev 1972 *incertae sedis*

Diagnosis: Microcotylid crochet en fleau in oncomiracidium (151); one middorsal vagina *(190); ovary elongate, double inverted U-shape (195); anchors present in oncomiracidium *(196); numerous pairs of haptoral suckers *(197).

Superfamily Microcotyloidea Taschenberg 1879 *incertae sedis*

Diagnosis: With characters of the suborder.

Family Microcotylidae Taschenberg 1879 *incertae sedis*

Diagnosis: With characters of the superfamily.

Superfamily Axinoidea Monticelli 1903

Diagnosis: Genital atrium with bilateral armed pads *(198).

Family Axinidae Monticelli 1903

Diagnosis: One dorsolateral vagina (199); angle between haptoral sucker rows greater than 103° (200); anterior pair of eyespots fused (201); ovary elongate, U-shaped *(202); four pairs of haptoral suckers *(203).

Family Diplasiocotylidae Hargis and Dillon 1965

Diagnosis: Two dorsal vaginae *(204).

Family Heteraxinidae Unnithan 1957

Diagnosis: Number of haptoral suckers in one row much greater than in the other (205).

Suborder Diclidophorinea Boeger and Kritsky in press

Diagnosis: Microcotylid crochet en fleau in oncomiracidium (151); one middorsal vagina *(190); ovary elongate, double inverted U-shape (195); anchors present in oncomiracidium *(196); numerous pairs of haptoral suckers *(197); midsclerite rod-shaped (206).

Superfamily Allopyragophoroidea Yamaguti 1963

Diagnosis: Haptoral sucker rows forming an angle of about 73° with body midline (207); lateral sclerite flattened bilaterally *(208).

Family Allopyragraphoridae Yamaguti 1963

Diagnosis: With characters of the superfamily.

Superfamily Diclidophoroidea Cerfontaine 1895

Diagnosis: Four pairs of haptoral suckers present *(209); anchors lacking in all stages of development *(210); two pairs of posterolateral sclerites *(211).

Family Diclidophoridae Cerfontaine 1895

Diagnosis: With characters of the superfamily.

Superfamily Pyragraphoroidea Yamaguti 1963

Diagnosis: Four pairs of haptoral suckers present *(209); firetongue sucker present (212).

Family Pterinotrematidae Caballero and Bravo-Hollis 1955

Diagnosis: Hooks in adult: two ventral, four in each of two lappets (213); subterminal buccal cavity present (214); cirrus sclerites present (215); single egg filament *(216); lateral sclerites much smaller than midsclerite (217).

Family Rhinecotylidae Lebedev 1979

Diagnosis: Cirrus sac present *(218); numerous pairs of haptoral suckers *(219); vaginae lacking *(220); rhinecotylid haptoral pad present (221).

Family Pyragaphoridae Yamaguti 1963

Diagnosis: Firetongue sucker present (212); cirrus sac present *(218); numerous pairs of haptoral suckers *(219); elongate cirrus present *(222).

Family Heteromicrocotylidae Unnithan 1961

Diagnosis: Firetongue sucker present (212); cirrus sac present *(218); numerous pairs of haptoral suckers *(219); genital atrium with bilateral armed pads *(223); heteromicrocotylid suckers present (224).

Infraclass Monogenea Van Beneden 1858: Infrafamilial relationships

Family Dactylogyridae Bychowsky 1933

Subfamily Ancyrocephalinae Bychowsky 1937 (Fig. 50)

References: Klassen and Beverly-Burton (1987, 1988a); Wheeler and Beverly-Burton (1989); Beverly-Burton and Klassen (1990).

Diagnosis: Two pairs of transverse bars present (1); 14 larval hooks, pair I median (2); cirrus sclerotized (3); accessory piece an incomplete sheath lateral to or partially surrounding cirrus shaft (in some cases lightly sclerotized longitudinal ribs present) (4); vas deferens looping dorsoventrally around left cecum (5).

"Group A" *incertae sedis*

Diagnosis: As above.

***Onchocleidus* Mueller 1936**

Diagnosis: Accessory piece with one ramus (7); spiral filament in cirrus present (8); vagina sinistral *(9).

***Leptocleidus* Mueller 1936**

Diagnosis: Accessory piece with one ramus (7); haptor small *(10); both roots of hamuli enlarged and heavily sclerotized *(11); one pair of transverse bars present (12).

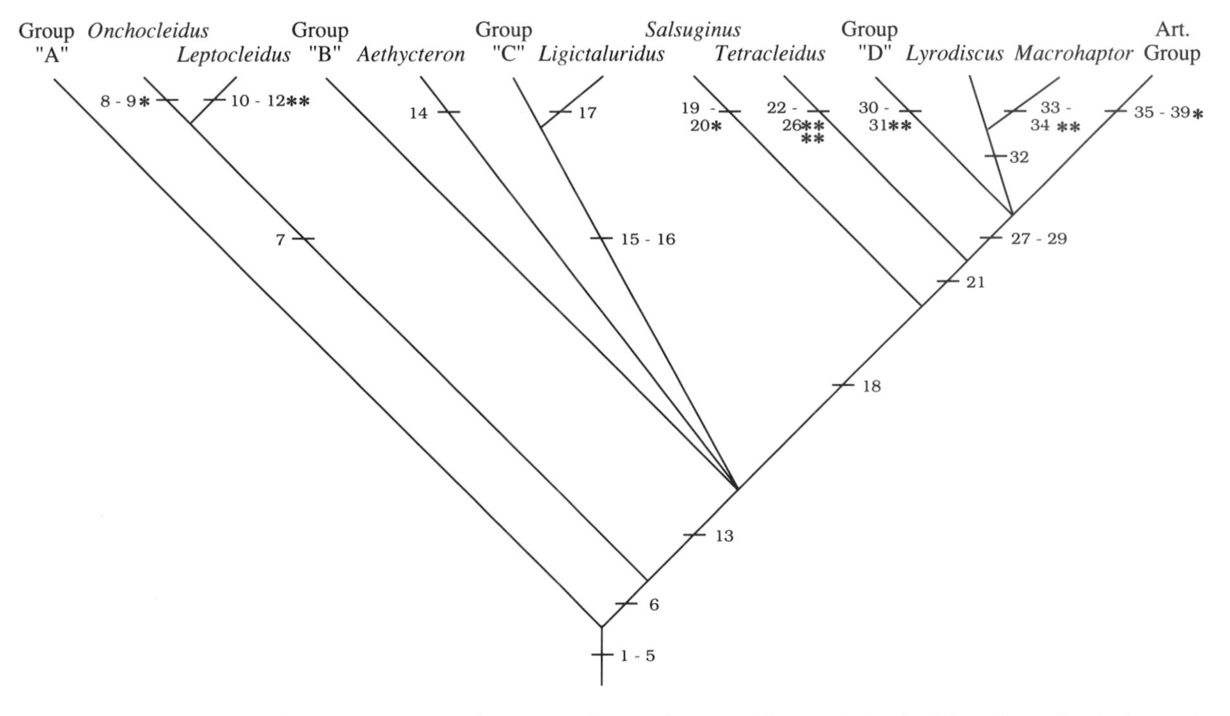

Figure 50. Phylogenetic tree for genera and generic groups of uncertain status (Groups A–D plus "Art. Group," referring to the species with "articulating bars") for the Ancyrocephalinae.

"Group B" *sedis mutabilis, incertae sedis*

Diagnosis: Accessory piece with two discrete rami (ribs) and sickle-shaped distal hook (6); one ramus of accessory piece attached proximally to base of cirrus (13).

Aethycteron* Suriano and Beverly-Burton 1982 *sedis mutabilis

Diagnosis: One ramus of accessory piece attached proximally to base of cirrus (13); one ramus and distal hook of accessory piece wrapping around cirrus to form spiral sheath (14).

"Group C" + *Ligictaluridus sedis mutabilis*

Diagnosis: Handle present at base of cirrus (15); base of sickle-shaped hook of accessory piece reduced, at least one ramus attached to base of cirrus shaft (16).

"Group C" *incertae sedis*

Diagnosis: As above.

***Ligictaluridus* Beverly-Burton 1984**

Diagnosis: Attached ramus of accessory piece separated distally from second ramus, which retains distal hook (17).

***Salsuginus* Beverly-Burton 1984**

Diagnosis: Accessory piece not attached, parallel to cirrus shaft (18); handle of larval hooks reduced *(19); base of sickle-shaped hook bifid (20).

***Tetracleidus* Mueller 1936**

Diagnosis: Accessory piece not attached, parallel to cirrus shaft (18); sickle-shaped hook of accessory piece reduced, recurved (21); accessory piece with one ramus (22); haptor small *(23); both roots of hamuli enlarged and heavily sclerotized *(24); handle of larval hooks reduced *(25); vagina sinistral *(26).

"Group D" *sedis mutabilis*

Diagnosis: Base of cirrus shaft inflated with curved distal end (27); proximal portion of accessory piece distinctly bifid (28); accessory piece proximally bifid, attached to basal third of cirrus shaft (29); haptor small *(30); both roots of hamuli enlarged and heavily sclerotized *(31).

Lyrodiscus* Rogers 1967 *incertae sedis

Diagnosis: Base of cirrus shaft inflated with curved distal end (27); proximal portion of accessory piece distinctly bifid (28); accessory piece proximally bifid, attached to basal third of cirrus shaft (29); hamulus shaft greatly elongate, roots lacking (32).

***Macrohaptor* Allison 1967**

Diagnosis: Base of cirrus shaft inflated with curved distal end (27); proximal portion of accessory piece distinctly bifid (28); accessory piece proximally bifid, attached to basal third of cirrus shaft (29); hamulus shaft elongate, both roots compressed and reduced *(33); ventral transverse bar V-shaped *(34).

"Articulating Bar Group" *sedis mutabilis*

Diagnosis: Base of cirrus shaft inflated with curved distal end (27); proximal portion of accessory piece distinctly bifid (28); accessory piece proximally bifid, attached to basal third of cirrus shaft (29); superficial root of hamulus prominent, deep root lacking (35); hamuli all oriented on ventral surface of haptor and directed posteriorly (36);

ventral transverse bar V-shaped *(37); dorsal transverse bar horizontal centrally with dorsally directed lateral struts (38); transverse bars articulating (39).

***Onchocleidus* Mueller 1936 (Fig. 51)**

Reference: Wheeler and Beverly-Burton (1989).

Diagnosis: Vagina dextral (1); dorsal hamuli reduced in size (2); spiral filament of cirrus present, sclerotized, encircling shaft (3).

***acer* Mueller 1936**

Diagnosis: Sclerotized spur of hamuli present, projecting from inner curvature of both dorsal and ventral hamuli *(4); pair II of larval hooks markedly smaller than others *(5); accessory piece lightly sclerotized sheath closely associated with spiral filament of cirrus (6).

***minimus* Mueller 1936**

Diagnosis: Accessory piece lightly sclerotized sheath with single sclerotized ramus (7).

***dispar* Group**

Diagnosis: Accessory piece with sclerotized blade with distal bifurcation closely associated with cirrus shaft, sheath lacking (8); dorsal hamuli markedly larger than ventral hamuli (9); deep root of dorsal hamuli reduced to tuberosity of hamular shaft (10).

furcatus* (Mueller 1937) Wheeler and Beverly-Burton 1989, *parvicirrus* (Mizelle and Jaskoski 1942) Wheeler and Beverly-Burton 1989 *sedis mutabilis incertae sedis

Diagnosis: With characters of the group.

dispar* Mueller 1936 *sedis mutabilis

Diagnosis: Accessory piece with sclerotized blade with distal ring through which cirrus protrudes *(11).

affinis* (Mueller 1937) Wheeler and Beverly-Burton 1989 *sedis mutabilis

Diagnosis: Cirrus helical *(12); cirrus spiral filament lacking *(13).

macropterus* (Harrises 1962) Wheeler and Beverly-Burton 1989 *sedis mutabilis

Diagnosis: Accessory piece with sclerotized blade with distal bifurcation closely associated with cirrus shaft, sheath lacking (8); no spiral filament associated with cirrus *(14).

***flieri* (Putz and Hoffman 1962) Wheeler and Beverly-Burton 1989**

Diagnosis: Accessory piece with sclerotized blade with distal bifurcation closely associated with cirrus shaft, sheath lacking (8); pair II of larval hooks markedly smaller than others *(15); hamuli shaft length reduced *(16).

***doloresae* (Hargis 1952) Wheeler and Beverly-Burton 1989**

Diagnosis: Pair II of larval hooks markedly smaller than others *(15); hamuli shaft length reduced *(16); accessory piece lightly sclerotized sheath with single sclerotized ramus *(17); distal aperture of cirrus flared, with enlarged opening *(18).

***nactus* (Mayes and Johnson 1975) Wheeler and Beverly-Burton 1989**

Diagnosis: Accessory piece with sclerotized blade with distal bifurcation closely associated with cirrus shaft, sheath lacking (8); pair II of larval hooks markedly smaller

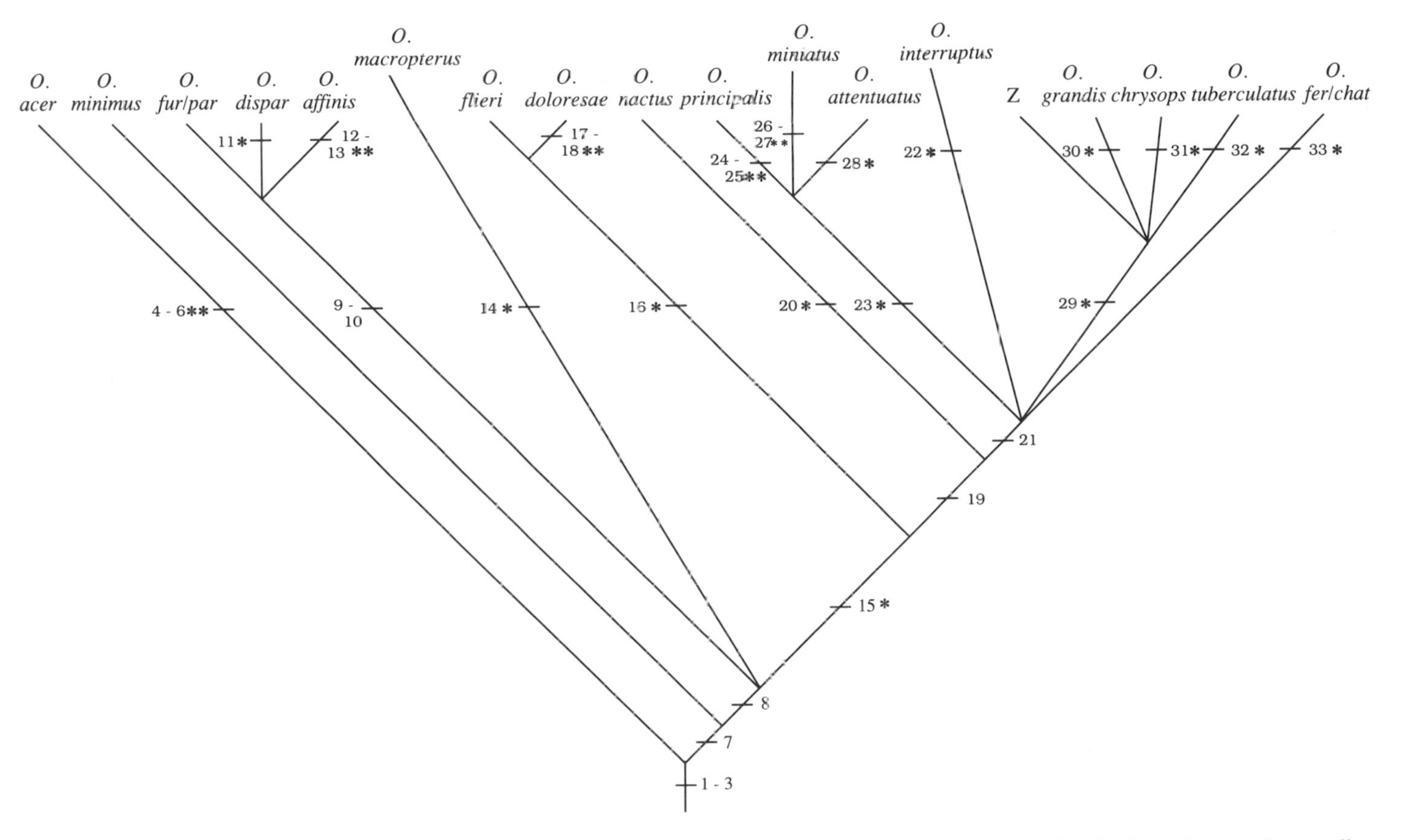

Figure 51. Phylogenetic tree for the species of *Onchocleidus*. *fur/par* = *O. furcatus* and *O. parvicirrus*; Z = *O. similis*, *O. chaenobryttus*, *O. cyanellus*, *O. distinctus*, and *O. variabilis*; *fer/chat* = *O. ferox* and *O. chatauquaensis*.

than others *(15); pairs III–VII of larval hooks directed anteriorly and concentrated in anterolateral region of haptor (19); sclerotized spur of hamuli present, projecting from inner curvature of both dorsal and ventral hamuli *(20).

interruptus* Mizelle 1936 *sedis mutabilis

Diagnosis: Accessory piece with sclerotized blade with distal bifurcation closely associated with cirrus shaft, sheath lacking (8); pair II of larval hooks markedly smaller than others *(15); pairs III–VII of larval hooks directed anteriorly and concentrated in anterolateral region of haptor (19); handle of larval hooks robust and elongate, extending almost to hook, shaft greatly reduced (21); distal aperture of cirrus flared, with enlarged opening *(22).

principalis* Group *sedis mutabilis

Diagnosis: Accessory piece with sclerotized blade with distal bifurcation closely associated with cirrus shaft, sheath lacking (8); pair II of larval hooks markedly smaller than others *(15); pairs III–VII of larval hooks directed anteriorly and concentrated in anterolateral region of haptor (19); handle of larval hooks robust and elongate, extending almost to hook, shaft greatly reduced (21); cirrus helical *(23).

principalis* Mizelle 1936 *sedis mutabilis

Diagnosis: Larval hook handle short, shaft elongate and slender *(24); all pairs of larval hooks approximately equal in size *(25).

miniatus* (Mizelle and Jaskoski 1942) Wheeler and Beverly-Burton 1989 *sedis mutabilis

Diagnosis: Sclerotized spur of hamuli present, projecting from inner curvature of both dorsal and ventral hamuli *(26); cirrus spiral filament lacking *(27).

attentuatus* (Mizelle 1941) Wheeler and Beverly-Burton 1989 *sedis mutabilis

Diagnosis: Hamuli shaft length reduced *(28).

similis* Group *sedis mutabilis

Diagnosis: Pair II of larval hooks markedly smaller than others *(15); pairs III–VII of larval hooks directed anteriorly and concentrated in anterolateral region of haptor (19); handle of larval hooks robust and elongate, extending almost to hook, shaft greatly reduced (21); accessory piece lightly sclerotized sheath with single sclerotized ramus *(29).

similis* Mueller 1936, *chaenobryttus* (Mizelle and Seamster 1939) Wheeler and Beverly-Burton 1989, *cyanellus* Mizelle 1936, *distinctus* Mizelle 1936, *variabilis* (Mizelle and Cronin 1943) Wheeler and Beverly-Burton 1989 *incertae sedis

Diagnosis: With characters of the species group.

***grandis* (Mizelle and Seamster 1939) Wheeler and Beverly-Burton 1989**

Diagnosis: Sclerotized spur of hamuli present, projecting from inner curvature of both dorsal and ventral hamuli *(30).

***chrysops* (Mizelle and Klucka 1953) Beverly-Burton 1984**

Diagnosis: Distal aperture of cirrus flared, with enlarged opening *(31).

***tuberculatus* (Allison and Rogers 1970) Cloutman 1988**

Diagnosis: Cirrus spiral filament lacking *(32).

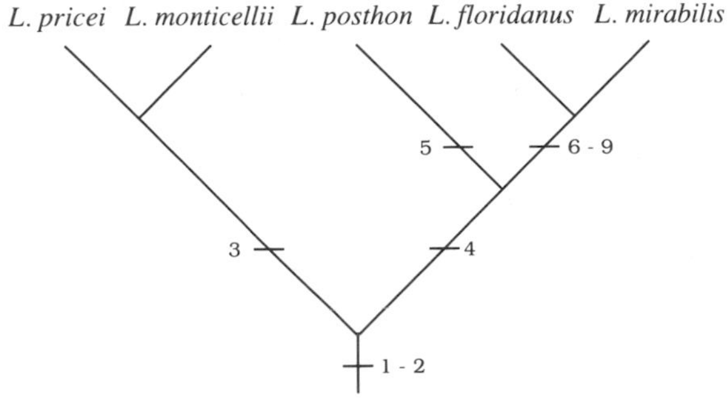

Figure 52. Phylogenetic tree for the species of *Ligictaluridus*.

ferox* (Mueller 1934) Mueller 1936, *chatauquaenesis* (Mueller 1938) Murith and Beverly-Burton 1984 *sedis mutabilis incertae sedis

Diagnosis: Pair II of larval hooks markedly smaller than others *(15); pairs III–VII of larval hooks directed anteriorly and concentrated in anterolateral region of haptor (19); handle of larval hooks robust and elongate, extending almost to hook, shaft greatly reduced (21); accessory piece with sclerotized blade with distal ring through which cirrus protrudes *(33).

***Ligictaluridus* Beverly-Burton 1984 (Fig. 52)**

Reference: Klassen and Beverly-Burton (1987).

Diagnosis: Basal handle of cirrus reduced (1); accessory piece consisting of two rami distally separated and aligned in parallel on one side of cirrus (2).

***pricei* (Mueller 1936) Klassen and Beverly-Burton 1985**

Diagnosis: Cirrus strongly curved with distal aperture oriented on same plane as proximal (3).

***monticellii* (Cognetti de Martiis 1924) Klassen and Beverly-Burton 1985**

Diagnosis: Cirrus strongly curved with distal aperture oriented on same plane as proximal (3).

***posthon* Klassen, Beverly-Burton, and Dechtiar 1985**

Diagnosis: Distally flaring funnel-like opening surrounding distal aperture of cirrus (4); entire cirrus shaft enlarged (5).

***floridanus* (Mueller 1936) Klassen and Beverly-Burton 1985**

Diagnosis: Distally flaring funnel-like opening surrounding distal aperture of cirrus (4); second basal handle on cirrus attached to opposite surface of cirrus base and directed toward distal cirrus aperture (6); accessory piece with a lightly sclerotized leaflike projection added to ramus containing hook (7); vaginal walls heavily sclerotized with centrally located conical projection (8); transverse bars of haptor greatly exceed hamuli in length (9).

***mirabilis* (Mueller 1937) Klassen and Beverly-Burton 1985**

Diagnosis: Distally flaring funnel-like opening surrounding distal aperture of cirrus (4); second basal handle on cirrus attached to opposite surface of cirrus base and directed toward distal cirrus aperture (6); accessory piece with a lightly sclerotized leaflike projection added to ramus containing hook (7); vaginal walls heavily sclerotized with centrally located conical projection (8); transverse bars of haptor greatly exceed hamuli in length (9).

Subfamily Ancyrocephalinae Bychowsky 1937: Clade of species with "articulating haptoral bars" (Fig. 53)

Reference: Klassen and Beverly-Burton (1988a).

Diagnosis: Hamuli oriented with blades projecting laterally (1); transverse bars articulated (2); ventral transverse bar V-shaped (3); dorsal transverse bar horizontal centrally with dorsal struts that may extend to dorsal margin of haptor (4).

***Crinicleidus crinicirrus* (Kritsky and Leiby 1973) Beverly-Burton 1986**

Diagnosis: Cirrus with additional sheath enveloping distal extremity of cirrus shaft (5).

***Crinicleidus longus* (Mizelle 1938) Beverly-Burton 1986**

Diagnosis: Cirrus shaft an elongate whiplike extension reflected back on itself (6).

***Syncleithrum fusiformis* (Mueller 1934) Price 1967**

Diagnosis: Ventral transverse bar with two fluted ventrolateral projections and medial constriction (8); dorsal transverse bar a solid shieldlike plate (9); cirrus with enlarged basal portion of shaft, distinguishing heavily sclerotized straight shaft from narrow curved distal ejaculatory limb (10); distal and proximal limbs of accessory piece constricted medially (11).

***Anchoradiscus triangularis* (Summers 1937) Mizelle 1941**

Diagnosis: Cirrus with basal shoulderlike enlargement *(7); larval hooks arranged around margin of haptor (12); transverse component of dorsal transverse bar with heavily slcerotized elongate, dorsolateral projections and lightly sclerotized lateral and medial expansions (13); shaft and extremities of accessory piece reduced *(14); enlargement of both superficial and deep roots of hamuli forming a triangular shape (15); larval hook shaft short, slender handle distinct, enlarged, and elongate (16); haptor wider than body (17); ventral bar with heavily sclerotized elongate ventrolateral projections and lightly sclerotized medial and lateral expansions (18); dorsal bar with heavily sclerotized elongate dorsolateral projections and lightly sclerotized ventral and medial expansions (19); flaring of enlarged basal portion of cirrus shaft (20).

***Anchoradiscoides serpentinus* Rogers 1967**

Diagnosis: Cirrus with basal shoulderlike enlargement *(7); larval hooks arranged around margin of haptor (12); transverse component of dorsal transverse bar with heavily sclerotized elongate, dorsolateral projections and lightly sclerotized lateral and medial expansions (13); shaft and extremities of accessory piece reduced *(14); enlargement of both superficial and deep roots of hamuli forming a triangular shape (15); larval hook shaft short, slender handle distinct, enlarged, and elongate (16); ventral bar with median, semicircular dorsal projection (21); dorsal bar crescentic *(22); sinusoidal shaft of accessory piece with reduced proximal bifurcation (23).

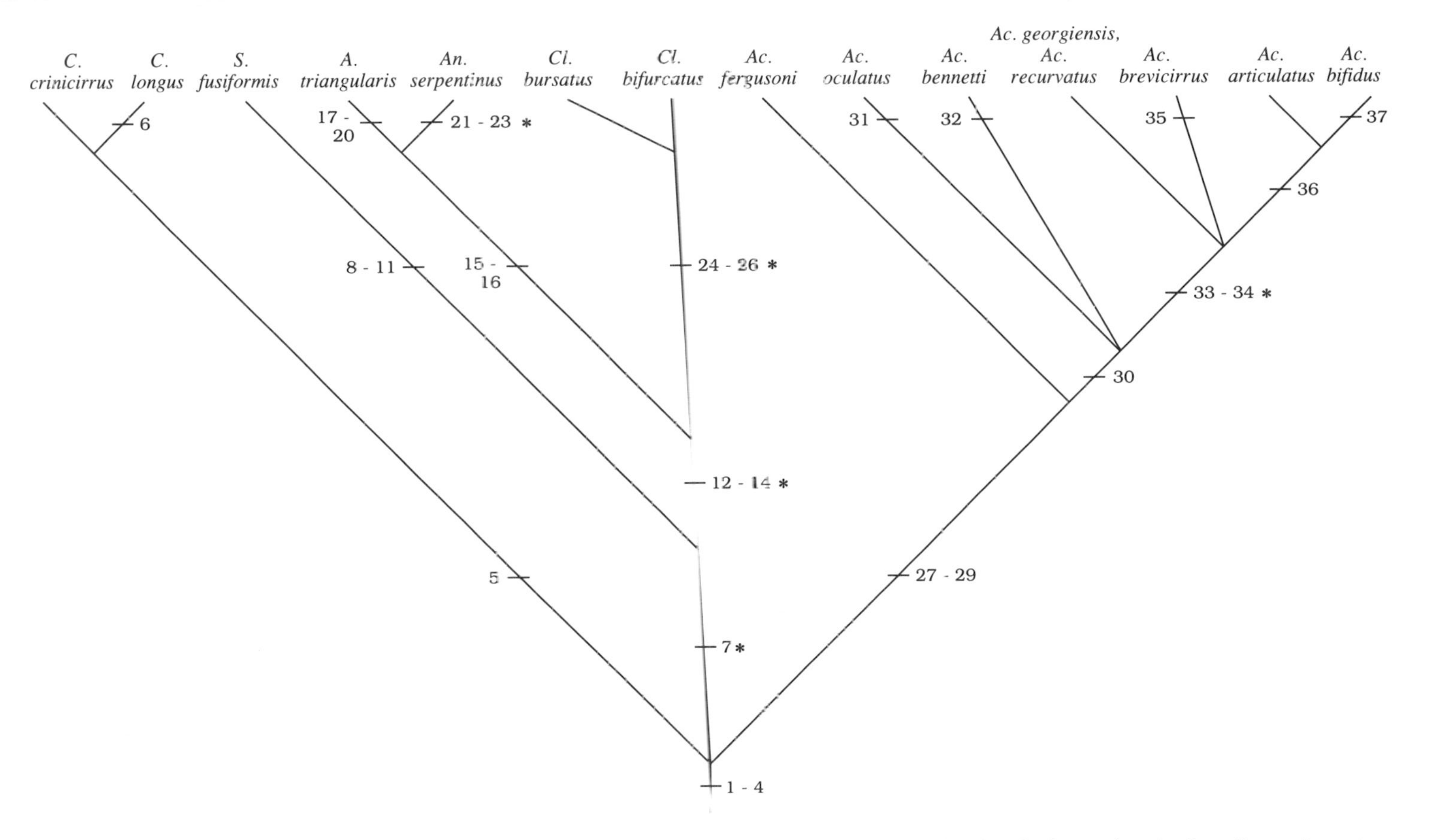

Figure 53. Phylogenetic tree for species of ancyrocephalines with articulating bars. *C.* = *Crinicleidus; S.* = *Syncleithrum; A.* = *Anchoradiscus; An.* = *Anchoradiscoides; Cl.* = *Clavunculus; Ac.* = *Actinocleidus.*

***Clavunculus bursatus* (Mueller 1936) Mizelle et al. 1956, *C. bifurcatus* (Mizelle 1941) Mizelle et al. 1956**

Diagnosis: Larval hooks arranged around margin of haptor (12); transverse component of dorsal transverse bar with heavily sclerotized elongate, dorsolateral projections and lightly sclerotized lateral and medial expansions (13); hamuli oriented posteriorly and reduced in size *(24); cirrus with expansion of entire shaft, including shoulderlike enlargement (25); proximal bifid extremities of accessory piece bent and touching to form a ringlike structure (26).

***Actinocleidus fergusoni* Mizelle 1938**

Diagnosis: Deep root of hamulus reduced (27); ventral transverse bar with heavily sclerotized ventrolateral projections and lightly sclerotized medial and lateral extensions (28); dorsal transverse bar centrally horizontal with two oblique dorsolateral struts (29).

***Actinocleidus oculatus* (Mueller 1934) Mueller 1937**

Diagnosis: Deep root of hamulus reduced (27); ventral transverse bar with heavily sclerotized ventrolateral projections and lightly sclerotized medial and lateral extensions (28); dorsal transverse bar horizontal centrally with two oblique dorsolateral struts, each with lateral inflation and small distal protruberance (30); accessory piece with distal bifid hook (31).

***Actinocleidus bennetti* Allison and Rogers 1967**

Diagnosis: Deep root of hamulus reduced (27); ventral transverse bar with heavily sclerotized ventrolateral projections and lightly sclerotized medial and lateral extensions (28); dorsal transverse bar horizontal centrally with two oblique dorsolateral struts, each with lateral inflation and small distal protruberance (30); cirrus with oblique ridge on sheath (32).

***Actinocleidus recurvatus* Mizelle and Donahue 1944, *A. georgiensis* Price 1966**

Diagnosis: Deep root of hamulus reduced (27); ventral transverse bar with heavily sclerotized ventrolateral projections and lightly sclerotized medial and lateral extensions (28); dorsal transverse bar horizontal centrally with two oblique dorsolateral struts, each with lateral inflation and small distal protruberance (30); cirrus with basal shoulderlike enlargement (33); accessory piece shaft and extremities reduced in size *(34).

***Actinocleidus brevicirrus* Mizelle and Jaskoski 1942**

Diagnosis: Deep root of hamulus reduced (27); ventral transverse bar with heavily sclerotized ventrolateral projections and lightly sclerotized medial and lateral extensions (28); dorsal transverse bar horizontal centrally with two oblique dorsolateral struts, each with lateral inflation and small distal protruberance (30); accessory piece shaft and extremities reduced in size *(34); cirrus with heavy sclerotization on shoulderlike enlargement (35).

***Actinocleidus articulatus* (Mizelle 1936) Mueller 1937**

Diagnosis: Deep root of hamulus reduced (27); ventral transverse bar with heavily sclerotized ventrolateral projections and lightly sclerotized medial and lateral extensions (28); dorsal transverse bar horizontal centrally with two oblique dorsolateral struts, each with lateral inflation and small distal protruberance (30); accessory piece shaft and extremities reduced in size *(34); cirrus with bifid distal aperture (36).

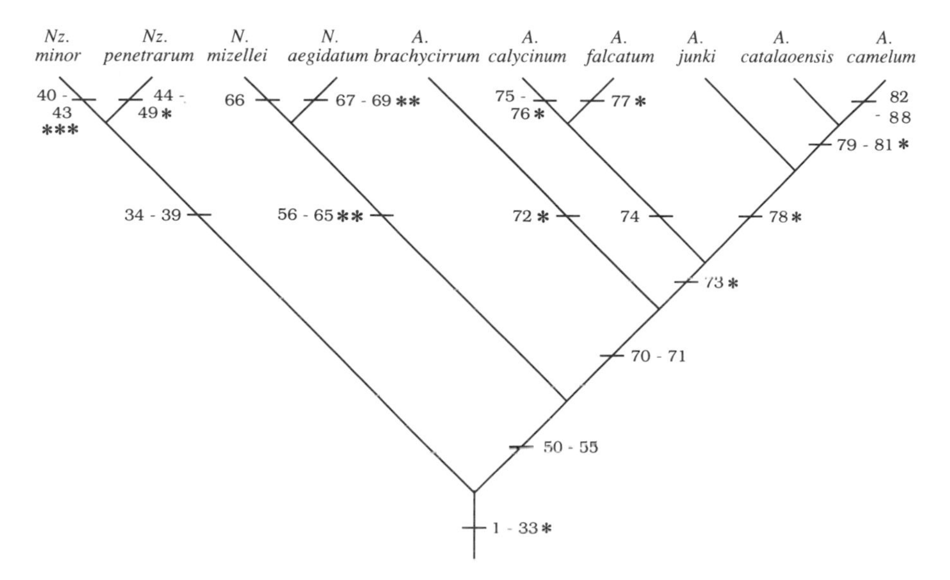

Figure 54. Phylogenetic tree for species of *Amphithecium* (*A.*), *Notothecium* (*N.*), and *Notozothecium* (*Nz.*) inhabiting Amazonian piranhas.

***Actinocleidus bifidus* Mizelle and Cronin 1943**

Diagnosis: Deep root of hamulus reduced (27); ventral transverse bar with heavily sclerotized ventrolateral projections and lightly sclerotized medial and lateral extensions (28); dorsal transverse bar horizontal centrally with two oblique dorsolateral struts, each with lateral inflation and small distal protruberance (30); cirrus with bifid distal aperture (36); accessory piece with conspicuously enlarged proximal bifurcation and additional medial sclerite (37).

***Amphithecium* Boeger and Kritsky 1988 + *Notothecium* Boeger and Kritsky 1988 + *Notozothecium* Boeger and Kritsky 1988 (Fig. 54)**

Reference: Boeger and Kritsky (1988).

Diagnosis: Body less than 1 mm long (1); body fusiform (2); dorsal body hump lacking (3); tegument scaled (4); esophagus lacking (5); eyes present (6); vaginal apertures bilateral (7); no sclerotized plate around vaginal apertures (8); vaginal duct sclerotized (9); seminal receptacle present (10); ovary with smooth margins (11); vitellaria randomly distributed (12); two prostatic reservoirs (13); seminal vesicle fusiform (14); cirrus coiled (15); one cirral ramus *(16); cirral aperture terminal (17); cirral tip not sclerotized (18); accessory piece T-shaped (19); distal end of accessory piece blunt (20); accessory piece flap lacking (21); ratio of fifth hook pair length to cirral length less than or equal to 0.1 (22); cleft on ventral bar lacking (23); anteromedial projection on ventral bar lacking (24); no projection on dorsal bar (25); anchor base without proximal sclerotization (26); ratio of ventral anchor length to dorsal anchor length less than 1:2 (27); ratio of anchor point length to anchor shaft length greater than 0.2 (28); ratio of hook pair 5 length to hook pair 2 length 1.0 (29); ratio of hook pair 5 length to hook pair 3 length less than 0.8 (30); ratio of hook pair 5 length to hook pair 4 length 0.46–0.61 (31); ratio of hook pair 5 length to hook pair 6 length less than 0.85 (32); ratio of hook pair 5 length to hook pair 7 length less than 0.8 (33).

Notozothecium **Boeger and Kritsky 1988**

Diagnosis: Anteromedial projection on ventral bar present (34); single dextral vaginal aperture (35); distal end of accessory piece ornamented (36); ratio of hook pair 5 length to hook pair 2 length less than 0.8 (37); ratio of hook pair 5 length to hook pair 4 length 1.0 (38); ratio of hook pair 5 length to hook pair 6 length less than 1.0 (39).

minor **Boeger and Kritsky 1988**

Diagnosis: Ratio of ventral anchor length to dorsal anchor length more than 1:4 *(40); sclerotized plate around vaginal apertures *(41); ratio of fifth hook pair length to cirral length 0.2–0.5 *(42); ratio of hook pair 5 length to hook pair 4 length 0.78 (43).

penetrarum **Boeger and Kritsky 1988**

Diagnosis: Body more than 1 mm long (44); tegument smooth *(45); one prostatic reservoir (46); anchor base with proximal sclerotization (47); ratio of hook pair 5 length to hook pair 3 length 1.0 (48); ratio of hook pair 5 length to hook pair 7 length less than 1.0 (49).

Notothecium **Boeger and Kritsky 1988**

Diagnosis: Vaginal duct not sclerotized (51); cirrus straight (52); two cirral rami (53); ratio of fifth hook pair length to cirral length 0.6–0.8 (55); one cirral rami tip pointed, one reduced (56); cirral rami unequal (57); single sinistral vaginal aperture (58); accessory piece flap small (59); eyes lacking (60); body strongly flattened (61); seminal vesicle C-shaped (62); distal end of accessory piece pointed (63); accessory piece flap present *(64); ratio of anchor point length to anchor shaft length less than 0.2 *(65).

mizellei **Boeger and Kritsky 1988**

Diagnosis: Projection on dorsal bar (66).

aegidatum **Boeger and Kritsky 1988**

Diagnosis: One cirral ramus *(67); sclerotized plate around vaginal apertures *(68); cirral aperture diagonal (69).

Amphithecium **Boeger and Kritsky 1988**

Diagnosis: Both cirral rami tips pointed (50); vaginal duct not sclerotized (51); cirrus straight (52); two cirral rami (53); cirral rami subequal (54); ratio of fifth hook pair length to cirral length 0.6–0.8 (55); esophagus present (70); seminal receptacle lacking (71).

brachycirrum **Boeger and Kritsky 1988**

Diagnosis: Accessory piece not T-shaped *(72).

calycinum **Boeger and Kritsky 1988**

Diagnosis: Ratio of fifth hook pair length to cirral length 0.2–0.5 *(73); distal end of accessory piece hooked (74); one cirral rami tip pointed, one funnel-shaped (75); tegument smooth *(76).

falcatum **Boeger and Kritsky 1988**

Diagnosis: Ratio of fifth hook pair length to cirral length 0.2–0.5 *(73); distal end of accessory piece hooked (74); accessory piece not T-shaped *(77).

junki **Boeger and Kritsky 1988**

Diagnosis: Ratio of fifth hook pair length to cirral length 0.2–0.5 *(73); ratio of anchor point length to anchor shaft length less than 0.2 *(78).

catalaoensis **Boeger and Kritsky 1988**

Diagnosis: Ratio of fifth hook pair length to cirral length 0.2–0.5 *(73); ratio of anchor point length to anchor shaft length less than 0.2 *(78); accessory piece flap about half length of distal portion of accessory piece (79); cirral tip sclerotized (80); accessory piece flap present *(81).

camelum **Boeger and Kritsky 1988**

Diagnosis: Ratio of fifth hook pair length to cirral length 0.2–0.5 *(73); ratio of anchor point length to anchor shaft length less than 0.2 *(78); body ventrally concave (82); dorsal body hump present (83); tegument papillated (84); ovary with irregular margins (85); vitellaria fimbriated (86); cleft on ventral bar present (87); ratio of ventral anchor length to dorsal anchor length more than 1:4 *(88).

Subfamily Anacanthorinae Price 1967

Anacanthorus **Mizelle and Price 1965: Clade of species inhabiting Amazonian piranhas (Fig. 55)**

Reference: Van Every and Kritsky (1992).

Diagnosis: No articulation of accesory piece to base of cirrus (1); cirrus coiled to some degree rather than straight or twisted (2); cirral "feather" lacking *(3); basal aperture of cirrus opening posterolaterally or laterally (4); distal aperture of cirrus subterminal *(5); cirral base lacking heel-like base (6); submedial (muscle) articulation point of accessory piece flat or minimally elevated *(7); subterminal expansion of accessory piece lacking (8); distal tip of accessory piece acute (9); peduncle surface smooth (10); "thumb" on haptoral hooks slightly depressed (11).

stachophallus **Boeger, Kritsky, and Van Every 1992,** ***thatcheri*** **Boeger and Kritsky 1988,** ***palamophallus*** **Boeger, Kritsky, and Van Every 1992**

Diagnosis: Cirral base with heel-like projection (12); distal tip of accessory piece blunt (13).

periphallus **Boeger, Kritsky, and Van Every 1992**

Diagnosis: Basal aperture of cirrus opening anteriorly or anterolaterally, cirrus J-shaped (14).

scapanus **Van Every and Kritsky 1992,** ***serrasalmi*** **Van Every and Kritsky 1992,** ***sciponophallus*** **Van Every and Kritsky 1992**

Diagnosis: Basal aperture of cirrus opening anteriorly or anterolaterally, cirrus J-shaped (14); distal aperture of cirrus terminal or diagonal (15).

jegui **Van Every and Kritsky 1992**

Diagnosis: Basal aperture of cirrus opening anteriorly or anterolaterally, cirrus J-shaped (14); distal aperture of cirrus terminal or diagonal (15); subterminal expansion

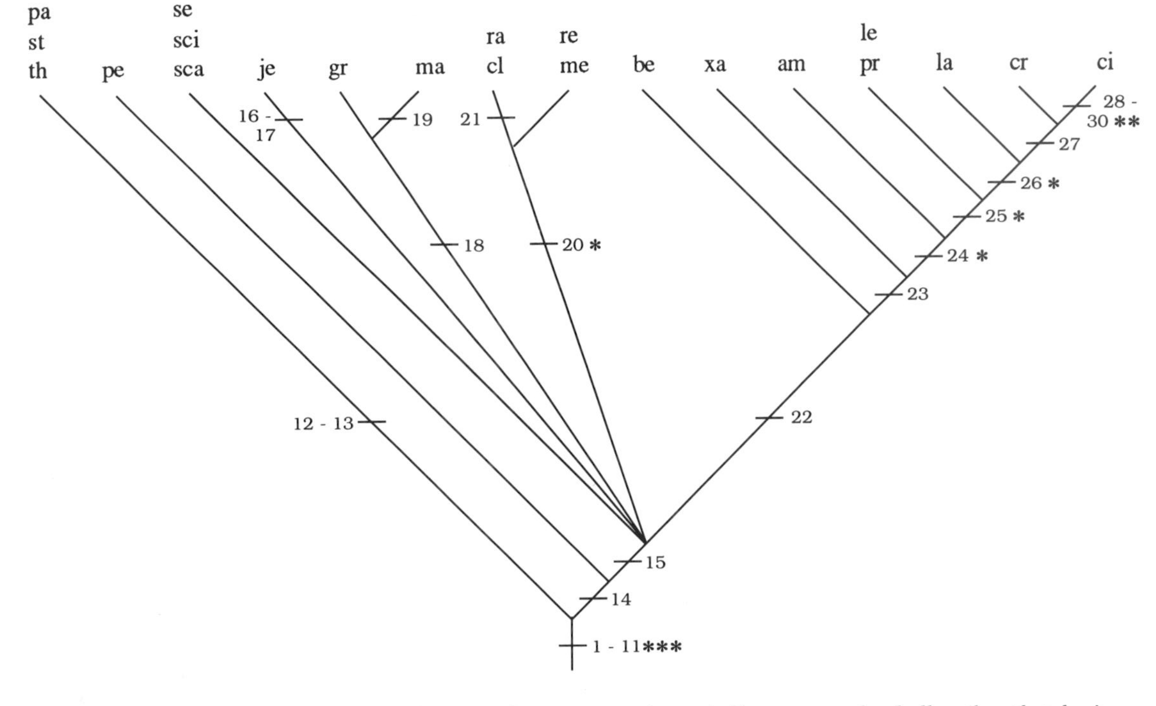

Figure 55. Phylogenetic tree for species of *Anacanthorus*. pa = *palamophallus;* st = *stachophallus;* th = *thatcheri;* pe = *periphallus;* se = *serrasalmi;* sci = *sciponophallus;* sca = *scapanus;* je = *jegui;* gr = *gravihamulatus;* ma = *mastigophallus;* ra = *ramosissimus;* cl = *cladophallus;* re = *reginae;* me = *mesocondylus;* be = *beleophallus;* xa = *xaniophallus;* am = *amazonicus;* le = *lepyrophallus;* pr = *prodigiosus;* la = *lasiophallus;* cr = *crytocaulus;* ci = *cinctus.*

of accessory piece present (16); subterminal expansion of accessory piece expanded from both margins (17).

***gravihamulatus* Van Every and Kritsky 1992**

Diagnosis: Basal aperture of cirrus opening anteriorly or anterolaterally, cirrus J-shaped (14); distal aperture of cirrus terminal or diagonal (15); "thumb" on haptoral hooks flattened (18).

***mastigophallus* Boeger, Kritsky, and Van Every 1992**

Diagnosis: Basal aperture of cirrus opening anteriorly or anterolaterally, cirrus J-shaped (14); distal aperture of cirrus terminal or diagonal (15); "thumb" on haptoral hooks flattened (18); basal aperture of cirrus opening posteriorly, cirrus curled to form one or more rings (19).

***reginae* Boeger and Kritsky 1988, *mesocondylus* Van Every and Kritsky 1992**

Diagnosis: Basal aperture of cirrus opening anteriorly or anterolaterally, cirrus J-shaped (14); distal aperture of cirrus terminal or diagonal (15); submedial (muscle) articulation point of accessory piece a small protuberance *(20).

***cladophallus* Van Every and Kritsky 1992, *ramosissimus* Van Every and Kritsky 1992**

Diagnosis: Basal aperture of cirrus opening anteriorly or anterolaterally, cirrus J-shaped (14); distal aperture of cirrus terminal or diagonal (15); submedial (muscle) articulation point of accessory piece a protruding branch (21).

***beleophallus* Boeger, Kritsky, and Van Every 1992**

Diagnosis: Basal aperture of cirrus opening anteriorly or anterolaterally, cirrus J-shaped (14); distal aperture of cirrus terminal or diagonal (15); cirral "feather" moderately developed, lacking terminal filaments (22).

***xaniophallus* Boeger, Kritsky, and Van Every 1992**

Diagnosis: Basal aperture of cirrus opening anteriorly or anterolaterally, cirrus J-shaped (14); distal aperture of cirrus terminal or diagonal (15); cirral "feather" well developed, with terminal filaments (23).

***amazonicus* Van Every and Kritsky 1992**

Diagnosis: Basal aperture of cirrus opening anteriorly or anterolaterally, cirrus J-shaped (14); distal aperture of cirrus terminal or diagonal (15); cirral "feather" well developed, with terminal filaments (23); subterminal expansion of accessory piece present *(24).

***lepyrophallus* Boeger, Kritsky, and Van Every 1992, *prodigiosus* Van Every and Kritsky 1992**

Diagnosis: Basal aperture of cirrus opening anteriorly or anterolaterally, cirrus J-shaped (14); cirral "feather" well developed, with terminal filaments (23); subterminal expansion of accessory piece present *(24); distal aperture of cirrus subterminal *(25).

***lasiophallus* Van Every and Kritsky 1992**

Diagnosis: Basal aperture of cirrus opening anteriorly or anterolaterally, cirrus J-shaped (14); cirral "feather" well developed, with terminal filaments (23); subterminal expansion of accessory piece present *(24); distal aperture of cirrus subterminal *(25); submedial (muscle) articulation point of accessory piece a small protuberance *(26).

***crytocaulus* Van Every and Kritsky 1992**

Diagnosis: Basal aperture of cirrus opening anteriorly or anterolaterally, cirrus J-shaped (14); cirral "feather" well developed, with terminal filaments (23); subterminal expansion of accessory piece present *(24); distal aperture of cirrus subterminal *(25); submedial (muscle) articulation point of accessory piece a small protuberance *(26); subterminal expansion of accessory piece thumblike, from a single margin (27).

***cinctus* Van Every and Kritsky 1992**

Diagnosis: Basal aperture of cirrus opening anteriorly or anterolaterally, cirrus J-shaped (14); subterminal expansion of accessory piece present *(24); distal aperture of cirrus subterminal *(25); subterminal expansion of accessory piece thumblike, from a single margin (27); cirral "feather" lacking (*28); submedial (muscle) articulation point of accessory piece flat or minimally elevated *(29); peduncle surface corrugated (30).

Family Monocotylidae Taschenberg 1879: Partial resolution of family relationships as part of analysis of species of *Monocotyle* (Fig. 56)

Reference: Measures et al. (1990).

Diagnosis: Eight loculi surrounding central loculus (1); male copulatory organ coiled or looping (2).

Anoplocotyoloides* Young 1967 *incertae sedis sedis mutabilis

Diagnosis: With characters of the family.

***Calicotyle* Diesing 1850 + *Merizocotyle* Cerfontaine 1894**

Diagnosis: Haptor with seven loculi surrounding central loculus (3); marginal membrane on haptor absent (4); egg appendages short *(5); vaginae doubled (6).

***Calicotyle* Diesing 1850**

Diagnosis: With characters of the subfamily.

***Merizocotyle* Cerfontaine 1894**

Diagnosis: Haptor with five loculi surrounding central loculus (7); egg appendages long *(8); male copulatory organ not coiled or looping *(9).

***Dendromonocotyle* Hargis 1955 + *Clemacotyle* Young 1967**

Diagnosis: Haptor with marginal papillae (10); haptor with ventral sclerites other than hamuli and larval hooks (11); hamuli absent (12); intestine dendritic (13); egg appendages short *(14); male copulatory organ not coiled or looping *(15).

***Dendromonocotyle* Hargis 1955**

Diagnosis: Egg appendages long *(16).

***Clemacotyle* Young 1967**

Diagnosis: Male copulatory organ not sclerotized (17).

Subfamily Monocotylinae Taschenberg 1879 *incertae sedis*

Diagnosis: Haptor with marginal papillae (10); haptor with ventral sclerites other than hamuli and larval hooks (11).

Monocotyle* Taschenberg 1878 *incertae sedis

Diagnosis: With characters of the subfamily.

granulata* Young 1967 *sedis mutabilis

Diagnosis: With characters of the genus.

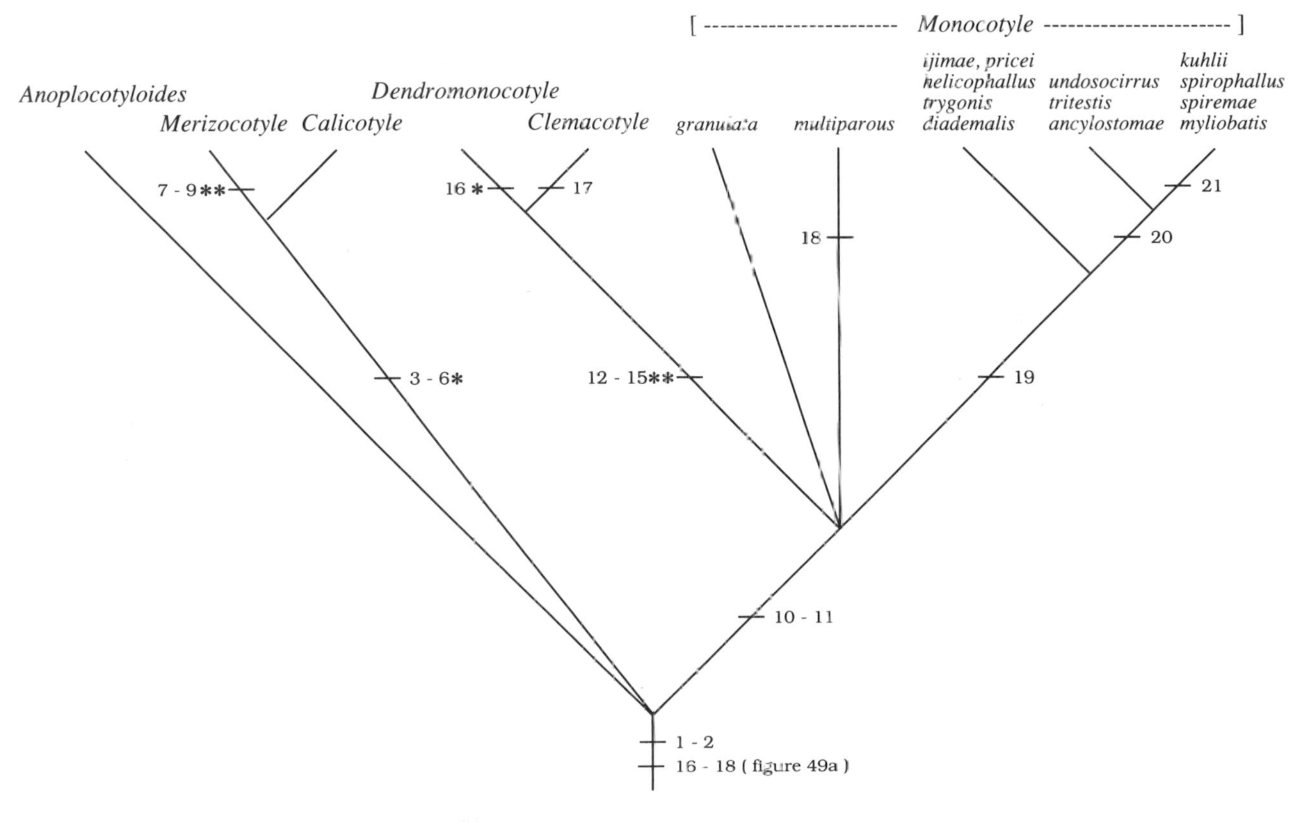

Figure 56. Phylogenetic tree for the Monocotylidae.

multiparous **Measures, Beverly-Burton, and Williams 1990** ***sedis mutabilis***

Diagnosis: Eggs retained to advanced stage of embryonation (egg containing oncomiracidium) (18).

diademalis **Hargis 1955,** ***trygonis*** **(Venkatanarsaiah and Kalkarni 1980) Timofeeva 1985,** ***helicophallus*** **Measures, Beverly-Burton, and Williams 1990,** ***pricei*** **Pearse 1949,** ***ijimae*** **Goto 1894**

Diagnosis: Accessory sclerotized structure present (19).

ancylostomae **Timofeeva 1984,** ***tritestis*** **Young 1967,** ***undosocirrus*** **Timofeeva 1984**

Diagnosis: Accessory sclerotized structure present (19); vaginal duct sclerotized (20).

spiremae **Measures, Beverly-Burton, and Williams 1990,** ***spirophallus*** **(Tripathi 1959),** ***myliobatis*** **Taschenberg 1878,** ***kuhlii*** **Young 1967**

Diagnosis: Accessory sclerotized structure present (19); vaginal duct sclerotized (20); egg appendages short *(21).

Family Hexabothriidae (Fig. 57)

Reference: Boeger and Kritsky (1989).

Diagnosis: Sclerite 1′ similar to sclerite 2′ (1); sclerite 1′ similar to sclerite 3′ (2); sclerite 1′ similar to sclerite 3 (3); sclerite 1 similar to sclerite 2′ (4); sclerite 1 similar to sclerite 3′ (5); sclerite 1 similar to sclerite 3 (6); sclerite 2′ similar to sclerite 2 (7); sclerite 2 similar to sclerite 3′ (8); sclerite 2 similar to sclerite 3 (9); sclerite 1 similar to sclerite 1′ (10); sclerite 1 similar to sclerite 2 (11); sclerite 1′ similar to sclerite 2 (12); sclerite 2′ similar to sclerite 3′ (13); vaginal duct differentiated in two portions (14); vaginal ducts parallel (15); vas deferens with thin distal wall (16); prostatic region lacking *(17); proximal portion of cirrus expanded along entire length (18); cirrus ovate (19); cirrus armed (20); distal portion of cirrus present (21); eggs with single elongate polar filament *(22); terminal vaginae muscular (23); terminal wall of vaginae delicate (24); ootype smooth (25); ovary lobate anteriorly (26); descending branch of ovary sinuous or coiled (27); ascending branch of ovary present (28); ascending branch of ovary straight (29); saclike seminal receptacle present (30); oviduct not dilated (31); ovary not dilated posteriorly (32); haptoral appendix marginal (33); haptor symmetrical (34); pair of anchors present (35); suckers on haptor adjacent (36).

Hexabothrium **Nordmann 1840**

Diagnosis: Vaginal ducts Y-shaped *(37).

Dasyonchocotyle **Hargis 1955**

Diagnosis: Proximal cirrus tubular (38); eggs with two elongate polar filaments (39); prostatic region present (40); sucker sclerite pair 2 displaced laterally (41); *sucker sclerite pair 3 large and robust, leading to:* sclerite 1′ more delicate than sclerite 3′ (42); sclerite 1′ more delicate than sclerite 3 (43); sclerite 1 more delicate than sclerite 3′ (44); sclerite 1 more delicate than sclerite 3 (45); sclerite 2′ more delicate than sclerite 3′ (46); sclerite 2′ more delicate than sclerite 3 (47); sclerite 2 more delicate than sclerite 3′ (48); sclerite 2 more delicate than sclerite 3 (49); distal cirrus elongate *(50).

Erpocotyle **van Beneden and Hesse 1863**

Diagnosis: Proximal cirrus tubular (38); eggs with two elongate polar filaments (39); prostatic region present (40); distal cirrus unarmed (51); descending branch of ovary straight

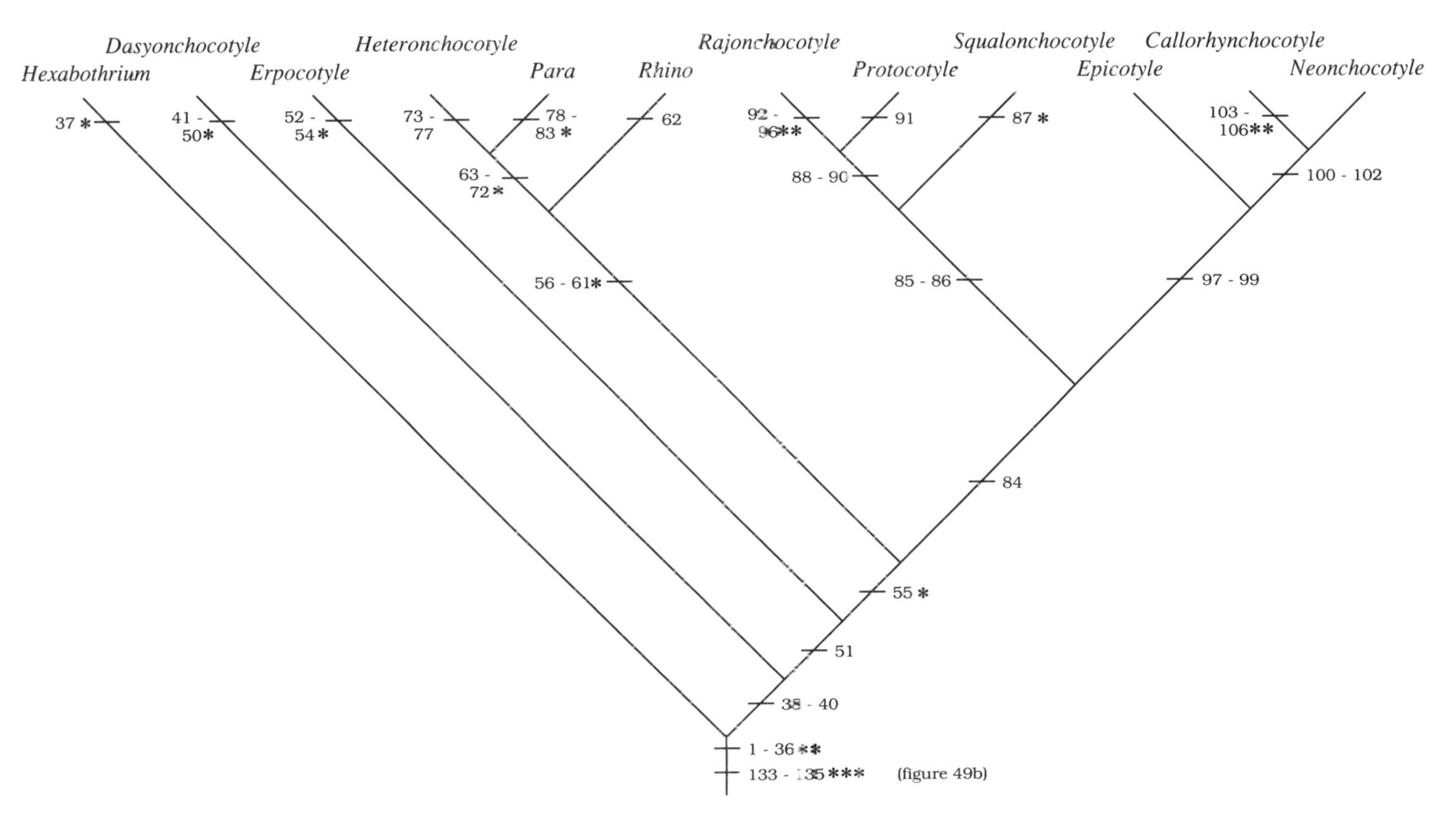

Figure 57. Phylogenetic tree for the genera of the Hexabothriidae. *Para* = *Paraheteronchocotyle*; *Rhino* = *Rhinobatonchocotyle*.

(52); terminal portion of vas deferens with thick wall (53); ovary branched anteriorly *(54).

***Rhinobatonchocotyle* Doran 1953**

Diagnosis: Proximal cirrus tubular (38); eggs with two elongate polar filaments (39); distal cirrus elongate *(55); oviduct dilated (presumably for sperm storage) (56); seminal receptacle lacking (57); terminal portion of vaginal duct thick-walled, muscular (58); prostatic region lacking *(59); ascending branch of ovary lacking (60); haptor asymmetrical, sucker complexes linear except complex 1 adjacent to base of haptoral appendix (61); vaginal duct X-shaped (62).

***Heteronchocotyle* Brooks 1934**

Diagnosis: Eggs with two elongate polar filaments (39); distal cirrus unarmed (51); oviduct dilated (presumably for sperm storage) (56); seminal receptacle lacking (57); terminal portion of vaginal duct thick-walled, muscular (58); prostatic region lacking *(59); ascending branch of ovary lacking (60); haptor asymmetrical, sucker complexes linear except complex 1 adjacent to base of haptoral appendix (61); proximal cirrus with thick wall *(63); *sclerites dissimilar in size and shape, leading to:* sclerite 1′ more robust than sclerite 2′ (64); sclerite 1′ more robust than sclerite 3′ (65); sclerite 1′ more robust than sclerite 3 (66); sclerite 1 more robust than sclerite 2′ (67); sclerite 1 more robust than sclerite 3′ (68); sclerite 1 more robust than sclerite 3 (69); sclerite 2′ more delicate than sclerite 2 (70); sclerite 2 more robust than sclerite 3′ (71); sclerite 2 more robust than sclerite 3 (72); sclerite 1 more robust than 1′ (73); sclerite 1 more delicate than sclerite 2 (74); haptoral appendix marginal, lateral to body midline (75); haptor asymmetrical, longitudinal haptoral axis forming an angle less than or equal to 45° with body midline (76); distal cirrus ovate (77).

***Paraheteronchocotyle* Mayes, Brooks, and Thorson 1981**

Diagnosis: Distal cirrus unarmed (51); distal cirrus elongate (55); oviduct dilated (presumably for sperm storage) (56); seminal receptacle lacking (57); terminal portion of vaginal duct thick-walled, muscular (58); prostatic region lacking *(59); ascending branch of ovary lacking (60); proximal cirrus with thick wall *(63); *sclerites dissimilar in size and shape, leading to:* sclerite 1′ more robust than sclerite 2′ (64); sclerite 1′ more robust than sclerite 3′ (65); sclerite 1′ more robust than sclerite 3 (66); sclerite 1 more robust than sclerite 2′ (67); sclerite 1 more robust than sclerite 3′ (68); sclerite 1 more robust than sclerite 3 (69); sclerite 2′ more delicate than sclerite 2 (70); sclerite 2 more robust than sclerite 3′ (71); sclerite 2 more robust than sclerite 3 (72); haptor asymmetrical, longitudinal haptoral axis congruent with body midline (78); sucker complexes linear except complex 1 adjacent to base of haptoral appendix (79); sclerites 1, 1′ similar (80); haptoral appendix marginal (81); anchors lacking (82); eggs with single elongate polar filament *(83).

***Squalonchocotyle* Cerfontaine 1899**

Diagnosis: Proximal cirrus tubular (38); eggs with two elongate polar filaments (39); prostatic region present (40); distal cirrus unarmed (51); distal cirrus elongate (55); vaginal ducts expanded, with gland cells surrounding base (84); ootype wall with longitudinal rows of large cells (85); terminal portion of vaginal ducts nonmuscular (86); ovary branched anteriorly *(87).

Protocotyle **Euzet and Maillard 1974**

Diagnosis: Proximal cirrus tubular (38); eggs with two elongate polar filaments (39); prostatic region present (40); distal cirrus unarmed (51); distal cirrus elongate (55); vaginal ducts expanded, with gland cells surrounding base (84); ootype wall with longitudinal rows of large cells (85); vaginal ducts parallel, terminal portion non-muscular (86); ascending branch of ovary sinuous (88); proximal cirrus expanded basally (89); vaginal ducts undifferentiated (90); ovary dilated posteriorly (91).

Rajonchocotyle **Cerfontaine 1899**

Diagnosis: Proximal cirrus tubular (38); vaginal ducts expanded, with gland cells surrounding base (84); ootype wall with longitudinal rows of large cells (85); ascending branch of ovary sinuous (88); vaginal ducts undifferentiated (90); vaginal ducts Y-shaped *(92); proximal cirrus completely expanded (93); distal cirrus lacking (94); eggs with two small polar filaments (95); prostatic region lacking *(96).

Epicotyle **Euzet and Maillard 1974**

Diagnosis: Proximal cirrus tubular (38); eggs with two elongate polar filaments (39); prostatic region present (40); distal cirrus unarmed (51); distal cirrus elongate (55); vaginal ducts expanded, with gland cells surrounding base (84); haptoral appendix lateral to midline (97); haptoral appendix dorsal (98); vaginal ducts parallel, distal portion tubular (99).

Neonchocotyle **Ktari and Maillard 1972**

Diagnosis: Proximal cirrus tubular (38); eggs with two elongate polar filaments (39); prostatic region present (40); distal cirrus unarmed (51); distal cirrus elongate (55); vaginal ducts expanded, with gland cells surrounding base (84); haptoral appendix lateral to midline (97); haptoral appendix dorsal (98); vaginal ducts parallel, distal portion tubular (99); sucker complex pair 1 on same side of longitudinal haptoral axis (100); haptor asymmetrical (101); longitudinal axis forming angle greater than 45° with body midline (102).

Callorhynchocotyle **Suriano and Incorvaia 1982**

Diagnosis: Eggs with two elongate polar filaments (39); distal cirrus unarmed (51); vaginal ducts expanded, with gland cells surrounding base (84); haptoral appendix lateral to midline (97); haptoral appendix dorsal (98); vaginal ducts parallel, distal portion tubular (99); sucker complex pair 1 on same side of longitudinal haptoral axis (100); haptor asymmetrical (101); longitudinal axis forming angle greater than 45° with body midline (102); prostatic region lacking *(103); distal cirrus ovate (104); vaginal ducts parallel with glandulomuscular terminal portion (105); proximal cirrus with thick wall *(106).

Cohort Gyrocotylidea Poche 1926 (Fig. 58)

Reference: Bandoni and Brooks (1987a).

Family Gyrocotylidae Benham 1901

Diagnosis: With characters of the order.

Genus *Gyrocotyle* Diesing 1850

Diagnosis: With characters of the family.

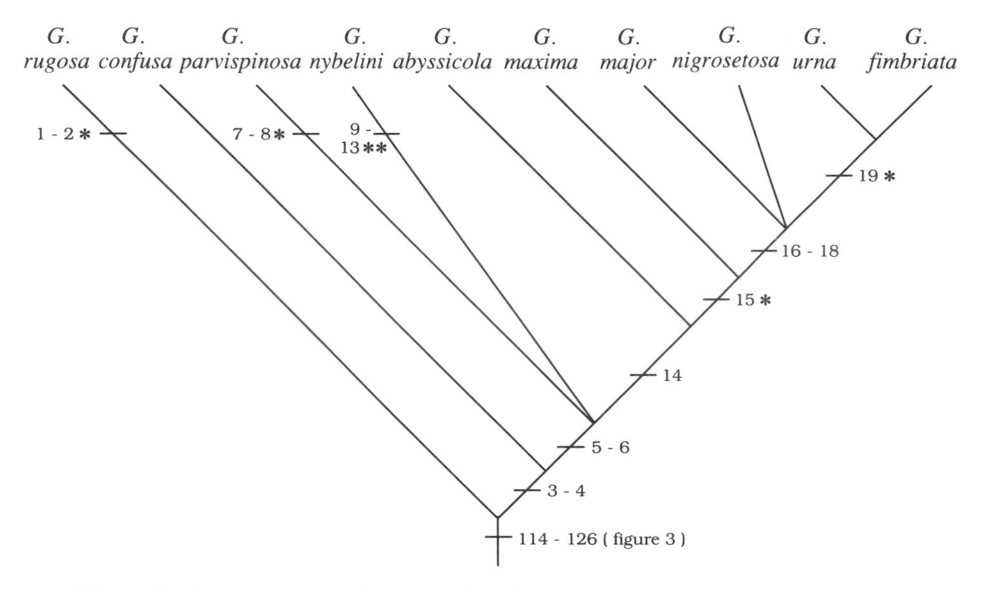

Figure 58. Phylogenetic tree for the species of *Gyrocotyle*.

***rugosa* Diesing 1850**

Diagnosis: *Gyrocotyle* with complex folding of uterus (1); short and thick tegumental spines *(2).

***confusa* van der Land and Dienske 1968**

Diagnosis: *Gyrocotyle* with crenate body margins, relatively small number of plications (3); large tegumental spines present on much of body surface (4).

parvispinosa* Lynch 1945 *sedis mutabilis

Diagnosis: *Gyrocotyle* with crenate body margins, with relatively small number of plications (3); large tegumental spines present on much of body surface (4); testes extending only to the anterior margin of the uterine sac (7); no copulatory papilla *(8).

nybelini* (Fuhrmann 1931) Bandoni and Brooks 1987 *sedis mutabilis

Diagnosis: *Gyrocotyle* with crenate body margins, relatively small number of plications (3); testes extending to posterior margin of uterine sac (6); posterior funnel elongate (9); uterine sac greatly expanded (10); no tegumental spines (11); highly crenulate body margins *(12); no genital notch *(13).

***abyssicola* van der Land and Templeman 1968**

Diagnosis: *Gyrocotyle* with crenate body margins, with relatively small number of plications (3); large tegumental spines present on much of body surface (4); uterine sac large (5); testes extending to posterior margin of uterine sac (6); pronounced genital notch (14).

***maxima* Macdonagh 1927**

Diagnosis: *Gyrocotyle* with crenate body margins, relatively small number of plications (3); uterine sac large (5); testes extending to posterior margin of uterine sac (6); pronounced genital notch (14); short and thick tegumental spines *(15).

major* van der Land and Templeman 1968, *nigrosetosa* Haswell 1902 *sedis mutabilis

Diagnosis: *Gyrocotyle* with large uterine sac (5); testes extending to posterior margin of uterine sac (6); pronounced genital notch (14); short and thick tegumental spines *(15); short and wide funnel (16); lateral body margins crenate with many plications (17); no tegumental spines on body margins (18).

urna* (Wagener 1852) Wagener 1858 *sedis mutabilis

Diagnosis: *Gyrocotyle* with large uterine sac (5); testes extending to posterior margin of uterine sac (6); pronounced genital notch (14); short and thick tegumental spines *(15); short and wide funnel (16); lateral body margins crenate with many plications (17); no tegumental spines on body margins (18); no copulatory papilla *(19).

fimbriata* Watson 1911 *sedis mutabilis

Diagnosis: *Gyrocotyle* with large uterine sac (5); testes extending to posterior margin of uterine sac (6); pronounced genital notch (14); short and thick tegumental spines *(15); short and wide funnel (16); lateral body margins crenate with many plications (17); no tegumental spines on body margins (18); no copulatory papilla *(19).

Molecular Data

Colin et al. (1986) suggested that *G. urna*, *G. confusa*, and *G. nybelini* all represented a single species. Bristow and Berland (1988) reported electrophoretic data covering 15 enzyme systems supporting the specific distinctness of *G. urna*, *G. confusa*, and *G. nybelini*. Subsequently, Berland et al. (1990) corroborated those findings using fatty acid chemometry patterns.

Cohort Cestoidea Rudolphi 1808

Subcohort Amphilinidea Poche 1922 (Fig. 59)

Reference: Bandoni and Brooks (1987b).

Family Amphilinidae Claus 1879

Diagnosis: With characters of the order.

Subfamily Amphilininae Claus 1879

Diagnosis: Amphilinids with elliptical body (1); testes partly scattered throughout parenchyma (2); vagina crossing male duct proximal to ejaculatory duct (3); supercoiling of third limb of uterus (4); ovary rotated ventral to seminal receptacle (5).

***Amphilina* Wagener 1858**

Diagnosis: With characters of the subfamily.

***foliacea* (Rudolphi 1819) Wagener 1858**

Diagnosis: *Amphilina* with testes completely scattered throughout parenchyma (6).

***japonica* Goto and Ishii 1936**

Diagnosis: *Amphilina* with vagina crossing male duct distal to ejaculatory duct (7).

Subfamily Schizochoerinae Poche 1922

Diagnosis: Amphilinids with small papilla in depression at posterior end (8); large seminal receptacle (9); kidney-shaped ovary, ovary entire, smooth margin (10).

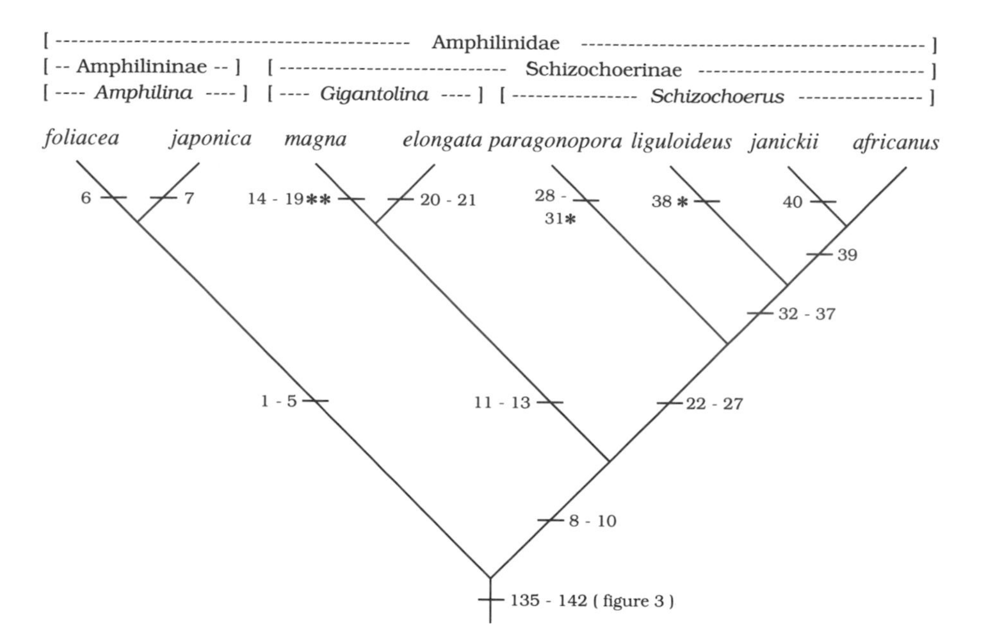

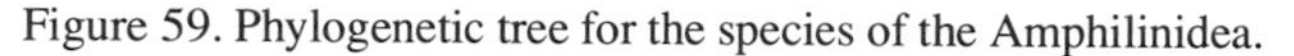
Figure 59. Phylogenetic tree for the species of the Amphilinidea.

***Gigantolina* Poche 1922**

Diagnosis: Schizochoerines with vas deferens not coiled (11); uterus looped (12); uterus dilated (13).

***magna* (Southwell 1915) Poche 1922**

Diagnosis: *Gigantolina* with large papilla in depression at posterior end (14); muscular vaginal pore (15); ovary bilobed, symmetrical, finely lobate *(16); ovary rotated ventral to seminal receptacle (17); vaginal extension *(18); ductus yamagutii present (19).

***elongata* (Johnston 1931) Bandoni and Brooks 1987**

Diagnosis: *Gigantolina* with common genital pore present (20); seminal duct dilated (21).

***Schizochoerus* Poche 1922**

Diagnosis: Schizochoerines with strongly muscular pharynx (22); paired testes (23); vagina coiled (24); uterine pore opening an anterior end of apical invagination (25); uterus reduced in size (26); accessory seminal receptacle formed as a dilation of seminal duct (27).

***paragonopora* (Woodland 1923) Bandoni and Brooks 1987**

Diagnosis: *Schizochoerus* with caudal lobes (28); reduced seminal receptacle *(29); ovary single, irregular (30); vagina bent (31).

***liguloideus* (Diesing 1850) Poche 1922**

Diagnosis: *Schizochoerus* with tegument with concentric annulations (32); doubled vagina (33); elongate vagina (34); atrophied uterine duct (35); descending limb of

N-shaped uterus adjacent to terminal limb (36); poorly developed vitellaria (37); vaginal extension *(38).

***janickii* (Poche 1922) Bandoni and Brooks 1987**

Diagnosis: *Schizochoerus* with tegument with concentric annulations (32); doubled vagina (33); elongate vagina (34); atrophied uterine duct (35); descending limb of N-shaped uterus adjacent to terminal limb (36); poorly developed vitellaria (37); fusiform body (39); unpaired testes, in wide bands (40).

***africanus* (Donges and Harder 1966) Bandoni and Brooks 1987**

Diagnosis: *Schizochoerus* with tegument with concentric annulations (32); doubled vagina (33); elongate vagina (34); atrophied uterine duct (35); descending limb of N-shaped uterus adjacent to terminal limb (36); poorly developed vitellaria (37); fusiform body (39).

Subcohort Eucestoda Southwell 1930: Ordinal relationships (Fig. 60)

References: Justine (1991a); Brooks et al. (1991).

Infracohort Pseudophylla infracohort n.

Diagnosis: Bilaterally symmetrical, bipartite scolices (the "difossate" condition), in which the modifications for attachment consist of leaflike longitudinal flaps (bothria) and their modifications (1); difossate plerocercoids (2).

Order Pseudophylliformes Carus 1863

Diagnosis: With characters of the group.

Infracohort Saccouterina infracohort n.

Diagnosis: Bilateral saccate uteri lacking permanent pores (there may be slitlike or porelike modifications of the tegument formed by an invagination of the subtegument that eventually fuse with an evagination of the uterine wall (a dehiscence), allowing the

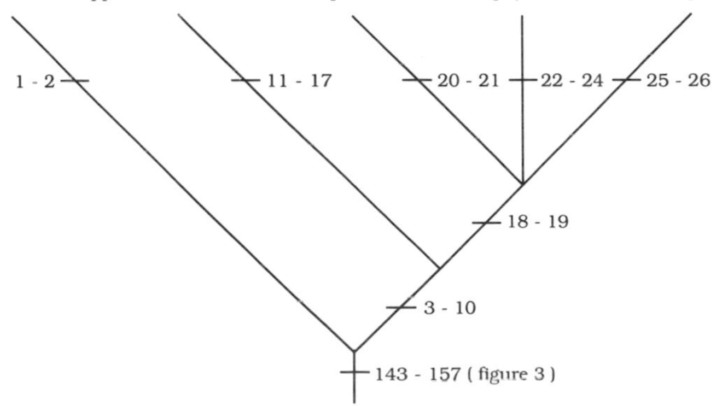

Figure 60. Phylogenetic tree for the orders of the Eucestoda.

release of eggs with the expansion of a gravid proglottid) (3); onchospheral flame cells are absent (4); hexacanths not ciliated (called oncospheres) (5); "oligolecithal" eggs (a minimal vitelline component, and a shell formed by the embryo) (6); two embryonic membranes (7); "ovoviviparous" development in which the hexacanth matures in utero (8); indeterminate growth (continuous budding from a growth zone) (9); apolytic (10).

Order Nippotaeniiformes Yamaguti 1939

Diagnosis: Fewer than six hooks on the detached cercomer (11); compact, preovarian vitellaria (12); euapolytic (13); ovary inverted-U-shaped (14); fewer than 13 proglottids per strobila (15); 25–32 testes per proglottid (16); single axoneme in sperm tails (17).

Tetrafossata Group

Diagnosis: Bilaterally symmetrical, quadripartite scolices (the "tetrafossate" condition), in which the modifications for attachment consist of four laterally positioned suckers and an apical sucker (18); medullary vitellaria (19).

Order Lecanicephaliformes Baylis 1920 *sedis mutabilis*

Diagnosis: Apical sucker greatly enlarged, called a *myzorhynchus* (20); myzorhynchus retractable (21).

Order Tetraphylliformes Carus 1863 *sedis mutabilis*

Diagnosis: Bothridia, arising as modifications of sucker margins (22); "X-shaped" ovaries when viewed in cross section (23); vagina anterior to cirrus sac (24).

Order Proteocephaliformes Mola 1928 *sedis mutabilis*

Diagnosis: Highly distinct medullary and cortical regions of the parenchyma, the latter being relatively extensive, that are defined by the longitudinal musculature (25); lateral branches and diverticula of the uterus (26).

Subcohort Eucestoda Southwell 1930: Infraordinal relationships

Order Pseudophylliformes Carus 1863 (Fig. 61)

Family Amphicotylidae Ariola 1899 *sedis mutabilis*

Diagnosis: Acaudate plerocercoids *(1); radially symmetrical saccate uterus (2); oncospheres *(3).

Family Philobythiidae Campbell 1977 *sedis mutabilis*

Diagnosis: Compact vitellaria (4).

Diphyllobothriidae Group *sedis mutabilis*

Diagnosis: Ventral genital pores (5).

Family Haplobothriidae Meggitt 1924

Diagnosis: Glanduloplerocercoids (6); tentacles on plerocercoid (7).

Family Cephalochlamydidae Blanchard 1908

Diagnosis: Circumcortical vitellaria *(8).

Family Diphyllobothriidae Luhe 1910

Diagnosis: Circumcortical vitellaria *(8); acaudate plerocercoid *(9); single sperm axoneme *(10).

Family Echinobothriidae Perrier 1897

Diagnosis: Scolex armed with two rows of apical spines (11).

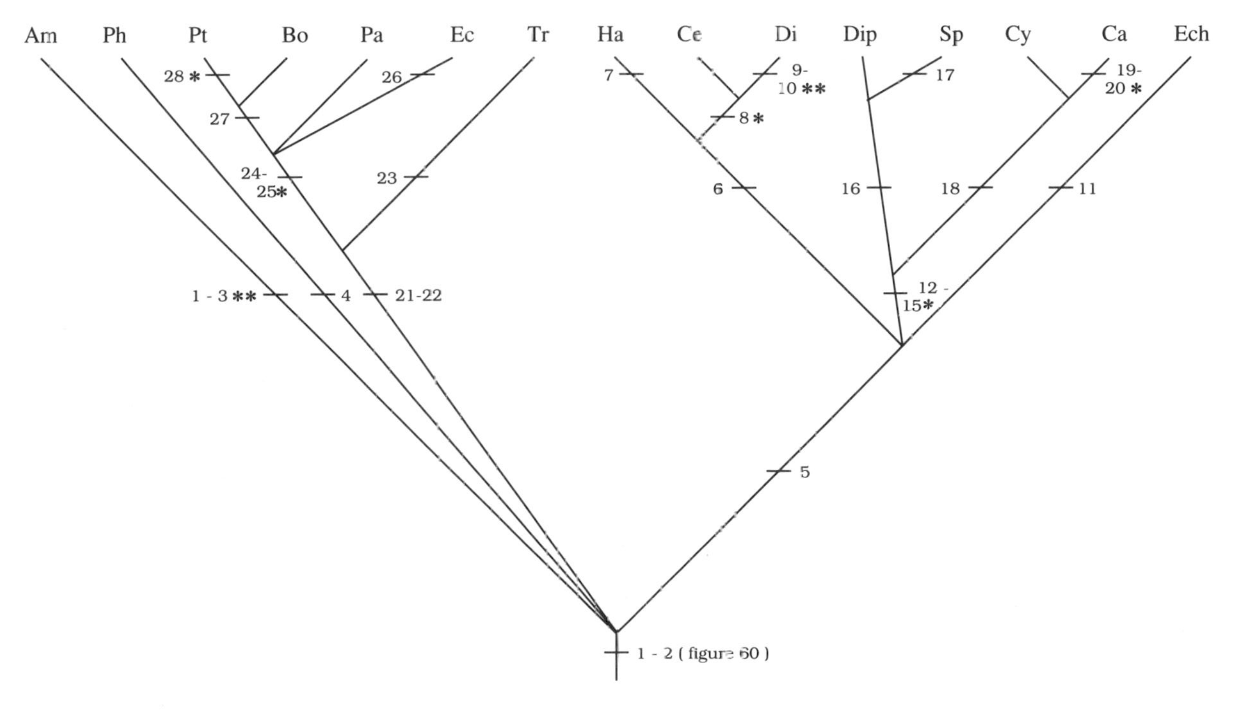

Figure 61. Phylogenetic tree for the families of the Pseudophylliformes. Am = Amphicotylidae; Ph = Philobythiidae; Pt = Ptychobothriidae; Bo = Bothriocephalidae; Pa = Parabothriocephalidae; Ec = Echinophallidae; Tr = Triaenophoridae; Ha = Haplobothriidae; Ce = Cephalochlamydidae; Di = Diphyllobothriidae; Dip = Diplocotylidae; Sp = Spathebothriidae; Cy = Cyathocephalidae; Ca = Caryophyllaeidae; Ech = Echinobothriidae.

Diplocotylidae Group

Diagnosis: Oncospheres *(12); indistinct segmentation (13); uterine pore between male and female genital pores (14); prominent uterine glands (radial glands) (15).

Family Diplocotylidae Monticelli 1892

Diagnosis: Egg filaments (16).

Family Spathebothriidae Wardle and McLeod 1952

Diagnosis: Genital pores dorsoventral (17).

Caryophyllaeidae Group

Diagnosis: Uterovaginal atrium (18).

Family Cyathocephalidae Nybelin 1922

Diagnosis: With characters of the group.

Family Caryophyllaeidae Leuckart 1878

Diagnosis: Monozoic (19); single sperm axoneme *(20).

Bothriocephalidae Group

Diagnosis: Scolex with apical disk (21); uterus with saccate distal portion (22).

Family Triaenophoridae Loennberg 1889

Diagnosis: Four hooks on scolex (23).

Family Parabothriocephalidae Yamaguti 1959 *sedis mutabilis, incertae sedis*

Diagnosis: Dorsolateral genital pores (24); circumcortical vitellaria *(25).

Family Echinophallidae Schumacher 1914 *sedis mutabilis*

Diagnosis: Dorsolateral genital pores (24); circumcortical vitellaria *(25); large spines on cirrus (26).

Family Bothriocephalidae Blanchard 1849

Diagnosis: Dorsolateral genital pores (24); circumcortical vitellaria *(25); dorsal genital pores (27).

Family Ptychobothriidae Luhe 1902

Diagnosis: Dorsolateral genital pores (24); circumcortical vitellaria *(25); dorsal genital pores (27); oncospheres *(28).

Order Tetraphylliformes Carus 1863 *sedis mutabilis* (Fig. 62)

Phyllobothriidae Braun 1900 Group I (e.g., *Calyptrobothrium* Monticelli 1893, *Crossobothrium* Linton 1889, *Anthobothrium* van Beneden 1850, *Orygmatobothrium* Diesing 1863, *Clydonobothrium* Euzet 1956, *Monorygma* Diesing 1863, *Rhodobothrium* Linton 1889, *Phyllobothrium lactuca*-group) *incertae sedis*.

Diagnosis: With characters of the order.

Phyllobothriidae Braun 1900 Group II *incertae sedis, sedis mutabilis*

Diagnosis: Medial loculi present (1); loculi present in plerocercoids (2); no sucker at apex of bothridium (3); single axoneme in sperm tails (4).

Trypanorhyncha Diesing 1863 Group *sedis mutabilis*

Diagnosis: Stiff muscularized bothridial margins (5); ciliated hexacanth larvae (6); circumcortical vitellaria (7); four tentacles and hydraulic system (8).

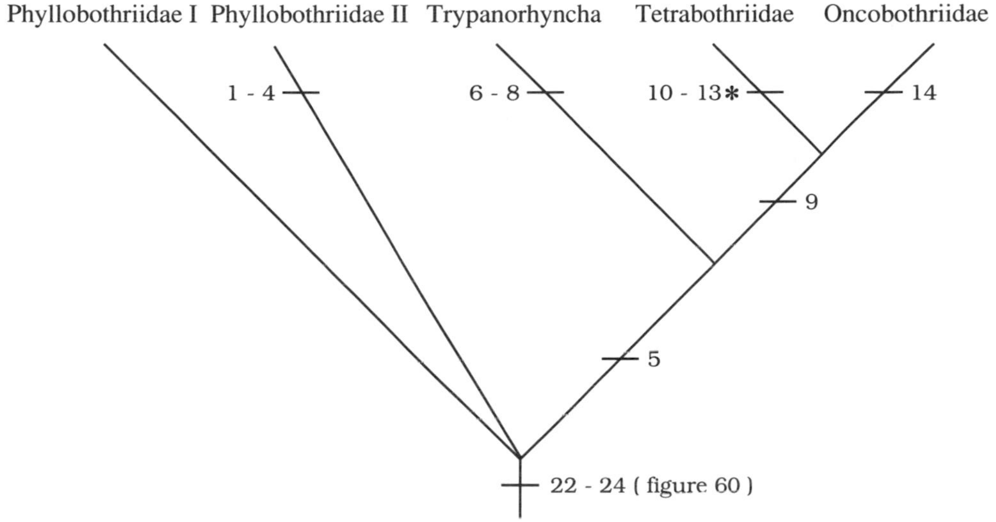

Phyllobothriidae I Phyllobothriidae II Trypanorhyncha Tetrabothriidae Oncobothriidae

1 - 4 6 - 8 10 - 13* 14

9

5

22 - 24 (figure 60)

Figure 62. Phylogenetic tree for the major groups of the Tetraphylliformes.

Family Tetrabothriidae Linton 1891

Diagnosis: Paedomorphic development of scolex (adult scolex form does not appear until definitive host is reached) (9); ovaries not X-shaped in cross section *(10); compact preovarian vitellaria (11); single uterine pore (12); "few" testes per proglottid (13).

Family Oncobothriidae Braun 1900 *sedis mutabilis*

Diagnosis: Paedomorphic development of scolex (adult scolex form does not appear until definitive host is reached) (9); hooks or cornified projections on each bothridium (14).

Order Proteocephaliformes Mola 1928 *sedis mutabilis* (Fig. 63)

Family Proteocephalidae LaRue 1911 Group I *incertae sedis, sedis mutabilis*

Diagnosis: With characters of the order.

Family Monticelliidae Mola 1929

Diagnosis: Forty to sixty testes per proglottid (1); vitellaria arranged like two parentheses on either side of the proglottid in cross section (2).

Cysticercoid Group *sedis mutabilis*

Diagnosis: Cysticercoid stage between procercoid and plerocercoid (3).

Proteocephalidae LaRue 1911 Group II *sedis mutabilis*

Diagnosis: With characters of the group.

Family Corallobothriidae Freze 1965 *sedis mutabilis*

Diagnosis: Metascolex formed by outgrowth of tissue immediately posterior to suckers (4).

Rostellate Group *sedis mutabilis*

Diagnosis: Apical sucker may be modified into a structure containing large spines, called a rostellum (5); rostellum not retractable (6); layer of peripheral microtubules in spermatozoa not interrupted at axoneme level (addition of lateral microtubules) (7).

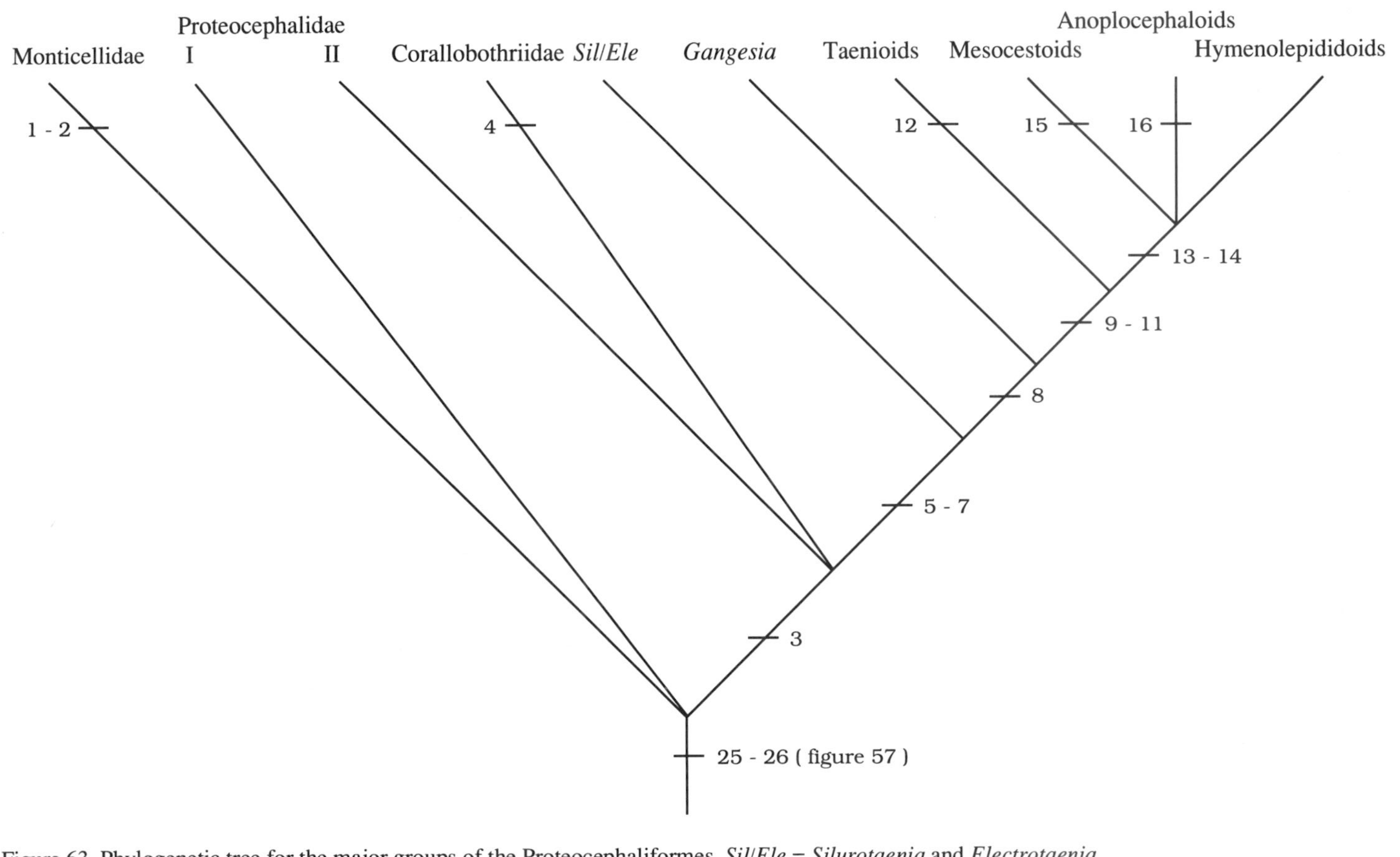

Figure 63. Phylogenetic tree for the major groups of the Proteocephaliformes. *Sil/Ele* = *Silurotaenia* and *Electrotaenia*.

***Silurotaenia* Nybelin 1942, *Electrotaenia* Nybelin 1942**

Diagnosis: With characters of the group.

Hooked Rostellate Group

Diagnosis: Rostellum armed with hooks (8).

***Gangesia* Woodland 1924**

Diagnosis: With characters of the group.

Cyclophyllidea van Beneden in Braun 1900 Group

Diagnosis: Compact, postovarian vitellarium (9); peripheral microtubules in spermatozoa twisted (10); single axoneme in sperm tail (11).

Taenioid Group

Diagnosis: Cysticerci and modifications (12).

Retractable Rostellum Group

Diagnosis: Rostellum retractable (13); uterus lacking lateral diverticula (14).

Hymenolepididoid Group *sedis mutabilis*

Diagnosis: With characters of the group.

Mesocestoid Group *sedis mutabilis*

Diagnosis: Parutcrine organs (15).

Anoplocephaloid Group *sedis mutabilis*

Diagnosis: Egg capsules (16).

Subcohort Eucestoda Southwell 1930: Infrafamilial relationships

Order Nippotaeniiformes Yamaguti 1939 (Fig. 64)

Family Nippotaeniidae Yamaguti 1939

Reference: Weekes (in press).

Diagnosis: With characters of the order.

***Nippotaenia* Yamaguti 1939**

Diagnosis: With characters of the family.

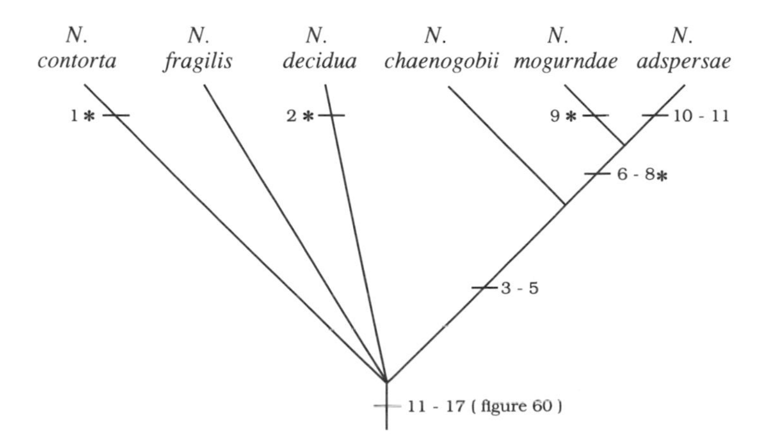

Figure 64. Phylogenetic tree for the species of *Nippotaenia*.

***fragilis* Hine 1977**

Diagnosis: With characters of the genus.

***contorta* Hine 1977**

Diagnosis: Up to 30 proglottids per strobila *(1).

***decidua* (Hine 1977) comb. n.**

Diagnosis: Hyperapolytic *(2).

***chaenogobii* Yamaguti 1939**

Diagnosis: Apical sucker surrounded by accessory muscular sphincter (3); cavity in cercomer (4); metacestode enclosed in membranous sac (5).

***mogurndae* Yamaguti and Miyata 1940**

Diagnosis: Apical sucker surrounded by accessory muscular sphincter (3); cavity in cercomer (4); metacestode enclosed in membranous sac (5); hyperapolytic *(6); 18–24 testes per proglottid (7); ovary flat to slightly U-shaped (8); up to 30 proglottids per strobila *(9).

***adspersae* Weekes in press**

Diagnosis: Apical sucker surrounded by accessory muscular sphincter (3); cavity in cercomer (4); metacestode enclosed in membranous sac (5); hyperapolytic *(6); 5–12 testes per proglottid (10); ovary U-shaped (11).

Order Lecanicephaliformes Baylis 1920 (Fig. 65)

Family Lecanicephalidae Braun 1900

Subfamily Lecanicephalinae Braun 1900 *incertae sedis*

Diagnosis: With characters of the family.

Discobothrium* van Beneden 1871, *Calycobothrium* Southwell 1911, *Echeneibothrium* van Beneden 1850, *Lecanicephalum* Linton 1890, *Hexacanalis* Perrenoud 1931 *sedis mutabilis, incertae sedis

Diagnosis: With characters of the subfamily.

***Polypocephalus* Braun 1878**

Diagnosis: Myzorhynchus comprising unarmed fleshy tentacles (1).

Subfamily Disculicipinae Joyeux and Baer 1935

Diagnosis: Nonretractable myzorhynchus (2).

***Staurobothrium* Shipley and Hornell 1905, *Tetragonocephalum* Shipley and Hornell 1905**

Diagnosis: Pedunculate suckers (3).

***Disculiceps* Joyeux and Baer 1935, *Adelobothrium* Shipley 1900**

Diagnosis: Circumcortical vitellaria (4); metascolex in the form of a collar (5).

***Prosobothrium* Cohn 1902**

Diagnosis: Circumcortical vitellaria (4); no myzorhynchus (6); large tegumental spines on neck (7).

***Cathetocephalus* Dailey and Overstreet 1973**

Diagnosis: Circumcortical vitellaria (4); no suckers (8); transversely elongate scolex (9).

Order Tetraphylliformes Carus 1863

Phyllobothriidae Group II *incertae sedis, sedis mutabilis* (Fig. 66)

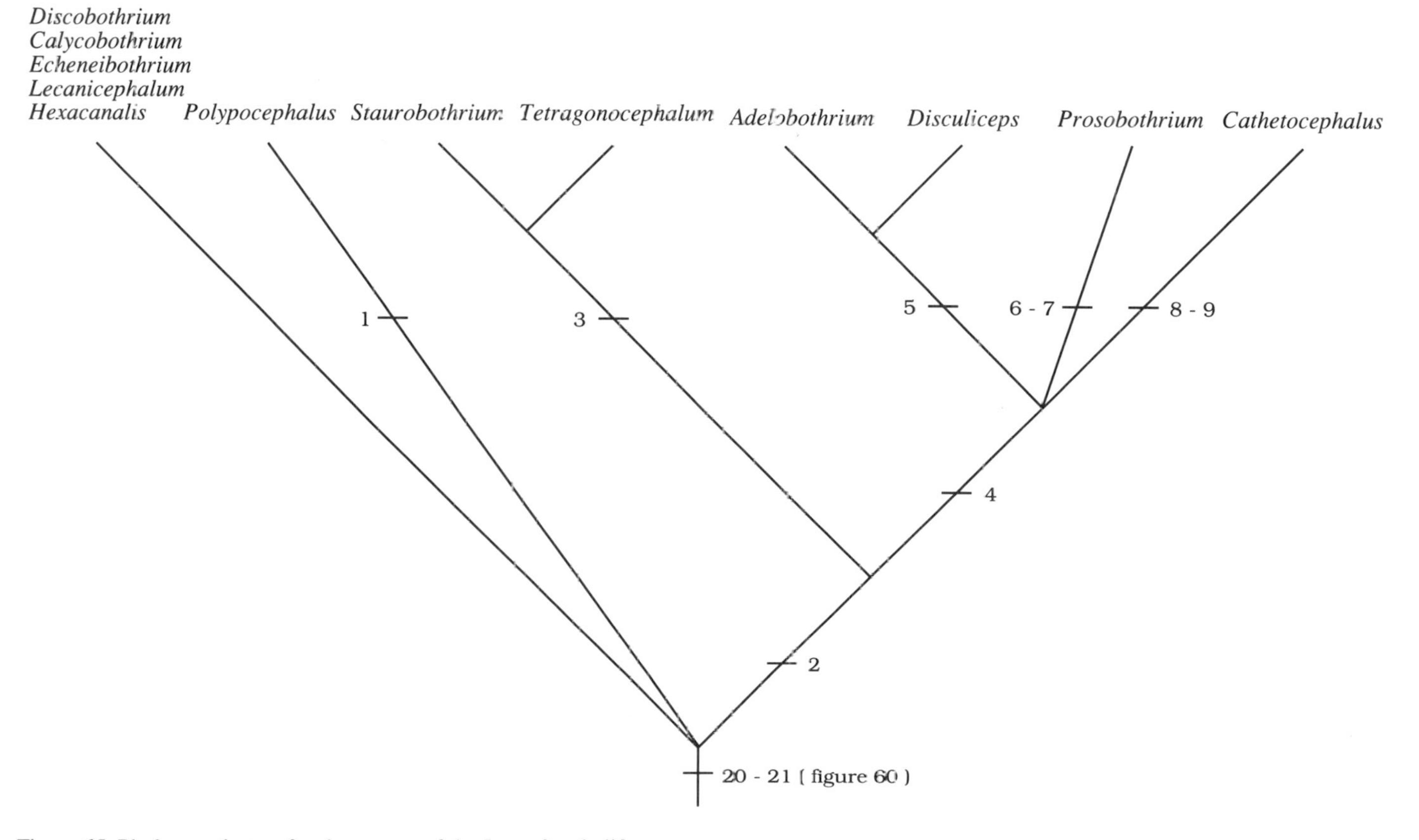

Figure 65. Phylogenetic tree for the genera of the Lecanicephaliformes.

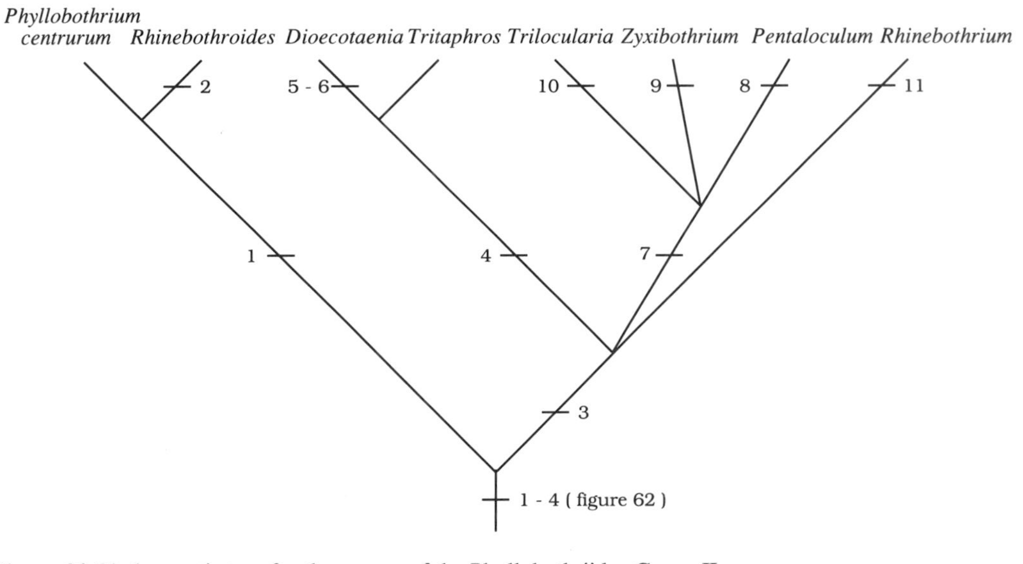

Figure 66. Phylogenetic tree for the genera of the Phyllobothriidae Group II.

Phyllobothrium centrurum* Group *incertae sedis

Diagnosis: With marginal loculi (1).

***Rhinebothroides* Mayes, Brooks, and Thorson 1981**

Diagnosis: With marginal loculi (1); large external seminal vesicle (2).

Family Dioecotaeniidae Schmidt 1969 *sedis mutabilis*

Diagnosis: Muscularized bothridial margins *(3) horizontal architecture of bothridial loculi (4).

***Tritaphros* Lonnberg 1889**

Diagnosis: Three loculi per bothridium.

***Dioecotaenia* Schmidt 1969**

Diagnosis: Multiple hexagonally shaped loculi (5); dioecious (6).

Family Triloculariidae Yamaguti 1959 *sedis mutabilis*

Diagnosis: Muscularized bothridial margins *(3); loculi irregularly arranged (7).

***Trilocularia* Olsson 1867**

Diagnosis: Three loculi per bothridium (10).

***Zyxibothrium* Hayden and Campbell 1981**

Diagnosis: Four loculi per bothridium (9).

***Pentaloculum* Alexander 1953**

Diagnosis: Five loculi per bothridium (8).

Rhinebothrium* Group *sedis mutabilis

Diagnosis: Muscularized bothridial margins *(3); bothridial loculi extending longitudinally along narrow bothridia (11).

***Rhinebothroides* Mayes, Brooks, and Thorson 1981 (Fig. 67)**

References: Brooks et al. (1981a); Brooks and Amato (1992).

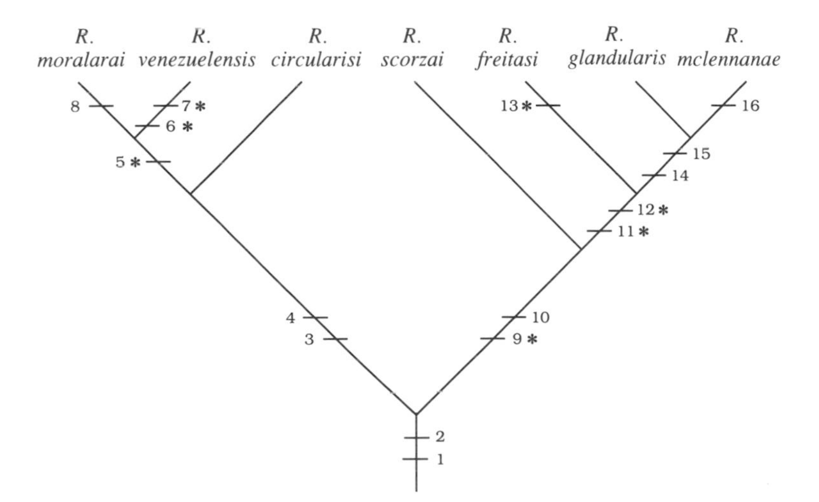

Figure 67. Phylogenetic tree for the species of *Rhinebothroides*.

Diagnosis: External seminal vesicle joining cirrus sac near poral end (1); elongate aporal ovarian lobes (2).

***circularisi* Mayes, Brooks, and Thorson 1981**

Diagnosis: Vitellaria on poral side interrupted in region of genital pore (3); poral ovarian lobe extending anteriorly only to posterior end of cirrus sac (4).

***venezuelensis* Brooks, Mayes, and Thorson 1981**

Diagnosis: Vitellaria on poral side interrupted in region of genital pore (3); poral ovarian lobe extending anteriorly only to posterior end of cirrus sac (4); 51–53 medial loculi per bothridium *(5); an average of 53 testes per proglottid *(6); proglottids craspedote *(7).

***moralarai* (Brooks and Thorson 1976) Brooks, Mayes, and Thorson 1981**

Diagnosis: Vitellaria on poral side interrupted in region of genital pore (3); poral ovarian lobe extending anteriorly only to posterior end of cirrus sac (4); an average of 53 testes per proglottid *(6); 45–49 medial loculi per proglottid (8).

***scorzai* (Lopez-Neyra and Diaz-Ungria 1958) Brooks, Mayes, and Thorson 1981**

Diagnosis: Vagina coiled (9); proglottids craspedote *(10).

***freitasi* Rego 1979**

Diagnosis: Vagina coiled (9); 49–59 medial loculi per bothridium *(11); an average of 55 testes per proglottid *(12); proglottids acraspedote *(13).

***glandularis* Brooks, Mayes, and Thorson 1981**

Diagnosis: Vagina coiled (9); 49–59 medial loculi per bothridium *(11); an average of 45 testes per proglottid (14); darkly staining glandular cells lying free in the parenchyma surrounding the terminal genitalia (15).

***mclennanae* Brooks and Amato 1992**

Diagnosis: Vagina coiled (9); 49–59 medial loculi per bothridium *(11); darkly staining glandular cells lying free in the parenchyma surrounding the terminal genitalia (15); an average of 31 testes per proglottid (16).

***Rhinebothrium* Species Group I (species with medial bothridial septa and medial bothridial constrictions) (Fig. 68)**

Reference: Brooks and Deardorff (1988).

Diagnosis: Preporal testes (1); medial bothridial septa (2); medial bothridial constrictions (3); 17–22 testes per proglottid (4); 42–56 loculi per bothridium (5).

***flexile* Linton 1890**

Diagnosis: V-shaped ovaries (6).

***burgeri* Baer 1948**

Diagnosis: V-shaped ovaries (6); muscular genital pore (7); 30–43 testes per proglottid (8).

***devaneyi* Brooks and Deardorff 1988**

Diagnosis: V-shaped ovaries (6); muscular genital pore (7); 30–43 testes per proglottid (8); 96–154 loculi per bothridium (9).

***himanturi* Williams 1964**

Diagnosis: Bothridial pedicels half bothridial length (10).

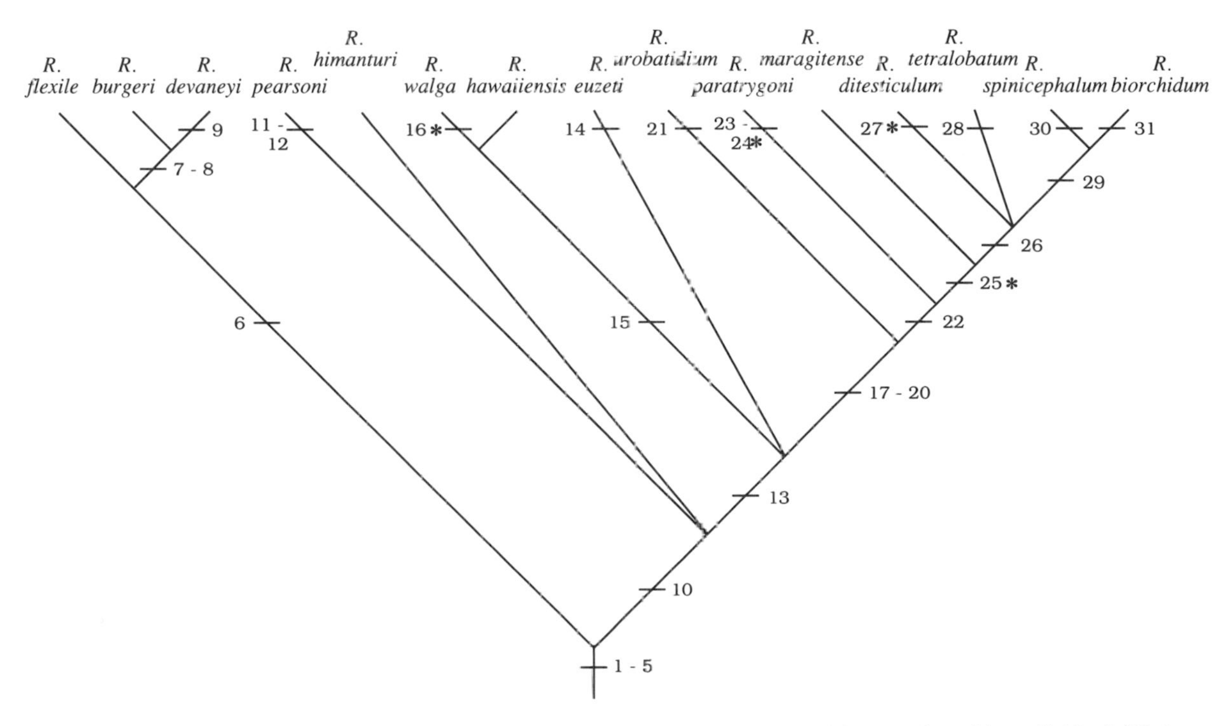

Figure 68. Phylogenetic tree for some species of *Rhinebothrium* species group containing species with medial bothridial septa and medial bothridial constrictions.

***pearsoni* Butler 1987**

Diagnosis: Bothridial pedicels half bothridial length (10); 20–33 testes per proglottid (11); 34–38 loculi per bothridium (12).

***euzeti* Williams 1958**

Diagnosis: Bothridial pedicels half bothridial length (10); 11–13 testes per proglottid (13); 78 loculi per bothridium (14).

***hawaiiensis* Cornford 1974**

Diagnosis: Bothridial pedicels half bothridial length (10); 11–13 testes per proglottid (13); compact vitelline bands (15).

***walga* (Shipley and Hornell 1905) Euzet 1958**

Diagnosis: Bothridial pedicels half bothridial length (10); compact vitelline bands (15); 4–6 testes per proglottid *(16).

***urobatidium* (Young 1955) Appy and Dailey 1978**

Diagnosis: Bothridial pedicels half bothridial length (10); 6–12 testes per proglottid (17); craspedote proglottids (18); more than 50 proglottids per strobila (19); square proglottids (20); 38–42 loculi per bothridium (21).

***paratrygoni* Rego and Dias 1976**

Diagnosis: Bothridial pedicels half bothridial length (10); craspedote proglottids (18); more than 50 proglottids per strobila (19); square proglottids (20); four to eight testes per proglottid (22); 72–76 loculi per bothridium (23); long cephalic peduncle *(24).

***margaritense* Mayes, and Brooks 1981**

Diagnosis: Bothridial pedicels half bothridial length (10); craspedote proglottids (18); more than 50 proglottids per strobila (19); square proglottids (20); three to six testes per proglottid *(25).

***ditesticulum* Appy and Dailey 1978**

Diagnosis: Bothridial pedicels half bothridial length (10); craspedote proglottids (18); more than 50 proglottids per strobila (19); square proglottids (20); two testes per proglottid (26); long cephalic peduncle *(27).

***tetralobatum* Brooks 1977**

Diagnosis: Bothridial pedicels half bothridial length (10); craspedote proglottids (18); more than 50 proglottids per strobila (19); square proglottids (20); two testes per proglottid (26); fragmented ovary (28).

***spinicephalum* Campbell 1970**

Diagnosis: Bothridial pedicels half bothridial length (10); craspedote proglottids (18); more than 50 proglottids per strobila (19); square proglottids (20); two testes per proglottid (26); 32–34 loculi per bothridium (29); spinose scolex (30).

***biorchidum* Huber and Schmidt 1985**

Diagnosis: Bothridial pedicels half bothridial length (10); craspedote proglottids (18); more than 50 proglottids per strobila (19); square proglottids (20); two testes per proglottid (26); 22–30 loculi per bothridium (31).

Family Tetrabothriidae Linton 1901 (Fig. 69)

Reference: Hoberg (1989).

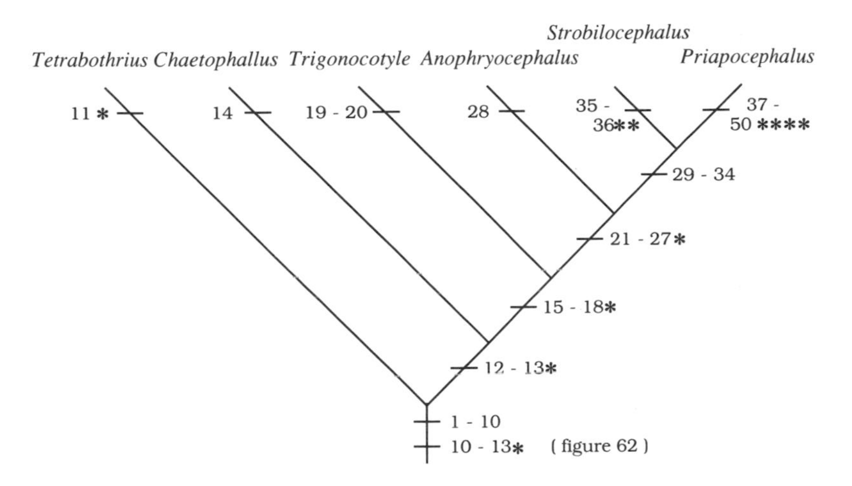

Figure 69. Phylogenetic tree for the genera of the Tetrabothriidae.

Diagnosis: Cirrus sac cylindrical (1); testes surrounding the ovary (2); scolex rectangular and flat (3); bothridia rectangular (4), ovary preequatorial (5); bothridial depth shallow (6); genital atrium with extensive muscular modification (7); fewer than 100 testes per proglottid (8); testes in dorsal fields (9); testes confined to space between osmoregulatory canals (10).

***Tetrabothrius* Rudolphi 1819**

Diagnosis: Ovoid cirrus sac *(11).

***Chaetophallus* Nybelin 1916**

Diagnosis: Scolex round and flat *(12); bothridia round (13); postovarian testes (14).

***Trigonocotyle* Baer 1932**

Diagnosis: Ovary postequatorial *(15); neck long (16); bothridia round (17); bothridial musculature moderate (18); scolex rectangular and cuboidal (19); three separate auricular appendages on margins of bothridia (20).

***Anophryocephalus* Baylis 1922**

Diagnosis: Ovary postequatorial *(15); neck long (16); bothridia round (17); bothridial musculature moderate (18); scolex round and flat *(21); genital pore ventrolateral (22); dorsal component of genital atrium reduced, ventral aspect with deep concavity (23); osmoregulatory canals atrophied (25); moderate apical development (26); pair of auricular structures directed laterally and medially on the anterior margin of each bothridium (27); uterus contained between osmoregulatory canals (28).

***Strobilocephalus* Baer 1932**

Diagnosis: Ovary postequatorial *(15); genital pore ventrolateral (22); dorsal component of genital atrium reduced, ventral aspect with deep concavity (23); male canal lacking (24); bothridia triangular (29); bothridia deep (30); bothridia highly muscularized (31); scolex globular (32); great apical development (33); single auricular appendage directed laterally from each bothridium (34); cirrus ovoid *(35); osmoregulatory system not atrophied *(36).

***Priapocephalus* Nybelin 1922**

Diagnosis: Genital pore ventrolateral (22); osmoregulatory canals atrophied (25); scolex globular (32); great apical development (33); ovary equatorial *(37); no auricles (38); genital atrium weakly developed with vestigial ventral concavity (39); testes lateral to ovary (40); testes extending beyond osmoregulatory canals (41); testes in dorsal and ventral fields (42); genital ducts between osmoregulatory canals (43); multiple uterine pores *(44); more than 100 testes *(45); vitelline gland follicular *(46); genital ducts ventral (47); osmoregulatory canals in scolex subtegumental and reticulate (48); scolex embedded in host mucosa (49); bothridia lacking (50).

***Anophryocephalus* Baylis 1922 (Fig. 70)**

Reference: Hoberg and Adams (1992).

***anophrys* Baylis 1922**

Diagnosis: Bothridial operculum longitudinal slitlike aperture not extending beyond bothridial musculature (1); cirrus sac thin-walled, weakly muscular *(2); atrial region aspinose *(3); no transverse ventral osmoregulatory canals (4); no dorsal osmoregulatory canals (5); lateral genital pore *(6).

***skrjabini* (Krotov and Deliamure 1955) Hoberg, Adams, and Rausch 1991**

Diagnosis: Genital papilla small (7); genital atrium pad ellipsoidal, massive (8); more than 34 testes per proglottid (9); neck extremely long (more than 16 mm) (10); lateral and medial auricles confluent along anterior margin of bothridia *(11).

***nunivakensis* Hoberg, Adams, and Rausch 1991**

Diagnosis: Genital papilla small (7); genital atrium pad ellipsoidal, massive (8); more than 34 testes per proglottid (9); bothridial operculum with longitudinal or diagonal aperture opening anterior to bothridia (12); bothridia contained in parenchymal envelope (13); cirrus sac ovoid (14); genital papilla directed slightly ventral (15); male canal decurved ventrally (16); atrial region aspinose *(17).

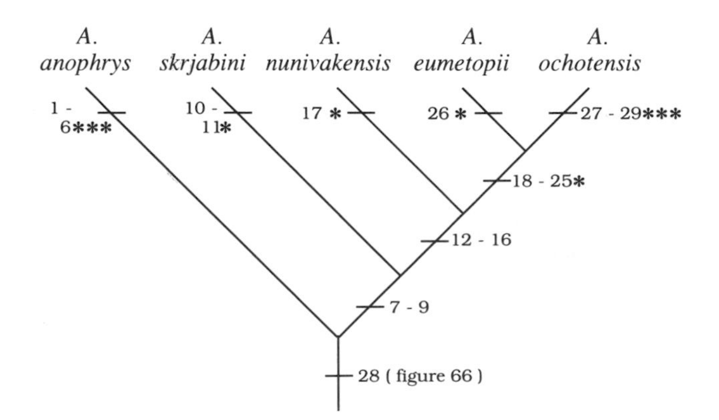

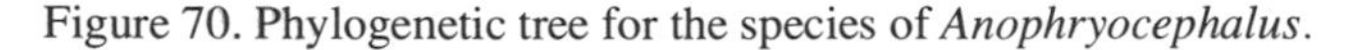

Figure 70. Phylogenetic tree for the species of *Anophryocephalus*.

***eumetopii* Hoberg, Adams, and Rausch 1991**

Diagnosis: More than 34 testes per proglottid (9); bothridial operculum with longitudinal or diagonal aperture opening anterior to bothridia (12); bothridia contained in parenchymal envelope (13); cirrus sac ovoid (14); genital papilla directed slightly ventral (15); male canal decurved ventrally (16); apical region hypertrophied (18); ventral osmoregulatory canals not hypertrophied (19); genital pore a protrusible suckerlike structure (20); genital papilla large (21); genital papilla strongly recurved ventrally (22); genital atrium pad ellipsoidal, small (23); testes completely overlapping ovary (24); neck short (2 mm) *(25); cirrus sac thin-walled, weakly muscular *(26).

***ochotensis* Deliamure and Krotov 1955**

Diagnosis: Bothridia contained in parenchymal envelope (10); genital papilla directed slightly ventral (12); cirrus sac ovoid (14); male canal decurved ventrally (16); apical region hypertrophied (18); ventral osmoregulatory canals not hypertrophied (19); genital pore a protrusible suckerlike structure (20); genital papilla large (21); genital papilla strongly recurved ventrally (22); genital atrium pad ellipsoidal, small (23); testes completely overlapping ovary (24); neck short (2 mm) *(25); no bothridial operculum *(27); 20–30 testes per proglottid *(28); lateral and medial auricles confluent along anterior margin of bothridia *(29).

Family Oncobothriidae Braun 1900

***Potamotrygonocestus* Brooks and Thorson 1976 (Fig. 71)**

Reference: Brooks et al. (1981a).

Diagnosis: Bothridial hooks more than 60 but less than 100 μm long (1); bothridial hooks join bases near middle (2); ovary shaped like an inverted A (3).

***amazonensis* Mayes, Brooks, and Thorson 1981**

Diagnosis: As above.

***orinocoensis* Brooks, Mayes, and Thorson 1981**

Diagnosis: Compact vitellaria (4); bothridial hooks up to 125 μm long (5); bothridial hook prongs join bases at one end (6); ovary Θ-shaped (7).

***magdalenensis* Brooks and Thorson 1976**

Diagnosis: Compact vitellaria (4); bothridial hooks never more than 60 μm long (8); genital pore posterolateral (9).

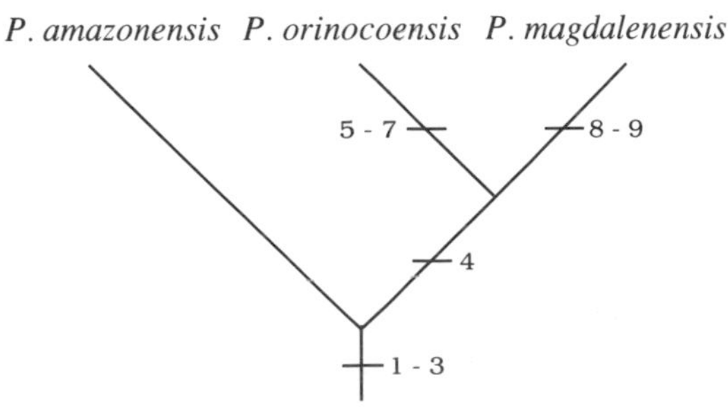

Figure 71. Phylogenetic tree for the species of *Potamotrygonocestus*.

Order Proteocephaliformes Mola 1928

Family Monticelliidae Mola 1929 (Fig. 72a,b)

References: Brooks and Rasmussen (1984); Brooks (in press).

Myzophorus* Woodland 1934 *incertae sedis

Diagnosis: With characters of the family.

***pirarara* Woodland 1934**

Diagnosis: With characters of the genus.

Nomimoscolex* Woodland 1934 *incertae sedis

Diagnosis: Ovary follicular (1); apical sucker/organ lacking (2); an average of 100–150 testes per proglottid (3).

***matogrossensis* Rego and Pavanelli 1990**

Diagnosis: Vagina posterior to cirrus sac only (4).

***piraeeba* Woodland 1934**

Diagnosis: Strongly developed internal muscle layer (5).

***mandube* (Woodland 1935) Brooks in press**

Diagnosis: Strongly developed internal muscle layer (5); vagina anterior to cirrus sac (6); trilobulate suckers (7).

Vaucheriella* Chambrier 1987 *sedis mutabilis

Diagnosis: Ovary follicular (1); no apical organ (2); vitellaria completely cortical (8); vitellaria not extending anterior to cirrus sac (9); vitellaria in two ventral bands on either side of proglottid *(10); 25–44 testes per proglottid (11).

***bicheti* Chambrier 1987**

Diagnosis: With characters of the genus.

Subfamily Rudolphiellinae Woodland 1935

Diagnosis: Ovary follicular (1); no apical organ (2); an average of 100–150 testes per proglottid (3); vitellaria completely cortical (8); type I metascolex (formed by growth of tissue around each sucker) (12); vitellaria in four quadrants, one dorsal and one ventral on each side of proglottid (13).

***Rudolphiella* Fuhrmann 1916**

Diagnosis: Ovary partly cortical *(14).

***lenha* (Woodland 1933) Brooks in press**

Diagnosis: With characters of the genus.

***microcephalus* (Diesing 1850) Brooks in press**

Diagnosis: Vitellaria in two ventral bands on either side of proglottid *(15); an average of 150–200 testes per proglottid *(16).

***lobosa* (Riggenbach 1896) Fuhrmann 1916**

Diagnosis: Vitellaria in two ventral bands on either side of proglottid *(15); an average of 150–200 testes per proglottid *(16); elongate eggs (17).

***myoides* (Woodland 1934) Woodland 1935**

Diagnosis: Vitellaria in two ventral bands on either side of proglottid *(15); elongate eggs (17); an average of 100 testes per proglottid *(18); eggs with long, thick filaments (19).

***piranubu* (Woodland 1934) Woodland 1935**

Diagnosis: Vitellaria in two ventral bands on either side of proglottid *(15); elongate eggs (17); an average of 100 testes per proglottid *(18); eggs with short, thin filaments (20).

***Amphoteromorphus* Diesing 1850**

Diagnosis: Uterus thick-walled, tubular (lateral diverticula reduced or lacking) (21).

***praeputialis* Rego, Santos, and Silva 1974**

Diagnosis: With characters of the genus.

***parkarmoo* Woodland 1935**

Diagnosis: Bilobulate suckers *(22); spinose scolex (23); 40–60 testes per proglottid *(24).

***peniculus* Diesing 1850**

Diagnosis: Bilobulate suckers *(22); spinose scolex (23); 40–60 testes per proglottid *(24); vitelline bands arranged like two parentheses on either side of proglottid *(25).

***piraeeba* Woodland 1934**

Diagnosis: Bilobulate suckers *(22); spinose scolex (23); 150–200 testes per proglottid *(26).

***megacephalum* (Diesing 1850) Brooks in press**

Diagnosis: Bilobulate suckers *(22); spinose scolex (23); 150–200 testes per proglottid *(26); tissue growth around suckers so pronounced that only an anterior and posterior opening remain (27).

Subfamily Monticelliinae Mola 1929

Diagnosis: Ovary follicular (1); no apical organ (2); an average of 100–150 testes per proglottid (3); vitellaria completely cortical (8); ovary at least partly cortical *(28).

Paramonticellia* Pavanelli and Rego 1991 *incertae sedis

Diagnosis: With characters of the subfamily.

piracatinga* (Woodland 1935) Brooks in press, *magna* (Rego, Santos, and Silva 1974) Brooks in press *sedis mutabilis

Diagnosis: With characters of the genus.

lenha* (Woodland 1933) Brooks in press, *sudobim* (Woodland 1935) Brooks in press, *admonticellia* (Woodland 1934) Brooks in press *sedis mutabilis

Diagnosis: An average of 200–250 testes per proglottid (29).

***lopesi* (Rego 1989) Brooks in press**

Diagnosis: An average of 300 testes per proglottid (30); vagina anterior to cirrus sac only *(31); about 25–35 uterine diverticula (32).

***itapipuensis* Pavanelli and Rego 1991**

Diagnosis: An average of 60 testes per proglottid *(33); about 14–20 uterine diverticula (34).

***Houssayela* Rego 1987**

Diagnosis: Two papillae on margins of each sucker (35).

***sudobim* (Woodland 1935) Rego 1987**

Diagnosis: Vagina anterior to cirrus sac only *(36); four papillae on margins of each sucker (37).

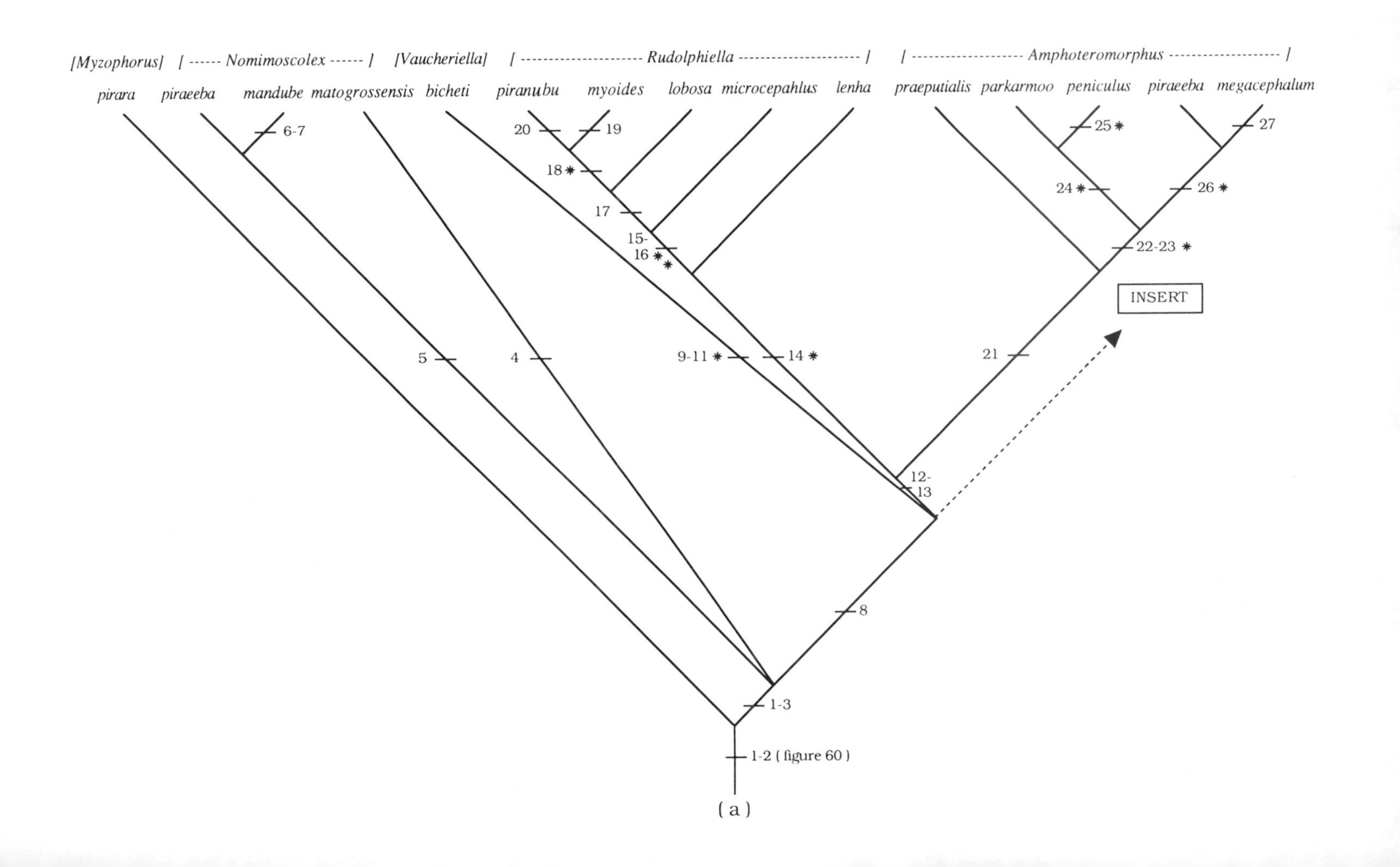
[Myzophorus]
[------ Nomimoscolex ------]
[Vaucheriella]
[-------------------- Rudolphiella --------------------]
[------------------ Amphoteromorphus ------------------]
pirara
piraeeba
mandube
matogrossensis
bicheti
piranubu
myoides
lobosa
microcepahlus
lenha
praeputialis
parkarmoo
peniculus
piraeeba
megacephalum
6-7
20
19
18 ✷
17
15-
16 ✷✷
25 ✷
27
24 ✷
26 ✷
22-23 ✷
INSERT
5
4
9-11 ✷
14 ✷
21
12-
13
8
1-3
1-2 (figure 60)
(a)

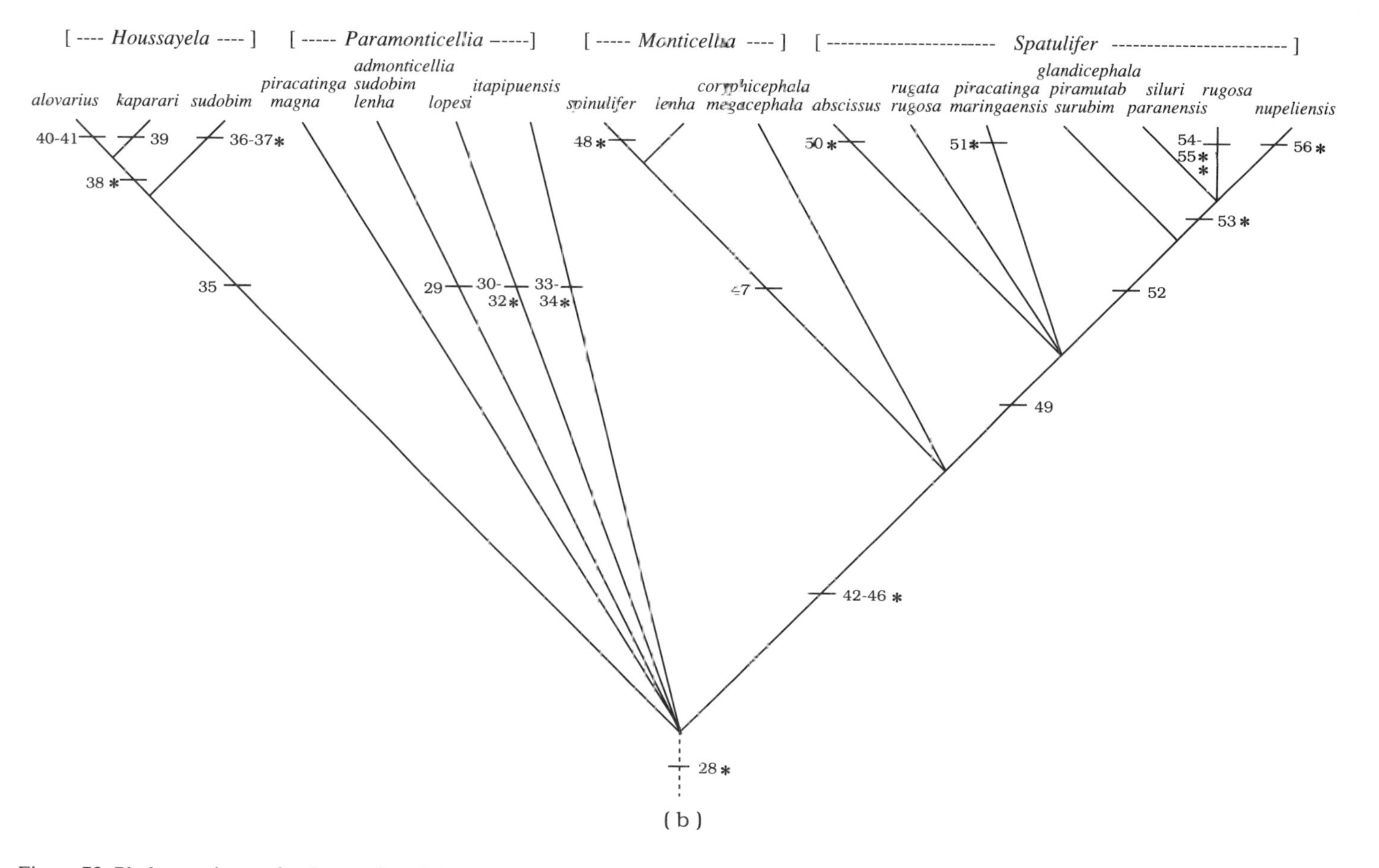

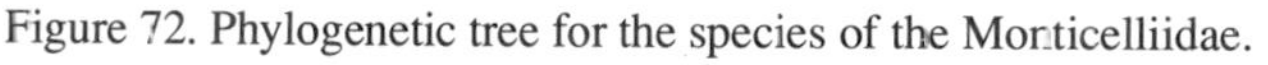

Figure 72. Phylogenetic tree for the species of the Monticelliidae.

***kaparari* (Woodland 1935) Brooks in press**

Diagnosis: An average of 100 testes per proglottid *(38); vagina posterior to cirrus sac only (39).

***alovarius* (Brooks and Deardorff 1980) Brooks in press**

Diagnosis: Forty to sixty testes per proglottid *(38); alate ovarian lobes (40); papillae on top of each sucker (41).

Monticellia* LaRue 1911 *incertae sedis

Diagnosis: Ovary completely cortical (42); testes cortical (43); uterus cortical (44); vitelline bands converging toward ventral midline of proglottid (45); an average of 150–200 testes per proglottid *(46).

***coryphicephala* (Monticelli 1891) LaRue 1911, *megacephala* Woodland 1934**

Diagnosis: With characters of the genus.

***lenha* Woodland 1933**

Diagnosis: Spinose suckers (47).

***spinulifer* Woodland 1934**

Diagnosis: Spinose suckers (47); 60–80 testes per proglottid *(48).

***Spatulifer* Woodland 1934**

Diagnosis: Ovary completely cortical (42); testes cortical (43); uterus cortical (44); vitelline bands converging toward ventral midline of proglottid (45); an average of 150–200 testes per proglottid *(46); type II metascolex (formed by expansion of neck posterior to scolex) (49).

***rugata* (Rego 1976) Brooks in press, *rugosa* (Woodland 1935) Brooks in press**

Diagnosis: With characters of the genus.

***abscissus* (Riggenbach 1896) Brooks in press**

Diagnosis: One hundred testes per proglottid *(50).

***piracatinga* (Woodland 1935) Brooks in press, *maringaensis* Pavanelli and Rego 1989**

Diagnosis: Forty to sixty testes per proglottid *(51).

***surubim* (Woodland 1933) Woodland 1934, *piramutab* (Woodland 1933) Brooks in press, *glandicephala* (Rego and Pavanelli 1985) Brooks in press**

Diagnosis: Two hundred fifty to four hundred testes per proglottid (52).

***siluri* (Fuhrmann 1916) Brooks in press, *paranensis* (Pavanelli and Rego 1989) Brooks in press**

Diagnosis: Two hundred fifty to four hundred testes per proglottid (52); bilobulate suckers *(53); egg filaments or not.

***rugosa* (Diesing 1850) Brooks in press**

Diagnosis: Two hundred fifty to four hundred testes per proglottid (52); bilobulate suckers *(53); ovary medullary *(54); uterus partly medullary *(55).

***nupeliensis* (Pavanelli and Rego 1991) Brooks in press**

Diagnosis: Bilobulate suckers *(53); 50–70 testes per proglottid *(56).

Anoplocephaloid Group *sedis mutabilis*

Family Dilepididae Railliet and Henry 1909

Alcataenia **Spasskaya 1971 (Fig. 73)**

Reference: Hoberg (1986).

Diagnosis: Genital atrium papillalike or muscular (1); uterine reticulations developed by posteriad extension (2).

larina **(Krabbe 1869) Hoberg 1984**

Diagnosis: Genital pores both between and dorsal to poral osmoregulatory canals (3).

fraterculae **Hoberg 1984**

Diagnosis: Genital ducts dorsal to osmoregulatory canals only (4).

cerorhincae **Hoberg 1984** ***sedis mutabilis***

Diagnosis: Genital ducts dorsal to osmoregulatory canals only (4); vagina thick-walled (5).

atlantiensis **Hoberg 1991** ***sedis mutabilis***

Diagnosis: Genital ducts dorsal to osmoregulatory canals only (4); vagina thick-walled (5); Mehlis's gland and vitellarium in poral portion of proglottid *(6).

pygamaeus **Hoberg 1984**

Diagnosis: Genital ducts dorsal to osmoregulatory canals only (4); vagina thick-walled (5); rostellar hooks about 40 μm in length (7); embedded in host mucosa (8); genital atrium muscular (9); cirrus sac thick-walled *(10); sphincter on seminal receptacle (11).

armillaris **(Rudolphi 1810) Spasskaia 1971**

Diagnosis: Genital ducts dorsal to osmoregulatory canals only (4); vagina thick-walled (5); rostellar hooks about 40 μm in length (7); embedded in host mucosa (8); rostellar hooks in two irregularly alternating rows (12); ovary initially reticulate (13); neck intermediate (about 800 μm) (14); proglottids longer than wide (15).

longicervica **Hoberg 1984**

Diagnosis: Genital ducts dorsal to osmoregulatory canals only (4); vagina thick-walled (5); rostellar hooks about 40 μm in length (7); embedded in host mucosa (8); rostellar hooks in two irregularly alternating rows (12); ovary initially reticulate (13); cirrus sac short, attaining but not crossing osmoregulatory canals (16); neck long (more than 1000 μm) (17); testes lateral to female organs in antiporal portion of proglottid *(18); Mehlis's gland and vitellarium in poral portion of proglottid *(19).

meinertzhageni **(Baer 1956) Spasskaia 1971**

Diagnosis: Vagina thick-walled (5); embedded in host mucosa (8); ovary initially reticulate (13); cirrus sac short, attaining but not crossing osmoregulatory canals (16); rostellar hooks about 30 μm in length (20); cirrus sac thick-walled *(21); genital ducts between osmoregulatory canals only *(22); rostellar hooks in two rows alternating in a mosaic pattern (23); testes lateral to female organs in antiporal portion of proglottid *(24); Mehlis's gland and vitellarium in poral portion of proglottid *(25).

campylacantha **(Krabbe 1869) Spasskaia 1971**

Diagnosis: Vagina thick-walled (5); rostellar hooks in two irregularly alternating rows (12); ovary initially reticulate (13); cirrus sac short, attaining but not crossing osmoregulatory canals (16); rostellar hooks about 30 μm in length (20); cirrus sac

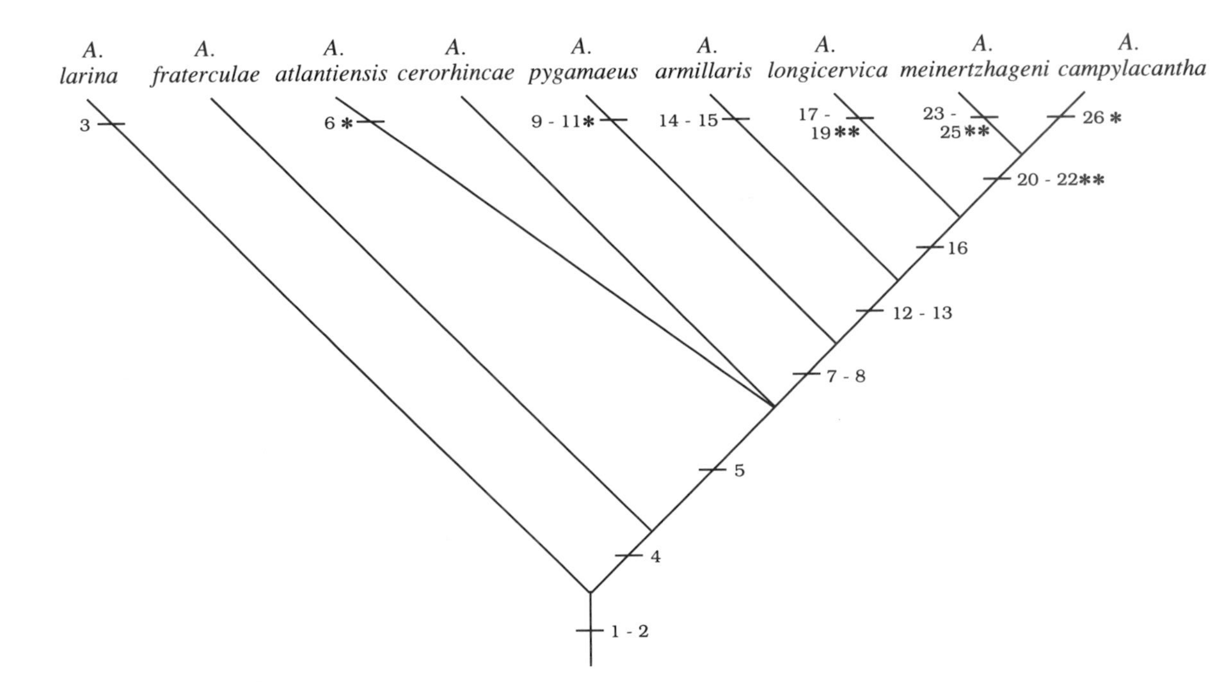

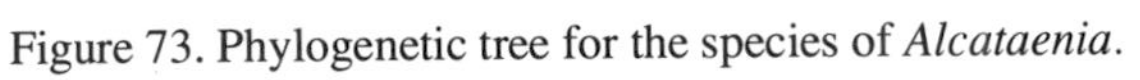
Figure 73. Phylogenetic tree for the species of *Alcataenia*.

thick-walled *(21); genital ducts between osmoregulatory canals only *(22); on surface of host mucosa *(26).

Phylum Nemata

Major Groups of Parasitic Nematodes and Their Relatives (Fig. 74)

References: Gadea (1973); Andrassy (1976); Adamson (1987).

Diagnosis: Ventral excretory pore present (1); unpaired preanal supplements in males (2); spicules present in males (3); amphids ventrally coiled (4); paired testes (5); 6 + 6 + 4 sensilla at anterior end (6); vulva present (7); postembryonic development goes through four stages (8); caudal glands present (9).

Class Enoplimorpha Gadea 1973 (also Penetrantia of Andrassy 1976; Enoplea of Inglis 1983 and Adamson 1987)

Diagnosis: Amphids pocketlike (10); five or more esophageal glands present (11).

Subclass Trefusiia *sensu* Adamson 1987 *sedis mutabilis, incertae sedis*

Diagnosis: As above.

Subclass Enoplia *sensu* Adamson 1987 *sedis mutabilis*

Diagnosis: Metanemes present (12).

Subclass Dorylaimia *sensu* Adamson 1987 *sedis mutabilis*

Diagnosis: Openings to all esophageal glands posterior to nerve ring (13); stylet present (14); bipartite esophagus (15).

Order Dorylaimida *sensu* Adamson 1987 *incertae sedis*

Diagnosis: As above.

Order Muspiceida *sensu* Adamson 1987

Diagnosis: Stichosome present (16); trophosome present (17); parasitic adult a protandrous hermaphrodite (18).

Order Mermithida *sensu* Adamson 1987

Diagnosis: Stichosome present (16); trophosome present (17); larval stages parasitic (19); adult or postparasite larva actively penetrating out of host to deposit eggs (20).

Order Trichurida Yamaguti 1961 *incertae sedis*

Diagnosis: Stichosome present (16); single or no spicules (21); eggs with bipolar plugs (22).

Order Dioctophymatida Yamaguti 1961

Diagnosis: Stichosome present (16); single or no spicules (21); eggs with bipolar plugs (22); caudal extremity of male modified to form a sucker (23); monodelphic (24).

Class Chromadorimorpha Gadea 1973 (also Rhabditea of Adamson 1987)

Diagnosis: Openings to subventral esophageal glands posterior, at or near base of esophageal corpus, dorsal esophageal gland opening at or near buccal cavity (25).

Subclass Leptolaimia *sensu* Adamson 1987 *sedis mutabilis, incertae sedis*

Diagnosis: As above.

Subclass Monhysteria *sensu* Adamson 1987 *sedis mutablilis*

Diagnosis: Ovaries outstretched (26).

Subclass Chromadoria *sensu* Adamson 1987 *sedis mutabilis*

Diagnosis: Cheilostome with 12 cheilorhabdions (27).

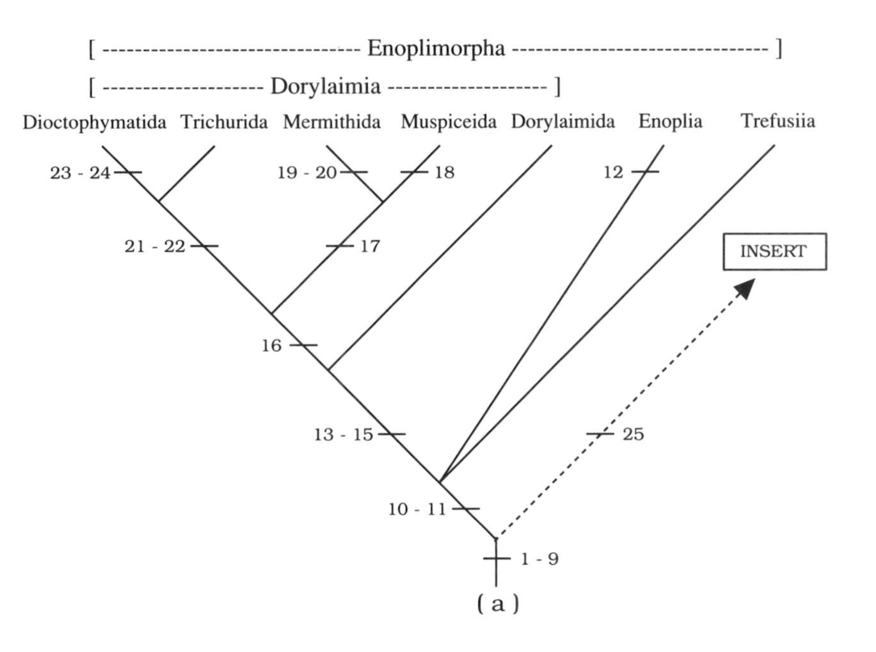
Enoplimorpha
Dorylaimia
Dioctophymatida
Trichurida
Mermithida
Muspiceida
Dorylaimida
Enoplia
Trefusiia
23 - 24
19 - 20
18
12
21 - 22
17
INSERT
16
13 - 15
25
10 - 11
1 - 9
(a)

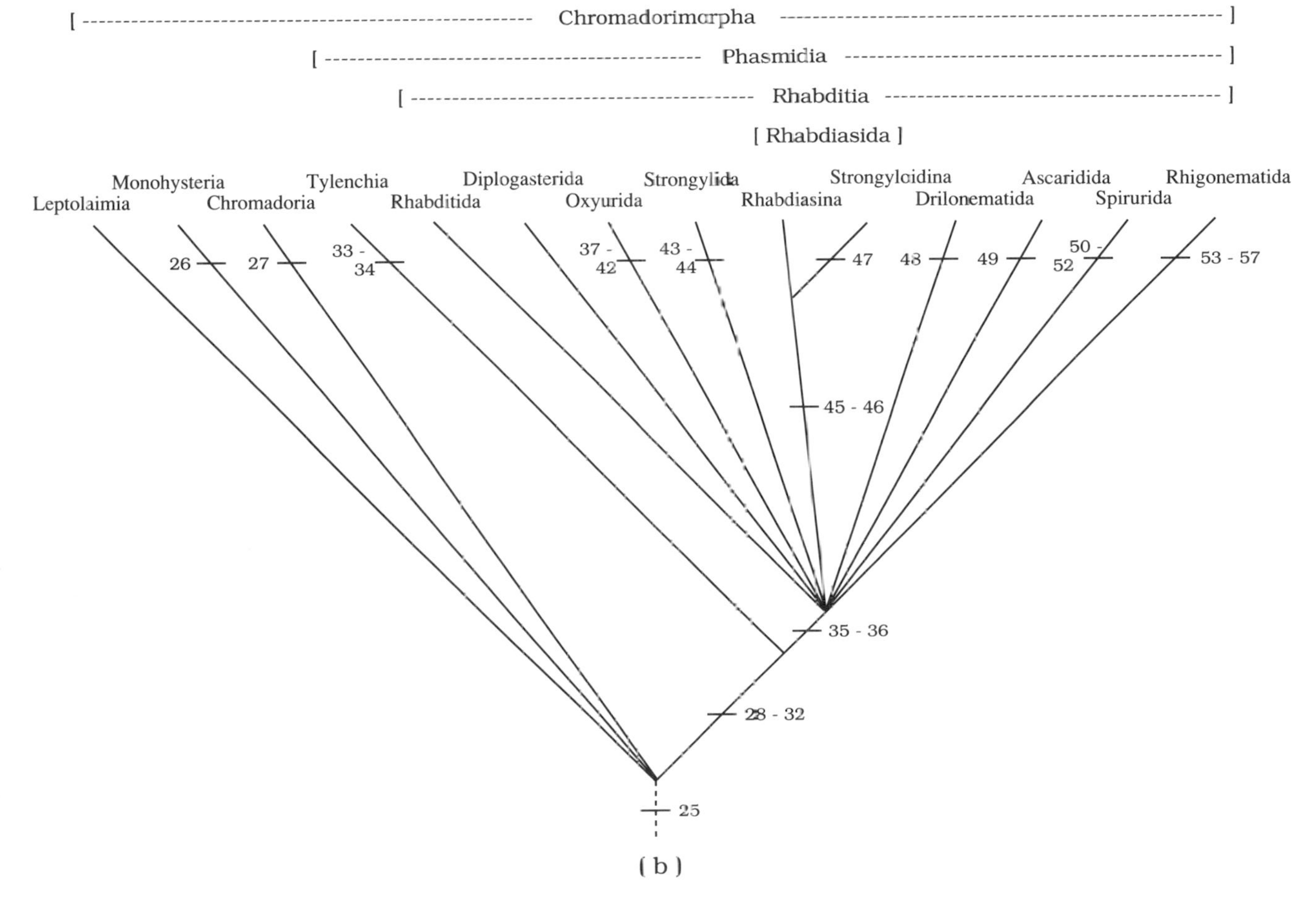

Figure 74. Phylogenetic tree for the major groups of parasitic nematodes and their closest relatives.

Subclass Phasmidia Chitwood 1933 (also Secernentea of Andrassy 1976; Rhabditea of Inglis 1983; Rhabditia of Adamson 1987) *sedis mutabilis*

Diagnosis: Amphids porelike (28); phasmids present (29); caudal glands lost (30); only anterior testis present (31); excretory duct lined with cuticle (32).

Superorder Tylenchia *sensu* Adamson 1987

Diagnosis: Buccal stylet present (33); inseminated female infective (34).

Superorder Rhabditia *sensu* Adamson 1987

Diagnosis: Males with ten paired and one upaired caudal papillae, three pairs of caudal papillae lateral (35); L_3 is dispersal stage (36).

Order Rhabditida *sensu* Adamson 1987 *sedis mutabilis, incertae sedis*

Diagnosis: As above.

Order Diplogasterida *sensu* Adamson 1987 *sedis mutabilis, incertae sedis*

Diagnosis: As above.

Order Oxyurida Weinland 1858 *sedis mutabilis*

Diagnosis: Single spicule (37); reduced number of caudal papillae (38); eggs subtriangular in end view (39); no externolateral papillae (40); first two molts in egg (41); haplodiploid reproduction (42).

Order Strongylida Diesing 1851 *sedis mutabilis*

Diagnosis: Caudal papillae at end of long bursal rays (43); L_1–L_3 free-living (44).

Order Rhabdiasida Yamaguti 1961 *sedis mutabilis*

Diagnosis: Parasitic adult a protandrous hermaphrodite (45); heterogonic development (parasitic generation alternating with free-living generation) (46).

Suborder Rhabdiasina *sensu* Adamson 1987 *incertae sedis*

Diagnosis: As above.

Suborder Strongyloidina *sensu* Adamson 1987 *sedis mutabilis*

Diagnosis: Parasitic adult parthenogenetic (47).

Order Drilonematida *sensu* Adamson 1987 *sedis mutabilis*

Diagnosis: Phasmids large and pocketlike (48).

Order Ascaridida Yamaguti 1961 *sedis mutabilis*

Diagnosis: One dorsal and two subventral lips present (49).

Order Spirurida Diesing 1861 *sedis mutabilis*

Diagnosis: Lateral pseudolabia present (50); esophagus divided into anterior muscular and posterior glandular portions (51); L_1–L_3 in arthropod host (52).

Order Rhigonematida *sensu* Adamson 1987 *sedis mutabilis*

Diagnosis: Buccal capsule modified to form three jawlike or three pennate cuticular projections (53); eggs extremely large (54); buccal capsule a cheilostome, protostome, and telostome, with telostome fused to at least part of the protostome (55); posterior portion of buccal capsule surrounded by esophageal tissue (56); vagina long and heavily muscled (57).

Phylum Nematoda: Infraordinal Relationships

Order Oxyurida

***Enterobius* Leach 1853 (Fig. 75)**

References: Brooks and Glen (1982); Hugot (1983); Hugot and Tourte-Schaefer (1985).

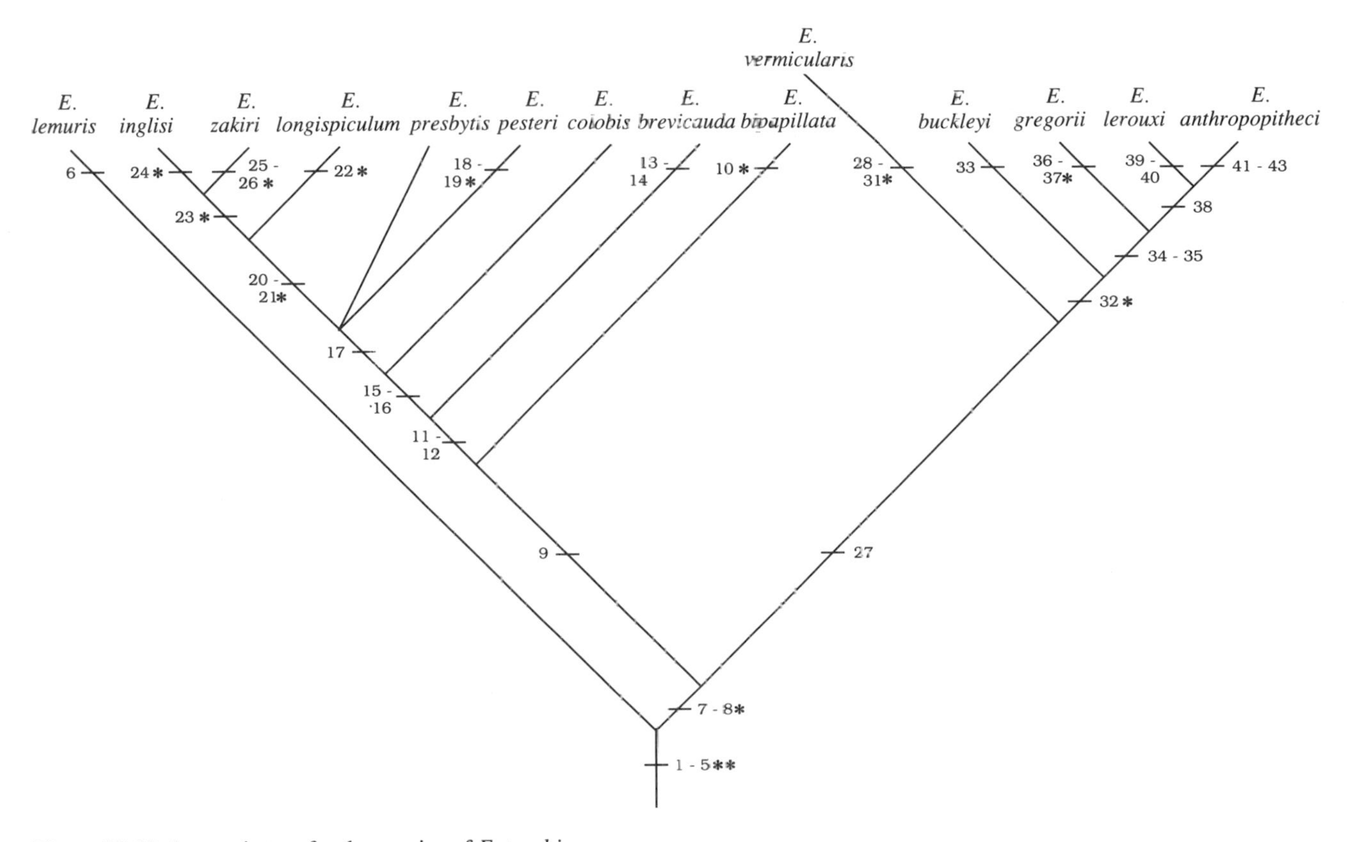

Figure 75. Phylogenetic tree for the species of *Enterobius*.

Diagnosis: Ratio of female tail length to total body length 1:5 *(1); esophageal bulb 70–100 μm long *(2); spicule 100–130 μm long (3); five pairs of caudal papillae in males (4); spicule type K of Inglis (1961) (5).

***lemuris* Baer 1935**

Diagnosis: Vagina directed anteriorly (6).

***bipapillata* (Gedoelst 1916) Yamaguti 1961**

Diagnosis: Esophagus 13–16% of total body length (7); esophageal bulb 100–130 μm long *(8); spicule type L of Inglis (1961) (9); ratio of female tail length to total body length 1:4 *(10).

***brevicauda* Sandosham 1950**

Diagnosis: Esophagus 13–16% of total body length (7); esophageal bulb 100–130 μm long *(8); ratio of vulva position to total body length 1:3 (11); ratio of female tail length to total body length 1:6–8 (12); spicule 77–81 μm long (13); spicule type M of Inglis (1961) (14).

colobis

Diagnosis: Esophagus 13–16% of total body length (7); esophageal bulb 100–130 μm long *(8); ratio of vulva position to total body length 1:3 (11); ratio of female tail length to total body length 1:10 (15); spicule type not one of Inglis's (1961) (16).

presbytis

Diagnosis: Esophagus 13–16% of total body length (7); esophageal bulb 100–130 μm long *(8); ratio of vulva position to total body length 1:3 (11); ratio of female tail length to total body length 1:10 (15); spicule type not one of Inglis's (1961) (16); spicule 200–240 μm long (17).

pesteri

Diagnosis: Esophagus 13–16% of total body length (7); esophageal bulb 100–130 μm long *(8); ratio of female tail length to total body length 1:10 (15); spicule type not one of Inglis's (1961) (16); spicule 200–240 μm long (17); esophageal bulb 150–165 μm long (18); ratio of vulva position to total body length 1:2 *(19).

longispiculum

Diagnosis: Esophagus 13–16% of total body length (7); ratio of vulva position to total body length 1:3 (11); ratio of female tail length to total body length 1:10 (15); spicule type not one of Inglis's (1961) (16); spicule 300–350 μm long (21); esophageal bulb 70–100 μm long *(22).

inglisi

Diagnosis: Esophagus 13–16% of total body length (7); spicule type not one of Inglis's (1961) (16); spicule 200–240 μm long (17); esophageal bulb 100–130 μm long *(20); ratio of female tail length to total body length 1:5 *(23); ratio of vulva position to total body length 1:2 *(24).

***zakiri* Siddiqi and Mirza 1956**

Diagnosis: Esophagus 13–16% of total body length (7); ratio of vulva position to total body length 1:3 (11); spicule type not one of Inglis's (1961) (16); spicule 300–350 μm long (21); ratio of female tail length to total body length 1:5 *(23); esophageal bulb 150–165 μm long (25); four pairs of caudal papillae in males *(26).

***vermicularis* (Linnaeus 1758) Leach 1853**

Diagnosis: Esophageal bulb 100–130 μm long *(8); spicule type P of Inglis (1961) (28); female worms more than 12 mm long (29); esophagus 8–10% of total body length (30); four pairs of caudal papillae in males *(31).

***buckleyi* Sandosham 1950**

Diagnosis: Esophagus 13–16% of total body length (7); esophageal bulb 100–130 μm long *(8); six pairs of caudal papillae in males (27); ratio of female tail length to total body length 1:4 *(32); spicule type N of Inglis (1961) (33).

***gregorii* Hugot 1983**

Diagnosis: Esophagus 13–16% of total body length (7); esophageal bulb 100–130 μm long *(8); ratio of female tail length to total body length 1:3 (34); spicule type Q of Inglis (1961) (35); spicule 70–80 μm long (36); four pairs of caudal papillae in males *(37).

***lerouxi* Sandosham 1950**

Diagnosis: Esophagus 13–16% of total body length (7); esophageal bulb 100–130 μm long *(8); ratio of female tail length to total body length 1:3 (34); spicule type Q of Inglis (1961) (35); spicule 66 μm long (38); seven pairs of caudal papillae in males (39); buccal ornamentation present (40).

***anthropopitheci* (Gedoelst 1916) Yamaguti 1961**

Diagnosis: Esophagus 13–16% of total body length (7); esophageal bulb 100–130 μm long *(8); six pairs of caudal papillae in males (27); ratio of female tail length to total body length 1:3 (34); esophageal bulb 67–80 μm long (41); spicule 52–56 μm long (42); spicule type O of Inglis (1961) (43).

Order Strongylida

Superfamily Strongyloidea

Tribe Oesophagostominea Lichtenfels 1980 *partem* (Fig. 76)

Reference: Glen and Brooks (1985).

Diagnosis: Ventral cervical groove present (1).

Daubneyia* LeRoux 1940 *incertae sedis

Diagnosis: As above.

***Oesophagostomum* (*Hysteracrum*) Railliet and Henry 1913**

Diagnosis: External corona with more than eight elements (2); cervical papillae posterior (3).

Oesophagostomum* (*Proteracrum*) Railliet and Henry 1913 *sedis mutabilis, incertae sedis

Diagnosis: External corona with more than eight elements (2); vagina short (4).

Oesophagostomum* (*Bosicola*) Sandground 1921 *sedis mutabilis, incertae sedis

Oesophagostomum* (*Oesophagostomum*) Molin 1861 *incertae sedis

Diagnosis: External corona with more than eight elements (2); vagina short (4); partially developed esophageal funnel (5).

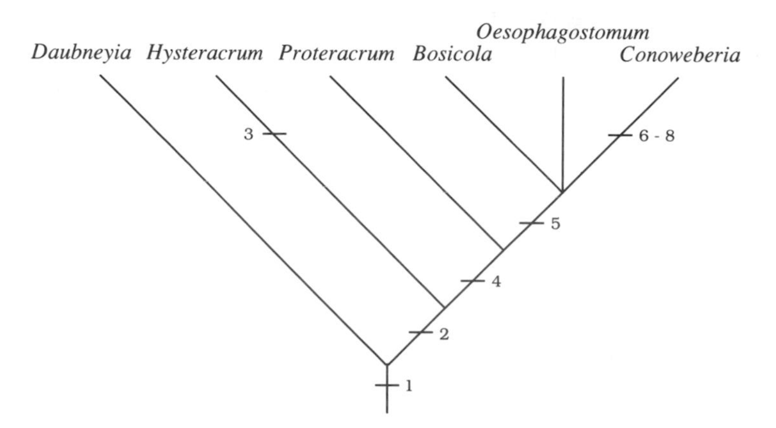

Figure 76. Phylogenetic tree for the subgenera of *Oesophagostomum.*

***Oesophagostomum* (*Conoweberia*) Ihle 1922**

Diagnosis: External corona with more than eight elements (2); vagina short (4); coronal elements small (6); three denticles (7); well-developed esophageal funnel with sclerotized plates (8).

***Oesophagostomum* (*Conoweberia*) Ihle 1922 (Fig. 77)**

Reference: Glen and Brooks (1985).

***zukowskyi* Travassos and Vogelsang 1932**

Diagnosis: Males less than 6 mm long (1); females less than 7 mm long (2); cephalic papillae 0.4–0.5 mm long *(3); buccal capsule 0.5–0.7 mm wide *(4); spicules short (5); male body less than 0.3 mm wide (6).

***bifurcum* Linstow 1879**

Diagnosis: Vulva close to posterior extremity (7); buccal capsule ratio 1:2.6–3.2 (8).

***brumpti* Railliet and Henry 1905**

Diagnosis: Vulva close to posterior extremity (7); buccal capsule ratio 1:2.6–3.2 (8); female tail short *(9).

***aculeatum* Linstow 1879**

Diagnosis: Vulva close to posterior extremity (7); buccal capsule ratio 1:2.6–3.2 (8); female tail short (9); female 13–20 mm long (10); ventral groove 0.23–0.30 mm long (11); esophageal funnel long (12).

***xeri* Ortlepp 1922, *susannae* Leroux 1940**

Diagnosis: Vulva close to posterior extremity (7); buccal capsule ratio 1:2.6–3.2 (8); female tail short (9); female 13–20 mm long (10); ventral groove 0.23–0.30 mm long (11); esophageal funnel long (12); cephalic papillae 0.4–0.5 mm long *(13); spicules long (14); female width 0.5–0.65 mm (15); transverse processes (16).

***blanchardi* Travassos and Vogelsang 1932**

Diagnosis: Buccal capsule ratio 1:2.6–3.2 (8); female 13–20 mm long (10); ventral groove 0.23–0.30 mm long (11); esophageal funnel long (12); spicules long (14); female

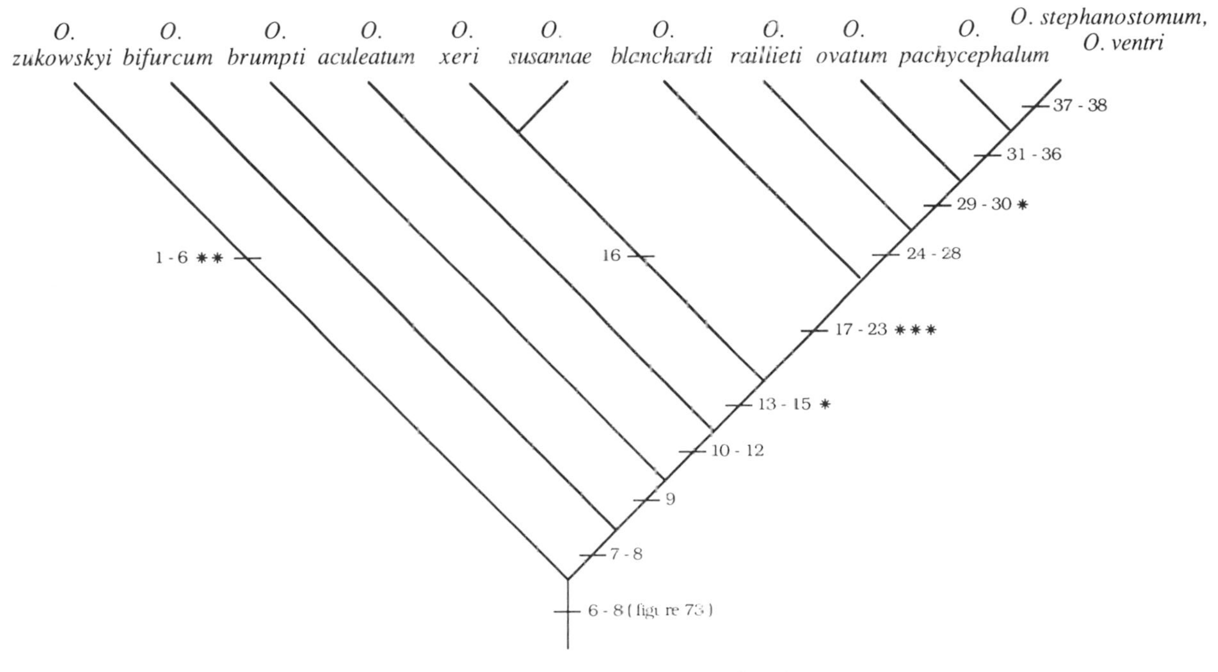

Figure 77. Phylogenetic tree for the species of *Oesophagostomum* (*Conoweberia*).

width 0.5–0.65 mm (15); male 13–18 mm long (17); cephalic papillae 0.5–0.6 mm long (18); esophagus 0.70–1.0 mm long (19); buccal capsule 0.50–0.70 mm wide *(20); female tail long *(21); esophagus 0.16–0.20 mm wide (22); vulva positioned anteriorly *(23).

***raillieti* Travassos and Vogelsang 1932**

Diagnosis: Buccal capsule ratio 1:2.6–3.2 (8); ventral groove 0.23–0.30 mm long (11); esophageal funnel long (12); female 13–20 mm long (13); spicules long (14); female width 0.5–0.65 mm (15); male 13–18 mm long (17); cephalic papillae 0.5–0.6 mm long (18); female tail long *(21); esophagus 0.16–0.20 mm wide (22); vulva positioned anteriorly *(23); ventral groove 0.3–0.45 mm long (24); esophagus 1.0–1.4 mm long (25); buccal capsule 0.8–1.4 mm wide (26); female body 0.65–1.1 mm wide (27); male body 0.6–1.2 mm wide (28).

***ovatum* Linstow 1906**

Diagnosis: Esophageal funnel long (12); female 13–20 mm long (13); spicules long (14); male 13–18 mm long (17); cephalic papillae 0.5–0.6 mm long (18); female tail long *(21); esophagus 0.16–0.20 mm wide (22); vulva positioned anteriorly *(23); ventral groove 0.3–0.45 mm long (24); esophagus 1.0–1.4 mm long (25); buccal capsule 0.8–1.4 mm wide (26); female body 0.65–1.1 mm wide (27); male body 0.6–1.2 mm wide (28); wide cephalic distension (29); buccal capsule ratio 1:3.6–4.1 *(30).

***pachycephalum* Molin 1861**

Diagnosis: Esophageal funnel long (12); spicules long (14); cephalic papillae 0.5–0.6 mm long (18); esophagus 0.16–0.20 mm wide (20); female tail long *(21); vulva positioned anteriorly *(23); ventral groove 0.3–0.45 mm long (24); esophagus 1.0–1.4 mm long (25); buccal capsule 0.8–1.4 mm wide (26); female body 0.65–1.1 mm wide (27); male body 0.6–1.2 mm wide (28); wide cephalic distension (29); buccal capsule ratio 1:3.6–4.1 *(30); male body 18–25 mm long (31); female body 20–30 mm long (32); 30–40 external elements (33); funnel V-shaped (34); external elements rounded (35); esophagus 0.2–0.3 mm wide (36).

***stephanostomum* Stossich 1904, *ventri* Thornton 1924**

Diagnosis: Esophageal funnel long (12); spicules long (14); cephalic papillae 0.5–0.6 mm long (18); esophagus 0.16–0.20 mm wide (20); female tail long *(21); vulva positioned anteriorly *(23); ventral groove 0.3–0.45 mm long (24); esophagus 1.0–1.4 mm long (25); buccal capsule 0.8–1.4 mm wide (26); female body 0.65–1.1 mm wide (27); male body 0.6–1.2 mm wide (28); wide cephalic distension (29); buccal capsule ratio 1:3.6–4.1 *(30); male body 18–25 mm long (31); female body 20–30 mm long (32); 30–40 external elements (33); funnel V-shaped (34); external elements rounded (35); esophagus 0.2–0.3 mm wide (36); six denticles (37); external elements not converging toward center of oral aperture (38).

Superfamily Trichostrongyloidea Leiper 1912

Family Heligmonellidae Durette-Desset and Chabaud 1977

Subfamily Pudicinae Skrjabin and Schikalobalova 1952 (Fig. 78)

Reference: Durette-Desset and Justine (1991).

Diagnosis: Orientation axis rotation 67–90° from sagittal axis (1); carene present (2).

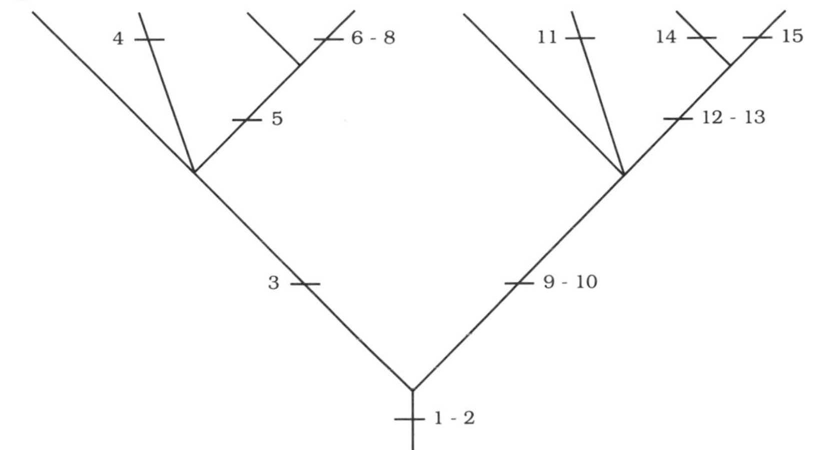

Figure 78. Phylogenetic tree for the genera of the Pudicinae. *Helig.* = *Heligmostrongylus; Pseudo.* = *Pseudoheligmosomum; Acanth.* = *Acanthostrongylus.*

***Heligmostrongylus* group**

Diagnosis: Ninth bursal ray shorter than tenth (3).

Heligmostrongylus* Travassos 1917 *incertae sedis, sedis mutabilis

Diagnosis: With characters of the group.

Fullebornema* Travassos 1937 *sedis mutabilis

Diagnosis: Caudal bursal ray type 1:3:1 (formula of Durette-Desset and Chabaud 1981) (4).

Pseudoheligmosomum* Subgroup *sedis mutabilis

Diagnosis: More than six ventral ridges (5).

Sciurodendrium* Durette-Desset 1971 *incertae sedis

Diagnosis: With characters of the subgroup.

***Pseudoheligmosomum* Travassos 1937**

Diagnosis: Cuticular protuberance supported by one left-developed ventral ridge and one left-developed dorsal ridge of equal length with other ridges (6); dorsal bursal ray divided into two independent stems without a common trunk (7); peculiar posterior synlophe present (8).

***Pudica* group**

Diagnosis: Comaretes present (9); at least some ridges discontinuous (10).

Pudica* Travassos and Dariba 1929 *incertae sedis, sedis mutabilis

Diagnosis: With characters of the group.

Acanthostrongylus* Travassos 1937 *sedis mutabilis

Diagnosis: Supernumerary spines present (11).

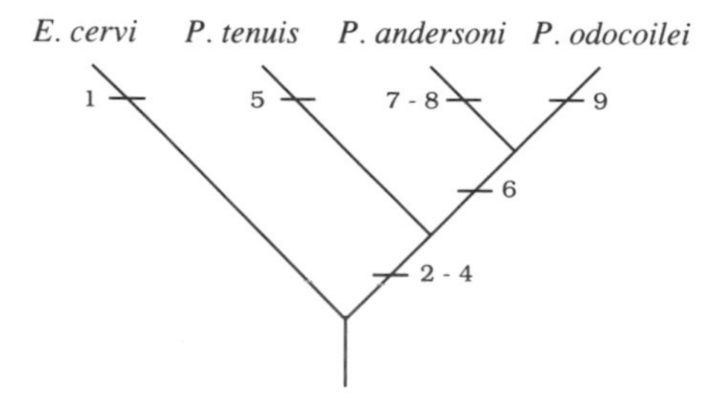

Figure 79. Phylogenetic tree for the species of the Elaphostrongylinae. *E.* = *Elaphostrongylus; P.* = *Parelaphostrongylus*.

Justinema* Subgroup *sedis mutabilis

Diagnosis: Carene absent and axis subfrontal (12); one ventral comarete, somewhat developed, present only in anterior half of body (13).

***Justinema* R'Kha and Durette-Desset 1991**

Diagnosis: Ratio of distance between distal extremity of tenth bursal ray and height of tenth bursal ray more than 60% (14).

***Durettestrongylus* Guerrero 1982**

Diagnosis: One ventral comarete, greatly developed, present along entire body length (15).

Superfamily Metastrongyloidea Lane 1917

Family Protostrongylidae Leiper 1926

Subfamily Elaphostrongylinae Boev and Shulz 1950 (Fig. 79)

Reference: Platt (1984).

***Elaphostrongylus cervi* Cameron 1931**

Diagnosis: Perityls lacking (1).

Parelaphostrongylus tenuis* (Dougherty 1945) Priadko and Boev 1971 *sedis mutabilis

Diagnosis: Gubernaculum crura present (2); corpus of gubernaculum split distally (3); branches of dorsal ray ventral (4); spicular foramen present (5).

Parelaphostrongylus andersoni* Prestwood 1972 *sedis mutabilis

Diagnosis: Gubernaculum crura present (2); corpus of gubernaculum split distally (3); branches of dorsal ray ventral (4); dorsal ray a compact bulb (6); distal ends of spicules bifid (7); dorsal ray ventral (8).

Parelaphostrongylus odocoilei* (Hobmaier and Hobmaier 1934) Boev and Schulz 1950 *sedis mutabilis

Diagnosis: Gubernaculum crura present (2); corpus of gubernaculum split distally (3); dorsal ray a compact bulb (6); dorsal ray dorsal (9).

Order Ascaridida

***Megalobatrachonema* Yamaguti 1941 (Fig. 80)**

Reference: Richardson (1988).

Diagnosis: Glandular esophageal bulb (1); ratio of spicule length to width more than 10:1 (2); subventral excretory gland cells extending anteriorly to near the level of the nerve ring (3).

***giganticum* (Olsen 1938) Baker 1980**

Diagnosis: Gubernaculum with rounded dorsal end (4); 11 pairs of caudal papillae (extra pair lateral) (5); three pairs of subventral papillae posterior to anus (6).

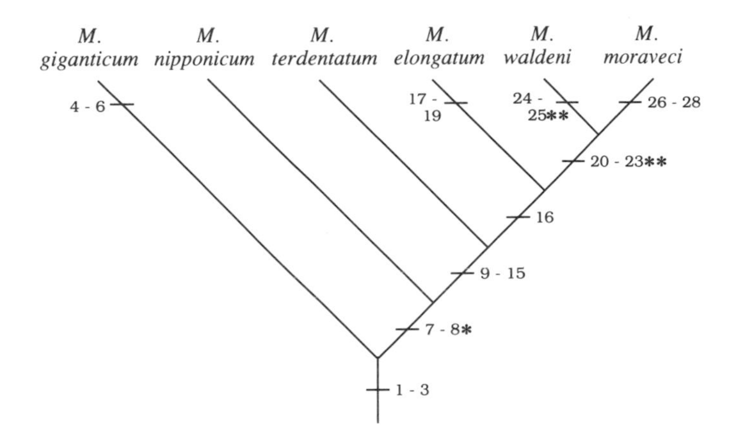

Figure 80. Phylogenetic tree for species of *Megalobatrachonema.*

***nipponicum* Yamaguti 1941**

Diagnosis: Elongate isthmus separating esophageal corpus and bulb (7); pseudosucker present *(8).

***terdentatum* (Linstow 1890) Hartwich 1960**

Diagnosis: Elongate isthmus separating esophageal corpus and bulb (7); pseudo sucker present *(8); circumoral tissue elevated (9); cheilostomal ring lacking (10); pharyngeal area showing little demarcation from rest of corpus of esophagus (11); esophageal bulb elongate (12); esophageal bulb without valves (13); ratio of spicule length to width 6–9:1 (14); 11 pairs of caudal papillae (extra pair subventral) (15).

***elongatum* (Baird 1858) Baker 1986**

Diagnosis: Elongate isthmus separating esophageal corpus and bulb (7); pseudosucker present *(8); circumoral tissue elevated (9); cheilostomal ring lacking (10); pharyngeal area showing little demarcation from rest of corpus of esophagus (11); esophageal bulb elongate (12); esophageal bulb without valves (13); 11 pairs of caudal papillae (extra pair subventral) (15); corpus swelling present (16); onchia three-pronged (17); ratio of spicule length to width less than 5:1 (18); eggs segmented in utero (19).

***waldeni* Richardson and Adamson 1988**

Diagnosis: Elongate isthmus separating esophageal corpus and bulb (7); cheilostomal ring lacking (10); pharyngeal area showing little demarcation from rest of corpus of esophagus (11); esophageal bulb elongate (12); esophageal bulb without valves (13); ratio of spicule length to width 6–9:1 (14); corpus swelling present (16); lips separate (20); lateral alae present (21); pseudosucker lacking *(22); nine pairs of caudal papillae (loss of one subventral and one lateral pair) (23); circumoral tissue not elevated *(24); cervical alae present (25).

***moraveci* Richardson and Adamson 1988**

Diagnosis: Elongate isthmus separating esophageal corpus and bulb (7); circumoral tissue elevated (9); cheilostomal ring lacking (10); pharyngeal area showing little demarcation from rest of corpus of esophagus (11); esophageal bulb elongate (12);

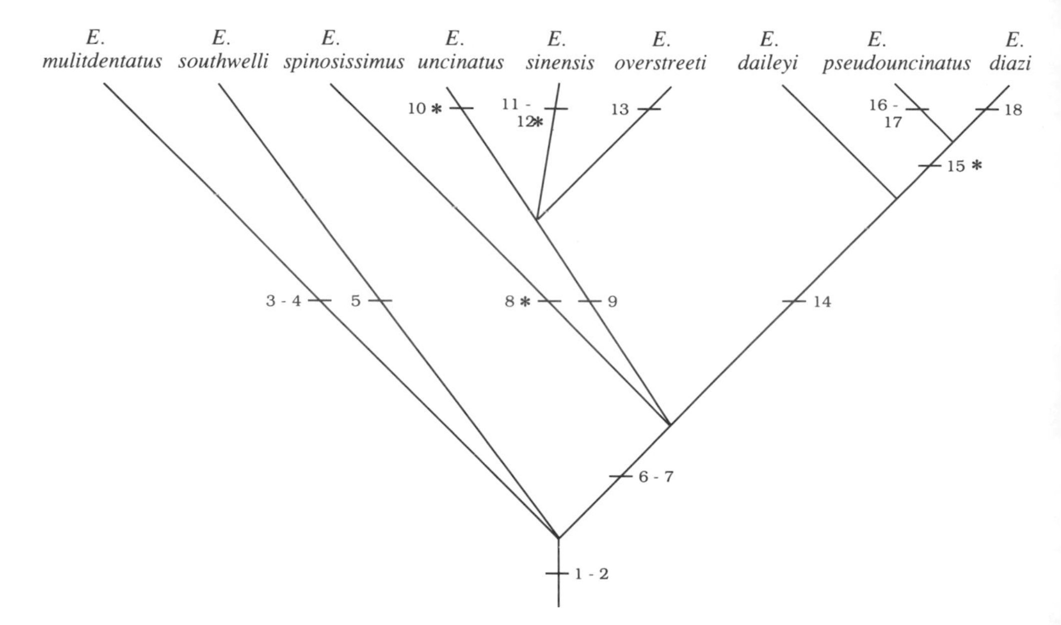

Figure 81. Phylogenetic tree for species of *Echinocephalus*.

esophageal bulb without valves (13); ratio of spicule length to width 6–9:1 (14); corpus swelling present (16); lips separate (20); lateral alae present (21); pseudosucker lacking *(22); cheilostomal thickening lacking, cuticular lining of lip surface sclerotized (26); hook-shaped pointed ventral end of gubernaculum (27); 12 pairs of caudal papillae (extra pairs both subventral) (28).

Order Spirurida

***Echinocephalus* Molin 1858 (Fig. 81)**

Reference: Brooks and Deardorff (1988).

Diagnosis: Five pairs of postanal papillae (1); three pairs of preanal papillae (2).

***multidentatus* Baylis and Lane 1920**

Diagnosis: Three pairs of preanal papillae (2); four pairs of postanal papillae (3); 15–18 rows of cephalic spines (4).

***southwelli* Baylis and Lane 1920**

Diagnosis: Five pairs of postanal papillae (1); three pairs of preanal papillae (2); 11–13 rows of cephalic spines (5).

***spinosissimus* Shipley and Hornell 1905**

Diagnosis: Three pairs of preanal papillae (2); 30–45 rows of cephalic spines (6); gubernaculum present (7); five pairs of postanal papillae in cluster *(8).

***uncinatus* Molin 1858**

Diagnosis: Three pairs of preanal papillae (2); 30–45 rows of cephalic spines (6); gubernaculum present (7); ventral rugose areas surrounding male cloaca (9); five pairs of postanal papillae in cluster *(10).

***sinensis* Ko 1975**

Diagnosis: Five pairs of postanal papillae (1); gubernaculum present (7); ventral rugose areas surrounding male cloaca (9); two pairs of preanal papillae (11); 26–29 rows of cephalic spines *(12).

***overstreeti* Deardorff and Ko 1983**

Diagnosis: Three pairs of preanal papillae (2); 30–45 rows of cephalic spines (6); gubernaculum present (7); ventral rugose areas surrounding male cloaca (9); three pairs of postanal papillae and one pair of adanal papillae (13).

***daileyi* Deardorff, Brooks, and Thorson 1981**

Diagnosis: Three pairs of preanal papillae (2); 30–45 rows of cephalic spines (6); gubernaculum present (7); six pairs of postanal papillae (14).

***pseudouncinatus* Milleman 1963**

Diagnosis: Three pairs of preanal papillae (2); gubernaculum present (7); three pairs of postanal papillae (16); 16–21 rows of cephalic spines (17).

***diazi* Troncy 1969**

Diagnosis: Gubernaculum present (7); six pairs of postanal papillae (14); 25–27 rows of cephalic spines *(15); two pairs of preanal papillae (18).

Order Rhigonematida (Fig. 82)

Reference: Adamson and van Waerebeke (1985).

Superfamily Rhigonematoidea *sensu* Adamson and van Waerebeke 1985

Diagnosis: Spines (1); cephalic cap present *(2); robust esophageal corpus (3); protostome and telostome completely fused (4); protostome and telostome completely surrounded by esophageal tissue (5); terminal duct of excretory system vesiculate (6); eggs with thick, smooth shell (7); no lips (8).

Family Rhigonematidae *sensu* Adamson and van Waerebeke 1985

Diagnosis: Cephalic collar present (9); esophageal isthmus represented only by constriction between corpus and bulb (10).

***Rhigonema* Cobb 1898**

Diagnosis: Esophageal isthmus about half as long as bulb width (11).

***Obainia* Adamson 1983**

Diagnosis: Esophageal corpus much shorter than bulb (12); dorsal jaw piece dissimilar to subventral piece and reduced in size (13).

***Glomerinema* van Waerebeke 1985**

Diagnosis: Esophageal corpus much shorter than bulb (12); jaw pieces with spine rows *(14); subdorsal and sublateral papillae pairs lacking *(15).

***Xustrostoma* Adamson and van Waerebeke 1984**

Diagnosis: Esophageal corpus much shorter than bulb (12); jaw pieces with spine rows *(14); oral opening dorsoventrally elongate (16); dorsal jaw piece dissimilar to subventral piece and well developed (17).

Family Ichthyocephalidae *sensu* Adamson and van Waerebeke 1985

Diagnosis: Oral opening laterally elongate (18); buccal capsule in the form of one dorsal and one ventral plate (19); jaw pieces with spine rows *(20); submedian outer

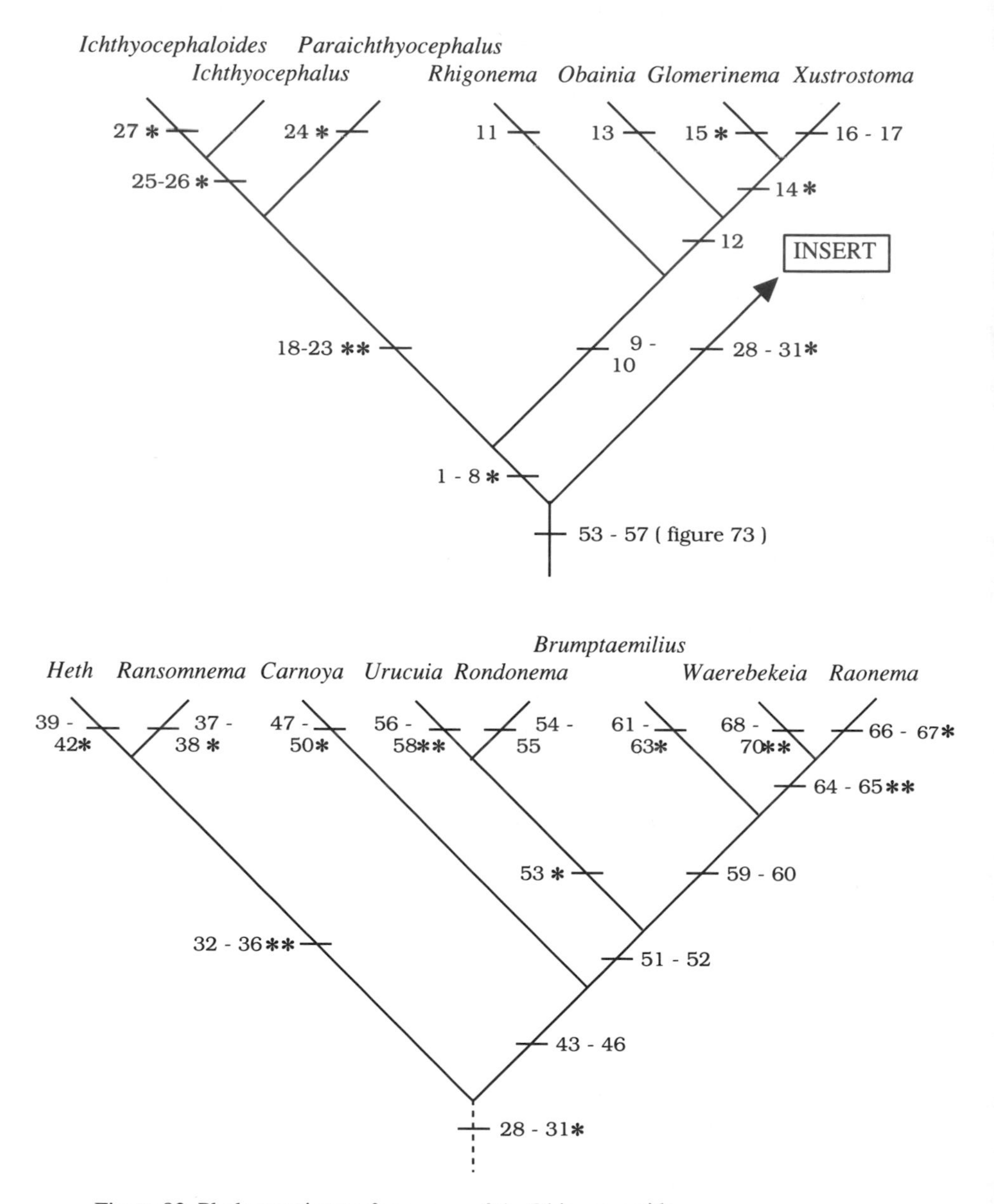

Figure 82. Phylogenetic tree for genera of the Rhigonematida.

papillae conical (21); subdorsal and sublateral papillae pairs lacking *(22); esophageal corpus ovoid (23).

***Paraichthyocephalus* Travassos and Kloss 1958**

Diagnosis: Spicules with surface sculpturing *(24).

Ichthyocephalus* Artigas 1926 *incertae sedis

Diagnosis: Spicules dissimilar *(25); esophageal isthmus less than half as long as bulb but forming distinct cylindrical portion (26).

***Ichthyocephaloides* Hunt and Sutherland 1984**

Diagnosis: Spicules dissimilar *(25); esophageal isthmus less than half as long as bulb but forming distinct cylindrical portion (26); female reproductive system monodelphic *(27).

Superfamily Ransomnematoidea

Diagnosis: Reduced number of caudal papillae (28); female reproductive system monodelphic *(29); three pennate cuticular metastomal projections (30); corpus spindle-shaped and sharply demarcated from isthmus and bulb (31).

Family Ransomnematidae *sensu* Adamson and van Waerebeke 1985

Diagnosis: Vulva markedly posterior (32); spermatids elongate (33); ventral sucker present (34); spicules dissimilar *(35); one dorsal and one ventral lip in males *(36); one subdorsal or subventral pair of caudal papillae (37); male lips lacking *(38).

***Ransomnema* Artigas 1926**

Diagnosis: With characters of the family.

Family Hethidae *sensu* Adamson and van Waerebeke 1985

Diagnosis: Vulva markedly posterior (32); spermatids elongate (33); ventral sucker present (34); spicules dissimilar *(35); one dorsal and one ventral lip in males *(36); esophageal corpus spindle-shaped in male, cylindrical in female (39); metastomal projections developed in male only, not of *Brumptaemilius* type (40); spicules fused (41); tail of male subulate *(42).

***Heth* Cobb 1989**

Diagnosis: With characters of the family.

Family Carnoyidae

Diagnosis: Esophageal corpus spindle-shaped in male, divided in female (43); metastomal projections reduced in both sexes (44); spicules parallel (45); outer submedian cephalic papillae of female forming subspherical saliences (46).

***Carnoya* Gilson 1898**

Diagnosis: Protostome long and tubular in male (47); no subdorsal or sublateral caudal papillae (48); one ventral, two dorsal lips in male *(49); annulations present in both sexes (50).

***Rondonema* Artigas 1926**

Diagnosis: Eggshell thin and inflexible (51); vagina directed anteriorly from vulva (52); male lips lacking *(53); 13 caudal papillae (54); one subdorsal or sublateral pair of papillae (55).

***Urucuia* Kloss 1961**

Diagnosis: Eggshell thin and inflexible (51); vagina directed anteriorly from vulva (52); male lips lacking (53); cephalic cap present *(56); male tail subulate *(57); ventral striae reinforced in caudal region of male (58).

***Brumptaemilius* Dollfus 1952**

Diagnosis: Vagina directed anteriorly from vulva (52); egg shell thick with surface sculpturing (59); annulations present in females (60); male tail truncate with caudal appendage (61); area rugosa present (62); one ventral, two dorsal lips in male *(63).

***Raonema* Kloss 1965**

Diagnosis: Vagina directed anteriorly from vulva (52); egg shell thick with surface sculpturing (59); annulations present in females (60); one dorsal and one ventral lip in males *(64); cephalic cap present *(65); male tail subulate *(66); gubernaculum reduced, dorsoventrally flattened (67).

***Waerebekeia* Adamson and Anderson 1985**

Diagnosis: Vagina directed anteriorly from vulva (52); egg shell thick with surface sculpturing (59); annulations present in females (60); cephalic cap present (65); metastomal projections developed in male only, of *Brumptaemilius* type (68); spicules with surface sculpturing *(69); one ventral, two dorsal lips in male *(70).

Superorder Tylenchia

References: Ferris (1979); Baldwin and Schouest (1990).

Family Heteroderidae Filip'ev and Schuurmans Stekhoven 1941 (Fig. 83)

Diagnosis: Female with irregularly swollen or lemon-shaped body (1); cuticle around vulva not fenestrated (2); phasmids in second-stage juveniles with lenslike structure in muscle layer (3); body wall of mature female lacking cyst (4); body wall of mature female not rugose (5); anus and vulva widely separated (6); spicules short (less than 30 μm long) (7); female cuticle annulated (8); vulva subequatorial (9); vulval cone present (10); spicule tips bifid (11); vulval slit short (less than 30 μm long) (12).

Meloidodera* Chitwood, Hannon, and Esser 1956 *incertae sedis

Diagnosis: With characters of the family.

Cryphodera* Colbran 1966 *incertae sedis

Diagnosis: Vulva subterminal (13).

***Atalodera* Wouts and Sher 1971, *Thecavermiculatus* Robbins 1978**

Diagnosis: Vulva subterminal (13); female cuticle annulated only at anterior end (14); spicules long (more than 30 μm) (15); anus and vulva located close together on prominence (17).

Hylonema* Luc, Taylor, and Cadet 1978 *incertae sedis

Diagnosis: Vulva subterminal (13); female cuticle annulated only at anterior end (14); spicules long (more than 30 μm) (15); anus and vulva located close together (16); body wall of mature female rugose (18).

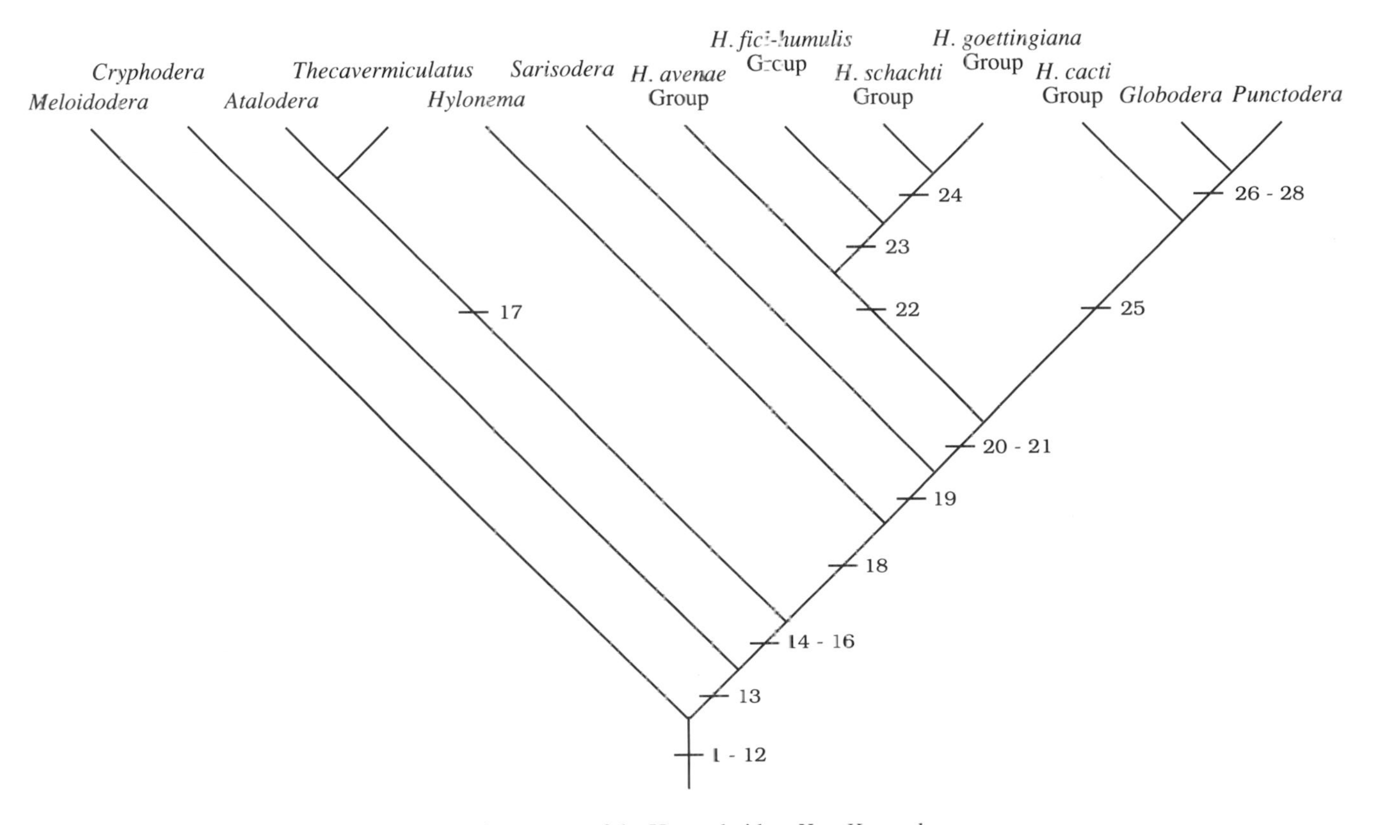

Figure 83. Phylogenetic tree for genera and species groups of the Heteroderidae. *H.* = *Heterodera*.

Sarisodera* Wouts and Sher 1971 *incertae sedis

Diagnosis: Vulva subterminal (13); female cuticle annulated only at anterior end (14); spicules long (more than 30 μm) (15); anus and vulva located close together (16); body wall of mature female with a cyst (19).

***Heterodera avenae* species group + *H. fici-humulis* species group + *H. goettingiana* species group + *H. schachti* species group group**

Diagnosis: Vulva subterminal (13); female cuticle annulated only at anterior end (14); spicules long (more than 30 μm) (15); anus and vulva located close together (16); body wall of mature female with a cyst (19); phasmids in second-stage juvenile without lenslike structure in muscle layer (20); cuticle around vulva bi- or ambifenestrated (21); lips with elongate disk (22).

***H. avenae* species group**

Diagnosis: With characters of the group.

***H. fici-humulis* species group**

Diagnosis: Vulval slit long (more than 30 μm) (23).

***H. goettingiana* species group + *H. schachti* species group**

Diagnosis: Vulval slit long (more than 30 μm) (23); lips with elongate disk and further development (24).

***Heterodera cacti* species group**

Diagnosis: Vulva subterminal (13); female cuticle annulated only at anterior end (14); spicules long (more than 30 μm) (15); anus and vulva located close together (16); body wall of mature female with a cyst (19); phasmids in second-stage juvenile without lenslike structure in muscle layer (20); cuticle around vulva circumfenestrated (25).

***Globodera* Skarbilovich 1959, *Punctodera* Mulvey and Stone 1976**

Diagnosis: Vulva subterminal (13); female cuticle annulated only at anterior end (14); spicules long (more than 30 μm) (15); anus and vulva located close together (16); body wall of mature female with a cyst (19); phasmids in second-stage juvenile without lenslike structure in muscle layer (20); cuticle around vulva circumfenestrated (25); female body round or pear-shaped (26); vulval cone lacking (27); spicule tips not bifid (28).

Phylum Acanthocephala

The phylum Acanthocephala represents a mysterious group of helminth parasites. All members of the phylum are endoparasites living as adults in the intestinal tract of vertebrates and as larvae and juveniles in the hemocoels of arthropods. Van Cleave (1941) summarized the current concepts about the phylogenetic affinities of the Acanthocephala, beginning with the following sentiment:

> The usual avenues for direct inquiries into phylogeny yield but scanty convincing evidence of direct relationship between the Acanthocephala and other animal groups. The reasons for this condition are found in the specialization which has accompanied perfect adaptation and complete organic adjustment to the par-

> asitic existence. Many other parasitic organisms, through either their anatomy or their ontogeny, reveal indisputable evidences of relationship with free-living forms which occupy a natural position in the scheme of classification. . . . Such possibilities are wholly lacking for the Acanthocephala, for the entire group stands apart in a condition of isolation that has baffled most investigators who have given thought to the problem.

Van Cleave felt that the only convincing character linking acanthocephalans with any other group was the absence of a digestive system, a trait shared with tapeworms. He suggested that the Acanthocephala should be accorded their own phylum, albeit allied with the Platyhelminthes in some manner. Although the acanthocephalan/tapeworm connection has not been adopted by biologists, even contemporary treatises on acanthocephalans (e.g., Brusca and Brusca 1990) find the group's placement among the "Aschelminthes" ambiguous. Conway Morris and Crompton (1982) presented a detailed discussion of the possible free-living sister groups of the Acanthocephala, concluding that the Priapulida is the most likely sister group, with the Kinorhyncha being the sister group of those two, the Rotifera possibly the sister group of those three, and the Gastrotricha, Nematomorpha, and Nemata (the "Aschelminthes") representing the sister group of those four (see Fig. 84).

Phylum Acanthocephala: Sister group Relationships (Fig. 84)

Diagnosis: Cement glands (see also rotifers and gastrotrichs) (1); separate sexes (widespread) (2); protonephridia (widespread) (3); cuticle (widespread and perhaps convergent) (4); pseudocoel (widespread and perhaps convergent) (5); circular muscle fibers (widespread) (6); eutely (widespread) (7); muscle tissues tubular in cross section (8); spiny retractable proboscis (9); trunk spines (10); single midventral nerve cord with circumpharyngeal proboscis ganglion (11).

There has never been a phylogenetic systematic analysis of the major groups of acanthocephalans. Conway Morris and Crompton (1982), however, discussed possible plesiomorphic conditions for a number of characters, including the cement glands, the ligament sacs, protonephridia, and definitive host type, in light of possible relationships among the three major groups of acanthocephalans recognized in most classifications:

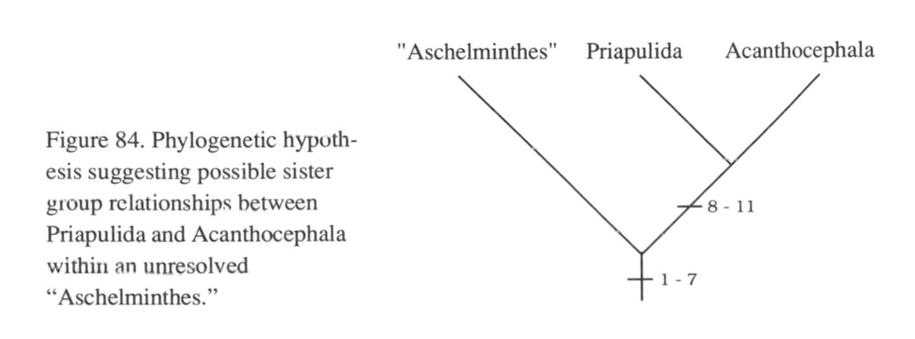

Figure 84. Phylogenetic hypothesis suggesting possible sister group relationships between Priapulida and Acanthocephala within an unresolved "Aschelminthes."

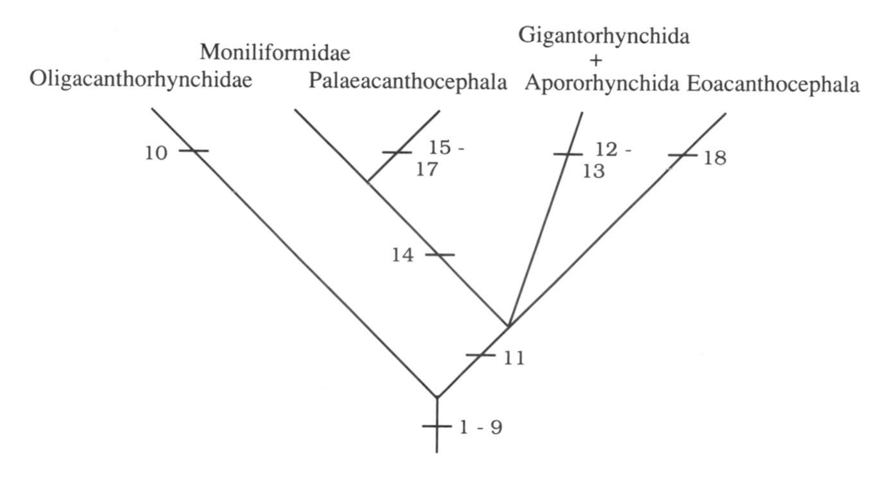

Figure 85. Phylogenetic tree for hypothesis I relationships among major groups of Acanthocephala.

the Archiacanthocephala Meyer 1933, Eoacanthocephala Van Cleave 1935, and Pale-acanthocephala Meyer 1933. We have attempted to use their data about plesiomorphic conditions to investigate phylogenetic hypotheses supported by apomorphic traits. Based on the data we have at present, the following three hypotheses are equally parsimonious. They are presented to stimulate interest in studying this enigmatic group.

Hypothesis I (Fig. 85)

Diagnosis: No gut (1); lacunar system (2); nonsyncytial cement glands (3); non-fragmented giant nuclei (4); persistent double ligament (5); proboscis hooks six per row in three rows (6); lacunar canals dorsoventrally oriented (7); male and female genital system (8); acanthor, acanthella, cystacanth developmental stages (9).

Oligacanthorhynchidae Southwell and Macfie 1925

Diagnosis: Proboscis hooks six per row in six rows (10).

Apororhynchida Thapar 1927 + Gigantorhynchida Southwell and Macfie 1925 *sedis mutabilis*

Diagnosis: No protonephridia (11); spines and hooks on proboscis (12); longitudinal muscle fibers in proboscis receptacle (13).

Palaeacanthocephala Meyer 1933 + Moniliformidae Van Cleave 1934 group *sedis mutabilis*

Diagnosis: No protonephridia (11); double wall of muscle around proboscis receptacle (14).

Moniliformidae Van Cleave 1934 *incertae sedis*

Diagnosis: With characters of the group.

Palaeacanthocephala Meyer 1933

Diagnosis: Lateral lacunar canals (15); fragmented giant nuclei (16); single ligament (17).

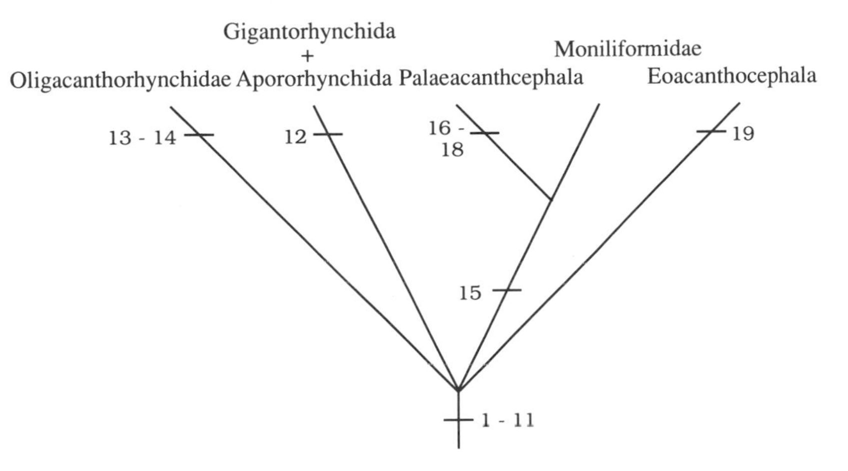

Figure 86. Phylogenetic tree for hypothesis II relationships among major groups of Acanthocephala.

Eoacanthocephala Van Cleave 1936 *sedis mutabilis*

Diagnosis: Syncytial cement glands (18).

Hypothesis II (Fig. 86)

Diagnosis: As hypothesis I, except assuming that no protonephridia (10) and longitudinal muscles restricted to proboscis receptacle (11) are plesiomorphic for Acanthocephala.

Apororhynchida Thapar 1927 + Gigantorhynchida Southwell and Macfie 1925 *sedis mutabilis*

Diagnosis: Spines and hooks on proboscis (12).

Oligacanthorhynchidae Southwell and Macfie 1925 *sedis mutabilis*

Diagnosis: Protonephridia present (13); proboscis hooks six per row in six rows (14).

Palaeacanthocephala Meyer 1933 + Moniliformidae Van Cleave 1934 group *sedis mutabilis*

Diagnosis: Double wall of muscle around proboscis receptacle (15).

Moniliformidae Van Cleave 1934

Diagnosis: With characters of the group.

Palaeacanthocephala Meyer 1933

Diagnosis: Lateral lacunar canals (16); fragmented giant nuclei (17); single ligament (18).

Eoacanthocephala Van Cleave 1936 *sedis mutabilis*

Diagnosis: Syncytial cement glands (19).

Hypothesis III (Fig. 87)

Diagnosis: As hypothesis II, except assuming that syncytial cement glands (12) are plesiomorphic for Acanthocephala.

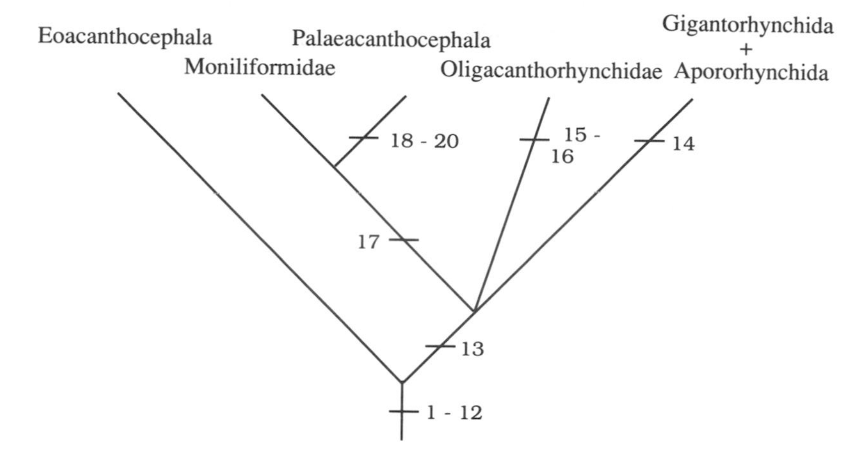

Figure 87. Phylogenetic tree for hypothesis III relationships among major groups of Acanthocephala.

Eoacanthocephala Van Cleave 1936 *sedis mutabilis*

Diagnosis: With characters of the phylum.

Apororhynchida Thapar 1927 + Gigantorhynchida Southwell and Macfie 1925 *sedis mutabilis*

Diagnosis: Nonsyncytial cement glands (13); spines and hooks on proboscis (14).

Oligacanthorhynchidae Southwell and Macfie 1925 *sedis mutabilis*

Diagnosis: Nonsyncytial cement glands (13); protonephridia present (15); proboscis hooks six per row in six rows (16).

Palaeacanthocephala Meyer 1933 + Moniliformidae Van Cleave 1934 group *sedis mutabilis*

Diagnosis: Nonsyncytial cement glands (13); double wall of muscle around proboscis receptacle (17).

Moniliformidae Van Cleave 1934

Diagnosis: With characters of the group.

Palaeacanthocephala Meyer 1933

Diagnosis: Lateral lacunar canals (18); fragmented giant nuclei (19); single ligament (20).

None of the three hypotheses supports the monophyly of the Archiacanthocephala Meyer 1933, comprising the Oligacanthorhynchidae, the Apororhynchida and Gigantorhynchida, and the Moniliformidae. In addition, none of the three hypotheses provides evidence that the Moniliformidae is monophyletic. In hypothesis I, four major monophyletic groups can be recognized and the Oligacanthorhynchidae is the sister group of the rest of the Acanthocephala. In hypothesis II, the four groups are monophyletic, but their relative relationships to each other are completely unresolved. In hypothesis III, the Eoacanthocephala represent the sister group of the rest of the Acanthocephala, but the Eoacanthocephala is not necessarily monophyletic. Clearly, more characters are needed to provide a robust starting point for elucidating the phylogenetic relationships of this fascinating group of parasites.

The only phylogenetic systematic treatment of any acanthocephalan group to date is the study by Amin (1986) for three members of the species-rich genus *Acanthocephalus* inhabiting North American freshwater fish.

Family Echinorhynchidae Cobbold 1897
***Acanthocephalus* Koelreuter 1771**

Reference: Amin (1986).

Plesiomorphic states for the genus proposed by Amin (1986): mean male body length greater than 5 mm (1); mean female body length more than 10 mm (2); body cylindrical (3); mean anterior testis length more than 800 µm (4); mean anterior testis length less than 15% of total body length (5); mean anterior testis width more than 50% of total body width (6); mean number of proboscis hooks per row in males less than nine (7); mean number of proboscis hooks per row in females less than nine (8); mean length of longest proboscis hooks in males more than 60 µm (9); mean length of longest proboscis hooks in females more than 70 µm (10).

***Acanthocephalus dirus* species group (Fig. 88)**

Diagnosis: Mean male body length 3–4 mm (1); mean female body length 7–9 mm (2); mean anterior testis length 450–800 µm (3); mean number of proboscis hooks per row in females 9–11 (4); mean length of longest hooks in males less than 50 µm (5); mean length of longest hooks in females less than 60 µm (6).

***dirus* (Van Cleave 1931) Van Cleave and Townsend 1936**

Diagnosis: Mean anterior testis length more than 19% of total body length *(7); mean number of proboscis hooks per row in males nine to ten (8); meristogram pattern C of Amin (1986) (9).

***alabamensis* Amin and Williams 1983**

Diagnosis: Mean male body length less than 3 mm (10); mean female body length less than 6 mm (11); mean anterior testis length less than 450 µm (12); mean length of largest proboscis hooks in males less than 50 µm (13); mean length of largest proboscis hooks in females less than 60 µm (14); mean anterior testis length more than 19% of total body length *(15); meristogram pattern D of Amin (1986) (16).

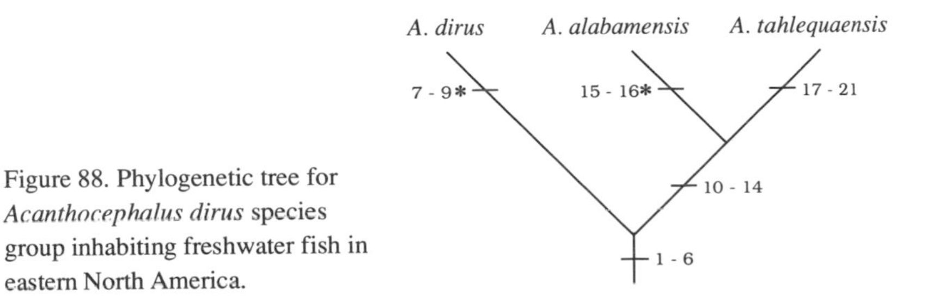

Figure 88. Phylogenetic tree for *Acanthocephalus dirus* species group inhabiting freshwater fish in eastern North America.

***tahlequaensis* Oetinger and Buckner 1976**

Diagnosis: Mean male body length less than 3 mm (10); mean female body length less than 6 mm (11); mean anterior testis length less than 450 μm (12); mean length of largest proboscis hooks in males less than 50 μm (13); mean length of largest proboscis hooks in females less than 60 μm (14); mean number of proboscis hooks per row in females greater than 11 (17); mean number of proboscis hooks per row in males more than 11 (18); mean anterior testis width less than 42% of total body width (19); body spindle-shaped (20); meristogram pattern E of Amin (1986) (21).

SUMMARY

The size of this appendix may suggest to some that we are overselling the point, but because we are advocating a fundamental change in the way biologists view the problem of reconstructing parasite phylogenies, we feel it is necessary to make as strong a case as possible. The data base presented in this appendix indicates clearly that phylogenetic relationships can be reconstructed on the basis of characteristics of the parasites themselves, without resorting to theories about coevolution or the effect of the host on the parasite. It also indicates that we can produce phylogenetic hypotheses for groups of parasites and use that information in studies of coevolution without introducing a crippling element of circular reasoning into the process. Thus, the stage is set for a revitalization of parasite evolutionary biology.

REFERENCES

Adamson, M. L.

1987. Phylogenetic analysis of the higher classification of the Nematoda. *Can. J. Zool.* 65:1478–1482.

Adamson, M. L., and D. van Waerebeke

1985. The Rhigonematida (Nematoda) of diplopods: Reclassification and its cladistic representation. *Ann. Parasitol. Hum. Comp.* 60:685–702.

Aho, J. M.

1990. Helminth communities of amphibians and reptiles: Comparative approaches to understanding patterns and processes. In *Parasite Communities: Patterns and Processes,* ed. G. Esch, A. Bush, and J. Aho, 157–195. Chapman and Hall, New York.

Alberch, P.

1980. Ontogenesis and morphological diversification. *Am. Zool.* 20:653–667.

1985. Problems with the interpretation of developmental sequences. *Syst. Zool.* 34:46–58.

Alberch, P., and J. Alberch

1981. Heterochronic mechanisms of morphological diversification and evolutionary change in the neotropical salamander, *Bolitoglossa occidentalis* (Amphibia: Plethodontidae). *J. Morphol.* 167:249–264.

Alberch, P., S. J. Gould, G. F. Oster, and D. B. Wake

1979. Size and shape in ontogeny and phylogeny. *Paleobiology* 5:296–315.

Allison, M. J., A. Pezzia, I. Hasigawa, and E. Gerszten

1974. A case of hookworm infection in a pre-Columbian American. *Am. J. Phys. Anthropol.* 41:103–106.

Ambros, V., and J. R. Horvitz

1984. Heterochronic mutants of the nematode *Caenorhabditis elegans*. *Science* 226:409–416.

Amin, O.

1986. On the species and populations of the genus *Acanthocephalus* (Acanthocephala: Echinorhynchidae) from North American freshwater fishes: A cladistic analysis. *Proc. Biol. Soc. Wash.* 99:574–579.

Anderson, R. C.

1957. The life cycles of dipetalonematid nematodes (Filaroidea: Dipetalonematidae): The problem of their evolution. *J. Helminthol.* 31:203–224.

1958. Possible steps in the evolution of filarial life cycles. *Proc. 6th Int. Congr. Trop. Med. Malaria* 2:444–449.

1982. Host-parasite relations and the evolution of the Metastrongyloidea (Nematoda). *Mem. Mus. Natl. Hist. Nat. Ser. A, Zool.* 123:129–133.

1984. The origins of zooparasitic nematodes. *Can. J. Zool.* 62:317–328.

Anderson, R. M., and R. M. May

1982. Coevolution of hosts and parasites. *Parasitology* 85:411–426.

1985. Epidemiology and genetics in the coevolution of parasites and hosts. *Proc. Roy. Soc. Lond. B* 219:281–283.

Anderson, S., and C. S. Anderson

1975. Three Monte Carlo models of faunal evolution. *Am. Mus. Novitates* 2563:1–6.

Andersson, M.

1982. Sexual selection, natural selection, and quality advertisement. *Biol. J. Linn. Soc.* 17:375–393.

1986. Evolution of condition-dependent sex ornaments and mating preferences: Sexual selection based on viability differences. *Evolution* 40:804–816.

Andrassy, I.

1976. *Evolution as a Basis for Systematization of Nematodes.* Putnam, London.

Araujo, A., L. F. Ferreira, U. Confalonieri, L. Nunez, and B. R. Filho

1985. The finding of *Enterobius vermicularis* eggs in pre-Columbian human coprolites. *Mem. Inst. Oswaldo Cruz* 80:141–143.

Archie, J. W.

1989. Homoplasy excess ratios: New indices for measuring levels of homoplasy in phylogenetic systematics and a critique of the consistency index. *Syst. Zool.* 38:253–269.

Arnold, S. J.

1983. Morphology, performance and fitness. *Am. Nat.* 23:347–361.

Avers, C.

1989. *Process and Pattern in Evolution.* Oxford University Press, Oxford.

Baer, J.-G.

1948. Les helminthes, parasites des vertébrés: Relations phylogénétiques entre leur évolution et celle de leurs hôtes. *Ann. Sci. Franche-Comte* 2:99–113.

1950. Phylogénie et cycles évolutifs des cestodes. *Rev. Suisse Zool.* 57:553–560.

1952. *Ecology of Animal Parasites.* University of Illinois Press, Urbana.

Baer, J.-G. (ed.)

1957. *Premier Symposium sur la specificité parasitaire des parasites des Vertébrés.* Paul Attinger, Neufchâtel.

Baldwin, J. G., and L. P. Schouest, Jr.

1990. Comparative detailed morphology of the Heteroderinae Filip'ev and Schuurmans Stekhoven, 1941, *sensu* Luc et al. (1988): Phylogenetic systematics and revised classification. *Syst. Parasitol.* 15:81–106.

Ball, G. H.
1943. Parasitism and evolution. *Am. Nat.* 77:345–364.

Bandoni, S. M., and D. R. Brooks
1987a. Revision and phylogenetic analysis of the Gyrocotylidea Poche, 1926 (Platyhelminthes: Cercomeria: Cercomeromorpha). *Can. J. Zool.* 65:2369–2389.
1987b. Revision and phylogenetic analysis of the Amphilinidea Poche, 1922 (Platyhelminthes: Cercomeria: Cercomeromorpha). *Can. J. Zool.,* 65:1110–1128.

Barta, J. R.
1989. Phylogenetic analysis of the class Sporozoea (Phylum Apicomplexa Levine, 1970): Evidence for the independent evolution of heteroxenous life cycles. *J. Parasitol.* 75:195–206.

Barton, N. H.
1989. Founder effect speciation. In *Speciation and its Consequences,* ed. D. Otte and J. A. Endler, 229–256. Sinauer, Sunderland, MA.

Barton, N. H., and B. Charlesworth
1984. Genetic revolutions, founder effects, and speciation. *Ann. Rev. Ecol. Syst.* 15:133–164.

Baum, D. A., and A. Larson
1991. Adaptation reviewed: A phylogenetic methodology for studying character macroevolution. *Syst. Zool.* 40:1–18.

Baverstock, P. R., R. Fielke, A. M. Johnson, R. A. Bray, and I. Beveridge
1991. Conflicting phylogenetic hypotheses for the parasitic platyhelminths tested by partial sequencing of 18S ribosomal RNA. *Int. J. Parasitol.* 21:329–339.

Begle, D. P.
1991. Relationships of the osmeroid fishes and the use of reductive characters in phylogenetic analysis. *Syst. Zool.* 40:33–53.

Behnke, J. M., and C. J. Barnard
1990. Coevolution of parasites and hosts: Host-parasite arms races and their consequences. In *Parasites: Immunity and Pathology. The Consequences of Parasitic Infection in Mammals,* ed. J. M. Behnke, 1–22. Taylor & Francis, London.

Bennett, G F., J. Blacou, E. M. White, and N. A. Williams
1978. Blood parasites of some birds from Senegal. *J. Wildl. Dis.* 14:67–73.

Benz, G. W., and G. B. Deets
1988. Fifty-one years later: An update on *Entepherus,* with a phylogenetic analysis of Cecropidae Dana, 1849 (Copepoda: Siphonostomatoidea). *Can. J. Zool.* 66:856–865.

Berenbaum, M. R.
1983. Coumarins and caterpillars: A case for coevolution. *Evolution* 37:163–179.

Berland, B., G. A. Bristow, and O. Grahl-Nielsen

1990. Chemotaxonomy of *Gyrocotyle* (Platyhelminthes: Cercomeria) species, parasites of chimaerid fishes (Holocephali), by chemometry of their fatty acids. *Mar. Biol.* 105:185–189.

Beverly-Burton, M., and G. J. Klassen

1990. New approaches to the systematics of the ancyrocephalid monogenea from Nearctic fishes. *J. Parasitol.* 76:1–21.

Bickham, J. W.

1981. Two-hundred-million-year-old chromosomes: Deceleration of the rate of karyotypic evolution in turtles. *Science* 212:1291–1293.

Blair, D., D. R. Brooks, J. Purdie, and L. Melville

1988. *Timoniella absita* n. sp. (Digenea: Cryptogonimidae) from the saltwater crocodile (*Crocodylus porosus* Schneider) from Australia. *Can. J. Zool.* 66:1763–1766.

Boeger, W. A., and D. C. Kritsky

1988. Neotropical Monogenea. 12. Dactylogyridae from *Serrasalmus nattereri* (Cypriniformes, Serrasalmidae) and aspects of their morphologic variation and distribution in the Brazilian Amazon. *Proc. Helminthol. Soc. Wash.* 55:188–213.

1989. Phylogeny, coevolution, and revision of the Hexabothriidae Price, 1942 (Monogenea). *Int. J. Parasitol.* 19:425–440.

In press. Phylogeny and a revised classification of the Monogenoidea Bychowsky, 1937 (Platyhelminthes). *Syst. Parasitol.*

Bonner, J. T., ed.

1982. *Heterochrony in Evolution. A Multidisciplinary Approach.* Plenum, New York.

Borgia, G.

1979. Sexual selection and the evolution of mating systems. In *Sexual Selection and Reproductive Competition in Insects,* ed. M. S. Blum and M. A. Blum, p. 19–80. Academic Press, New York.

1986. Satin bowerbird parasites: A test of the bright male hypothesis. *Behav. Ecol. Sociobiol.* 19:355–358.

Bowler, P. J.

1983. *The Eclipse of Darwinism.* Johns Hopkins University Press, Baltimore.

1989. *Evolution: The History of an Idea.* Revised ed. University of California Press, Berkeley and Los Angeles.

Brandt, B. B.

1936. Parasites of certain North Carolina salientia. *Ecol. Monogr.* 6:491–532.

Bray, R. A.

1986. A revision of the family Zoogonidae Odhner, 1902 (Platyhelminthes: Digenea): Introduction and subfamily Zoogoninae. *Syst. Parasitol.* 9:3–28.

1987. A revision of the family Zoogonidae Odhner, 1902 (Platyhelminthes: Digenea): Subfamily Lepidophyllinae and comments on some aspects of biology. *Syst. Parasitol.* 9:83–123.

1988. A discussion of the status of the subfamily Baccigerinae Yamaguti, 1958 (Digenea) and the constitution of the family Fellodistomidae Nicoll, 1909. *Syst. Parasitol.* 11:97–112.

Bremer, B.

1987. The sister-group of the paleotropical tribe Argostemmateae: A redefined neotropical tribe Hamelieae (Rubiaceae, Rubioidea). *Cladistics* 3:35–51.

Bremer, B., and K. Bremer

1989. Cladistic analysis of blue-green procaryote interrelationships and chloroplast origin based on 16S rRna oligonucleotide catalogues. *J. Evol. Biol.* 2:13–30.

Bristow, G. A., and B. Berland

1988. A preliminary electrophoretic investigation of the gyrocotylid parasites of *Chimaera monstrosa* L. *Sarsia* 73:75–77.

Brooks, D. R.

1976. Parasites of amphibians of the Great Plains, II. Platyhelminths of amphibians in Nebraska. *Bull. Univ. Nebr. State Mus.* 10:65–92.

1977. Evolutionary history of some plagiorchioid trematodes of anurans. *Syst. Zool.* 26:277–289.

1978a. Systematic status of proteocephalid cestodes from reptiles and amphibians in North America with descriptions of three new species. *Proc. Helminthol. Soc. Wash.* 45:1–28.

1978b. Evolutionary history of the cestode order Proteocephalidea. *Syst. Zool.* 27:312–323.

1979a. Testing the context and extent of host-parasite coevolution. *Syst. Zool.* 28:299–307.

1979b. Testing hypotheses of evolutionary relationships among parasitic helminths: The digeneans of crocodilians. *Am. Zool.* 19:1225–1238.

1981a. Revision of the Acanthostominae (Digenea: Cryptogonimidae). *Zool. J. Linn. Soc.* 70:313–382.

1981b. Hennig's parasitological method: A proposed solution. *Syst. Zool.* 30:229–249.

1982. Higher level classification of parasitic platyhelminths and fundamentals of cestode classification. In *Parasites: Their World and Ours,* ed. D. F. Mettrick and S. S. Desser, 189–193. Elsevier, Amsterdam.

1985. Phylogenetics and the future of helminth systematics. *J. Parasitol.* 71:719–727.

1988. Macroevolutionary comparisons of host and parasite phylogenies. *Ann. Rev. Ecol. Syst.* 19:235–259.

1989a. A summary of the database pertaining to the phylogeny of the major groups of parasitic platyhelminths, with a revised classification. *Can. J. Zool.* 67:714–720.

1989b. The phylogeny of the Cercomeria (Platyhelminthes: Rhabdocoela) and general evolutionary principles. *J. Parasitol.* 75:606–616.

1990a. Phylogenetic systematic evaluation of the classification of the Zoogonidae Odhner, 1902 (Cercomeria: Trematoda: Digenea: Plagiorchiformes). *Syst. Parasitol.* 19:127–137

1990b. Parsimony analysis in historical biogeography and coevolution: Methodological and theoretical update. *Syst. Zool.* 39:14–30.

1992a. Origins, diversification, and historical structure of the helminth fauna inhabiting neotropical freshwater stingrays (Potamotrygonidae). *J. Parasitol.* 78:588–595.

1992b. New distribution records for *Bunodera eucaliae* (Miller, 1936) Miller, 1940 and *Bunodera inconstans* (Lasee, Font, and Sutherland, 1988) comb. n. (Digenea: Plagiorchiformes: Allocreadiidae: Bunoderinae) with discussion of their phylogenetic relationships. *J. Parasitol.* 78:779–783.

In press. Phylogenetic systematic evaluation of the Monticelliidae (Eucestoda: Proteocephalidea). *Rev. Brasil. Biol.*

Brooks, D. R., and J. F. R. Amato

1992. Cestode parasites in *Potamotrygon motoro* (Muller and Henle) (Chondrichthyes: Potamotrygonidae) from southwestern Brazil, including *Rhinebothroides mclennanae* sp. n. (Tetraphyllidea: Phyllobothriidae) and a revised host-parasite checklist for helminths inhabiting neotropical freshwater stingrays. *J. Parasitol.* 78:393–398.

Brooks, D. R., and S. M. Bandoni

1988. Coevolution and relicts. *Syst. Zool.* 37:19–33.

Brooks, D. R., and J. N. Caira

1982. *Atrophecaecum lobacetabulare* sp. n. (Digenea: Cryptogonimidae: Acanthostominae) with discussion of the generic status of *Paracanthostomum* Fischthal and Kuntz, 1965 and *Ateuchocephala* Coil and Kuntz, 1960. *Proc. Biol. Soc. Wash.* 95:223–231.

Brooks, D. R., and T. L. Deardorff

1988. *Rhinebothrium devaneyi* n. sp. (Eucestoda: Tetraphyllidea) and *Echinocephalus overstreeti* Deardorff and Ko, 1983 (Nematoda: Gnathostomatidae) in a thorny back ray, *Urogymnus asperrimus,* from Enewetak Atoll, with phylogenetic analysis of both species groups. *J. Parasitol.* 74:459–465.

Brooks, D. R., and D. R. Glen

1982. Pinworms and primates: A case study in coevolution. *Proc. Helminthol. Soc. Wash.* 49:76–85.

Brooks, D. R., and B. Holcman

In press. Phylogenetic revision and new classification of the Acanthostominae (Platyhelminthes: Cercomeria: Trematoda: Digenea: Opisthorchiformes: Cryptogonimidae) with description of a new species. *Proc. Biol. Soc. Wash.*

Brooks, D. R., and C. A. Macdonald

1986. A new species of *Phyllodistomum* Braun, 1899 (Digenea: Gorgoderidae) in a neotropical catfish, with a discussion of the generic relationships of the Gorgoderidae. *Can. J. Zool.* 64:1326–1330.

Brooks, D. R., and D. A. McLennan

1990. Searching for a general theory of evolution. *J. Ideas* 1:35–46.

1991. *Phylogeny, Ecology, and Behavior: A Research Program in Comparative Biology.* University of Chicago Press, Chicago.

1992. Historical ecology as a research program in macroevolution. In *Systematics, Historical Ecology, and North American Freshwater Fishes,* ed. R. L. Mayden, 76–113. Stanford University Press, Stanford.

In press. Historical ecology and community ecology. In *Historical and Geographic Determinants of Diversity,* ed. R. Ricklefs and D. Schluter. University of Chicago Press, Chicago.

Brooks, D. R., and C. Mitter

1984. Analytical basis of coevolution. In *Fungus/Insect Relationships: Perspectives in Ecology and Evolution,* ed. Q. Wheeler and M. Blackwell, 42–53. Columbia University Press, New York.

Brooks, D. R., and R. T. O'Grady

1989. Crocodilians and their helminth parasites: Macroevolutionary considerations. *Am. Zool.* 29:873–883.

Brooks, D. R., and R. M. Overstreet

1978. The family Liolopidae (Digenea) including a new genus and two new species from crocodilians. *Int. J. Parasitol.* 8:267–273.

Brooks, D. R., and J. R. Palmieri

1979. *Neopronocephalus orientalis* sp. n. (Digenea: Pronocephalidae) and *Spirhapalum elongatum* Rohde, Lee and Lim, 1968 (Digenea: Spirorchiidae) from *Cuora amboinensis* (Daudin) in Malaysia. *Proc. Helminthol. Soc. Wash.* 46:55–57.

Brooks, D. R., and G. Rasmussen

1984. Proteocephalid cestodes from Venezuelan catfish, with a new classification of the Monticelliidae. *Proc. Biol. Soc. Wash.* 97:748–760.

Brooks, D. R., and E. O. Wiley

1988. *Evolution as Entropy: Toward a Unified Theory of Biology.* University of Chicago Press, Chigaco.

Brooks, D. R., M. A. Mayes, and T. B. Thorson

1981a. Systematic review of cestodes infecting freshwater stingrays (Chondrichthyes: Potamotrygonidae) including four new species from Venezuela. *Proc. Helminthol. Soc. Wash.* 48:43–64.

Brooks, D. R., T. B. Thorson, and M. A. Mayes

1981b. Freshwater stingrays (Potamotrygonidae) and their helminth parasites: Testing hypotheses of evolution and coevolution. In *Advances in Cladistics: Proceedings of the First Meeting of the Willi Hennig Society,* ed.

V. A. Funk and D. R. Brooks, 147–175. New York Botanical Garden, New York.

Brooks, D. R., R. T. O'Grady, and D. R. Glen

1985a. The phylogeny of the Cercomeria Brooks, 1982 (Platyhelminthes). *Proc. Helminthol. Soc. Wash.* 52:1–20.

1985b. Phylogenetic analysis of the Digenea (Platyhelminthes: Cercomeria) with comments on their adaptive radiation. *Can. J. Zool.* 63:411–443.

Brooks, D. R., S. M. Bandoni, C. A. Macdonald, and R. T. O'Grady

1989a. Aspects of the phylogeny of the Trematoda Rudolphi, 1808 (Platyhelminthes: Cercomeria). *Can. J. Zool.* 67:2609–2624.

Brooks, D. R., J. Collier, B. A. Maurer, J. D. H. Smith, and E. O. Wiley

1989b. Entropy and information in evolving biological systems. *Biol. Philos.* 4:407–432.

Brooks, D. R., E. P. Hoberg, and P. J. Weekes

1991. Preliminary phylogenetic systematic analysis of the Eucestoda Southwell, 1930 (Platyhelminthes: Cercomeria). *Proc. Biol. Soc. Wash.* 104:651–668.

Brooks, D. R., J. B. Catto, and J. F. R. Amato

1992. A new phylogenetic classification of the genera of the Proterodiplostomidae Dubois, 1936 (Digenea: Strigeiformes). *Proc. Biol. Soc. Wash.* 105: 143–147.

Brusca, R. C., and G. J. Brusca

1990. *Invertebrates.* Sinauer, Sunderland, MA.

Burt, M. D. B., and L. Jarecka

1982. Phylogenetic host specificity of cestodes. *Mem. Mus. Natl. Hist. Nat. Ser. A, Zool.* 123:47–51.

Bush, A. O., J. M. Aho, and C. R. Kennedy

1990. Ecological versus phylogenetic determinants of helminth parasite community richness. *Evol. Ecol.* 4:1–20.

Butler, P. M.

1982. Directions of evolution in the mammalian dentition. In *Problems of Phylogenetic Reconstruction,* ed. K. A. Joysey and A. E. Friday, 235–244. Systematics Association Special Vol. 21. Academic Press, London.

Buttner, A.

1951. La progénese chez les trématodes digénétiques (fin): Etude de quelque metacercaires à evolution inconnue et de certaines formes de développement voisines de la progénese. Conclusions générales. *Ann. Parasitol. Hum. Comp.* 26:279–322.

Byrd, E. E.

1939. Studies on the blood flukes of the family Spirorchidae. Part II. Revision of the family and description of new species. *J. Tenn. Acad. Sci.* 14:116–161.

Byrd, E. E., and J. F. Denton

1938. New trematodes of the subfamily Reniferinae, with a discussion of the systematics of the genera and species assigned to the subfamily group. *J. Parasitol.* 24:379–385.

Byrd, E. E., and R. J. Reiber

1942. Strigeid trematodes of the alligator, with remarks on the prostate gland and terminal portions of the genital ducts. *J. Parasitol.* 28:51–73.

Cable, R. M.

1974. Phylogeny and taxonomy of trematodes with reference to marine species. in *Symbiosis in the Sea,* ed. W. B. Vernberg, 173–193. University of South Carolina Press, Columbia.

Caira, J. N.

1989. A revision of the North American papillose Allocreadiidae with independent cladistic analyses of larval and adult forms. *Bull. Univ. Neb. State Mus.* 11:1–58.

Cameron, T. W. M.

1929. The species of *Enterobius* in primates. *J. Helminthol.* 7:161–182.

1956. *Parasites and Parasitism.* Wiley, New York.

1964. Host specificity and evolution of helminthic parasites. *Adv. Parasitol.* 2:1–34.

Carmichael, A. C.

1984. Phylogeny and historical biogeography of the Schistosomatidae. PhD dissertation, Michigan State University, East Lansing

Carney, J. P., and D. R. Brooks

1991. Phylogenetic analysis of *Alloglossidium* Simer, 1929 (Digenea: Plagiorchiformes: Macroderoididae) with discussion of the origin of truncated life cycle patterns in the genus. *J. Parasitol.* 77:890–900.

Carson, H. L.

1975. The genetics of speciation at the diploid level. *Am. Nat.* 109:83–92.

1982. Speciation as a major reorganization of polygenic balances. In *Mechanisms of Speciation,* ed. C. Barigozzi, 411–433. Liss, New York.

Carson, H. L., and K. Y. Kaneshiro

1976. *Drosophila* of Hawaii: Systematics and evolutionary genetics. *Annu. Rev. Ecol. Syst.* 7:311–346.

Carson, H. L., and A. R. Templeton

1984. Genetic revolutions in relation to speciation phenomena: The founding of new populations. *Annu. Rev. Ecol. Syst.* 15:97–131.

Case, S. M.

1978. Biochemical systematics of some members of the genus *Rana* native to North America. *Syst. Zool.* 27:299–311.

Caullery, M.

1952. *Parasitism and Symbiosis.* Sidgwick and Jackson, London.

Chabaud, A. G.

1954. Sur le cycle évolutif des spirurides et nématodes ayant une biologie comparable: Valeur systèmatique des caractères biologiques. *Ann. Parasitol. Hum. Comp.* 29:42–88, 206–249, 358–425.

1955. Essai d'interpretation phylétique des cycles évolutifs chez les Nématodes parasites de vertebres. *Ann. Parasitol. Hum. Comp.* 30:83–126.

1965a. Specificité parasitaire. In *Traite de Zoologie,* ed. P. P. Grasse. Tome 4 Fascicule 2, 548–557. Masson et Cie, Paris.

1965b. Cycles évolutifs des nématodes parasites de vertebres. In *Traite de Zoologie,* ed. P. P. Grasse. Tome 4, Fascicule 2, 437–463. Masson et Cie, Paris.

1982. Evolution et taxonomie des nématodes: Revue. In *Parasites: Their World and Ours,* ed. D. F. Mettrick and S. S. Desser, 216–221. Elsevier, Amsterdam.

Chandler, A. C.

1940. *Introduction to Parasitology.* Wiley, New York.

Chandler, A. C., and C. P. Read

1961. *Introduction to Parasitology.* Wiley, New York.

Charlesworth, B., and S. Rouhani

1988. The probability of peak shifts in a founder population. II. An additive polygenic trait. *Evolution* 42:1129–1145.

Cheng, T. C.

1986. *General Parasitology,* 2nd ed. Academic Press, New York.

Chitwood, B. G., and M. B. Chitwood

1974. *An Introduction to Nematology.* Monumental, Baltimore.

Clayton, D. H.

1991. The influence of parasites on host sexual selection. *Parasitol. Today* 7:329–334.

Coddington, J. A.

1988. Cladistic tests of adaptational hypotheses. *Cladistics* 4:3–22.

Colbert, E. H., and M. Morales

1991. *Evolution of the Vertebrates,* 4th ed. Wiley, New York.

Colin, J. A., H. H. Williams, and O. Halvorsen

1986. One or three gyrocotylideans (Platyhelminthes) in *Chimaera monstrosa* (Holocephali)? *J. Parasitol.* 72:10–21.

Collette, B. B.

1982. South American needlefishes of the genus *Potamorrhaphis* (Beloniformes: Belonidae). *Proc. Biol. Soc. Wash.* 95:714–747.

Combes, C.

1972. Influence of the behaviour of amphibians on helminth life-cycles. *Zool. J. Linn. Soc.* Suppl. 1: 151–170.

Conway Morris, S., and D. W. T. Crompton

1982. The origins and evolution of the Acanthocephala. *Biol. Rev.* 57:85–115.

Cort, W. W.

1917. Homologies of the excretory system of fork-tailed cercariae. *J. Parasitol.* 4:49–57.

Cracraft, J.

1982a. A nonequilibrium theory for the rate-control of speciation and extinction and the origin of macroevolutionary patterns. *Syst. Zool.* 31:348–365.

1982b. Geographic differentiation, cladistics, and vicariance biogeography: Reconstructing the tempo and mode of evolution. *Am. Zool.* 22:411–424.

1984. Conceptual and methodological aspects of the study of evolutionary rates, with some comments on bradytely in birds. In *Living Fossils,* ed. N. Eldredge and S. M. Stanley, 95–104. Springer-Verlag, New York.

1985. Biological diversification and its causes. *Ann. Miss. Bot. Garden* 72:794–822.

Crane, P. R.

1985. Phylogenetic relationships in seed plants. *Cladistics* 1:329–348.

Crawley, H.

1923. Evolution in the ciliate family Ophryoscolecidae. *Proc. Acad. Nat. Sci. Philadelphia* 75:393–412.

Cressey, R. F., B. Collette, and J. Russo

1983. Copepods and scombrid fishes: A study in host parasite relationships. *Fish. Bull.* 81:227–265.

Cribb, T. H.

1988. Life cycle and biology of *Prototransversotrema steeri* Angel, 1969 (Digenea: Transversotrematidae). *Austral. J. Zool.* 36:111–129.

Cribb, T. H., and R. A. Bray

In press. A review of the Tandanicolinae (=Monodhelminthinae) (Digenea: Fellodistomidae) with a description of *Prosogonarium angelae* n. sp. *Syst. Parasitol.*

Crisci, J. V., I. J. Gamundi, and M. N. Cansello

1988. A cladistic analysis of the genus *Cyttaria* (Fungi-Ascomycotina). *Cladistics* 4:279–290.

Crompton, D. W. T.

1973. The sites occupied by some parasitic helminths in the alimentary tract of vertebrates. *Biol. Rev.* 48:27–83.

Crompton, D. W. T., and S. M. Joyner

1980. *Parasitic Worms.* Wykeham, London.

Darling, S. J.

1920. Observations on the geographical and ethnological distribution of hookworms. *Parasitology* 12:217–233.

1921. The distribution of hookworms in the zoological regions. *Science* 53:323–324.

Darwin, C.

1859. *On the Origin of Species by Means of Natural Selection.* 1st ed. John Murray, London.

1871. *The Descent of Man, and Selection in Relation to Sex.* John Murray, London.

1872. *The Origin of Species*. 6th ed. John Murray, London.

Deblock, S.

1974. Contribution à l'étude des Microphallidae Travassos, 1920 (Trematoda). XXVII. *Microphallus abortivus* n. sp. espèce à cycle évolutif abrégé originaire d'Oleron. *Ann. Parasitol. Hum. Comp.* 49:175–184.

Deets, G. B.

1987. Phylogenetic analysis and revision of *Kroeyerina* Wilson, 1932 (Siphonostomatoidea: Kroyeriidae), copepods parasitic on chondrichthyans, with descriptions of four new species and the erection of a new genus, *Prokroyeria*. *Can. J. Zool.* 65:2121–2148.

Deets, G. B., and J.-S. Ho

1988. Phylogenetic analysis of the Eudactylinidae with descriptions of two new genera. *Proc. Biol. Soc. Wash.* 101:317–339.

Densmore, L. D., III, and R. D. Owen

1989. Molecular systematics of the order Crocodilia. *Am. Zool.* 29:831–841.

Densmore, L. D., III, and P. S. White

1991. The systematics and evolution of the Crocodilia as suggested by restriction endonuclease analysis of mitochondrial and nuclear ribosomal DNA. *Copeia* 1991:602–615.

Dickinson, H., and J. Antonovics

1973. Theoretical consideration of sympatric divergence. *Am. Nat.* 107:256–274.

Dial, K. P., and J. M. Marzluff

1989. Nonrandom diversification within taxonomic assemblages. *Syst. Zool.* 38:26–37.

Diehl, S. R., and G. L. Bush

1989. The role of habitat preference in adaptation and speciation. In *Speciation and Its Consequences,* ed. D. Otte and J. A. Endler, 345–365. Sinauer, Sunderland, MA.

Dietz, R. S., and J. C. Holden

1966. Miogeoclines (Miogeosynclines) in space and time. *J. Geol.* 74:566–583.

Dobzhansky, T.

1937. *Genetics and the Origin of Species*. Columbia University Press, New York.

Dodson, E. O., and P. Dodson

1985. *Evolution: Process and Product.* Prindle, Weber, and Schmidt, Boston.

Dogiel, V. A.

1962. *General Parasitology.* 3rd ed. Oliver and Boyd, London.

Dojiri, M., and G. B. Deets

1988. *Norkus cladocephalus,* new genus, new species (Siphonostomatoida: Sphyriidae), a copepod parasitic on an elasmobranch from southern California waters, with a phylogenetic analysis of the Sphyriidae. *J. Crust. Biol.* 8:679–687.

Dougherty, E. C.

1949. The phylogeny of the nematode family Metastrongylidae Leiper (1909): A correlation of host and symbiote evolution. *Parasitology* 39:222–234.

1951. Evolution of zooparasitic groups in the phylum Nematoda with special reference to host distribution. *J. Parasitol.* 37:353–378.

Dubois, G.

1945. A propos de la spécificité parasitaire des Strigeida. *Bull. Soc. Neuchat. Sci. Nat.* 69:5–103.

Duellman, W. E., and L. Trueb

1986. *Biology of Amphibians.* McGraw-Hill, New York.

Dunn, E. R.

1925. The host-parasite method and the distribution of frogs. *Am. Nat.* 59:370–375.

Dunn, F. L.

1966. Patterns of parasitism in primates: Phylogenetic and ecological interpretations, with particular reference to the Hominoidea. *Folia Primatol.* 4:329–345.

Durette-Desset, M. C.

1985. Trichostrongyloid nematodes and their vertebrate hosts: Reconstruction of the phylogeny of a parasitic group. *Adv. Parasitol.* 24:239–306.

Durette-Desset, M. C., and A. G. Chabaud

1981. Nouvel essai de classification des Nématodes Trichostrongyloidea. *Ann. Parasitol. Hum. Comp.* 56:297–312.

Durette-Desset, M. C., and J.-L. Justine

1991. A cladistic analysis of the genera in the subfamily Pudicinae (Nematoda, Trichostrongyloidea, Heligmonellidae). *Int. J. Parasitol.* 21:579–587.

Ehlers, U.

1984. Phylogenetisches System der Plathelminthes. *Verhandl. Naturwiss. Ver. Hamburg* 27:291–294.

1985a. *Das phylogenetische System der Plathelminthes.* Gustav Fischer Verlag, Stuttgart.

1985b. Phylogenetic relationships within the Platyhelminthes. In *The Origins and Relationships of Lower Invertebrates,* ed. S. Conway Morris, J. D. George, R. Gibson, and H. M. Platt, 143–158. Oxford University Press, Oxford.

1986. Comments on a phylogenetic system of the Platyhelminthes. *Hydrobiologia* 132:1–12.

Ehrlich, P. R., and P. H. Raven

1964. Butterflies and plants: A study in coevolution. *Evolution* 18:586–608.

Ehrlich, P. R., and E. O. Wilson

1991. Biodiversity studies: Science and policy. *Science* 253:758–762.

Eichler, W.

1941a. Wirtsspezifitat und stammesgeschichtliche Gleichläufigkeit (Fahrenholz'sche Regel) bei Parasiten im allgemeinen und bei Mallophagen im besonderen. *Zool. Anz.* 132:254–262.

1941b. Korrelation in der Stammesentwicklung von Wirten und Parasiten. *Zeitschr. Parasitenk. (Berlin)* 12:94.

1942. Die Entfaltungsregel und andere Gesetzmässigkeiten in den parasitologischen Beziehungen der Mallophagen und anderer ständiger Parasiten zu ihren Wirten. *Zool. Anz.* 137:77–83.

1948a. Evolutionsfragender Wirtsspezifitat. *Biol. Zentralbl.* 67:373–406.

1948b. Some rules in ectoparasitism. *Ann. Mag. Nat. Hist.* 1:588–598.

1966. Two new evolutionary terms for speciation in parasitic animals. *Syst. Zool.* 15:216–218.

1973. Neuere Überlegungen zu den parasitophyletischen Regeln. *Helminthologia* 14:1–4.

1982. Les règles parasitophylétiques comme phénomène de la théorie de l'évolution. 2nd Symposium on Host Specificity among Parasites of Vertebrates. *Mem. Mus. Natl. Hist. Nat., Nouv. Ser.* 123:69–71.

Ejsmont, L.

1927. *Spirhapalum polesianum* n. g., n. sp. trematode du sang d'*Emys orbicularis* L. *Ann. Parasitol. Hum. Comp.* 5:220–235.

Eldredge, N.

1976. Differential evolutionary rates. *Paleobiology* 2:174–177.

Eldredge, N., and J. Cracraft

1980. *Phylogenetic Patterns and the Evolutionary Process*. Columbia University Press, New York.

Emerson, S. B.

1986. Heterochrony and frogs: The relationship of a life history trait to morphological form. *Am. Nat.* 127:167–183.

Eshel, I., and E. Aiken

1983. On the evolutionary instability of inner Nash solutions of the coevolving populations. *J. Math. Biol.* 18:123–133.

Eshel, I., and W. D. Hamilton

1984. Parent–offspring correlation in fitness under fluctuating selection. *Proc. Roy. Soc. Lond. B* 222:1–14.

Estrada, E., and B. J. Meggers.

1961. A complex of traits of probable transpacific origin on the coast of Ecuador. *Am. Anthropol.* 63:913–939.

Euzet, L., and C. Combes

1980. Les problèmes de l'espèce chez les animaux parasites. *Mem. Soc. Zool. France* 40:239–285.

Ewing, H. E.

1924. On the taxonomy, biology and distribution of the biting lice of the family Gyropidae. *Proc. U.S. Natl. Mus.* 63:1–42.

1928. A revision of the American lice of the genus *Pediculus,* together with consideration of the significance of their geographical and host distribution. *Proc. U.S. Natl. Mus.* 68:1–38.

Fahrenholz, H.

1909. Aus dem Myobien-Nachlass des Herrn Poppe. *Abhandl. Naturwiss. Vereins Bremen* 19:359–370.

1913. Ectoparasiten und Abstammungslehre. *Zool. Anz. (Leipzig)* 41:371–374.

Farrell, B. D., C. Mitter, and D. J. Futuyma

1992. Diversification at the insect–plant interface. *BioScience* 42:34–42.

Farris, J. S.

1989. The retention index and the rescaled consistency index. *Cladistics* 5:417–419.

Faulkner, C. T., S. Patton, and S. S. Johnson

1989. Prehistoric parasitism in Tennessee: Evidence from the analysis of dessicated fecal material collected from Big Bone Cave, Van Buren County, Tennessee. *J. Parasitol.* 75:461–463.

Feeny, P. P.

1976. Plant apparency and chemical defense. *Recent Adv. in Phytochem.* 10:1–40.

Felsenstein, J.

1978. The number of evolutionary trees. *Syst. Zool.* 27:27–33.

1981. Skepticism towards Santa Rosalia, or why are there so few kinds of animals? *Evolution* 35:124–138.

Ferreira, L. F., A. Araujo, and U. Confalonieri (eds.)

1988a. *Paleoparasitologia no Brasil.* Manguinhos, Rio de Janeiro.

Ferreira, L. F., A. Araujo, U. Confalonieri, M. Chame, and B. R. Filho

1988b. Encontro de ovos de ancilostomoideos em coprolitos humanos datados de 7.230 80 anos, no estado do Piaui, Brasil. In *Paleoparasitologia no Brasil,* ed. L. F. Ferreira, A. Araujo, and U. Confalonieri, 37. Manguinhos, Rio de Janeiro.

Ferreira, L. F., A. Araujo, U. Confalonieri, and L. Nunez

1989. Infeccao por *Enterobius vermicularis* em populacoes agro-pastoris pre-Colombianos de San Pedro de Atacama, Chile. *Mem. Inst. Oswaldo Cruz* 84:197–199.

Ferris, V. R.

1979. Cladistic approaches in the study of soil and plant parasitic nematodes. *Am. Zool.* 19:1195–1215.

Filipchenko, A. A.

1937. Ecological concept of parasitism. *Uch. Zap. Leningr. Gos. Univ. Ser. Biol. Nauk.* 13:4–14.

Fink, W. L.

1982. The conceptual relationships between ontogeny and phylogeny. *Paleobiology* 8:254–264.

Fisher, R.

1930. *The Genetical Theory of Natural Selection.* Clarendon Press, Oxford.

Fitzpatrick, J. W.

1988. Why so many passerine birds? A response to Raikow. *Syst. Zool.* 37:71–76.

Flessa, K. W., and J. S. Levinton

1975. Phanerozoic diversity patterns: Tests for randomness. *J. Geol.* 83: 239–248.

Font, W. F.

1980. The effect of progenesis on the evolution of *Alloglossidium* (Trematoda, Plagiorchiida, Macroderoididae). *Acta Parasitol. Polon.* 27:173–183.

Foster, W. D.

1965. *A History of Parasitology.* Livingstone, London.

Frailey, C. D.

1986. Late Miocene and Holocene mammals, exclusive of Notoungulata, of the Rio Acre region, western Amazonia. *Contr. Sci. Nat. Hist. Mus., Los Angeles Co.* 364:1–46.

Freeman, R. S.

1973. Ontogeny of cestodes and its bearing on their phylogeny and systematics. In *Advances in Parasitology,* Vol. 11, ed. B. Dawes, 481–517. Academic Press, New York.

Freire, S. E.

1987. A cladistic analysis of *Lucilia* Cass. (compositae, Inuleae). *Cladistics* 3:254–272.

Frey, E., J. Reiss, and S. F. Tarsitano

1989. The axial tail musculature of recent crocodiles and its phyletic implications. *Am. Zool.* 29:857–862.

Fry, G. F.

1977. *Analysis of prehistoric coprolites from Utah.* University of Utah Anthropological Papers, No. 97. University of Utah Press, Salt Lake City.

Fry, G. F., and H. J. Hall

1969. Parasitological examination of prehistoric coprolites from Utah. *Utah Acad. Sci. Arts Lett. Proc.* 46:102–105.

1975. Human coprolites from Antelope House: Preliminary analysis. *Kiva* 41:87–96.

Fry, G. F., and J. G. Moore

1969. *Enterobius vermicularis:* 10,000 year old human infection. *Science* 166:1620.

Funk, V. A., and D. R. Brooks

1990. Phylogenetic systematics as the basis of comparative biology. *Smithsonian Contr. Bot.* 73:1–45.

Futuyma, D. J.

1986. *Evolutionary Biology,* 2nd ed. Sinauer, Sunderland, MA.

Futuyma, D. J., and G. C. Mayer

1980. Non-allopatric speciation in animals. *Syst. Zool.* 29:254–271.

Futuyma, D. J., and G. Moreno

1988. The evolution of ecological specialization. *Annu. Rev. Ecol. Syst.* 19:207–233.

Futuyma, D. J., and M. Slatkin (eds.)
1983. *Coevolution.* Sinauer, Sunderland, MA.

Gadea, E.
1973. Sobre la filogenia interna des nematodos. *Publ. Inst. Biol. Apl. Barcelona* 54:87–92.

Gaffney, E. S.
1975. A phylogeny and classification of the higher categories of turtles. *Bull. Am. Mus. Nat. Hist.* 55:388–436.

Gaffney, E. S., and P. A. Meylan
1988. A phylogeny of turtles. In *The Phylogeny and Classification of the Tetrapods, Vol. 1: Amphibians, Reptiles, Birds,* ed. M. J. Benton, 157–219. Systematics Association Special Vol. 35A. Clarendon Press, Oxford.

Gaffney, E. S., J. H. Hutchinson, F. A. Jenkins, Jr., and L. J. Meeker
1987. Modern turtle origins: The oldest known cryptodire. *Science* 237:289–291.

Gardner, S. L.
1991. Phyletic coevolution between subterranean rodents of the genus *Ctenomys* (Rodentia: Hystricognathi) and nematodes of the genus *Paraspidodera* (Heterakoidea: Aspidodeidae) in the Neotropics: Temporal and evolutionary implications. *Zool. J. Linn. Soc.* 102:169–201.

Gardner, S. L., and M. L. Campbell
1992. Parasites as probes for biodiversity. *J. Parasitol.* 78:596–600.

Garnham, P. C. C.
1973. Distribution of malaria parasites in primates, insectivores, and bats. *Symp. Zool. Soc. Lond.* 33:377–404.

Gibson, D. I.
1981. Evolution of digeneans. *Parasitology* 82:161–163.
1987. Questions in digenean systematics and evolution. *Parasitology* 95:429–460.

Gibson, D. I., and R. A. Bray
1979. The Hemiuroidea: Terminology, systematics, and evolution. *Bull. Brit. Mus. (Nat. Hist.) (Zool.)* 36:35–146.

Gilinsky, N. L.
1981. Stabilizing selection in the Archaeogastropoda. *Paleobiology* 7:316–331.

Gittenberger, E.
1988. Sympatric speciation in snails: A largely neglected model. *Evolution* 42:826–828.

Glen, D. R., and D. R. Brooks
1985. Phylogenetic relationships of some strongylate nematodes of primates. *Proc. Helminthol. Soc. Wash.* 52:227–236.
1986. Parasitological evidence pertaining to the phylogeny of the hominoid primates. *Biol. J. Linn. Soc.* 27:331–354.

Goodchild, C. G.

1943. The life history of *Phyllodistomum solidum* Rankin, 1937 with observations on the morphology, development, and taxonomy of the Gorgoderinae (Trematoda). *Biol. Bull.* 84:59–86.

Goodnight, C. J.

1987. On the effect of founder events on epistatic genetic variance. *Evolution* 41:80–91.

Gosliner, T. M., and M. T. Ghiselin

1984. Parallel evolution in opisthobranch molluscs and its implications for phylogenetic methodology. *Syst. Zool.* 33:255–274.

Gould, S. J.

1977. *Ontogeny and Phylogeny.* Harvard University Press, Cambridge, MA.

Gould, S. J., and E. S. Vrba

1982. Exaptation: A missing term in the science of form. *Paleobiology* 8:4–15.

Gould, S. J., D. M. Raup, J. J. Sepkoski, T. J. M. Schopf, and D. S. Simberloff

1977. The shape of evolution: A comparison of real and random clades. *Paleobiology* 3:23–40.

Grabda-Kazubska, B.

1976. Abbreviation of the life cycles in plagiorchid trematodes: General remarks. *Acta Parasitol. Polon.* 24:125–141.

Grant, P. R., and B. R. Grant

1989. Sympatric speciation and Darwin's finches. In *Speciation and Its Consequences,* ed. D. Otte and J. A. Endler, 433–457. Sinauer, Sunderland, MA.

Greene, H. W.

1986. Diet and arboreality in the emerald monitor, *Varanus prasinus,* with comments on the study of adaptation. *Field. Zool. N. Series* 31: 1–12.

Greiner, E. C., G. F. Bennett, E. M. White, and R. F. Coombs

1975. Distribution of the avian hematozoa of North America. *Can. J. Zool.* 53:1762–1787.

Griffith, H.

1991. Heterochrony and evolution of sexual dimorphism in the *fasciatus* group of the Scincid genus *Eumeces. J. Herpetol.* 25:24–30.

Guegan, J.-F., and J.-F. Agnese

1991. Parasite evolutionary events inferred from host phylogeny: The case of *Labeo* species (Teleostei, Cyprinidae) and their dactylogyrid parasites (Monogenea, Dactylogyridae). *Can. J. Zool.* 69:595–603.

Guerrant, E. O., Jr.

1982. Neotenic evolution of *Delphinium nudicaule* (Ranunculaceae): A hummingbird-pollinated larkspur. *Evolution* 36:699–712.

Haeckel, E.

1866. *Generelle Morphologie der Organismen.* Reiner, Berlin.

Hafner, M. S., and S. A. Nadler

1988. Phylogenetic trees support the coevolution of parasites and their hosts. *Nature* 332:258–259.

1990. Cospeciation in host-parasite assemblages: Comparative analysis of rates of evolution and timing of cospeciation events. *Syst. Zool.* 39: 192–204.

Hall, H. J.

1972. *Diet and Disease at Clyde's Cavern, Utah.* M.Sc. thesis, University of Utah, Salt Lake City.

Hamilton, W. D.

1982. Pathogens as causes of genetic diversity in their host populations. In *Population Biology of Infectious Diseases,* ed. R. M. Anderson and R. M. May, 269–296. Springer-Verlag, New York.

Hamilton, W. D., and M. Zuk

1982. Heritable true fitness and bright birds: A role for parasites? *Science* 218: 384–387.

Harant, H.

1955. *Histoire de la Parasitologie*. Les Conférences du Palais de la Découverte. Série D, No. 35. Université de Paris, Paris.

Harrison, L.

1914. The Mallophaga as a possible clue to bird phylogeny. *Austral. Zool.* 1:7 11.

1915a. Mallophaga from *Apteryx,* and their significance: With a note on *Rallicola. Parasitology* 8:88–100.

1915b. The relationship of the phylogeny of the parasite to that of the host. *Rep. Br. Assoc. Adv. Sci.* 85:476–477.

1916. Bird-parasites and bird-phylogeny. *Ibis* 10:254–263.

1922. On the Mallophagan family Trimenoponidae, with a description of a new genus and species from an Australian marsupial. *Austral. Zool.* 2:154–159.

1924. The migration route of the Australian marsupial fauna. *Austral. Zool.* 3:247–263.

1926. Crucial evidence for antarctic radiation. *Am. Nat.* 60:374–383.

1928a. On the genus *Stratiodrilus* (Archiannelida: Histriobdellidae), with a description of a new species from Madagascar. *Rec. Austral. Mus.* 16:116–121.

1928b. Host and parasite. *Proc. Linn. Soc. New South Wales* 53:ix–xxxi.

1929. The composition and origins of the Australian fauna, with special reference to the Wegener hypothesis. *Rep. Mtgs. Australas. Assoc. Adv. Sci. Perth* 1926:332–396.

Harvey, P. H., and M. Pagel

1991. *The Comparative Method in Evolutionary Biology.* Oxford University Press, Oxford.

Harvey, P. H., and L. Partridge
1982. Bird coloration and parasites: A task for the future? *Nature* 300:480–481.

Hecht, M. K., and J. L. Edwards
1976. The determination of parallel or monophyletic relationships: The proteid salamanders: A test case. *Am. Nat.* 110:653–677.

Hegner, R. W.
1926. The biology of host-parasite relationships among protozoa living in man. *Q. Rev. Biol.* 1:393–418.

Hegner, R. W., F. M. Root, D. L. Augustine, and C. G. Huff
1938. *Parasitology.* Appleton-Century, New York.

Hennig, W.
1950. *Grundzüge einer Theory der phylogenetischen Systematik.* Deutscher Zentralverlag, Berlin.
1966. *Phylogenetic Systematics.* University of Illinois Press, Urbana.

Henry, S. M. (ed.)
1966. Foreword. In *Symbiosis,* Vol. 1. Academic Press, New York.

Hillis, D. M., and S. K. Davis
1986. Evolution of ribosomal DNA: Fifty million years of recorded history in the frog genus *Rana. Evolution* 40:1275–1288.

Hillis, D. M., and C. Moritz (eds.)
1990. *Molecular Systematics.* Sinauer, Sunderland, MA.

Hillis, D. M., J. S. Frost, and D. A. Wright
1983. Phylogeny and biogeography of the *Rana pipiens* complex: A biochemical evaluation. *Syst. Zool.* 32:132–143.

Ho, J.-S.
1988. Cladistics of *Sunaristes,* a genus of harpacticoid copepods associated with hermit crabs. *Hydrobiologia* 167/168:555–560.

Ho, J.-S., and T. T. Do
1985. Copepods of the family Lernanthropidae parasitic on Japanese marine fishes, with a phylogenetic analysis of the Lernanthropid genera. *Rep. Sado Mar. Biol. Stat. Niigata Univ.* 15:31–76.

Hoberg, E. P.
1986. Evolution and historical biogeography of a parasite-host assemblage: *Alcataenia* spp. (Cyclophyllidea: Dilepididae). *Can. J. Zool.* 64:2576–2589.
1987. Recognition of larvae of the Tetrabothriidae (Eucestoda): Implications for the origins of tapeworms in marine homeotherms. *Can. J. Zool.* 65:997–1000.
1989. Phylogenetic relationships among genera of the Tetrabothriidae (Eucestoda). *J. Parasitol.* 75:617–626.
1992. Congruent and synchronic patterns in biogeography and speciation among seabirds, pinnipeds, and cestodes. *J. Parasitol.* 78:601–615.

Hoberg, E. P., and A. M. Adams

1992. Phylogeny, historical biogeography and ecology of *Anophryocephalus* spp. (Tetrabothriidae) among pinnipeds of the Holarctic during the late Tertiary and Pleistocene. *Can. J. Zool.* 70:703–719.

Høiland, K.

1987. A new approach to the phylogeny of the order Boletales (Basidiomycotina). *Nord. J. Bot.* 7:705–718.

Holmes, J. C.

1983. Evolutionary relationships between helminth parasites and their hosts. In *Coevolution,* ed. D. J. Futuyma and M. Slatkin, 161–185. Sinauer, Sunderland, MA.

1990. Helminth communities in marine fishes. In *Parasite Communities: Patterns and Processes,* ed. G. Esch, A. Bush, and J. Aho, 101–130. Chapman and Hall, New York.

Holmes, J. C., and W. M. Bethel

1972. Modification of intermediate host behaviour by parasites. In *Behavioural Aspects of Parasitic Transmission,* ed. E. V. Canning and C. A. Wright, 123–143. Academic Press, New York.

Hopkins, G. H. E.

1942. The Mallophaga as an aid to the classification of birds. *Ibis* 6:94–106.

Horne, P. D.

1985. A review of the evidence of human endoparasitism in the pre-Columbian New World through the study of coprolites. *J. Archaeol. Sci.* 12:299–310.

Hugot, J. P.

1983. *Enterobius gregorii* (Oxyuridae Nematoda): Un nouveau parasite humaine: Note preliminaire. *Ann. Parasitol. Hum. Comp.* 58:403–404.

1988. Les nématodes Syphaciinae, parasites des rongeurs et de lagomorphes: Taxonomie, Zoogéographie, Evolution. *Mem. Mus. Natl. Hist. Nat. Ser. A. Zool.* 141:94–120.

Hugot, J. P., and C. Tourte-Schaefer

1985. Etude morphologique des deux Oxyures parasites de l'homme: *Enterobius vermicularis* et *E. gregorii. Ann. Parasitol. Hum. Comp.* 60:57–64.

Hull, D. L.

1988. *Science as a Process.* University of Chicago Press, Chicago.

Humphery-Smith, I.

1989. The evolution of phylogenetic specificity among parasitic organisms. *Parasitol. Today* 5:385–387.

Ihering, H. von

1891. On the ancient relations between New Zealand and South America. *Trans. Proc. New Zealand Inst.* 24:431–445.

1902. Die Helminthen als Hilfsmittel der zoogeographischen Forschung. *Zool. Anz. (Leipzig)* 26:42–51.

Inglis, W. G.

1961. The oxyurid parasites (nematodes) of primates. *Proc. Zool. Soc. Lond.* 136:103–122.

1983. An outline classification of the phylum Nematoda. *Austral. J. Zool.* 31: 243–255.

Jcrmy, T.

1976. Insect–host plant relationships: Coevolution or sequential evolution? *Symp. Biol. Hung.* 16:109–113.

1984. Evolution of insect/host plant relationships. *Am. Nat.* 124:609–630.

Johnston, S. J.

1912. On some trematode parasites of Australian frogs. *Proc. Linn. Soc. New South Wales* 37:285–362.

1913. Trematode parasites and the relationships and distribution of their hosts. *Rep. Austral. Assoc. Adv. Sci. Melbourne* 14:272–278.

1914a. Australian trematodes and cestodes: A preliminary study in zoogeography. *Proc. Brit. Assoc. Adv. Sci. Australia* 84:424.

1914b. Australian trematodes and cestodes. *Med. J. Austral.* 1:243–244.

1916. On the trematodes of Australian birds. *J. Proc. Roy. Soc. New South Wales* 50:187–261.

Jordan, D. S., and V. Kellogg

1900. *Animal Life: A First Book of Zoology.* Appleton, New York.

1908. The law of geminate species. *Am. Nat.* 42:73–80.

Jordan, D. S., V. Kellogg, and H. Heath

1909. *Animal Studies: A Textbook of Elementary Zoology for Use in High Schools and Colleges.* Appleton, New York.

Justine, J.-L.

1991a. Phylogeny of parasitic platyhelminthes: A critical study of synapomorphies proposed from the ultrastructure of spermiogenesis and spermatozoa. *Can. J. Zool.* 69:1421–1440.

1991b. Cladistic study in the Monogenea (Platyhelminthes), based upon a parsimony analysis of spermiogenic and spermatozoal ultrastructural characters. *Int. J. Parasitol.* 21:821–838.

Kearn, G. C.

1986. The eggs of monogeneans. *Adv. Parasitol.* 25:175–273.

Keith, A.

1931. *New Discoveries Relating to the Antiquity of Man.* Norton, New York.

Kéler, S.

1938. Zur Geschichte der Mallophagen-forschung. *Z. Parasitenkd.* 10:31–66.

1939. Zur Kenntnis der Mallophagen-fauna Polens. 2. Beitrag. *Z. Parasitenkd.* 11:47–57.

Kellogg, V. L.

1896a. New Mallophaga, I, with special reference to a collection from maritime birds of the Bay of Monterey, California. *Proc. Cal. Acad. Sci. 2nd Ser.* 6:31–168.

1896b. New Mallophaga, II, from land birds; together with an account of the mallophagous mouthparts. *Proc. Cal. Acad. Sci. 2nd Ser.* 6:431–548.

1913a. Distribution and species-forming of ectoparasites. *Am. Nat.* 47:129–158.

1913b. Ectoparasites of the monkeys, apes, and man. *Science, N. S.* 38:601–602.

1914. Ectoparasites of mammals. *Am. Nat.* 48:257–279.

Kellogg, V. L., and I. Kuwana

1902. Mallophaga from birds. *Proc. Wash. Acad. Sci.* 4:457–499.

Kennedy, C. E. J.

1990. Helminth communities in freshwater fish: Structured communities or stochastic assemblages? In *Parasite Communities: Patterns and Processes,* ed. G. Esch, A. Bush, and J. Aho, 131–156. Chapman and Hall, New York.

Kethley, J. B., and D. E. Johnston

1975. Resource tracking patterns in bird and mammal ectoparasites. *Misc. Publ. Entomol. Soc. America* 9:231–236.

Kim, K. C. (ed.)

1985. *Coevolution of Parasitic Arthropods and Mammals.* Wiley, New York.

Kirby, J., Jr.

1937. Host-parasite relations in the distribution of protozoa in termites. *Univ. Calif. Publ. Zool.* 41:189–212.

Kirkpatrick, C. E., and T. B. Smith

1988. Blood parasites of birds in Cameroon. *J. Parasitol.* 74:1009–1013.

Kirkpatrick, C. E., and H. B. Suthers

1988. Epizootiology of blood parasite infections in passerine birds from central New Jersey. *Can. J. Zool.* 66:2374–2382.

Kirkpatrick, M.

1982. Sexual selection and the evolution of female choice. *Evolution* 36:1–12.

1986. Sexual selection and cycling parasites: A simulation study of Hamilton's hypothesis. *J. Theor. Biol.* 119:263–271.

Klassen, G. J.

1992. Coevolution: A history of the macroevolutionary approach to studying host-parasite associations. *J. Parasitol.* 78:573–587.

Klassen, G. J., and M. Beverly-Burton

1987. Phylogenetic relationships of *Ligictaluridus* spp. (Monogenea: Ancyrocephalidae) and their ictalurid (Siluriformes) hosts: An hypothesis. *Proc. Helminthol. Soc. Wash.* 54:84–90.

1988a. North American freshwater ancyrocephalids (Monogenea) with articulating haptoral bars: Phylogeny reconstruction. *Syst. Parasitol.* 11:49–57.

1988b. North American freshwater ancyrocephalids (Monogenea) with articulating haptoral bars: Host-parasite coevolution. *Syst. Zool.* 37:179–189.

Kluge, A. G.

1983. Cladistics and the classification of the Great Apes. In *New Interpretations of Apes and Human Ancestry,* ed. R. L. Ciochon and R. S. Corruccini, 151–177. Plenum Press, New York.

Kochmer, J. P., and R. H. Wagner

1988. Why are there so many kinds of passerine birds? Because they are so small. *Syst. Zool.* 37:68–69.

Kodric-Brown, A., and J. H. Brown.

1984. Truth in advertising: The kinds of traits favored by sexual selection. *Am. Nat.* 124:309–323.

Kuhn, H. J.

1967. Parasites and phylogeny of the catarrhine primates. In *Taxonomy and Phylogeny of the Old World Primates with References to the Origin of Man,* ed. B. Chiarelli, 187–195. Rosenberg and Sellier, Torino.

Lande, R.

1981. Models of speciation by sexual selection on polygenic traits. *Proc. Nat. Acad. Sci. U.S.A.* 78:3721–3725.

Larson, A., D. B. Wake, L. R. Maxson, and R. Highton

1981. A molecular phylogenetic perspective on the origins of morphological novelties in the salamanders of the Tribe Plethodontini (Amphibia, Plethodontidae). *Evolution* 35:405–422.

LaRue, G. R.

1951. Host-parasite relations among the digenetic trematodes. *J. Parasitol.* 37:333–342.

Lauder, G. V.

1981. Form and function: Structural analysis in evolutionary biology. *Paleobiology* 7:430–442.

Lauder, G. V., and K. F. Liem

1989. The role of historical factors in the evolution of complex organismal functions. In *Complex Organismal Functions: Integration and Evolution in Vertebrates,* ed. D. B. Wake and G. Roth., 63–78. Wiley, London.

Layzer, D.

1978. A macroscopic approach to population genetics. *J. Theor. Biol.* 73:769–788.

1980. Genetic variation and progressive evolution. *Am. Nat.* 115:809–826.

Leuckart, R.

1879. *Allgemeine Naturgeschichte der Parasiten mit besonderer Berücksichtigung der bei Menschen schmarotzenden Arten.* C. F. Winter, Leipzig.

Leviton, A. E., and M. L. Aldrich (eds.)

1985. Plate tectonics and biogeography. *J. Hist. Earth Sci. Soc.* 4:91–201.

Lichtenfels, J. R., and P. A. Pilitt
1983. Cuticular ridge patterns of *Nematodirella* (Nematoda: Trichostrongyloidea) of North American ruminants, with a key to species. *Syst. Parasitol.* 5:271–285.

Liem, K. F.
1973. Evolutionary strategies and morphological innovations: Cichlid pharyngeal jaws. *Syst. Zool.* 22:424–441.

Liem, K. F., and D. B. Wake
1985. Morphology: Current approaches and concepts. In *Functional Vertebrate Morphology,* ed. M. Hildebrand, D. M. Bramble, K. F. Liem, and D. B. Wake, 366–377. Harvard University Press, Cambridge, MA.

Llewellyn, J.
1968. Larvae and larval development of monogeneans. *Adv. Parasitol.* 6:373–383.
1981. Biology of monogeneans. *Parasitology* 82:57–68.

Lockwood, R.
1872. A new entozoon from the eel. *Am. Nat.* 6:449.

Lotz, J. M.
1986. On the phylogeny of the Lecithodendriidae (Trematoda). *J. Parasitol.* suppl.: 81.

Lynch, J. D.
1989. The gauge of speciation: On the frequencies of modes of speciation. In *Speciation and Its Consequences,* ed. D. Otte and J. Endler, 527–553. Sinauer, Sunderland, MA.

MacCallum, G. A.
1921. Studies in helminthology: Part 1—Trematodes, Part 2—Cestodes, Part 3—Nematodes. *Zoopathologica* 1:135–284.
1926. Review du genre *Spirorchis* MacCallum. *Ann. Parasitol. Hum. Comp.* 4:97–103.

Macdonald, C. A., and D. R. Brooks
1989a. Redescription of *Pseudotelorchis compactus* (Cable and Sanborn, 1970) Yamaguti, 1971 (Cercomeria: Trematoda: Digenea: Plagiorchiformes) with discussion of its phylogenetic affinities. *Can. J. Zool.* 67:1421–1424.
1989b. Revision and phylogenetic analysis of the North American species of *Telorchis* Luhe, 1899 (Cercomeria: Trematoda: Digenea: Telorchiidae). *Can. J. Zool.* 67:2301–2320.

McIntosh, A.
1935. A progenetic metacercaria of a *Clinostomum* in a West Indian land snail. *Proc. Helminthol. Soc. Wash.* 2:79–80.

Mackiewicz, J. S.
1981. Caryophyllidea (Cestoidea): Evolution and classification. *Adv. Parasitol.* 19:139–206.

McKinney, M. L.
1986. Ecological causation of heterochrony: A test and implications for evolutionary theory. *Paleobiology* 12:282–289.
McKinney, M. L. (ed.)
1988. *Evolution and Development.* Dahlem Conference. Springer-Verlag, New York.
McLennan, D. A., and D. R. Brooks
1991. Parasites and sexual selection: A macroevolutionary perspective. *Q. Rev. Biol.* 66:255–286.
McMullen, D. B.
1938. Observations on precocious metacercarial development in the trematode superfamily Plagiorchioidea. *J. Parasitol.* 24:273–280.
McNamara, K. J.
1982. Heterochrony and phylogenetic trends. *Paleobiology* 8:130–142.
1986. A guide to the nomenclature of heterochrony. *J. Paleontol.* 60:4–13.
McVicar, A. H.
1979. The distribution of cestodes within the spiral intestine of *Raja naevis* Muller & Henle. *Int. J. Parasitol.* 9:165–176.
Maddison, W. P., M. J. Donoghue, and D. R. Maddison
1984. Outgroup analysis and parsimony. *Syst. Zool.* 33:83–103.
Manter, H. W.
1940. The geographical distribution of digenetic trematodes of marine fishes of the tropical American Pacific. *Allan Hancock Pac. Exped. Rep.* 2:531–547.
1954. Some digenetic trematodes from fishes of New Zealand. *Trans. Roy. Soc. New Zealand* 82:475–568.
1955. The zoogeography of trematodes of marine fishes. *Exp. Parasitol.* 4:62–86.
1963. The zoogeographical affinities of trematodes of South American freshwater fishes. *Syst. Zool.* 12:45–70.
1966. Parasites of fishes as biological indicators of recent and ancient conditions. In *Host-Parasite Relationships,* ed. J. E. McCauley. 59–71. Oregon State University Press, Corvallis.
1967. Some aspects of the geographical distribution of parasites. *J. Parasitol.* 53:3–9.
Maurer, B. A.
1989. Diversity dependent species dynamics: Incorporating the effects of population level processes on species dynamics. *Paleobiology* 15:133–146.
Maurer, B. A., and D. R. Brooks.
1991. Energy flow and entropy production in biological systems. *J. Ideas* 2:48–53.
Mayden, R. L.
1986. Speciose and depauperate phylads and tests of punctuated and gradual evolution: Fact or artifact? *Syst. Zool.* 35:591–602.
1992. *Phylogeny, Historical Ecology, and North American Freshwater Fish.* Stanford University Press, Palo Alto.

Maynard Smith, J.

1966. Sympatric speciation. *Am. Nat.* 100:637–650.

1978. *The Evolution of Sex.* Cambridge University Press, Cambridge.

1985. Sexual selection, handicaps, and true fitness. *J. Theor. Biol.* 115:1–8.

Mayr, E.

1942. *Systematics and the Origin of Species.* Columbia University Press, New York.

1954. Change of genetic environment and evolution. In *Evolution as a Process.* ed. J. Huxley, A. C. Hardy, and E. B. Ford, 157–180. Allen & Unwin, London.

1957. Evolutionary aspects of host specificity among parasites of vertebrates. In *Premier Symposium sur la spécificité parasitaire des parasites des Vertébrés,* ed. J.-G. Baer, 7–14. Paul Attinger, Neuchâtel.

1960. The emergence of evolutionary novelties. In *Evolution after Darwin,* ed. S. Tax, 349–380. University of Chicago Press, Chicago.

1963. *Animal Species and Evolution.* Belknap Press of Harvard University, Cambridge, MA.

1969. *Principles of Systematic Zoology.* McGraw-Hill, New York.

1982. Processes of speciation in animals. In *Mechanisms of Speciation,* ed. C. Barigozzi, 1–19. Liss, New York.

Measures, L. N., M. Beverly-Burton, and A. Williams

1990. Three new species of *Monocotyle* (Monogenea: Monocotylidae) from the stingray, *Himantura uarnak* (Rajiformes: Dasyatidae) from the great Barrier Reef: Phylogenetic reconstruction, systematics, and emended diagnoses. *Int. J. Parasitol.* 20:755–767.

Meggers, B. J., and C. Evans

1966. A transpacific contact in 3000 B.C. *Sci. Am.* 214:28–35.

Meier, R., P. Kores, and S. Darwin

1991. Homoplasy slope ratio: A better measurement of observed homoplasy in cladistic analyses. *Syst. Zool.* 40:74–88.

Metcalf, M. M.

1920. Upon an important method of studying problems of relationship and geographical distribution. *Proc. Natl. Acad. Sci. U.S.A.* 6:432–433.

1922. The host parasite method of investigation and some problems to which it gives approach. *Anat. Rec.* 23:117.

1923a. The opalinid ciliate infusorians. *Bull. U.S. Natl. Mus.* 120:1–484.

1923b. The origin and distribution of the Anura. *Am. Nat.* 57:385–411.

1926. Larval stages in a protozoon. *Proc. U.S. Natl. Acad. Sci.* 12:734–737.

1928a. The bell-toads and their opalinid parasites. *Am. Nat.* 62:5–21.

1928b. Trends in evolution: A discussion of data bearing upon 'orthogenesis.' *J. Morphol. Physiol.* 45:1–45.

1929. Parasites and the aid they give in problems of taxonomy, geographic distribution, and paleogeography. *Smithsonian Misc. Coll.* 81:1–36.

1934. Frogs and opalinids. *Science* 79:213–214.

1940. Further studies on the opalinid ciliate infusorians. *Proc. U.S. Natl. Mus.* 87:465–634.

Miller, J. S.

1992. Host-plant associations among prominent moths. *BioScience* 42:50–57.

Mishler, B. D., and S. P. Churchill

1984. A cladistic approach to the phylogeny of the "Bryophytes." *Brittonia* 36:406–424.

Mitter, C., and D. R. Brooks

1983. Phylogenetic aspects of coevolution. In *Coevolution,* ed. D. J. Futuyma and M. Slatkin, 65–98. Sinauer, Sunderland, MA.

Mitter, C., B. Farrell, and B. Wiegemann

1988. The phylogenetic study of adaptive zones: Has phytophagy promoted insect diversification? *Am. Nat.* 132:107–128.

Mitter, C., B. Farrell, and D. J. Futuyma

1991. Phylogenetic studies of insect–plant interactions: Insights into the genesis of diversity. *Trends Ecol. Evol.* 6:290–293.

Mode, C. J.

1958. A mathematical model for the co-evolution of obligate parasites and their hosts. *Evolution* 12:158–165.

Møller, A. P.

1990. Parasites and sexual selection: Current status of the Hamilton and Zuk hypothesis. *J. Evol. Biol.* 3:319–328.

Mooi, R. J.

1987. Analysis of the Sand Dollars (Clypeasteroida: Scutellina) and the interpretation of heterochronic phenomena. Ph.D. dissertation, University of Toronto, Toronto, Canada.

Moore, J.

1981. Asexual reproduction and environmental predictability in cestodes (Cyclophyllidea: Taeniidae). *Evolution* 35:723–741.

1984. Altered behavioral responses in intermediate hosts: An acanthocephalan parasite strategy. *Am. Nat.* 123:572–577.

Moore, J., and D. R. Brooks

1987. Asexual reproduction in cestodes (Cyclophyllidea: Taeniidae): Ecological and phylogenetic influences. *Evolution* 41:882–891.

Moore, J. G., G. F. Fry, and E. Englert, Jr.

1969. Thorny-headed worm infection in North American prehistoric man. *Science* 163:1324–1325.

Morely Davies, A.

1920. *An Introduction to Palaeontology.* Van Nostrand, New York.

Moshkovski, Sh. D.

1946. Functional parasitology, Parts. 1, 2, and 3. *Med. Parazit.* 15(4):22–36, (5):28–42, (6):3–19.

Myers, A. A.
1988. A cladistic and biogeographic analysis of the Aorinae subfamily nov. *Crustaceana Suppl.* 13:167–192.

Nelson, G.
1984. Identity of the anchovy *Hildebrandichthys setiger* with notes on relationships and biogeography of the genera *Engraulis* and *Cetengraulis*. *Copeia* 1984:422–427.

Noble, E. R., and G. A. Noble
1961. *Parasitology: The Biology of Animal Parasites,* 1st ed. Lea & Febiger, Philadelphia.
1976. *Parasitology: The Biology of Animal Parasites,* 4th ed. Lea & Febiger, Philadelphia.

Noble, E. R., G. A. Noble, G. A. Schad, and A. J. MacInnes
1989. *Parasitology: The Biology of Animal Parasites,* 6th ed. Lea & Febiger, Philadelphia.

Novacek, M. J.
1984. Evolutionary stasis in the elephant-shrew, *Rhynchocyon*. In *Living Fossils,* ed. N. Eldredge and S. M. Stanley, 4–22. Springer-Verlag, New York.

Odening, K.
1974a. Ontogenese und Lebenszyklus bei Helminthen und ihre Widerspiegelung in der Wirtssklassifikation. *Zool. Anz. Jena* 194:43–55.
1974b. Verwandtschaft, System und zyklo-ontogenetische Besonderheiten der Trematoden. *Zool. Jahrb. Syst. Bd.* 101:345–396.

Odhner, T.
1913. Zum natürlichen System der digenen Trematoden. VI. *Zool. Anz.* 42:287–319.

O'Donald, P.
1962. The theory of sexual selection. *Heredity, London* 17:541–552.
1967. A general model of sexual and natural selection. *Heredity, London* 22:499–518.

O'Grady, R. T.
1985. Ontogenetic sequences and the phylogenetics of parasitic flatworm life cycles. *Cladistics* 1:159–170.
1987. Phylogenetic systematics and the evolutionary history of some intestinal flatworm parasites (Trematoda: Digenea: Plagiorchioidea) of anurans. Ph.D. dissertation, University of British Columbia, Vancouver, Canada.
1989. Distribution and biogeography. Parasite-host specificity. Evolution of parasitism. In *The Biology of Animal Parasites,* 6th ed., ed. E. R. Noble, G. A. Noble, G. A. Schad, and A. J. MacInnes, 482–544. Lea & Febiger, Philadelphia.

Osche, G.
1958. Beiträge zur Morphologie, Ökologie, und Phylogenie der Ascaridoidea. Parallelen in der Evolution von Parasit und Wirt. *Z. Parasitenk.* 18:479–572.

1960. Systematische, morphologische und parasitophyletische Studien an parasitischen Oxyuroidea (Nematoda) exotischer Diplopoden (ein Beitrag zur Morphologie des Sexualdimorphismus). *Zool. Jahrb.* 87:395–440.

1963. Morphological, biological, and ecological considerations in the phylogeny of parasitic nematodes. In *The Lower Metazoa: Comparative Biology and Phylogeny,* ed. E. C. Dougherty, 283–302. University of California Press, Berkeley.

Otte, D., and J. A. Endler (eds.)

1989. *Speciation and Its Consequences.* Sinauer, Sunderland, MA.

Owen, H. G.

1983. *Atlas of Continental Displacement: 200 million years to the Present.* Cambridge University Press, Cambridge.

Patrucco, R., R. Tello, and D. Bonavia

1983. Parasitological studies of coprolites of pre-hispanic Peruvian populations. *Current Anthropol.* 24:393–394.

Patterson, C., and H. G. Owen

1991. Indian isolation or contact? A response to Briggs. *Syst. Zool.* 40:96–100.

Patton, J. L., and M. F. Smith

1989. Population structure and the genetic and morphologic divergence among pocket gopher species (genus *Thonomys*). In *Speciation and its Consequences,* ed. D. Otte and J. A. Endler, 284–304. Sinauer, Sunderland, MA.

Pearson, J. C.

1972. A phylogeny of life cycle patterns of the Digenea. *Adv. Parasitol.* 10:153–189.

Pence, D.

1990. Helminth communities of mammalian hosts: Concepts at the infracommunity, component, and compound community levels. in *Parasite Communities: Patterns and Processes,* ed. G. Esch, A. Bush, and J. Aho, 233–260. Chapman and Hall, New York.

Peters, W., P. C. C. Garnham, R. Killick-Kendrick, N. Rajapaksa, W. Cheong, and F. Cadigan

1976. Malaria of the orang-utan (*Pongo pygmaeus*) in Borneo. *Proc. Roy. Soc. Lond., Ser. B* 275:439–482.

Pilbeam, D.

1984. The descent of hominoids and hominids. *Sci. Am.* 250:84–96.

Platt, T. R.

1984. Evolution of the Elaphostrongylinae (Nematoda: Metastrongyloidea: Protostrongylidae) parasites of cervids (Mammalia). *Proc. Helminthol. Soc. Wash.* 51:196–204.

1988. Phylogenetic analysis of the North American species of the genus *Hapalorhynchus* Stunkard, 1922 (Trematoda: Spirorchiidae), blood flukes of freshwater turtles. *J. Parasitol.* 74:870–874.

1991. Notes on the genus *Hapalorhynchus* (Digenea: Spirorchidae) from African turtles. *Trans. Am. Microsc. Soc.* 110:182–184.

1992. A phylogenetic and biogeographic analysis of the genera of Spirorchinae (Digenea: Spirorchidae) parasitic in freshwater turtles. *J. Parasitol.* 78: 616–629.

Pomiankowski, A.

1987a. Sexual selection: The handicap principle does work—sometimes. *Proc. R. Soc. Lond., Ser. B* 231:123–145.

1987b. The costs of choice in sexual selection. *J. Theor. Biol.* 128:195–218.

Price, E. W.

1934. New genera and species of blood flukes from a marine turtle with a key to the family Spirorchidae. *J. Wash. Acad. Sci.* 24:132–141.

Price, P. W.

1980. *Evolutionary Biology of Parasites.* Princeton University Press, Princeton.

1986. Evolution in parasite communities. *Int. J. Parasitol.* 17:209–214.

Pritchard, M. H.

1966. Studies on digenetic trematodes of Hawaiian fishes: Family Opecoelidae Ozaki, 1925. *Zool. Jahrb. Syst. Bd.* 93:173–202.

Prokopic, J., and K. Krivanec

1975. Helminths of amphibians: Their interaction and host-parasite relationships. *Acta Sci. Nar. Brno* 9:1–48.

Raff, R. A., and E. C. Raff (eds.)

1987. *Development as an Evolutionary Process: Proceedings of a Meeting Held at the Marine Biological Laboratory in Woods Hole, Massachusetts, August 23 & 24, 1985.* Liss, New York.

Raikow, R. J.

1986. Why are there so many kinds of passerine birds? *Syst. Zool.* 35:255–259.

1988. An analysis of evolutionary success. *Syst. Zool.* 37:76–79.

Rannala, B. H.

1990. An electrophoretic perspective on the relationship between *Haplometrana* Lucker, 1931 and *Glypthelmins* Stafford, 1905 (Digenea: Plagiorchiiformes). *J. Parasitol.* 76:746–748.

Raup, D. M.

1984. Mathematical models of cladogenesis. *Paleobiology* 11:42–52.

Raup, D. M., and S. J. Gould

1974. Stochastic simulation and evolution of morphology: Towards a nomothetic paleontology. *Syst. Zool.* 23:305–322.

Raup, D. M., S. J. Gould, T. J. M. Schopf, and D. S. Simberloff

1973. Stochastic models of phylogeny and the evolution of diversity. *J. Geol.* 81:525–542.

Read, A. F., and P. H. Harvey

1989. Read and Harvey reply. *Nature* 340:105.

Read, C. P.

1972. *Animal Parasitism.* Prentice Hall, Englewood Cliffs, NJ.

Reduker, D. W., D. W. Duszynski, and T. L. Yates

1987. Evolutionary relationships among *Eimeria* spp. (Apicomplexa) infecting cricetid rodents. *Can. J. Zool.* 65:722–735.

Reinhard, K. J.

1990. Archaeoparasitology in North America. *Am. J. Phys. Anthropol.* 82:145–163.

Reinhard, K. J., R. H. Hevly, and G. A. Anderson

1987. Helminth remains from prehistoric Indian coprolites on the Colorado Plateau. *J. Parasitol.* 73:630–639.

Reiss, J. O.

1989. The meaning of developmental time: A metric for comparative embryology. *Am. Nat.* 134:170–189.

Remane, A.

1956. *Die Grundlagen des natürlichen Systems der vergleichenden Anatomie und Phylogenetik.* 2. Geest und Portig, Leipzig.

Rennie, J.

1992. Living together. *Sci. Am.* 258:122–133.

Richardson, J. P. M.

1988. Phylogenetic analysis of the genus *Megalobatrachonema* Yamaguti, 1941 (Ascaridida: Cosmocercoidea: Kathlaniidae), with field and laboratory observations of *M. waldeni*. M.Sc. thesis, University of British Columbia, Vancouver, Canada.

Ricklefs, R., and D. Schluter

In press. *Historical and Geographic Determinants of Diversity,* University of Chicago Press, Chicago.

Riggs, M. R., and M. J. Ulmer

1983a. Host-parasite relationships of helminth parasites of the genus *Haemopis.* I. Associations at the individual host level. *Trans. Am. Microsc. Soc.* 102: 213–226.

1983b. Host-parasite relationships of helminth parasites of the genus *Haemopis.* II. Associations at the host–species level. *Trans. Am. Microsc. Soc.* 102: 227–239.

Ringo, J. M.

1977. Why 300 species of Hawaiian *Drosophila*? *Evolution* 31:695–754.

Rogers, W. P.

1962. *The Nature of Parasitism: The Relation of Some Metazoan Parasites to Their Hosts.* Academic Press, New York.

Rohde, K.

1979. A critical evaluation of intrinsic and extrinsic factors responsible for niche restriction in parasites. *Am. Nat.* 114:648–671.

1981. Niche width of parasites in species-rich and species-poor communities. *Experientia* 37:359–361.

1984. Ecology of marine parasites. *Helgolander Meeresuntersuchungen* 37:5–33.

1989. At least eight types of sense receptors in an endoparasitic flatworm: A counter-trend to sacculinization. *Naturwissenschaften* 76:383–385.

Rohde, K., and R. P. Hobbs
1986. Species segregation: Competition of reinforcement of reproductive barriers? In *Parasite Lives: Papers on Parasites, Their Hosts, and Their Associations to Honour J. F. A. Sprent,* ed. M. Cremin, C. Dobson, and D. E. Moorehouse, 189–199. University of Queensland Press, St. Lucia, Australia.

Rohde, K., S. K. Lee, and H. W. Lim
1968. Ueber drei malayische Trematoden. *Ann. Parasitol.* 43:33–43.

Ronquist, F., and S. Nylin
1990. Process and pattern in the evolution of species associations. *Syst. Zool.* 39:323–344.

Rosen, D. E.
1978. Vicariant patterns and historical explanation in biogeography. *Syst. Zool.* 27:159–188.

Roza, L.
1989. [Host-parasite relationships: Coevolution or sequential evolution?] *Parasitol. Hung.* 22:29–33.

Russell, E. S.
1916. *Form and Function.* John Murray, London.

Saether, O. A.
1977. Female genitalia in Chironimidae and other Nematocera: Morphology, phylogenies, keys. *Bull. Fish. Res. Bd. Canada* 197:1–211.

Samuels, R.
1965. Parasitological study of long-dried fecal samples. *Mem. Soc. Am. Archaeol.* 19:175–179.

Sanderson, M. J., and M. J. Donoghue
1989. Patterns of variation in levels of homoplasy. *Evolution* 43:1781–1795.

Sandground, J. H.
1926. Speciation and specificity in the nematode genus *Strongyloides. J. Parasitol.* 12:59–80.

Sandosham, A. A.
1950. On *Enterobius vermicularis* (Linnaeus, 1758) and some related species from primates and rodents. *J. Helminthol.* 24:171–204.

Schad, G. A.
1963. Niche diversification in a parasitic species flock. *Nature* 198:404–406.

Scharpilo, V. P.
1960. [Species of reptilian helminthes new for the fauna of the USSR.] *Dopov. Akad. Nauk Ukr. R.S.R.* 8:1120–1123. [in Ukrainian]

Schmidt, G. D., and L. S. Roberts
1985. *Foundations of Parasitology,* 3rd ed. Times Mirror/Mosby College: St. Louis, MO.

Schultz, A. H.
1930. The skeleton of the trunk and limbs of higher primates. *Hum. Biol.* 2:303–438.

Schwartz, J. H.
1984. The evolutionary relationships of man and orang-utans. *Nature* 308:501–505.

Seurat, L. G.
1920. Histoire naturelle des nématodes de la Berbèrie, premiere partie: Morphologie, développement, éthologie, et affinités des nématodes. Université d'Alger, Faculté Sciences Fondation. Joseph Azoubib, Algiers.

Sey, O.
1991. *CRC Handbook of the Zoology of Amphistomes.* CRC Press, Boca Raton, FL.

Shea, B. T.
1983. Allometry and heterochrony in the African apes. *Am. J. Phys. Anthropol.* 62:275–289.

Shoop, W. L.
1988. Trematode transmission patterns. *J. Parasitol.* 74:46–59.
1989. Systematic analysis of the Diplostomidae and Strigeidae. *J. Parasitol.* 75:21–32.

Siddall, M. E., and S. S. Desser
1991. Merogonic development of *Haemogregarina balli* (Apicomplexa: Adeleina: Haemogregarinidae) in the leech *Placobdella ornata* (Glossiphoniidae), its transmission to a chelonian intermediate host, and phylogenetic implications. *J. Parasitol.* 77:426–436.

Siddall, M. E., H. Hong, and S. S. Desser
1992. Phylogenetic analysis of the Diplomonadida (Wenyon, 1926) Brugerolle, 1975: Evidence for heterochrony in protozoa and against *Giardia lamblia* as a "missing link." *J. Protozool.* 39:361–367.

Siddall, M. E., D. R. Brooks, and S. S. Desser
In press. Phylogenetic reversibility of parasitism. *Evolution.*

Siegel-Causey, D.
1991. Systematics and biogeography of North Pacific shags, with a description of a new species. *Occ. Pap. Mus. Nat. Hist. Kansas* 140:1–17.

Simpson, G. G.
1944. *Tempo and Mode in Evolution.* Columbia University Press, New York.
1953. *The Major Features of Evolution.* Columbia University Press, New York.

Smiley, J.
1978. Plant chemistry and the evolution of host specificity: New evidence from *Heliconius* and *Passiflora*. *Science* 201:745–746.

Soper, F. L.
1927. The report of a nearly pure *Ancylostoma duodenale* infestation in native South American Indians and a discussion of its ethnological significance. *Am. J. Hyg.* 7:174–184.

Spieth, H. T.

1974. Mating behavior and evolution of the Hawaiian *Drosophila*. In *Genetic Mechanisms of Speciation in Insects,* ed. M. J. D. White, 94–101. Australia and New Zealand Book, Sydney.

Sprent, J. F. A.

1982. Host-parasite relationships of ascaridoid nematodes and their vertebrate hosts in space and time. *Mem. Mus. Natl. Hist. Nat. Paris, Ser. A, Zoologie* 123:255–263.

Stammer, H. J.

1955. Ökologische Wechselbeziehungen zwischen Insekten und anderen Tiergruppen. *Wand. Versamml. dtsch. Ent.* 7:12–61.

1957. Gedanken zu den parasitophyletischen Regeln und zur Evolution der Parasiten. *Zool. Anz.* 159:255–267.

Stanley, S. M.

1979. *Macroevolution: Pattern and Process*. Freeman, San Francisco.

Stanley, S. M., P. W. Signor, S. Lidgard, and A. F. Karr

1981. Natural clades differ from "random" clades: Simulations and analysis. *Paleobiology* 7:115–127.

Stein, B. A.

1992. Sicklebill hummingbirds, ants, and flowers. *BioScience* 42:27–33.

Stiassny, M. L. J., and J. Jensen

1987. Labroid interrelationships revisited: Morphological complexity, key innovations, and the study of comparative diversity. *Bull. Mus. Comp. Zool.* 151:269–319.

Stiger, M. A.

1977. *Anasazi Diet: The Coprolite Evidence,* M.Sc. thesis. University of Colorado, Boulder.

Stock, T. M., and J. C. Holmes

1987. *Dioecocestus asper* (Cestoda: Dioecocestidae): An interference competitor in an enteric helminth community. *J. Parasitol.* 73:1116–1123.

Strauss, R. E.

1990. Heterochronic variation in the developmental timing of cranial ossifications in Poeciilid fishes (Cyprinodontiformes). Evolution 44:1558–1567.

Stunkard, H. W.

1921. Notes on North American blood flukes. *Am. Mus. Novitates* 12:1–5.

1923. Studies on North American blood flukes. *Bull. Am. Mus. Nat. Hist.* 48:165–221.

1940. Life history studies and the development of parasitology. *J. Parasitol.* 26:1–15.

1959. The morphology and life history of the digenetic trematode, *Asymphylodora amnicolae* n. sp.: The possible significance of progenesis for the phylogeny of the Digenea. *Biol. Bull.* 117:562–581.

1970. Trematode parasites of insular and relict vertebrates. *J. Parasitol.* 56: 1041–1054.

Sundberg, P.

1989. Phylogeny and cladistic classification of the paramonostiliferous family Plectonemertidae (phylum Nemertea). *Cladistics* 5:87–100.

Szidat, L.

1940. Beiträge zum Aufbau eines natürlichen Systems der Trematode. I. Die Entwicklung von *Echinocercariea choanophila* U. Szidat zu *Cathaemasia hians* und die Ableitung der Fasciolidae von den Echinostomidae. *Zeitschr. Parasitenk.* 11:239–283.

1956a. Über den Entwicklungszyklus mit progenetischen Larvenstadien (Cercariaeen) von *Genarchella genarchella* Travassos, 1928 (Trematoda, Hemiuridae) und die Möglichkeit einer hormonalen Beeinflussung der Parasiten durch ihre Wirtstiere. *Zeitschr. Tropenmed. Parasitol.* 7:132–153.

1956b. Der marine Charakter der Parasitenfauna der Süsswasserfische des Stromssystems des Rio de la Plata und ihre Deutung als Reliktfauna des Tertiaren Tethys-Meeres. *Proc. 14th Int. Congr. Zool.* 1953:128–138.

1956c. Geschichte, Anwendung und einige Folgerungen aus den parasitogenetischen Regeln. *Zeitschr. Parasitenk.* 17:237–268.

1961. Versuch einer Zoogeographie des Sud-Atlantic mit Hilfe von Leitparasiten der Meeresfische. *Parasitol. Schriften.* 13:1–97.

Taplin, L. E., and G. C. Grigg

1989. Historical zoogeography of eusuchian crocodilians: A physiological perspective. *Am. Zool.* 29:885–901.

Tarsitano, S. F., E. Frey, and J. Reiss

1989. The evolution of the Crocodilia: A conflict between morphological and biochemical data. *Am. Zool.* 29:843–856.

Tauber, C. A., and M. J. Tauber

1989. Sympatric speciation in insects: Perception and perspective. In *Speciation and Its Consequences,* ed. D. Otte and J. A. Endler, 307–344. Sinauer, Sunderland, MA.

Taylor, G., and G. Williams

1982. The lek paradox not resolved. *Theor. Pop. Biol.* 22:392–404.

Templeton, A. R.

1979. Once again, why 300 species of Hawaiian *Drosophila*? *Evolution* 33:513–517.

1980. The theory of speciation by the founder principle. *Genetics* 92:1011–1938.

1981. Mechanisms of speciation: A population genetic approach. *Ann. Rev. Ecol. Syst.* 12:23–48.

Thorson, T. B., D. R. Brooks, and M. A. Mayes

1983. The evolution of freshwater adaptation in stingrays. *Nat. Geogr. Soc. Rep.* 15:663–694.

Throckmorton, L. H.

1965. Similarity versus relationship in *Drosophila*. *Syst. Zool.* 14:221–236.

Tinsley, R. C.

1982. The reproductive strategy of a polystomatid monogenean in a desert environment. *Parasitology* 85:xv.

1983. Ovoviviparity in platyhelminth life-cycles. *Parasitology* 86:161–196.

Trivers, R. L.

1972. Parental investment and sexual selection. In *Sexual Selection and the Descent of Man, 1871–1971,* ed. B. Campbell, 136–179. Aldine, Chicago.

Van Cleave, H. J.

1941. Relationships of the Acanthocephala. *Am. Nat.* 75:31–47.

Van Every, L. R., and D. C. Kritsky

1992. Neotropical Monogenoidea. 18. *Anacanthorus* Mizelle and Price 1965 (Dactylogyrindae, Anacanthorinae) of piranha (Characoidea, Serrasalmidae) from the Central Amazon, their phylogeny, and aspects of host-parasite coevolution. *J. Helminthol. Soc. Wash.* 59:52–75.

Vermeij, G. J.

1988. The evolutionary success of passerines: A question of semantics? *Syst. Zool.* 37:69–71.

Vrba, E. S.

1980. Evolution, species, and fossils: How does life evolve? *S. Afr. J. Sci.* 76: 61–84.

1984a. What is species selection? *Syst. Zool.* 33:318–328.

1984b. Evolutionary pattern and process in the sister-group Alcelaphini-Aepycerotini (Mammalia: Bovidae). In *Living Fossils,* ed. N. Eldredge and S. M. Stanley, 62–79. Springer-Verlag, New York.

Wagner, M.

1868. *Die Darwin'sche Theorie und das Migrationsgesetz der Organismen.* Duncker & Humboldt, Leipzig.

Wallace, B.

1955. Inter-population hybrids in *Drosophila melanogaster. Evolution* 9:302–316.

Walton, A. C.

1942. Some oxyurids from a Galapagos tortoise. *Proc. Helminthol. Soc. Wash.* 9:1–17.

Ward, H. B.

1921. A new blood fluke from turtles. *J. Parasitol.* 7:114–128.

Washburn, S. L.

1973. Primate studies and human evolution. In *Nonhuman Primates and Medical Research,* ed. G. H. Browne, 467–485. Academic Press, New York.

Weekes, P. J.

In press. A review of the order Nipotaeniidea Yamaguti (Eucestoda) with the description of a new genus and a new species. *New Zeal. J. Zool.*

Wegener, A.

1912. Die Entstehung der Kontinente. *Peterm. Mitt.* 1912:185–195.

West-Eberhard, M. J.

1983. Sexual selection, social competition, and speciation. *Q. Rev. Biol.* 58:155–183.

1989. Phenotypic plasticity and the origins of diversity. *Annu. Rev. Ecol. Syst.* 20:249–278.

Wheeler, T. A., and M. Beverly-Burton

1989. Systematics of *Onchocleidus* Mueller, 1936 (Monogenea: Ancyrocephalidae): Phylogenetic relationships, evolution, and host associations. *Can. J. Zool.* 67:706–713.

Whitfield, P. J.

1979. *The Biology of Parasitism: An Introduction to the Study of Interacting Organisms*. Edward Arnold, London.

Wiley, E. O.

1981. *Phylogenetics: The Theory and Practice of Phylogenetic Systematics*. Wiley, New York.

1988. Parsimony analysis and vicariance biogeography. *Syst. Zool.* 37:271–290.

Wiley, E. O., and R. L. Mayden

1985. Species and speciation in phylogenetic systematics, with examples from the North American fish fauna. *Ann. Mo. Bot. Garden* 72:596–635.

Wiley, E. O., D. Siegel-Causey, D. R. Brooks, and V. A. Funk

1991. *The Compleat Cladist: A Primer of Phylogenetic Procedures*. University of Kansas Museum of Natural History Press, Lawrence, Kansas.

Wilkinson, M.

1991. Homoplasy and parsimony analysis. *Syst. Zool.* 40:105–109.

Williams, H. H.

1960. The intestine in members of the genus *Raja* and host-specificity in the Tetraphyllidea. *Nature* 188:514–516.

1966. The ecology, functional morphology, and taxonomy of *Echeneibothrium* Beneden, 1849 (Cestoda: Tetraphyllidea), a revision of the genus and comments on *Discobothrium* Beneden, 1870, *Pseudanthobothrium* Baer, 1956, and *Phormobothrium* Alexander, 1963. *Parasitology* 56:227–285.

1968. *Phyllobothrium piriei* sp. nov. (Cestoda: Tetraphyllidea) from *Raja naevus* with a comment on its habit and mode of attachment. *Parasitology* 58:929–937.

Williams, H. H., A. H. McVicar, and R. Ralph

1970. The alimentary canal of fish as an environment for helminth parasites. *Symp. Brit. Soc. Parasitol.* 8:43–77.

Wilson, A. C., G. L. Bush, S. M. Case, and M. C. King

1975. Social structuring of mammalian populations and rate of chromosomal evolution. *Proc. Natl. Acad. Sci. U.S.A.* 72:5061–5065.

Wilson, R. A., G. Smith, and M. R. Thomas
1982. Fascioliasis. In *Population Dynamics of Infectious Diseases: Theory and Applications,* ed. R. M. Anderson, 262–319. Chapman and Hall, London.

Windley, B. F.
1986. *The Evolving Continents,* 2nd ed. Wiley, New York.

Winterbottom, R.
1990. The *Trimmatom nanus* species complex (Actinopterygii, Gobiidae): Phylogeny and progenetic heterochrony. *Syst. Zool.* 39:253–265.

Wirth, U.
1984. Die Struktur der Metazoen-Spermien und ihre Bedeutung für die Phylogenetik.*Verhandl. naturwiss. Ver. Hamburg* 27:295–362.

Xylander, W. E. R.
1986. Ultrastrukturelle Befunde zur Stellung von *Gyrocotyle* im System der parasitischen Plathelminthen. *Verhandl. Deutsch. Zool. Ges.* 79:193.
1987a. Ultrastructure of the lycophore larva of *Gyrocotyle urna* (Cestoda, Gyrocotylidea). I. Epidermis, neodermis anlage and body musculature. *Zoomorphology* 106:352–360.
1987b. Ultrastructural studies on the reproductive system of Gyrocotylidea and Amphilinidea (Cestoda). II. Vitellaria, vitellocyte development and vitelloduct of *Gyrocotyle urna. Zoomorphology* 107:293–297.
1987c. Ultrastructure of the lycophora larva of *Gyrocotyle urna* (Cestoda, Gyrocotylidea). II. Receptors and nervous system. *Zool. Anz.* 219:239–255.
1987d. Das Protonephridialsystem der Cestoda: Evolutive Veränderungen und ihre mögliche funktionelle Bedeutung. *Verhandl. Deutsch. Zool. Ges.* 80:257–258.
1988. Ultrastructural studies on Udonellidae: Evidence for a position within the Neodermata. *Fortschr. Zool.* 36:51–57.
1989. Ultrastructural studies on the reproductive system of Gyrocotylidea and Amphilinidea (Cestoda): Spermatogenesis, spermatozoa, testes, and vas deferens of *Gyrocotyle. Int. J. Parasitol.* 19:897–905.
1990. Ultrastructure of the lycophore larva of *Gyrocotyle urna* (Cestoda, Gyrocotylidea). *Zoomorphology* 109:319–328.

Yamaguti, S.
1971. *Synopsis of the Digenetic Trematodes of Vertebrates.* Keigaku, Tokyo.

Zahavi, A.
1975. Mate selection: A selection for a handicap. *J. Theor. Biol.* 53:205–214.
1977. The cost of honesty. *J. Theor. Biol.* 67:603–605.

Zimmerman, M. R., and R. E. Morilla
1983. Enterobiasis in pre-Columbian America. *Paleopathol. News* 42:8.

Zschokke, F.
1904. Die Darmcestoden der amerikanischen Beuteltiere. *Centralbl. Bakt. Parasit.* 36:51–61.
1933. Die Parasiten als Zeugen für die geologische Vergangenheit ihrer Träger. *Forsch. Fortschr.* 9:466–467.

AUTHOR INDEX

The Appendix has not been indexed.

SUBJECT INDEX

The Appendix has not been indexed.